Visual Information Processing in Wireless Sensor Networks:

Technology, Trends, and Applications

Li-minn Ang
University of Nottingham (Malaysia Campus), Malaysia

Kah Phooi Seng
Sunway University, Malaysia

Information Science
REFERENCE

Managing Director:	Lindsay Johnston
Senior Editorial Director:	Heather Probst
Acquisitions Editor:	Erika Gallagher
Development Manager:	Joel Gamon
Development Editor:	Michael Killian
Book Production Manager:	Sean Woznicki
Typesetters:	Mackenzie Snader and Milan Vracarich
Print Coordinator:	Jamie Snavely
Cover Design:	Nick Newcomer

Published in the United States of America by
Information Science Reference (an imprint of IGI Global)
701 E. Chocolate Avenue
Hershey PA 17033
Tel: 717-533-8845
Fax: 717-533-8661
E-mail: cust@igi-global.com
Web site: http://www.igi-global.com

Library of Congress Cataloging-in-Publication Data

Visual information processing in wireless sensor networks: technology, trends, and applications / Li-minn Ang and Kah Phooi Seng, editors.
p. cm.
Summary: "This book provides a central source of reference on visual information processing in wireless sensor network environments and its technology, application, and society issues"-- Provided by publisher.
Includes bibliographical references.
ISBN 978-1-61350-153-5 (hardcover) -- ISBN 978-1-61350-154-2 (ebook) -- ISBN 978-1-61350-155-9 (print & perpetual access) 1. Wireless sensor networks. 2. Optical detectors. I. Ang, Li-minn, 1971- II. Seng, Kah Phooi, 1974-
TK7872.D48V57 2012
681'.2--dc23

2011026269

British Cataloguing in Publication Data
A Cataloguing in Publication record for this book is available from the British Library.

Table of Contents

Preface

The combination of image sensors with wireless sensor network (WSN) technology has resulted in a new network technology called a visual sensor network (VSN). The use of image sensors in VSNs increases the range of potential applications because compared to scalar sensors, image sensors are able to provide more information which can be used for visual processing tasks such as detection, identification, and tracking. On the one hand, VSNs can be seen as an extension of traditional WSNs where image sensors have replaced scalar sensors. On the other hand, the use of image sensors in VSNs brings with it a whole different set of practical and research challenges. The successful development of VSNs will require research contributions from different disciplines such as visual information processing, wireless networks, hardware and embedded systems, and power engineering.

The objective of this book is to assemble together the technology, trends, and applications for research on VSNs to serve as a comprehensive source of reference and to play an influential role in setting the scopes and directions of this emerging field of research. It contains open-solicited and invited chapters written by leading academics, researchers, and practitioners in the field. The target audience would be researchers, professionals, and students in engineering and information technology that engage with visual information processing and wireless systems. While there are other books on wireless sensor networks, few provide a primary focus for discussing the specific challenges for visual information processing in wireless sensor network environments.

This book contains sixteen chapters including an opening chapter to introduce the reader to this area. It is structured into four sections which are: foundations of visual information processing in wireless sensor networks, energy efficient information processing and transmission in VSN, collaborative information processing in VSN, and hardware technology and applications for VSN. The first section on foundations of visual information processing in wireless sensor networks contains an introductory chapter. Chapter 1 by Li-minn Ang and Kah Phooi Seng, presents an introduction to VSN technology and its applications. The authors provide an overview of research issues and trends. Issues related to energy efficient processing, collaborative processing, and hardware technology will be highlighted. This chapter will also give a brief introduction to the other chapters in the book with a focus on showing how the topics covered in each chapter relate to the overall picture of visual information processing in wireless sensor network environments.

Section II is devoted to energy efficient information processing and transmission in VSNs. Image sensors generate a very high amount of data that would have to be processed and transmitted within the network. There is a trade-off between the energy required for processing and the energy required for transmission. To save energy in transmission, image and video compression algorithms can be applied to reduce the amount of data to be transmitted. However, the implementation of compression algorithms would require higher computational complexity, leading to higher energy consumption for processing. For energy efficient implementation in VSNs, the algorithms would need to have low computational and memory complexity. This section contains six chapters.

The first four chapters in the section present various approaches for energy efficient compression techniques to reduce the data. The techniques range from current standards like JPEG, MPEG-x and H.26x to newer techniques using distributed video coding and compressive sensing. Chapter 2 by Yinhao Ding and Cheng-Chew Lim, presents compressive sampling techniques to achieve information compression in VSNs. The authors focus on energy-efficiency and reliability issues such as the relationships between sparsity control and compression ratio, the effect of block-based sampling on reconstruction quality, the complexity consideration of the reconstruction process for real-time applications, and the compensation for packets missing in network flows. The effectiveness of the approach on recovered image quality is evaluated using Euclidean distance and variance analysis. The next chapter by A. Elamin, Varun Jeoti and Samir Belhouari, describes the use of distributed video coding for VSNs. The authors review important developments such as the Stanford Wyner-Ziv coding architecture and discuss the latest research trends highlighting a Region-Based-Wyner-Ziv video codec that enables low-complexity encoding while achieving high compression efficiency.

The fourth chapter by Mammeri Abdelhamid, Brahim Hadjou, Ahmed Khoumsi and Djemel Ziou, presents the design of wavelet filters for use in power constrained VSNs. While designing a wavelet-based coder (WBC) in the context of VSN, engineers and designers must respect the strict constraint on power consumption. The authors evaluate and select the best set of wavelet filters. Their results provide a good reference for designers of WBC for power-constrained applications such as VSN. Chapter 5 by Ruth Aguilar-Ponce, J. Luis Tecpanecatl-Xihuitl, and Alfonso Alba-Cadena, presents approaches for low level representation in VSNs. The authors introduce algorithms for data encoding and data filtering. Data encoding is performed by means of predictive video encoding using a phase-only correlation function for motion estimation. Data filtering is performed in three phases: pixel classification, background update, and detection. The algorithms involved in each phase are low in terms of complexity and memory resources.

The remaining two chapters in the section present various approaches for energy efficient transmission in VSNs. Visual information requires higher bandwidth and lower delay and delay jitter to provide the required Quality of Service (QoS) for multimedia transmission. Chapter 6 by Nalin Sharda, presents an overview of the evolution of WSNs and discusses factors for multimedia transmission in such networks such as bandwidth, delay and delay jitter and how these influence the QoS in the delivery of multimedia information. The author explores various transmission issues and limitations, such as limited energy, computational power, and bandwidth. The final chapter in the section by S. Mehta and K. S. Kwak, presents the use of energy-efficient backoff algorithms in MAC protocols. The authors present a study of various backoff algorithms and propose an improved backoff (IB) algorithm for an energy efficient MAC protocol in WSNs. Their results show that the IB algorithm gives improvements in throughput, channel access delay, and energy efficiency under varying traffic conditions over the widely used binary exponential backoff based algorithm.

Section III is devoted to collaborative information processing in VSNs. To realize the full potential of the VSN, the visual nodes can collaborate with each other to achieve the goals of the application. The issues for collaborative processing are similar to that involving multi-camera networks. Typical visual processing tasks to be performed collaboratively are to detect, track, and recognize objects in the scene. Object detection refers to a visual processing task to locate a certain class of objects in the image frame (for example, faces and vehicles). The goal of object tracking is to associate target objects in different video frames. Object recognition refers to a visual processing task where the general class of the object is known and the objective is to recognize an object's exact identity. This section contains four chapters.

Chapter 8 by Yi Zhou, Hichem Snoussi, Shibao Zheng and Fethi Smach, presents an approach in wireless camera networks for visual human tracking. The authors exploit the use of distinctive and fast to compute local features to represent the non-rigid targets. Transmission of feature descriptors between cameras is done without any calibration. They show that their proposed local features succeed to re-identify and relocate the target among the distributed cameras and also propose an efficient interest point detection and matching scheme for the visual tracking under real-time constraints. The next chapter by Shung Han Cho, Yunyoung Nam and Sangjin Hong, presents an object association method through multiple cameras collaboration for a large-scale surveillance system. The objects association is achieved by locally generating homographic lines on targets in collaborating cameras. Their proposed method is verified with real video sequences. The simulation result demonstrates that the proposed method is robust against false association because it considers all the possible pairing cases of occluded targets.

Chapter 10 by Juan Gómez-Romero, Jesús García, Miguel A. Patricio, José M. Molina, and James Llinas, presents an approach for high-level fusion in VSNs. Information fusion techniques combine data from multiple sensors, along with additional information and knowledge, to obtain better estimates of the observed scenario than could be achieved by the use of single sensors or information sources alone. The authors study the architectural and functional issues of applying context information to improve high-level fusion procedures, with a particular focus on visual data applications. The next chapter by Majdi Mansouri, Hichem Snoussi, Jing Teng, Ouachani Ilham, and Cédric Richard, describes a technique using quantized bayesian filtering in WSNs. The focus of the chapter is to study the Bayesian inference problem in distributed WSNs with particular emphasis on the trade-off between estimation precision and energy-awareness. The authors propose a variational approach to approximate the particle distribution to a single Gaussian distribution, while respecting the communication constraints of WSNs. Simulation results demonstrate the significantly improved performance of the approach.

The final section (Section IV) of the book deals with the hardware technology and applications for VSNs. This section contains five chapters. Important issues for the hardware technology are related to the image sensor, processor hardware and power considerations. The authors in chapters 11 and 12 present approaches using line scan cameras in place of two-dimensional cameras to reduce data during the acquisition process. Chapter 12 by Jiang Yu Zheng, describes the use of line sensors for monitoring and surveillance in VSNs using linear CCD sensors. It reads temporal data from a CCD array continuously and stores them to form a 2D image profile. The method delivers more information such as color, shape, and event of a flowing scene. It abstracts passing objects in the profile without heavy computation and transmits much less data than a video from normal cameras. This chapter focuses on several unsolved issues of line sensors in capturing targets in the 3D space such as sensor setting, shape analysis, robust object extraction, and real time background adapting to ensure long-term sensing and visual data collection via networks.

Chapter 13 by Mangesh Chitnis, Claudio Salvadori, Matteo Petracca, Paolo Pagano, Giuseppe Lipari, and Luca Santinelli, present an innovative technique of using line sensor based image capturing and processing in order to detect moving objects such as vehicles. The authors show that line sensor techniques may achieve faster processing results with less storage and bandwidth requirements while conserving node energy. Their solution presents itself as a low-cost candidate for Intelligent Transport Systems (ITS) to monitor and control urban traffic. The next chapter by Wai Chong Chia, Christopher Ngau, Lee Seng Yeong, Li Wern Chew, Li-minn Ang, and Kah Phooi Seng, presents hardware technology for VSNs using field programmable gate array technology (FPGA). The authors propose a FPGA implementation of a low-complexity and strip-based microprocessor architecture for image processing. The captured image data is processed in a strip-by-strip manner to reduce the local memory requirement. This allows an image with higher resolution to be captured and processed with the limited hardware resources. The approach is illustrated with a visual saliency application.

The remaining two chapters are concerned with applications for VSNs. Chapter 15 by Johannes Karlsson, Tim Wark, Keni Ren, Karin Fahlquist, and Haibo Li, presents a practical application of VSNs – the Digital Zoo. The authors describe their work to set up a large scale VSN in a Swedish zoo. It is located close to the Arctic Circle making the environment very hard for this type of deployment. To reach this goal the sensed data will be processed and semantic information will be used to support interaction design, which is a key component to provide a new type of experience for the visitors. The final chapter by Julien Sebastien Jainsky, and Deepa Kundur, presents a security application for VSNs. In particular, attention is brought to steganographic security and how it can be discouraged without challenging the primary objectives of the network. Preventative steganalysis aims at discouraging any potential steganographic attacks by processing the collected data such that the possibility of steganography becomes very small and the steganalysis leads to high rate of success. The authors motivate the development and implementation of more lightweight steganalytic solutions that take into account the resources made available by the network's deployment and its application in order to minimize the steganalysis impact on the workload.

We would like to express our deepest appreciation to the authors of this book and to the reviewers for their tremendous service by critically reviewing the chapters and offering useful suggestions. Most of the authors of chapters included in this book also served as referees for chapters written by other authors. We hope that the reader will share our enthusiasm to present this volume and will find it useful.

Li-minn Ang
University of Nottingham (Malaysia Campus), Malaysia

Kah Phooi Seng
Sunway University, Malaysia

Acknowledgment

This book is made possible through the efforts of many people. We especially want to thank the authors of each chapter for contributing the results of their research work and expanding our understanding of aspects related to visual information processing in wireless sensor networks. We also wish to acknowledge the support of the editorial advisory board and the help of all who were involved in the collection and review process of this book. A special note of thanks goes to the staff at IGI Global. In particular, we would like to express our deepest appreciation to Michael Killian and Christine Bufton. Their editorial assistance and professional support, from the inception of the initial idea to final publication, have been invaluable.

Li-minn Ang
University of Nottingham (Malaysia Campus), Malaysia

Kah Phooi Seng
Sunway University, Malaysia

Section 1
Foundations of Visual Information Processing in Wireless Sensor Networks

Chapter 1
Visual Sensor Network Technology and its Applications

Li-minn Ang
University of Nottingham (Malaysia Campus), Malaysia

Kah Phooi Seng
Sunway University, Malaysia

ABSTRACT

The combination of image sensors with wireless sensor network (WSN) technology has resulted in a new network technology called a visual sensor network (VSN). On the one hand, VSNs can be seen as an extension of traditional WSNs where image sensors have replaced scalar sensors. On the other hand, the use of image sensors in VSNs brings with it a different set of practical and research challenges. This is because image sensors generate a very high amount of data that would have to be processed and transmitted within the network. In this chapter, we present an introduction to VSN technology and provide an overview of research issues and trends. Issues related to energy efficient processing, collaborative processing, and hardware technology will be highlighted. This chapter will also give a brief introduction to the other chapters in the book with a focus on showing how the topics covered in each chapter relate to the overall picture of visual information processing in wireless sensor network environments.

INTRODUCTION

The field of wireless sensor networks (WSNs) has seen much success in diverse applications ranging from environmental monitoring, smart homes, healthcare, to industrial and defence applications. Traditional wireless sensor networks capture scalar data such as temperature, vibration, pressure, or humidity. Motivated by the success of WSNs and also with the emergence of new technology in the form of low-cost image sensors, researchers have proposed combining image sensors with WSNs to form visual sensor networks (VSNs). The use of image sensors in VSNs increases the range of potential applications because compared to scalar sensors, image sensors are able to provide

DOI: 10.4018/978-1-61350-153-5.ch001

more information which can be used for visual processing tasks such as detection, identification, and tracking. On the one hand, VSNs can be seen as an extension of traditional WSNs where image sensors have replaced scalar sensors. On the other hand, the use of image sensors in VSNs brings with it a different set of practical and research challenges. This is because image sensors generate a very high amount of data that would have to be processed and transmitted within the network. The main issue is that sensor nodes have limited battery power and hardware resources. The energy efficient processing and transmission of the data within the VSN is of primary importance to maximize the lifetime of the overall network. Depending on the application, the network should be active for a duration of time ranging from weeks to years without the need for battery replacement.

Researchers have used several different terms such as camera sensor networks, wireless image sensor networks (WISNs), and wireless visual sensor networks (WVSNs) to describe these image-based sensor networks. When combined with other sensors such as audio sensors, the networks are termed wireless multimedia sensor networks (WMSNs). Throughout this chapter, we will use the term visual sensor networks. A brief description of a VSN can be stated as follows. A VSN comprises of many visual camera nodes (VCNs) working collaboratively to detect useful events in the overall scene data to be conveyed to a base station for higher-level processing. Each VCN is an embedded device which has the capability to perform sensing, processing and communication within the node itself. The VCNs are connected using a wireless sensor network which relays the scene event data from one node of the network to another until it arrives at the base station.

The successful development of VSNs will require research contributions from different disciplines such as visual information processing, wireless networks, hardware and embedded systems, and power engineering. Several survey papers on different issues of VSNs and WMSNs can be found in the papers by (Akyildiz et al., 2007; Almalkawi, et al., 2010; Soro et al., 2009). The objective of this chapter is to present a brief introduction to VSN technology and to provide an overview of research issues and trends to assist the reader in the journey throughout this book. We will focus our discussions on visual information processing issues for wireless sensor network environments. The end of each section will also draw the reader's attention towards further information and case studies which can be obtained from the later chapters in this book.

VSN TECHNOLOGY

Researchers and authors have used different terms to describe the various components in the VSN. Figure 1 shows a general structure of a VSN consisting of four main components: Visual Camera Node (VCN), Visual Cluster Head (VCH), Network Node, and Base Station. Figure 1 shows that there a decreasing information flow from the VCNs to the base station. The captured scene data are processed and transformed to useful event data. The focus for the upper levels of the network (VCN and VCH) is on visual information processing and the focus for the lower levels of the network (network node) is on wireless network communications. The primary theme for both upper and lower network levels is to achieve energy efficiency within the constraints of the battery powered nodes.

The VCNs form the end points of the network. Each VCN consist of a camera sensor, processing unit, communication unit, and power unit. Each camera sensor has a field of view (FOV) of the scene. A captured scene is called an image frame. The processing unit performs the visual processing to reduce the high amount of scene data. Two approaches can be used. The first approach uses event detectors to identify useful events in the scene data. If an event is not detected, then the image frame is discarded and there is no need to

Figure 1. General structure of a VSN

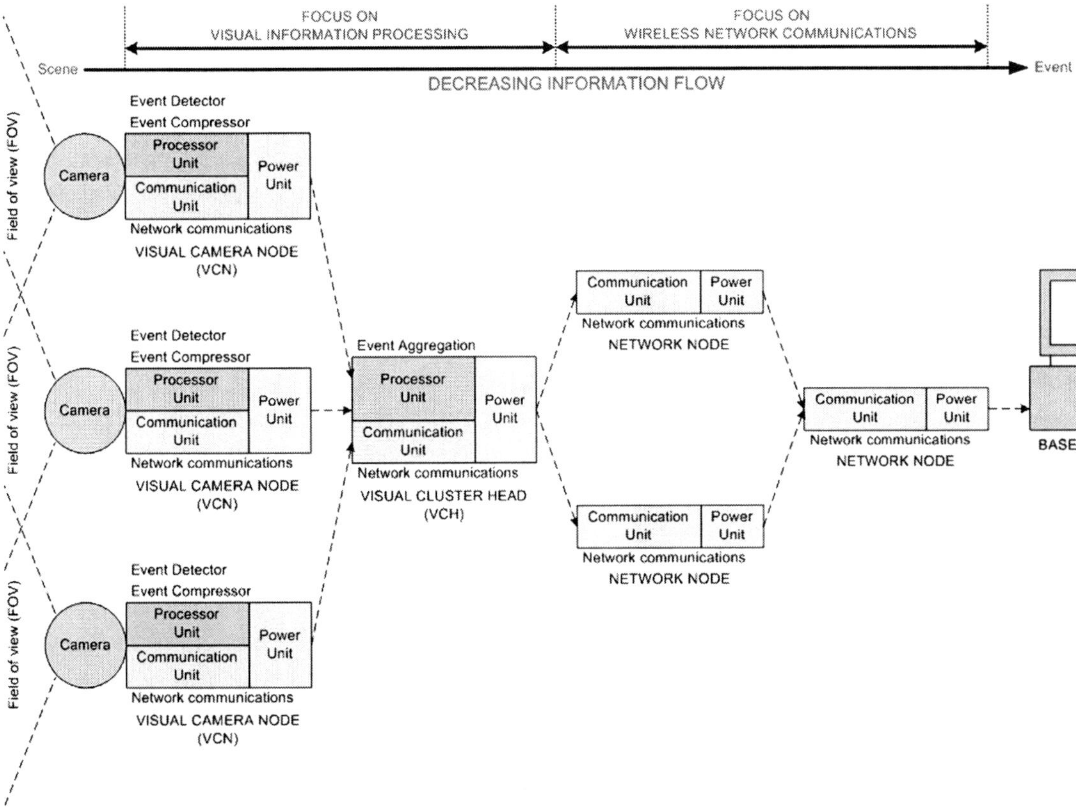

transmit the frame through the network. The second approach uses event compressors to reduce the data for image frames that have to be transmitted through the network. Various image and video compression techniques can be used and this will be discussed in a later section. The communication unit transmits the compressed data to other nodes. Each VCN receives its power supply from a power unit which is mostly battery powered.

The role of the VCHs is to receive data from several VCNs. Each VCH consist of a processing unit, communication unit, and power unit. Each VCH receives data from several VCNs. The FOVs of VCNs may overlap and the VCH can further reduce the data by performing event aggregation by consolidating the information from the VCNs. For example, the image frames from different VCNs may be stitched into a single frame to remove the overlapping data. This will be further discussed in a later section. The communication unit and power unit performs a similar role as for the VCN. The network node performs the same role as for a traditional wireless sensor network and consists of a communication unit and power unit. The communication unit relays the data from node to node until it arrives at the base station. The base station is the destination of all the data gathered throughout the network. This is likely to be a conventional computer capable of powerful processing and connected to a main power supply. Thus, energy efficiency and power issues are not important here.

Researchers have classified VSNs into three models: single-tier flat architecture, single-tier clustered architecture, and multi-tier architecture (Akyildiz et al., 2007). Using the terminology above, a single-tier flat architecture would be a VSN without the VCHs. In this model, the data is directly relayed from VCNs to network nodes. The single-tier clustered architecture would be a WVSN utilizing cluster heads to perform visual event aggregation. If there are different types of VCNs or other sensor nodes in the network, then this is termed a multi-tier architecture. Typically, there would be a lower-resolution VCN working together with a higher-resolution VCN which is inactive most of the time. When the lower-resolution VCN detects a significant event, it instructs the higher-level VCN to wake-up to perform higher-level processing. This achieves energy efficiency because most of the time, only the lower-level and lower power VCN nodes are active. The higher-level and higher power VCN nodes are only activated when significant events are detected. An example of a multi-tier VSN architecture organized as a hierarchy of levels to achieve energy efficiency can be found in the work by (Lopes et al., 2007). This work uses a three tier architecture to detect, identify, and track objects for an environmental monitoring application. The first level uses passive infrared (PIR) sensors to detect objects in the sensing field. The PIR sensors are constantly active. When the PIR sensors detect an object, they activate the camera sensors in the second level. The camera sensors in the second level use a scheme to identify the object based on its predominant colour. When an object is identified, then the third level of powerful visual sensors is activated to track the movements of the object. Other examples of multi-tier VSN approaches can be found in the papers by (Prati et. al, 2005; Kim et al., 2008).

ENERGY EFFICIENT INFORMATION PROCESSING

Image sensors generate a very high amount of data that would have to be processed and transmitted within the network. There is a trade-off between the energy required for processing and the energy required for transmission. To save energy in transmission, image and video compression algorithms can be applied to reduce the amount of data to be transmitted. However, the implementation of compression algorithms would require higher computational complexity, leading to higher energy consumption for processing. For energy efficient implementation in VSNs, the algorithms would need to have low computational and memory complexity.

Two different visual information processing approaches can be employed to reduce the image data. The approaches can be divided into single-view approaches and multi-view approaches. Single-view approaches attempt to reduce the image data from each individual VCN. Two techniques can be used here: event detection and event compression. The first technique uses event detection to reduce data by only transmitting frames in the network when significant events are detected. Image frames which are not significant are discarded by the VCN. The second technique is to perform event compression by performing compression on the image frames which have to be transmitted. Event compression approaches rely on image and video compression algorithms to remove redundancy from the data. These algorithms range from current standards like JPEG, MPEG-x and H.26x to newer techniques using distributed video coding and compressive sensing. While single-view approaches attempt to reduce the scene data from each individual VCN, multi-view approaches perform the data reduction by aggregating the data from different VCNs. This is possible because of the overlapping FOVs of the different VCNs. These techniques will be briefly discussed in this section.

Event Detection

Event detection approaches reduce scene data by only transmitting image frames when significant events are detected for the application. For example, a surveillance application could use a face detector to decide which image frames to send to the base station. However, the face detector would need to have low computational complexity to meet the energy requirements in the VCN. There is a trade-off between energy required for processing and energy required for transmission. On the one hand, using an event detector in the VCN requires more computational power. On the other hand, this could result in a saving of transmission power when frames are discarded. The other advantage of an event detector is that it could also serve as an early stage for visual preprocessing. To perform the facial recognition process, the central computer would need to perform at least two stages. The first stage is to locate the face location in the image and the second stage would then perform the recognition task by comparing the facial features with a stored database. To reduce the large amount of image data for processing by the central computer, the event detector performs the face detection task and the location(s) of the face(s) is then communicated to the central computer to perform the facial recognition task.

An example of a VSN employing an event detector can be found in the paper by (Czarlinska & Kundur, 2008). The authors propose an event detector using simple image processing at the camera nodes based on difference frames and the chi-squared detector. Their algorithm performs well on indoor surveillance sequences and some, but not all outdoor sequences. Event detection is a mature field in computer vision. A comprehensive review of event detection can be found in the book by (Niels & Niels, 2001). The research challenge is to find suitable detectors which are reliable and can be efficiently implemented within the hardware constraints of the VCN.

Event Compression

There is a wide range of image compression techniques. The research challenge is to identify algorithms which are suitable for implementation in VSNs. As a main characteristic, the algorithms should have low computational complexity and memory requirements while maintaining high coding performance to achieve energy efficiency. Figure 2 shows a summary of image compression and several representative algorithms (Chew et al., 2008). In general, image coding can be categorized under first generation and second generation image coding approaches (Kunt et al., 1985). The authors observe that image and video coding is basically carried out in two steps. The first step converts the image data into a sequence of messages. In the second step, code words are assigned to the messages. First generation approaches put the emphasis on the second step whereas second generation approaches put it on the first step and use available results for the second step. Examples of first generation approaches are algorithms based on the discrete cosine transform (DCT) and discrete wavelet transform (DWT). Figure 2 shows the basic components of an image coding process. An image captured by the visual sensor nodes is first transformed to a domain that is more suitable for processing. Quantization is then carried out on the transformed image data. The image compression algorithm is then applied followed by entropy coding. This generates a bit stream that is now ready for transmission. At the decoder side, the above processes are reversed which finally leads to the reconstruction of the original image.

Second generation approaches require more computational processing compared to first generation approaches. Thus, researchers have focused on adapting and modifying first generation algorithms to lower computational and memory requirements while still maintaining the coding performance. The work in (Lu et al., 2008) proposes using the lapped biorthogonal transform (LBT) in place of the CDF 9/7 filter used in JPEG2000. This reduces the computational requirements by half and only requires 1/15 of the memory. The work in (Lee et al., 2009) proposes a design approach based on precision-optimized

Figure 2. Image compression algorithms and coding process

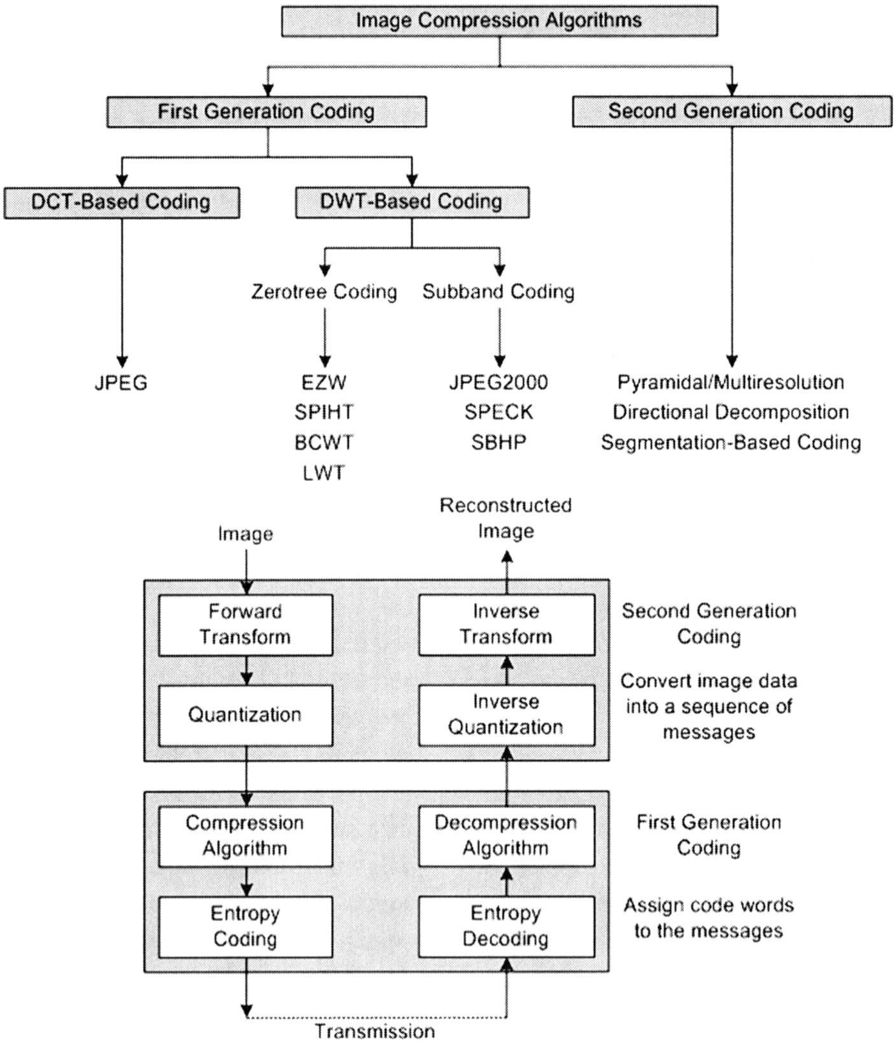

custom arithmetic to obtain an energy-minimal, platform-specific implementations of the DCT and quantization steps in the JPEG algorithm. Their approach shows speed and energy improvements ranging from factors of 2 to 5 compared to the traditional JPEG algorithm. A comprehensive tutorial on the application of low memory wavelet transforms for wireless sensor networks can be found in (Rein & Reisslein, 2011). The authors also propose a fractional wavelet transform which require only 1.5 kByte of RAM to transform a grayscale 256x256 image.

New Approaches

Researchers have proposed new approaches for image and video compression specifically for implementation in hardware constrained environments like VSNs. We will briefly describe three recent approaches: distributed video coding (DVC), compressive sensing (CS), and distributed image compression (DIC). Traditional approaches like MPEG-x and H.26x were designed for broadcast type applications where the encoder complexity is high and the decoder complexity is low. This is in contrast to the requirements for VSN which require that the encoder has low complexity. DVC and CS approaches aim to have a low complexity encoder at the expense of a higher complexity decoder. The low complexity encoder can be implemented in the VCN and the base station which has fewer constraints on energy efficiency and power issues can implement the high complexity decoder. Distributed video coding is based on two major results in information theory: the Slepian-Wolf and Wyner-Ziv theorems. In DVC, two correlated sources are independently encoded using separate encoders and the bit streams associated with each source are jointly decoded exploiting the correlation between them. The authors in (Pereira et al., 2008) examine different application scenarios for DVC including its use in VSNs. The authors conclude that the potential advantages of DVC in VSN are lower encoder complexity, higher coding efficiency, improved error resilience, multiview correlation exploitation and codec/resolution independent scalability. The drawbacks are lower compression efficiency if more complex encoders are not allowed and higher decoding complexity which could affect the scalability of the system in terms of the number of sensors it can support.

Compressive sensing (CS) is a recent development in signal processing. The Nyquist-Shannon sampling theory states that a signal has to be sampled at twice the bandwidth to be reconstructable. The CS technique allows for a lower sampling rate when the signal has a sparse structure. CS is able to compress the input data in the sampling process. For sparse signals, CS provides a great reduction in the sampling rate, power consumption and computational complexity. Examples of the application of CS for VSN can be found in (Han et al., 2010; Kang & Lu, 2009). Another new approach for compression in VSN is based on DIC which reduces the computation and energy requirements of individual VCNs by sharing the processing of tasks. The paper by (Wu & Abouzeid, 2005) proposes a DIC approach by exploiting the characteristics of the Discrete Wavelet Transform (DWT). In the scheme, nodes compress an image while forwarding it to the destination subject to a specific image quality requirement.

Multi-View Compression

VSNs with VCHs can employ multi-view approaches to further reduce the data. Each VCH receive data from several VCNs. Note that the role of a VCH could be played by VCNs as well. For example, the VCN with the highest remaining energy level could be designated as the current VCH and when its energy level decreases to a certain level, another VCN could take over the role of the VCH. As shown in Figure 1, the FOVs of VCNs may overlap and the VCH can further reduce the data by performing event aggregation by consolidating the information from the VCNs. This requires solving what is known in computer vision as the correspondence problem. This is to find a set of points in one image which can be identified as the same points in another image. The application of multi-view compression approaches for WVSN is a very recent development. We will illustrate this approach using the work by (Chia et al., 2011) which proposes an approach using image stitching. The computational requirements for image stitching are relatively high. To reduce the complexity, the authors utilize the processing power available at the base station. Initially, the two images to be stitched are sent to the base station. The base station performs the stitching and sends back the relevant information to the nodes on the pixel locations of the overlapping regions. This allows the nodes to discard the overlapping regions during transmission.

Figure 3 shows an overview of the approach. There are eight steps in the algorithm. The first step is feature extraction where feature points are extracted from the two image views. This is denoted as images A and B in the figure. Different feature detectors can be adopted to complete the task provided that the same features or objects can be found within the two images. The purpose of doing this is to determine the region where the two images overlap. The second step is feature point description. A feature point descriptor is required to describe the feature points and generate an identity for each of them. Without the identity, each of the feature points is just a coordinate that indicates where the point is located. The identity is needed to perform the matching of feature points. In the third step, feature matching is performed. The matching of feature points can be achieved by exhaustively comparing feature points to feature points in the two images. Techniques such as using the k-d tree can be used to improve the processing speed (Brown & Lowe, 2007). After a set of matching feature points is obtained, the transformation parameters are estimated using the Random Sample Consensus (RANSAC) algorithm (Fischler & Bolles, 1981). Based on the matched feature points, the RANSAC algorithm finds a model that fits best and is then used to estimate the transformation parameter, T, as shown in Eq. (1).

In Step five, the transformation parameter, T is then used to determine the source location for every pixel in the stitched image using Eq. (2) and (3). The coordinate (x', y') is the source location where the data has the corresponding point in the original image. After one of the images (A or B) has been transformed into the coordinate system of the other, the two images are ready to be stitched together. This is performed in step six. Multi-band blending is performed in the seventh step. Ideally the images that are to be stitched together would have the same intensity. However, this is difficult to achieve in reality and can result in a visible seam in the stitched image. This problem can be resolved using techniques such as multi-band blending (Burt & Adelson, 1983). To further reduce the data, image compression is performed on the stitched image to remove spatial redundancy in the image. This also reduces the bandwidth requirements for transmission.

$$T = \begin{bmatrix} t_0 & t_1 & t_2 \\ t_3 & t_4 & t_5 \\ t_6 & t_7 & 1 \end{bmatrix} \tag{1}$$

$$x' = \frac{\tilde{x}'}{\tilde{z}'} = \frac{t_0 x + t_1 y + t_2}{t_6 x + t_7 y + 1} \tag{2}$$

$$y' = \frac{\tilde{y}'}{\tilde{z}'} = \frac{t_3 x + t_4 y + t_5}{t_6 x + t_7 y + 1} \tag{3}$$

FURTHER INFORMATION

Further information on energy efficient processing and transmission techniques in VSN can be found in Section II. Four chapters present various approaches for energy efficient compression techniques to reduce the data. The techniques range from current standards like JPEG, MPEG-x and H.26x to newer techniques using distributed video coding and compressive sensing. Chapter 2 by Yinhao Ding and Cheng-Chew Lim, presents compressive sampling techniques to achieve information compression in VSNs. The authors focus on energy-efficiency and reliability issues such as the relationships between sparsity control and compression ratio, the effect of block-based sampling on reconstruction quality, the complexity consideration of the reconstruction process for real-time applications, and the compensation for packets missing in network flows. The effectiveness of the approach on recovered image quality is evaluated using Euclidean distance and variance analysis. The next chapter by A. Elamin, Varun Jeoti and Samir Belhouari, describes the use of distributed video coding for VSNs. The authors review important developments such as the Stanford Wyner-Ziv coding architecture and discuss the latest research trends highlighting a Region-Based-Wyner-Ziv video codec that enables low-complexity encoding while achieving high compression efficiency.

Figure 3. Multi-view compression

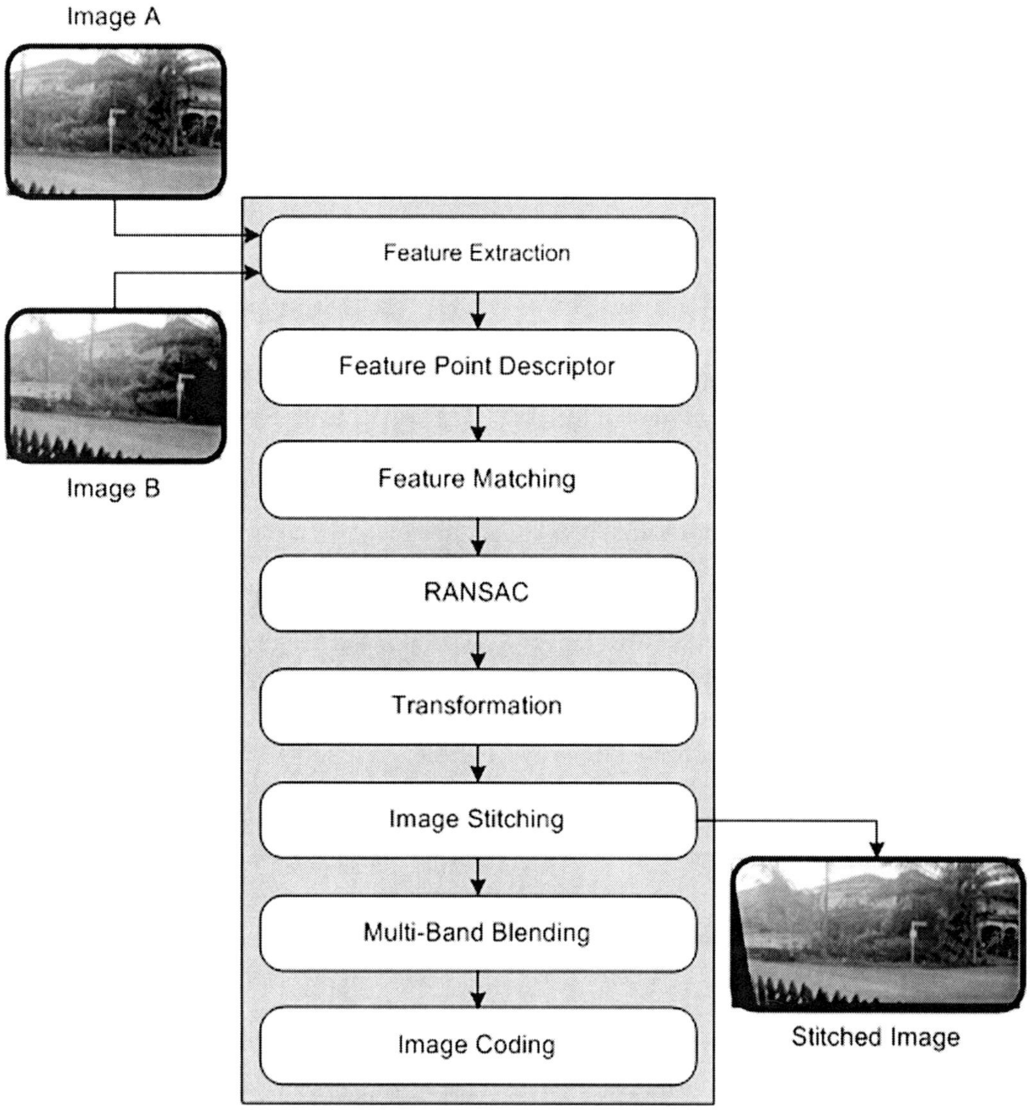

The fourth chapter by Mammeri Abdelhamid, Brahim Hadjou, Ahmed Khoumsi and Djemel Ziou, presents the design of wavelet filters for use in power constrained VSNs. While designing a wavelet-based coder (WBC) in the context of VSN, engineers and designers must respect the strict constraint on power consumption. The authors evaluate and select the best set of wavelet filters. Their results provide a good reference for designers of WBC for power-constrained applications such as VSN. Chapter 5 by Ruth Aguilar-Ponce, J. Luis Tecpanecatl-Xihuitl, and Alfonso Alba-Cadena, presents approaches for low level representation in VSNs. The authors introduce algorithms for data encoding and data filtering. Data encoding is performed by means of predictive video encoding using a phase-only correlation function for motion estimation. Data filtering is performed in three phases: pixel classification, background update, and detection. The algorithms involved in each phase are low in terms of complexity and memory resources.

The next two chapters present various approaches for energy efficient transmission in VSNs. Visual information requires higher bandwidth and lower delay and delay jitter to provide the required Quality of Service (QoS) for multimedia transmission. Chapter 6 by Nalin Sharda, presents an overview of the evolution of WSNs and discusses factors for multimedia transmission in such networks such as bandwidth, delay and delay jitter and how these influence the QoS in the delivery of multimedia information. The author explores various transmission issues and limitations, such as limited energy, computational power, and bandwidth. Chapter 7 by S. Mehta and K. S. Kwak, presents the use of energy-efficient backoff algorithms in MAC protocols. The authors present a study of various backoff algorithms and propose an improved backoff (IB) algorithm for an energy efficient MAC protocol in WSNs. Their results show that the IB algorithm gives improvements in throughput, channel access delay, and energy efficiency under varying traffic conditions over the widely used binary exponential backoff based algorithm.

COLLABORATIVE INFORMATION PROCESSING

To realize the full potential of the VSN, the visual nodes can collaborate with each other to achieve the goals of the application. The issues for collaborative processing are similar to those involving multi-camera and distributed networks.

Multi Camera Network Models

Researchers have classified multi-camera networks into three models: centralized processing, distributed processing, and clustered processing (Aghajan & Cavallaro, 2009). Figure 4 shows the centralized and clustered processing models. The centralized processing model resembles a traditional VSN architecture where the data sensed by the VCNs are sent to the base station for specific application processing. The processing performed in the VCNs are only for energy efficiency requirements, for example compression. The clustered processing model allows the VCNs to perform the visual processing at the node level. Each cluster may also collaborate and share data with its neighbouring clusters. The distributed processing model is a clustered processing model where each cluster only consists of a single VCN.

Collaborative Object Detection, Tracking and Recognition

Typical visual processing tasks to be performed collaboratively are to detect, track, and recognize objects in the scene. Object detection refers to a visual processing task to locate a certain class of objects in the image frame (for example, humans and vehicles). The goal of object tracking is to associate target objects in different video frames. Object recognition refers to a visual processing task where the general class of the object is known and the objective is to recognize an object's exact identity. The paper by (Sankaranarayanan et al., 2008) provides a good overview of these different tasks for multi-camera networks.

There are two approaches which can be used for object detection: detection of objects using static features and detection of moving objects. The first approach, where the object is detected using features from a single frame, involves maintaining a model of the objects of interests. The detection process is then decomposed into two steps of first finding the features in the image and then validating whether the features sufficiently explains the presence of the object. This approach has the disadvantages of being computationally intensive and also requiring training of the detection system beforehand. Researchers in (Frigo et al., 2010) present a vehicle license plate system using the Viola-Jones object detection algorithm (Viola & Jones, 2001). This algorithm uses Haar-like rectangular features and a number of classi-

Figure 4. Multi-camera network models

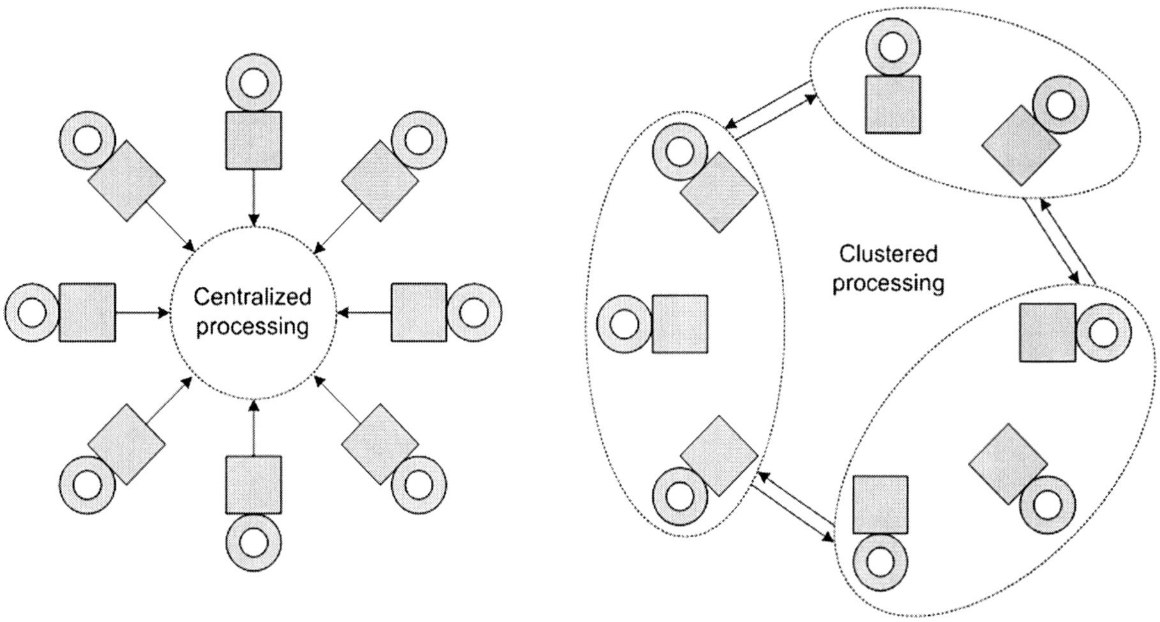

fiers arranged as a cascade. The beginning of the cascade uses simpler classifiers with a smaller number of features and more complex classifiers are employed later on. A region which is rejected is discarded and not processed any further. The idea is to discard early on in the processing regions which are not of the object type and to use the more complex classifiers only for the remaining regions. The authors report that the algorithm is very robust to misdetections, both in terms of missing regions on the license plate and false detections in the background.

The second approach involves detection of moving objects. A common approach is to use a technique termed background subtraction. Figure 5 shows a general model for background subtraction (Cheung & Kamath, 2007). There are four stages: preprocessing, background modeling, foreground detection, and data validation. In the initial stage, smoothing operations for temporal or spatial preprocessing can be performed to reduce camera and environmental noise such as rain and snow. Frame size reduction can also be performed to reduce the data processing rate. Background modeling is then performed to use the new video frame to calculate and update the background model. Various models can be used. These ranges from simple frame differencing to more sophisticated models using mixture of Gaussians (Stauffer & Grimson, 1999). Frame differencing uses the video frame at time t-1 as the background model for the frame at time t. Since it uses only a single previous frame, this model has a weakness that it is not able to identify the interior pixels of a uniformly coloured moving object. Also, this model does not account for changes in the environment such as illumination, shadows, and weather changes. The mixture of Gaussians approach is based on the assumption that the foreground and the background can be modeled as a mixture of Gaussians (MOG), where the Gaussians with low probability represent moving objects. The third stage of foreground detection identifies pixels in the video frame that cannot be sufficiently explained by the background model and outputs them as a binary mask. The final stage of data validation examines the candidate mask and removes those that do not correspond

Figure 5. General model for background subtraction

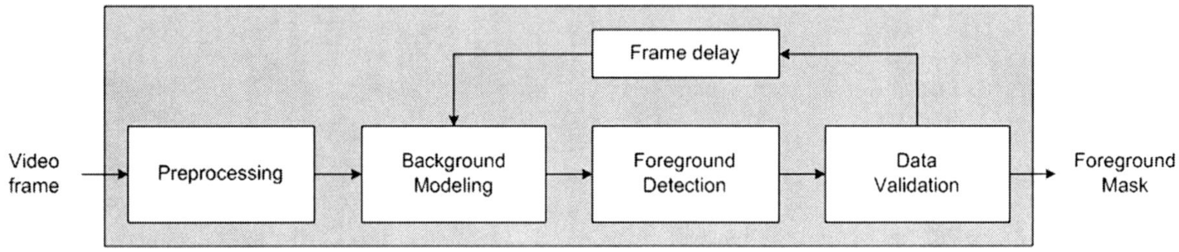

to moving objects. This stage often involves using domain knowledge. A survey of background subtraction techniques can be found in the paper by (Piccardi, 2004).

The aim of object tracking is to generate the trajectory of an image over time by locating its position in every frame of the video. A survey of object tracking can be found in (Yilmaz et al., 2006). There are two approaches which can be used for object tracking: deterministic approaches or stochastic approaches. Deterministic approaches pose tracking as an optimization problem. Stochastic approaches estimate the posterior distribution of the object location using Kalman filters or particle filters. The authors in (Medeiros, 2008) present an object tracking algorithm suitable for a network of wireless cameras. The algorithm uses a clustering protocol to establish connections among cameras that detect the target and to enable the propagation of the state of the target as it moves. Data aggregation is carried out using a decentralized Kalman filter. They carry out a series of experiments and show that the algorithm is capable of tracking a target with accuracy comparable to that achieved by a centralized approach wherein every measurement is transmitted to the base station. There are two approaches which can be used for object recognition: feature-based approaches and global approaches. Feature-based approaches detect points of interest in each image of an object and describe the object using descriptors of these local feature points. Global approaches describe the object using their global properties such as shape, silhouette, texture, or colour.

Further Information

Further information on collaborative information processing techniques in VSN can be found in Section III. The authors address various issues in object detection and tracking. Chapter 8 by Yi Zhou, Hichem Snoussi, Shibao Zheng and Fethi Smach, presents an approach in wireless camera networks for visual human tracking. The authors exploit the use of distinctive and fast to compute local features to represent the non-rigid targets. Transmission of feature descriptors between cameras is done without any calibration. They show that their proposed local features succeed to re-identify and relocate the target among the distributed cameras and also propose an efficient interest point detection and matching scheme for the visual tracking under real-time constraints. The next chapter by Shung Han Cho, Yunyoung Nam and Sangjin Hong, presents an object association method through multiple cameras collaboration for a large-scale surveillance system. The objects association is achieved by locally generating homographic lines on targets in collaborating cameras. Their proposed method is verified with real video sequences. The simulation result demonstrates that the proposed method is robust against false association because it considers all the possible pairing cases of occluded targets.

Chapter 10 by Juan Gómez-Romero, Jesús García, Miguel A. Patricio, José M. Molina, and

James Llinas, presents an approach for high-level fusion in VSNs. Information fusion techniques combine data from multiple sensors, along with additional information and knowledge, to obtain better estimates of the observed scenario than could be achieved by the use of single sensors or information sources alone. The authors study the architectural and functional issues of applying context information to improve high-level fusion procedures, with a particular focus on visual data applications. The next chapter by Majdi Mansouri, Hichem Snoussi, Jing Teng, Ouachani Ilham, and Cédric Richard, describes a technique using quantized bayesian filtering in WSNs. The focus of the chapter is to study the Bayesian inference problem in distributed WSNs with particular emphasis on the trade-off between estimation precision and energy-awareness. The authors propose a variational approach to approximate the particle distribution to a single Gaussian distribution, while respecting the communication constraints of WSNs. Simulation results demonstrate the significantly improved performance of the approach.

HARDWARE TECHNOLOGY AND APPLICATIONS

This section will present an overview concerning the hardware technology and applications for VSNs. Important issues for the hardware technology are related to the processor, storage and communications components. In WSNs, these components are often integrated together to form a device called a wireless mote. A key consideration for the wireless mote is the energy that it consumes because this will determine its lifetime. VSN technology has the potential for developing new applications in diverse areas such as smart homes and highways, environmental and habitat monitoring, industry, agriculture and security. In smart home environments, VSNs can be used to assist and improve independent living for elderly and disabled people.

Mote Technology and Energy System

Researchers have classified wireless mote platforms into three categories depending on their processing power and storage: lightweight-class platforms, intermediate-class platforms, and PDA-class platforms (Almalkawi, et al., 2010). Lightweight-class platforms have low processing power capability, small storage, and are usually equipped with basic communications only. These motes consume a very low amount of energy. An example of a lightweight-class mote platform is the FireFly (Mangharam et al., 2007). Intermediate-class platforms have better computational processing power and a larger storage memory than lightweight devices, but are also usually equipped with basic communications. An example of a intermediate-class mote platform is the TelosB (TelosB, 2011). PDA-class platforms are designed to process multimedia content and have more powerful processing capability and large storage. It supports different operating systems and multiple radios. The drawback is that they consume more energy. An example of a PDA-class mote platform is the Stargate platform (Stargate, 2011). Table 1 shows a comparison of the different mote platforms in terms of their processing power and memory.

To minimize the mote energy requirements, most commercial manufacturers allow the motes to be configured for different operational modes. For example, the Waspmote from Libelium has four operational modes: On, Sleep, Deep Sleep, and Hibernate (Waspmote, 2011). Figure 6 shows a state diagram of the Waspmote operational modes. It is important to utilize the mote in the correct mode to maximize energy efficiency. Table 1 shows the power consumption for the Waspmote modes. The normal operation mode is 'On'. The consumption in this state is 9 mA. In the 'Sleep' mode, the microcontroller is put into a latent state. It can be woken up by asynchronous interrupts and by the synchronous interrupts generated by the watchdog timeout. The duration

Table 1. Comparison of mote platforms

Platform	Mote	Microcontroller	RAM	Flash Memory
Lightweight	FireFly	8-bit ATmega128L	8 KB	128 KB
Intermediate	TelosB	16-bit TI MSP430	10 KB	1 MB
PDA-class	Stargate	32-bit PXA255 XScale	64 MB	32 MB

Figure 6. State diagram for Waspmote operational modes

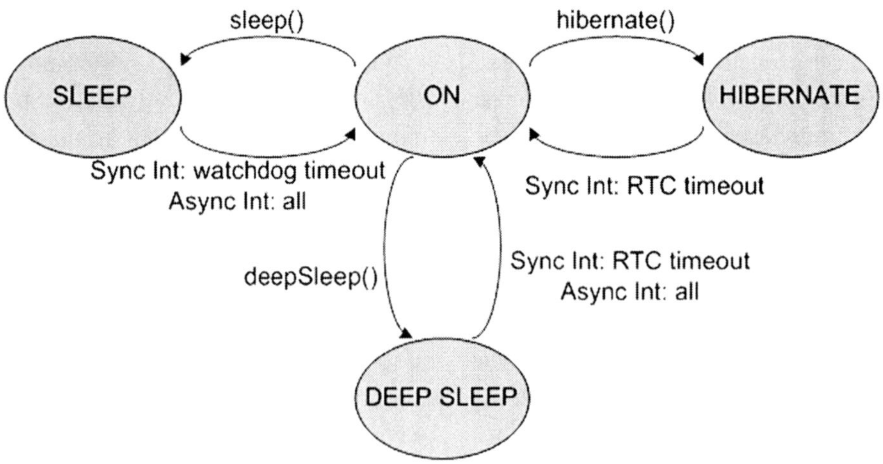

interval of this state is from 32 ms to 8 seconds. The consumption in this state is 62 μA. In the 'Deep Sleep' mode, the microcontroller is put into a latent state and can be woken up by asynchronous interrupts and by the synchronous interrupts triggered by the real time clock (RTC). The duration interval of this state is from 8 seconds to minutes, hours, or days. The consumption in this state is 62 μA. In the 'Hibernate' mode, the microcontroller and all the Waspmote modules are completely disconnected. In this mode, it can only be woken up through the programmed alarm in the RTC. The duration interval of this state is from 8 seconds to minutes, hours, or days. The consumption in this state is only 0.7 μA.

Applications of Visual Sensor Networks

A general description of various applications for VSNs and WMSNs can be found in the survey papers by (Akyildiz et al., 2007; Almalkawi, et al., 2010; Soro et al., 2009). We will focus our discussion on a recent trend of applying VSNs for assistive technology in smart environments like homes and communities. The work in (Incel et al., 2010) presents an application of VSN for ambient assistive living (AAL) in home environments. The authors propose a system framework for tracking and monitoring the health status and well-being of the elderly. The system consists of scalar sensors (temperature, motion, and biosensors) together with embedded cameras, audio sensors, and RFID tags. The embedded cameras provide monitoring of the activities of the residents. RFID tags are used for location tracking and

Table 2. Operational modes for Waspmote

	Consumption	Microprocessor	Cycle	Accepted interruptions
ON	9 mA	ON	-	Synchronous and Asynchronous
Sleep	62 μA	ON	32 ms – 8 s	Synchronous (watchdog timeout) and Asynchronous
Deep Sleep	62 μA	ON	8 s – minutes / hours / days	Synchronous (real time clock) and Asynchronous
Hibernate	0.7 μA	OFF	8 s – minutes / hours / days	Synchronous (real time clock)

to activate the associated embedded cameras in the room when an emergency is detected. Biosensors measure the blood pressure and heart beat rate and provide information on the health status and well-being of the residents. The authors propose an approach to detect falls. Their algorithm uses background subtraction, blob formation, and silhouette identification. The silhouette is then fitted on an ellipsoid model and tracked to give the speed and acceleration of the major axis. Falls are then identified and differentiated from other activities. When a fall is detected, the system sends an alarm over the network and does not transmit the video image.

Applications for assistive living need visual processing approaches to detect basic human movements. An application using audio and visual sensors to detect human movements and events like falling, walking, and standing for a smart home environment was presented in (Silva, 2008). The author defined an event and spatial relationship based approach to the problem and reported that the video sensors detected the actions at 94.44% accuracy while the audio sensors detected the actions at 83.35%. The video based fall detection accuracy was 93.3%. VSNs can also be used for application in smart communities. The work in (Yang et al., 2003) proposes an application to count the number of people in a crowd. Counting people in a crowd is a difficult problem because there would be many occlusions when people are hidden behind other objects in the scene. The authors propose an approach using a VSN to form clusters according to the geometry so that each cluster can count the number of people at a local checkpoint. The clusters communicate with each other to determine the global count of the crowd in the area. The authors propose a geometric algorithm that computes bounds on the number and possible locations of people using silhouettes computed by each sensor through background subtraction. Their system requires no initialization and operates in real-time. Their prototype cluster consists of eight networked cameras which counts people in a 144 square feet area.

For applications in private spaces, such as homes and communities, it is important to consider the social and privacy issues for the people being monitored. Traditional camera-based surveillance using people to monitor screens causes intrusion of privacy and high demands for continuous monitoring. VSNs are unobtrusive in the monitoring and can be requested to send information only when an event such as an emergency is detected. Furthermore, cameras which are "blind" and cannot take pictures can be used to minimize monitoring privacy. An example of this approach to maintaining privacy is presented in the paper by (Teixeira et al., 2006) using the concept of address event image sensors (AER). AER imagers measure a scene and looks for the most relevant data for the processing job. The imagers do not communicate pixel values explicitly, like traditional cameras. Instead, they produce a stream of addresses consisting of the addresses of the pixels that meet a certain criteria. Information about the intensity of each event is contained in the frequency of the recurring addresses.

Further Information

Further information on hardware technology and applications for VSN can be found in Section IV. The authors in chapters 11 and 12 present approaches using line scan cameras in place of two-dimensional cameras to reduce data during the acquisition process. Chapter 12 by Jiang Yu Zheng, describes the use of line sensors for monitoring and surveillance in VSNs using linear CCD sensors. It reads temporal data from a CCD array continuously and stores them to form a 2D image profile. The method delivers more information such as color, shape, and event of a flowing scene. It abstracts passing objects in the profile without heavy computation and transmits much less data than a video from normal cameras. This chapter focuses on several unsolved issues of line sensors in capturing targets in the 3D space such as sensor setting, shape analysis, robust object extraction, and real time background adapting to ensure long-term sensing and visual data collection via networks.

Chapter 13 by Mangesh Chitnis, Claudio Salvadori, Matteo Petracca, Paolo Pagano, Giuseppe Lipari, and Luca Santinelli, present an innovative technique of using line sensor based image capturing and processing in order to detect moving objects such as vehicles. The authors show that line sensor techniques may achieve faster processing results with less storage and bandwidth requirements while conserving node energy. Their solution presents itself as a low-cost candidate for Intelligent Transport Systems (ITS) to monitor and control urban traffic. The next chapter by Wai Chong Chia, Wing Hong Ngau, Li-minn Ang, Kah Phooi Seng, Li Wern Chew, and Lee Seng Yeong, presents hardware technology for VSNs using field programmable gate array (FPGA) technology. The authors present a FPGA implementation of a low-complexity and strip-based microprocessor architecture for image processing. The captured image data is processed in a strip-by-strip manner to reduce the local memory requirement. This allows an image with higher resolution to be captured and processed with the limited hardware resources. The approach is illustrated with a visual saliency application.

The remaining two chapters are concerned with applications for VSNs. Chapter 15 by Johannes Karlsson, Tim Wark, Keni Ren, Karin Fahlquist, and Haibo Li, presents a practical application of VSNs – the Digital Zoo. The authors describe their work to set up a large scale VSN in a Swedish zoo. It is located close to the Arctic Circle making the environment very hard for this type of deployment. To reach this goal the sensed data will be processed and semantic information will be used to support interaction design, which is a key component to provide a new type of experience for the visitors. The final chapter by Julien Sebastien Jainsky, and Deepa Kundur, presents a security application for VSNs. In particular, attention is brought to steganographic security and how it can be discouraged without challenging the primary objectives of the network. Preventative steganalysis aims at discouraging any potential steganographic attacks by processing the collected data such that the possibility of steganography becomes very small and the steganalysis leads to high rate of success. The authors motivate the development and implementation of more lightweight steganalytic solutions that take into account the resources made available by the network's deployment and its application in order to minimize the steganalysis impact on the workload.

CONCLUSION

Visual sensor networks (VSNs) are an emerging enabling technology with the potential of supporting new novel vision-based applications. There is a vast landscape of research areas in algorithms, architectures, and implementations which have to be studied and explored to meet its requirements and constraints. In this chapter, we gave a brief introduction to VSN technology and presented a high level overview of issues in energy efficiency, collaborative processing, and hardware technology. The rest of this book will take the reader on a journey into further depth on the technology, trends, and applications involving the inter-disciplinary areas of visual information processing, network communications, and hardware technology, which researchers have proposed to realize the full potential of VSNs in human society.

REFERENCES

Aghajan, H., & Cavallaro, A. (2009). *Multi-camera networks: Principles and applications*. Waltham, MA: Academic Press.

Akyildiz, I., Melodia, T., & Chowdhury, K. (2007). A survey on wireless multimedia sensor networks. [Amsterdam, The Netherlands: Elsevier.]. *Computer Networks*, *51*(4), 921–960. doi:10.1016/j.comnet.2006.10.002

Almalkawi, I. T., Guerrero Zapata, M., Al-Karaki, J. N., & Morillo-Pozo, J. (2010). Wireless multimedia sensor networks: Current trends and future directions. *Sensors (Basel, Switzerland)*, *10*(7), 6662–6717. doi:10.3390/s100706662

Brown, M., & Lowe, D. G. (2006). Automatic panoramic image stitching using invariant features. [Berlin/Heidelberg, Germany: Springer.]. *International Journal of Computer Vision*, *74*(1), 59–73. doi:10.1007/s11263-006-0002-3

Burt, P. J., & Adelson, E. H. (1983). A multiresolution spline with application to image mosaics. [New York, NY: ACM.]. *ACM Transactions on Graphics*, *2*(4), 217–236. doi:10.1145/245.247

Chew, L. W., Ang, L. M., & Seng, K. P. (2008). Survey of image compression algorithms in wireless sensor networks. *Proceedings of International Symposium on Information Technology (ITSim '08)* (pp. 1–9).

Chia, W. C., Chew, L. W., Ang, L. M. & Seng, K. P. (2011). Low memory image stitching and compression for WMSN using strip-based processing. *International Journal of Sensor Networks*. Geneva, Switzerland: Inderscience Publisher.

Czarlinska, A., & Kundur, D. (2008). Reliable event-detection in wireless visual sensor networks through scalar collaboration and game-Theoretic consideration. *IEEE Transactions on Multimedia*, *10*(5), 675–690. doi:10.1109/TMM.2008.922775

Fischler, M. A., & Bolles, R. C. (1981). Random sample consensus: A paradigm for model fitting with applications to image analysis. *Communications of the ACM*, *24*(6), 381–395. doi:10.1145/358669.358692

Haering, N., & Lobo, N. D. V. (2001). *Visual event detection*. Berlin/Heidelberg, Germany: Springer.

Han, B., Wu, F., & Wu, D. (2010). Image representation by compressive sensing for visual sensor networks. [Amsterdam, The Netherlands: Elsevier.]. *Journal of Visual Communication and Image Representation*, *21*(4), 325–333. doi:10.1016/j.jvcir.2010.02.007

Incel, O. D., Alemdar, H. A., & Ersoy, E. (2010). Multi-modal wireless sensor networks for ambient assisted living. *Proceedings AALIANCE European Conference on AAL* (poster paper).

Kang, L.-W., & Lu, C.-S. (2009). Distributed compressive video sensing. *Proceedings of IEEE International Conference on Acoustics Speech and Signal Processing* (pp. 1169-1172).

Kim, H., Rahimi, M., Estrin, D., & Villasenor, J. D. (2008). Energy-aware high resolution image acquisition via heterogeneous image sensors. *IEEE Journal of Selected Topics in Signal Processing*, *2*(4), 526–537. doi:10.1109/JSTSP.2008.925985

Kunt, M., Ikonomopoulos, A., & Kocher, M. (1985). Second generation image coding techniques. *Proceedings of the IEEE*, *73*(4), 549–575. doi:10.1109/PROC.1985.13184

Lee, D.-U., Kim, H., Rahimi, M., Estrin, D., & Villasenor, J. D. (2009). Energy-efficient image compression for resource-constrained platforms. *IEEE Transactions on Image Processing*, *18*(9), 2100–2113. doi:10.1109/TIP.2009.2022438

Lopes, C. E. R., Linhares, F. D., Santos, M. M., & Ruiz, L. B. (2007). A multi-tier, multimodal wireless sensor network for environmental monitoring, *Proceedings of the 4th International conference on Ubiquitous Intelligence and Computing (UIC'07)* (pp. 589-598).

Lu, Q., Luo, W., Wang, J., & Chen, B. (2008). Low-complexity and energy efficient image compression scheme for wireless sensor networks. *Computer Networks*, *52*(13), 2594–2603. doi:10.1016/j.comnet.2008.05.006

Mangharam, R., Rowe, A., & Rajkumar, R. (2007). FireFly: A cross-layer platform for real-time embedded wireless networks. *Real-Time Systems*, (37): 183–231. doi:10.1007/s11241-007-9028-z

Medeiros, H., Park, J., & Kak, A. (2008). Distributed object tracking using a cluster-based Kalman Filter in wireless camera networks. *IEEE Journal of Selected Topics in Signal Processing*, *2*(4), 448–463. doi:10.1109/JSTSP.2008.2001310

Pereira, F., Torres, L., Guillemot, C., Ebrahimi, T., Leonardi, R., & Klomp, S. (2008). Distributed video coding: Selecting the most promising application scenarios. *Signal Processing Image Communication*, *23*(5), 339–352. doi:10.1016/j.image.2008.04.002

Piccardi, M. (2004). Background subtraction techniques: A review. *2004 IEEE International Conference on Systems Man and Cybernetics* (pp. 3099-3104).

Prati, A., Vezzani, R., Benini, L., Farella, E., & Zappi, P. (2005). An integrated multi-modal sensor network for video surveillance. *Proceedings of the 3rd ACM International Workshop on Video Surveillance Sensor Networks VSSN* (pp. 95-102). New York, NY: ACM Press.

Rein, S., & Reisslein, M. (2011). *Low-memory wavelet transforms for wireless sensor networks: A tutorial*. IEEE Communications Surveys & Tutorials.

Sankaranarayanan, A. C., Veeraraghavan, A., & Chellappa, R. (2008). Object detection, tracking and recognition for multiple smart cameras. *Proceedings of the IEEE*, *96*(10), 1606–1624. doi:10.1109/JPROC.2008.928758

Silva, L. C. D. (2008). Audiovisual sensing of human movements for home-care and security in a smart environment. *International Journal On Smart Sensing and Intelligent Systems*, *1*(1), 220–245.

Soro, S., & Heinzelman, W. (2009). A survey of visual sensor networks. [Cairo/New York: Hindawi Publishing Corporation.]. *Advances in Multimedia*, *2009*, 1–22. doi:10.1155/2009/640386

Stargate X-scale Processor Platform. (2011). Retrieved from www.willow.co.uk/ Stargate_Datasheet.pdf

Stauffer, C., & Grimson, W. E. L. (1999). Adaptive background mixture models for real-time tracking. *Proceedings of IEEE International Conference on Computer Vision and Pattern Recognition*, *2*(c), 246–252.

Teixeira, T., Lymberopoulos, D., Culurciello, E., Aloimonos, Y., & Savvides, A. (2006). A lightweight camera sensor network operating on symbolic information. *Proceedings of the 1st ACMIEEE International Conference on Distributed Smart Cameras DSC.*

TelosB Mote Platform. (2011). Retrieved from www.willow.co.uk/ TelosB_Datasheet.pdf

Viola, P., & Jones, M. (2001). Rapid object detection using a boosted cascade of simple features. *Proceedings of IEEE International Conference on Computer Vision and Pattern Recognition CVPR, 1*(C), I-511–I-518.

Waspmote Technical Guide. (2011). Retrieved from http://www.libelium.com/ documentation/ waspmote/ waspmote-datasheet_eng.pdf

Wu, H., & Abouzeid, A. A. (2005). Energy efficient distributed image compression in resource-constrained multi-hop wireless networks. *Elsevier Computer Communications, 28*, 1658–1668.

Yang, D. B., González-Baños, H. H., & Guibas, L. J. (2003). Counting people in crowds with a real-time network of simple image sensors. *Proceedings 9th IEEE International Conference on Computer Vision* (pp. 122-129).

Yilmaz, A., Javed, O., & Shah, M. (2006). Object tracking: A survey. [ACM.]. *ACM Computing Surveys, 38*(4), 13. doi:10.1145/1177352.1177355

Section 2
Energy Efficient Information Processing and Transmission in VSN

Chapter 2
Energy Efficiency and Reliability Considerations of a Compressive Sensing Technique in Wireless Visual Sensor Networks

Yinhao Ding
The University of Adelaide, Australia

Cheng-Chew Lim
The University of Adelaide, Australia

ABSTRACT

This chapter focuses on the energy efficiency and reliability issues when applying the novel compressive sensing technique in wireless visual sensor networks. An explanation is given for why compressive sensing is useful for visual sensor networks. The relationships between sparsity control and compression ratio, the effect of block-based sampling on reconstruction quality, complexity consideration of reconstruction process for real-time applications, and compensation for packets missing in network flows are discussed. We analyse the effectiveness of using the 2-dimensional Haar wavelet transform for sparsity control, the difference between compressive sampling in spatial and frequency domains, and the computation of the prime-dual optimisation method and the log barrier algorithm for reconstruction. The effectiveness of the approach on recovered image quality is evaluated using Euclidean distance and variance analysis.

DOI: 10.4018/978-1-61350-153-5.ch002

INTRODUCTION

A wireless visual sensor network (WVSN) is a self-organised wireless network widely used for surveillance purposes. The network normally has a large number of sensor nodes to monitor and detects critical events in a broad physical area. The development of image sensors and wireless technologies has greatly increased the surveillance capacity of wireless visual sensor networks. In recent years, the wireless sensor networks have become more and more attractive and have a wide range of applications, such as green-house environment surveillance (Chi, Chen & Gao, 2008), weather monitoring (Chen, Zhong, Chen & Liao, 2006), road traffic surveillance and control (Xie, Zhang & Chen, 2007), health monitoring for aged care (Choi, Choi & Cha, 2008) and broad situational surveillance for defence systems. For different kinds of applications, the networks need to handle diverse sensor data, ranging from scalar data, image data (the generalised meaning of any 2-D array of numbers), and video data (from monitoring and surveillance) to the data from sophisticated medical sensors, e.g. compressed magnetic resonance imaging data (Lustig, Donoho, Santos & Pauly, 2008) for wireless network transmission. Modern image and video sensors are capable of generating a huge amount of data in order to improve the effectiveness of visual information. Therefore, efficient visual data compression techniques are essential to WVSN (Han, Wu and Wu, 2010).

The purpose of this chapter is to examine energy efficiency and reliability issues of employing a compressive sensing (CS) technique in WVSN. These issues are listed below.

1. The methods for image quality measures in human visual based monitoring system.
2. The relationship between compressive ratio and sparsity control.
3. The effectiveness of the 2-dimensional Haar wavelet for image sparsity enhancement.
4. The differences and energy efficient consideration of applying CS in spatial and frequency domains and the corresponding reconstruction techniques.
5. The effectiveness of sampling block size adjustment and network flow control to compensate for lost packets in a wireless network.

The block-based compression method used in JPEG for image and MPEG for video can significantly reduce the data rate for network transmission. However, conventional compression techniques are not completely suitable for wireless visual sensor networks. A wireless sensor network capability is severely constrained by the limited memory, battery life and micro-processing capability of sensor nodes. Conventional image compression techniques with high sampling rates generate a huge amount of data for further processing in visual sensor nodes, and the sensors' battery lives are depleted quickly.

Compressive sensing is a recent development. The CS technique acquires and reconstructs signals at a sampling rate much lower than that required by the Nyquist-Shannon criteria (Ash, 1965). During the CS sampling process, the original visual data is projected to a subspace. The energy to handle the data required by a sensor node using CS will be less than that of conventional compression techniques using a high sampling rate. The heavy sensor node computation load arising from conventional image compression, such as the use of discrete cosine transform (DCT), is not energy efficient. Furthermore, the sensor nodes may not be capable of managing large numbers of data generated by a high sample rate due to their limited memories and micro-processing capability.

Compressive sensing enables high computational tasks to be pushed to the back-end of the wireless sensor network, such as the base station, which offers an effective way for the sensor network to reduce traffic load and prolong battery life. At the back-end, the reconstruction process aims to recover original sparse or nearly sparse signals from undersampled data. This process involves solving an underdetermined matrix equation. Many existing methods can deal with these kinds of convex optimisation problems. An example is l_1-MAGIC (Candès & Romberg, 2005), which uses interior point methods to solve these linear and second-order cone programs. We will consider the energy efficiency issue of CS by focusing on solving the sparsity control problem prior to CS, and testing the complexity of a variety of methods that solve convex optimisation problems in the reconstruction process.

Compressive sensing can be applied in both the spatial domain and the frequency domain. Spatial domain sampling is more energy efficient, as it allows for the elimination of any image transform steps. Data compression is completely incorporated into the data acquisition process without further data processing. However, the compression ratio and reconstruction quality depend heavily on the sparsity of the original sensed image. We will explore their relationship further in the "Energy efficient sparsity control" section. Frequency domain compressed sampling requires more computation, but gives more flexibility in compression ratio control and provides extremely high reconstruction quality (Candès, Romberg & Tao, 2006). Frequency domain compressive sensing can be used as an extra stage in a conventional compression technique for a higher data compression ratio.

We will also consider the problems and effects of packets missing during the data transmission, and the methods to compensate for the packets lost in a CS-based WVSN. The effect of a lost packet in CS-based WVSN is significantly different from that in the case of conventional image compression, since the distortion is spread over the whole image. We consider the adjustment of sampling block size and network flow control collaborating with the Haar wavelet transform to reduce the effect of a lost packet.

The reconstructed image quality is evaluated by variance analysis in the sense of the Euclidean metric (l_2). We will introduce the technique of applying Euclidean distance and human visual system (HVS) correction in the reconstructed image quality evaluation.

The chapter presents the background to the compressive sensing technique and explores the linear relationship between sparsity and compression ratio. Test results are given to explain how the 2D Haar transform can effectively reduce the number of non-zero pixels in an image. We show that CS in the spatial domain is more efficient in a WVSN of limited energy and data processing capability. Meanwhile, frequency domain sparse sampling is another useful tool for further compressing visual signals in JPEG-based networks. The mathematical model of primal-dual and log-barrier methods for image reconstruction is explained briefly, together with some tests. All simulation results given in this chapter are tested in MATLAB 7.9 with the image processing toolbox V6.4.

BACKGROUND ON COMPRESSIVE SENSING

The idea behind compressive sensing can be traced back to the 1970s, when the images of reflective layers within the earth were constructed by seismologists based on incomplete data that did not seem to satisfy the Nyquist-Shannon criterion. A comprehensive description and explanation about compressive sensing was conducted by Candès, Romberg and Tao (2006) and Donoho (2006). The central idea in compressive sensing (or compressive sampling, or sparse sampling) is that the number of samples needed to capture a signal depends primarily on the data structural content, rather than its bandwidth (as described by Nyquist-Shannon theory). The data structural content refers to the sparsity level and energy distribution in an image. Compressive sensing

Figure 1. An example of compressive sensing in a sparse 2D image

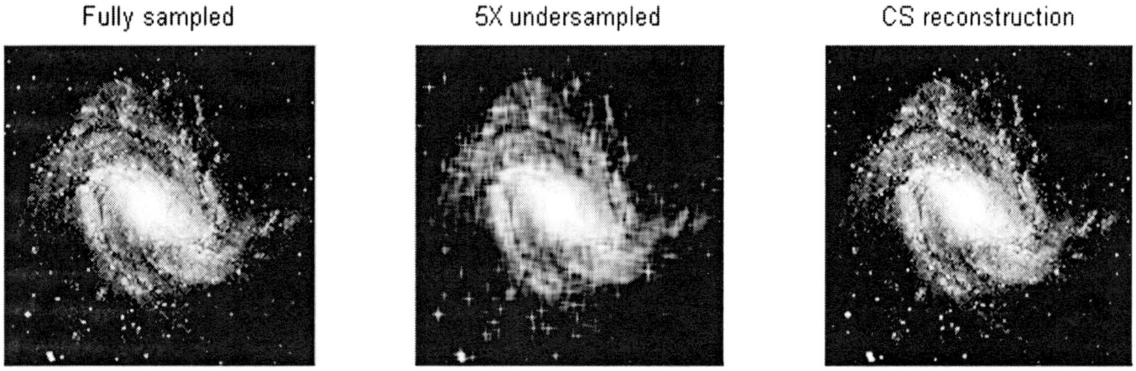

is a signal acquisition and reconstruction theory asserting that one can exploit the sparsity property in a compressible image when capturing signals (Chartrand, Baraniuk, Eldar, Figueiredo & Tanner, 2010). That is, one can design sensing techniques that condense the information or energy in a sparse signal into a small amount of data (Donoho, 2006). This fact may change the traditional way of data acquisition based on signal bandwidth. As an example, the following images show how a sparse image can be nearly exactly reconstructed (0.15% difference in the Euclidean sense) from incomplete data acquisition (5 times undersampled). The original data is undersampled in the frequency domain and reconstructed using the log-barrier algorithm provided in the l_1 MAGIC MATLAB routines (Candès & Romberg, 2005).

Let us consider the simplest case: time domain undersampling of a vector. The compressive sensing process can be modelled as a random linear measurement of a sparse vector $x_0 \in \mathbf{R}^N$. Denote the measurement by a $K \times N$ matrix $A \in \mathbf{R}^{K \times N}$. Then the compressively sampled data can be represented by a small number of linear measurements $b = Ax_0 \in \mathbf{R}^K$, where $K \ll N$ (or, $b = Ax_0 + \varepsilon \in \mathbf{R}^K$, , when there is measurement and channel noise in WVSN). The central result of CS states that the data can be recovered from this highly incomplete information by solving a convex optimisation program (Candès, Romberg and Tao, 2006). This process raises two important questions. First, is it possible to design $K \ll N$ sensing schemes to capture almost all the information about x_0? Second, how can one approximate x_0 from this incomplete information?

The answer to the first question is that the signal must be sparse or compressible to allow the CS process to condense the information by undersampling. How much the data can be compressed directly depends on how sparse the sensed data is (Candès, Romberg & Tao, 2006; Blu, Dragotti, Vetterli, Marziliano & Coulot, 2008). The following tests show how sparsity affects the data reconstruction quality.

The original signals are 512 length vectors, and contain 20 and 80 spikes in Figure 2 (a) and Figure 2 (b), respectively. After a random linear measurement by a 100 × 512 matrix, the compressively sampled signals contain 100 sampling points. The primal-dual method is used to recover the signal from these 100 incomplete sampling points. The first test shows an exact reconstruction. As the number of non-zero points increases, the data recovery quality declines.

The second question relates to the data reconstruction methods. The reconstruction process aims to solve a convex program to find a vector $x_r \in \mathbf{R}^N$. that approximates the original sparse signal x_0. One has to find the optimal or near-optimal 'candidate', which is supposed to be the

Figure 2. (a) Exact reconstruction of a sparse signal; (b) Reconstruction of a less sparse signal

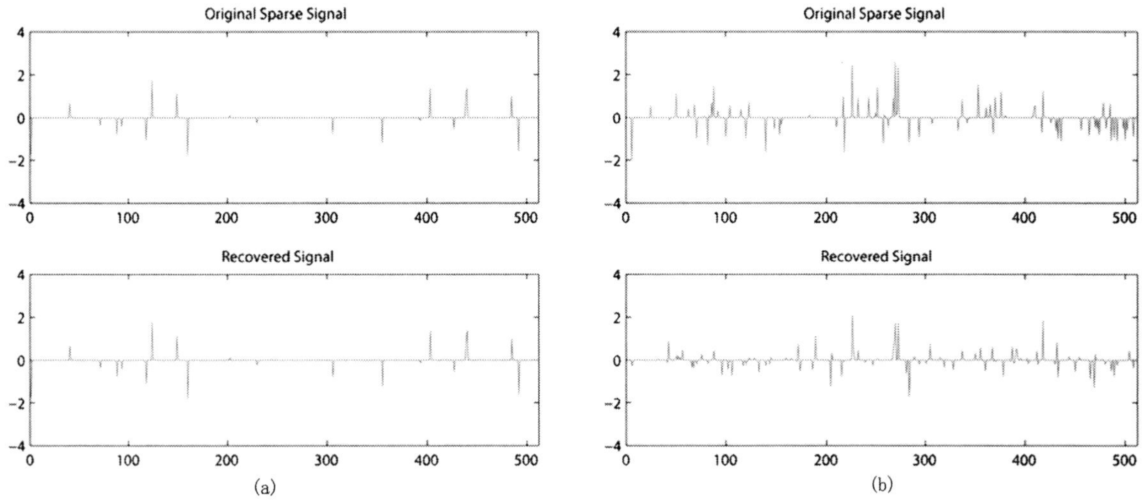

same as the original image. The recovered vector x_r should be sufficient for the incomplete measurement (i.e., $Ax_r=b$), and must also meet certain optimisation constraints, such as minimising the l_1 norm or minimising the total variation of x_r (the l_2 norm). In the above example, given these 100 observed samples, we search for optimal solutions in the set of length-512 signals that have samples that match our observations. The searching set is an affine subspace of dimension $512-100=412$. The optimisation problems are:

$$\min \|x_r\|_1 \text{ subject to } Ax_r=b, \tag{1}$$

$$\min \mathrm{TV}(x_r) \text{ subject to } Ax_r=b, \tag{2}$$

where TV stands for total variance and is defined as $\mathrm{TV}(x_r)\sum_{i=2}^{N}(\boldsymbol{x}_i-\boldsymbol{x}_{i-1})^2$. The optimisation algorithms named l_1-MAGIC (Candès & Romberg, 2005) use interior point methods to solve these linear and second-order cone programs (SOCP). The first optimisation program aims to minimise the l_1 norm and can be solved by primal-dual methods. The second optimisation program is to minimise the l_2 norm and the log-barrier algorithm is a powerful method for the solution. The "Spatial domain compressive sensing" section and "Frequency domain compressive sensing" section will give further details.

Compressive sensing normally gives a better result in frequency domain undersampling. Many natural signals have coherent representations when expressed in the spatial basis, and nearly all the image pixels have nonzero values. The Fourier coefficients offer a condensed representation of images: a majority of coefficients close to or equal to zero, which gives a better condition for compressive sensing. Most Fourier coefficients are small, and only a few large coefficients capture most of the information. This is the same insight used in many image compression techniques, e.g., JPEG ignores the large number of small terms resulting from DCT.

Compressive sensing shows us how data compression can be implicitly incorporated into the data acquisition process, which provides an energy-efficient way of data compression for WVSN. Sensor nodes only need to capture a small amount of data, and the complexity during data compression is minimised, which can avoid the computational bottleneck in sensor nodes. High computation tasks are pushed to the back-end such as the base station, where enough energy and processing capability are readily available.

IMAGE QUALITY EVALUATION

The quality of all the images tested in this chapter is evaluated in the Euclidean sense (l_2). This section briefly explains the algorithms used for image quality evaluation. Image quality measures are figures of merit used for the evaluation of imaging systems. The most frequently used measures are the deviations between original and processed image with mean square errors (MSE) or signal to noise ratio (SNR) (Avcibas, Sankar & Sayood, 2002). Mean square errors refer to the Euclidean distance and SNR is the normalised variation. Given the original image $I(i, j)$ and the reconstructed one $I_r(i, j)$, the normalised error between these two images is:

$$\varepsilon = \frac{\sqrt{\sum_{i,j} (I(i,\ j) - I_r(i,\ j))^2}}{\sqrt{\sum_{i,j} (I(i,\ j))^2}} \tag{3}$$

In order to evaluate the performance of an image compression and reconstruction algorithm, we use analysis of variance (ANOVA) as a statistical tool to evaluate the merits of the image quality on a set of images. Avcibas, Sankar and Sayood (2002) pointed out that the statistical analysis results in an F test with $N - 1$ degrees of freedom, where N is the total number of compressed images. Two statistical variables are of the most importance in the analysis:

1. **Sample mean:** F distribution has a positive mean. The smaller the mean is, the better the quality of the compressed image.
2. **Sample skewness:** The probability distribution function (PDF) of an F distribution has a long tail on the right. Thus the skewness is positive. A larger sample skewness corresponds to a better image quality, since the bulk of the values lie to the left of the mean (close to the original image).

Image quality measures describe the differences between a compressed image and the original image. In the application of human visual monitoring system, image quality measures need to be modified by human visual system (HVS) distortion, as minor image differences may cause significant visible artifacts (such as Figure 9 (b) and Figure 9 (c)). In order to obtain a closer relationship to the assessment by the human visual system, both the original and compressed images can be processed via filters that simulate human vision (Nill, 1985). One simple model for the human visual system proposed by Nill is a band-pass filter with a transfer function:

$$H(\rho) = \begin{cases} 0.05e^{\rho^{0.554}}, \rho < 7 \\ e^{-9(|\log(\rho) - \log(9)|)^{2.3}}, \rho \geq 7 \end{cases} \tag{4}$$

where $\rho = \sqrt{i^2 + j^2}$. The original image and compressed image are processed through such a spectral mask and then the inverse 2D DCT. After this human visual system modification, normal image quality measures can be applied. We use this HVS modification filter in all the tests and image quality measures.

ENERGY EFFICIENT SPARSITY CONTROL

Compressive sensing exploits sparsity natures when acquiring signals (Candès & Romberg, 2005). Thus, sparsity is an important prior condition for compressive sensing. A sparse vector in a strict sense means that all but a few of its entries are zero or near zero. Sparsity expresses the idea that the "information rate" of a discrete time signal is much smaller than the "data rate" suggested by its bandwidth. Thus, reducing of sampling points without losing effectiveness in visual information is possible. More precisely, compressive sensing exploits the fact that many natural signals are compressible due to their sparse nature (Ma, Baraniuk, Candes & Elad, 2010). Unlike JPEG (Dufaux & Nicholson, 2004), which suggests that discarding a large fraction of small coefficients would not cause much visible information in an image to be lost, compressive sensing takes the benefit of data structural content and provides a better reconstruction quality with the same compression ratio as using JPEG. In this section, we will examine the relationship between sparsity and compression ratio in compressive sensing, and propose suitable image pre-conditioning methods for sparsity control in WVSN with limited energy and micro-processing capability in sensor nodes.

Linear Relationship between Sparsity and Compression Ratio

We now consider the number of time domain samples that are required for exact reconstruction, or for reconstruction with some sufficiently small errors).

When reconstructing a signal exactly, there exists a lower bound on the number of observed samples. This minimum number is directly linked to the sparsity of original data. The purpose of this section is to analyse the relationship between signal sparsity and compression ratio. The compression ratio here is defined as the compressed image size divided by the uncompressed image size.

First, we consider a simple data set which is sparse in the frequency domain. Suppose that x is a length-100 discrete signal composing five complex sinusoids whose frequencies, phases and amplitudes are unknown. The discrete Fourier transform (DFT) of x has only five nonzero spikes. Below is a plot of DFT of x *(Figure 3)*.

Figure 3. Magnitude of DFT of five complex sinusoids

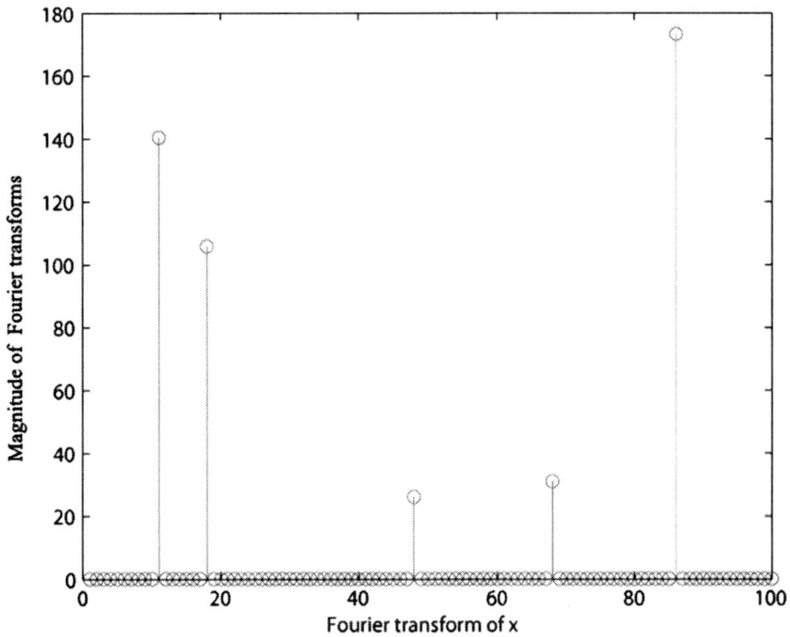

Suppose that we are given only 30 random observations. Figure 4 shows the plot of $\boldsymbol{x}$ in the time domain. The observed samples are in ⊗.

We are unable to estimate the missing samples by interpolation. However, since the signal is sparse in the frequency domain, the optimum recovered signal can be found by minimising the l_1 norm of its DFT. The problem becomes finding the optimum solution from a set of length-100 vectors. The candidates are a set of vectors in a subspace of dimension 100 − 30 (= 70), and the optimal solution is the candidate that has the smallest sum of the magnitude of the DFT. The primal-dual algorithm is the solver of this optimisation problem, and the details are in the later section on "Spatial domain compressive sensing". Through the method just described, the signal is recovered exactly.

In this example, we observe that a signal of 95% zero coefficients in the frequency domain can be exactly recovered from 30% of the random time domain samples. Indeed, the signal can be reconstructed with sufficiently small errors from as low as 20% of the random samples. Expressed mathematically, if there exist L sinusoids in the signal, we can exactly recover all the sinusoids using l1 minimisation from the order of Llog(N) samples. See Candès, Romberg and Tao (2006) for a full exposition.

So far, we have considered the compressive sensing of random time point sampling and the sparsity in the frequency domain. The result can be extended to more general compressive sampling methods and the more general sense of sparsity (sparsity in the time or spatial domain). Rather than taking K random samples in the time domain,

Figure 4. Fully sampled signal x and observed samples in the time domain

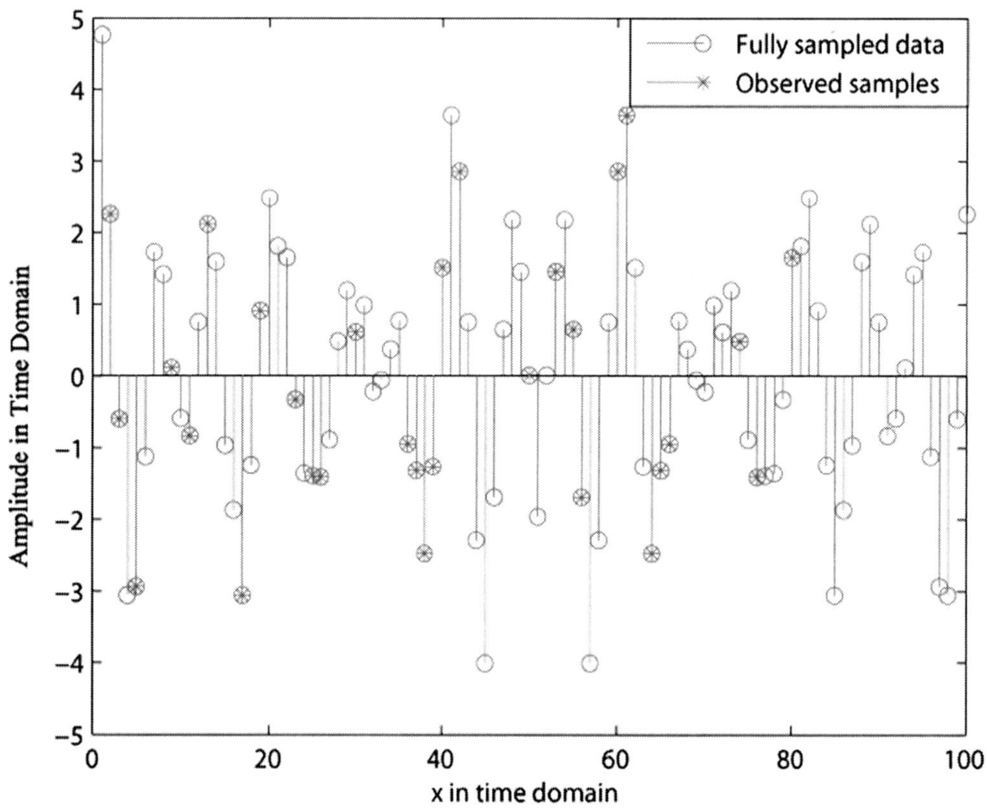

we randomly measure the signal by a $K \times N$ matrix $A \in \mathbf{R}^{K\times N}$, that is, to project the signal onto a randomly chosen K dimensional subspace. Again, the candidates form a subspace of dimension $N - K$. Then, if x is sparse in an L-dimensional orthogonal basis, and $K \geq O(Llog(N))$, we can recover x without error by minimising the l_1 norm (Candès, Romberg & Tao, 2006). The sparsity level is defined by $\xi = \frac{L}{N}$. Then, $\xi \ll \frac{1}{\log(N)}$ is required for a better compression ratio. Note that the sparsity level also means the percentage of nonzero entries in DFT, or the dimension of the orthogonal basis that can fully express the sparse signal in the spatial domain.

We conclude that as N increases, a smaller sparsity level is required in order to maintain the same compression ratio. This is usually the case in natural signals (objects of the same size in a larger background are more sparse). Since the compression ratio is $\rho = L\log(N)/N$, the relationship between the compression ratio ρ and the sparsity level ξ is linear with the slope $\log(N)$. The ratio increases as the sampling block size increases (larger N). Therefore, smaller block size sampling is more efficient for compressive sensing. We will further analyse the effect of block size on image reconstruction quality in the "Reducing the effect of missing packets" section.

Haar Wavelet Sparsity Control

Many natural images have coherent information when expressed in the spatial domain, which means that there are very few zero pixels in images. Frequency domain representation gives more condensed information. However, frequency transform requires more computation tasks in sensor nodes and consumes more energy. Image

Figure 5. Exact reconstruction of x

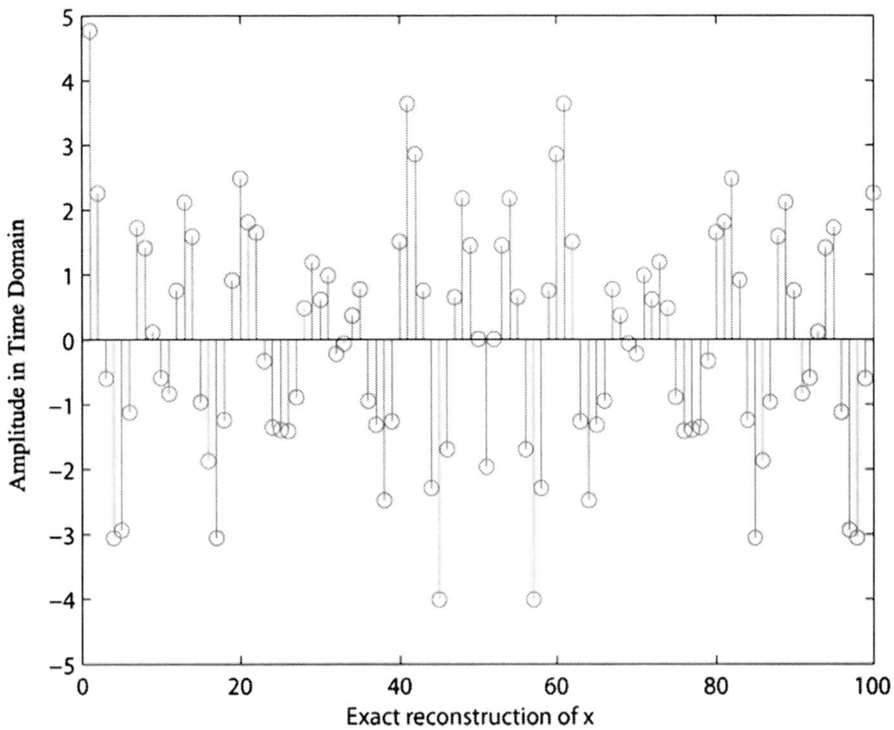

Haar wavelet transform gives an efficient way to reduce the sparsity level (proportion of nonzero pixels) in the spatial domain. A basic 4×4 Haar wavelet pattern is given below:

$$\begin{bmatrix}1 & 1\\1 & 1\end{bmatrix}, \begin{bmatrix}1 & -1\\1 & -1\end{bmatrix}, \begin{bmatrix}1 & 1\\-1 & -1\end{bmatrix}, \begin{bmatrix}1 & -1\\-1 & 1\end{bmatrix}.$$

The transform can be performed as a combination of down-sampling steps. At each step, the image is divided into four parts: highlighting the average value, vertical edges, horizontal edges, and diagonal edges. This process can greatly increase the number of zero or near zero points with minimum computations. The computation for the first step requires $3N$ fixed point additions, where N is the total number of pixels. For n step transform, the number of fixed point additions is

$$3N\times(1+1/4+1/16+...+(1/4)^{n-1}) = 4N\times(1-(1/4)^{n}) < 4N. \quad (5)$$

The reconstruction process involves repeated up-sampling and wavelet inversion steps. The figure below shows the three-step transform of the camera man image. The number of zeros pixels and near zero pixels (less than 1% of full gray scale level) in the original image is 4.8%, which increases to 77.8% after the two-step Haar transform.

This Haar wavelet transform uses a minimum number of operations, less than $4N$ fixed point additions, to greatly increase the proportion of zero points in images. This process can be used prior to compressive sensing to make the image ready for a much lower compression ratio. Since the part on the top left corner is of the most importance for reconstruction, this transform also provides a good way to control network flow. Details are discussed further in the "Reducing the effect of missing packets" section.

SPATIAL DOMAIN COMPRESSIVE SENSING

Compressed sensing of images in the spatial domain is energy efficient. The compression is performed as random measurement during sampling. This section focuses on an under-sampling

Figure 6. (a) Original camera man image; (b) Two-step Haar transform

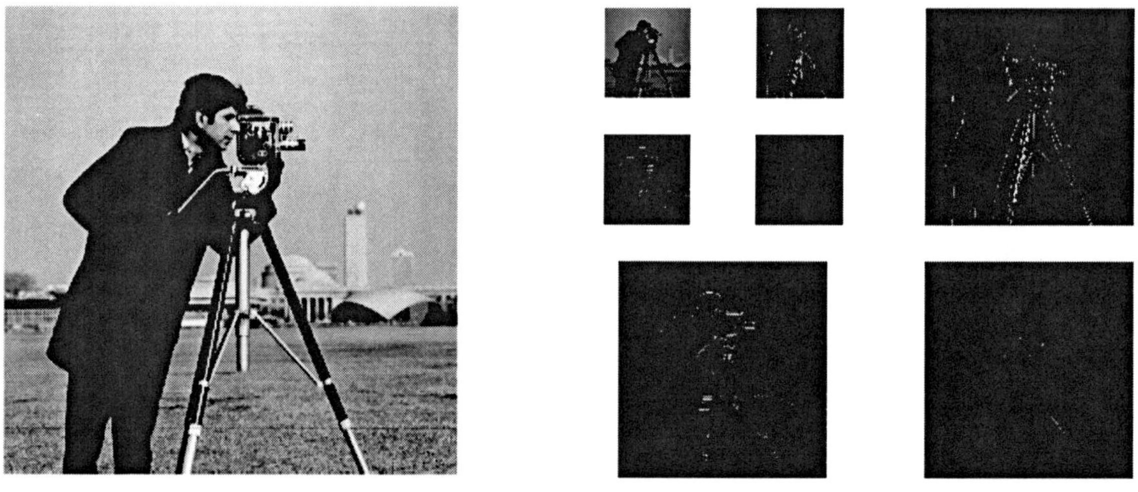

(a) (b)

method in a random K-dimensional subspace and the primal-dual method for reconstruction.

Problem Formulation of Time Domain Undersampling

We will use a randomly selected $N \times K$ matrix A for the compressed measurement (rank(A)=K, and $K \ll N$). The first issue is how to choose the compression ratio $\frac{K}{N}$. In a previous section, we have shown that the relationship between compression ratio and sparsity level is linear. The compression ratio can be chosen according to the sparse nature of the original image. Haar wavelet transform can be used to increase the number of zero or near zero pixels and provide more space for compressive sensing. The central idea of CS-based WVSN is to release the complexity in sensor nodes and relinquish most of the computations to the base station for the reconstruction task. Since we know that the source signal is sparse, a straightforward recovery method is to minimise the l_1 norm. A standard form of the linear program is expressed as:

$$\min \|x\|_1$$

$$\text{subject to } A_x = b$$

$$0 \le x_i \le 1,\ i=1,2,\ldots,N, \tag{6}$$

where $x_1, x_2, \ldots, x_N$ are the entries of vector x. Here we restrict the gray scale level to within [0, 1]. The primal-dual algorithm is a fast interior point method to solve this convex linear problem.

Primal-Dual Algorithm

The primal problem and its corresponding dual problem are shown in Table 1.

For any feasible primal-dual solution pair (x,y), we have $y^T b \le c^T x$. If x is optimal for the primal, and y is optimal for the dual, then $y^T b = c^T x$. The solution procedure is the classical Newton method. We will not go through the detailed procedure here. Instead, we will focus on using the primal-dual algorithm to solve the l_1 minimisation linear programming (LP) problem. In a standard LP form:

$$\min c^T$$

$$\text{subject to } Ax=b$$

$$f_i(x) \le 0,\ i=1,2,\ldots,2N, \tag{7}$$

where $c=1$, and $f_i(\boldsymbol{x}) = \boldsymbol{c}_i^T \boldsymbol{x} + \boldsymbol{d}_i \le 0$ is the same constraint as $0 \le x_1, x_2, \ldots, x_N \le 1$ for properly chosen $c_i \in \mathbf{R}^N$ and $d_i \in \mathbf{R}$. Then, at the optimal point x^*, we will have an optimal dual solution y^*. The Karush-Kuhn-Tucker (KKT) conditions at the optimal point are:

$$\begin{aligned} & c + A^T y^* + \sum_i \lambda_i^* c_i = 0, \\ & \lambda_i^* f_i(x^*) = 0, i = 1,2,\ldots,2N, \\ & Ax^* = b, \\ & f_i(x^*) \le 0,\ i = 1,2,\ldots,2N. \end{aligned} \tag{8}$$

The primal-dual algorithm finds optimal x^* along with optimal y^*. At every Newton iteration, we start from the previous interior point ($\boldsymbol{x}^l$, $\boldsymbol{y}^l$,

Table 1. Primal and dual programs

Primal	Dual
$\min c^T x$ subject to $Ax \ge b$ $x \ge 0$	$\max y^T b$ subject to $y^T A \le c^T$ $y \ge 0$

λ^l), and find a new interior point ($\boldsymbol{x}^{l+1}$, $\boldsymbol{y}^{l+1}$, λ^{l+1}). In practice, the condition $\lambda_i^l f_i\left(x^l\right) = 0$ to $\lambda_i^l f_i\left(x^l\right) = -\frac{1}{\tau^l}$ can be relaxed (Boyd & Vandenberghe, 2004; Candès & Romberg, 2005). By increasing τ^l as we progress through the primal-dual procedure, there will be a smooth central path from an interior point to the solution on the boundary. At each step, after finding the step direction ($\Delta\boldsymbol{x}$, $\Delta\boldsymbol{y}$, $\Delta\lambda$), we choose the maximum step size that just keeps the feasible solutions in the interior. For more detail of the proof of the primal-dual algorithm used to solve l_1 minimisation LP, see the l_1 MAGIC user guide (Candès & Romberg, 2005).

An example is shown in Figure 7. After the two-step Haar wavelet transform, the camera man image becomes a perfectly sparse expression for compressive sensing. We take 5,000 random measurements of this 65,536 pixel image. That is, the image is projected onto a 5,000 subspace. We search for the image which has the minimum l_1 norm using the primal-dual procedure introduced above. Figure 7 (b) shows that the original image has been perfectly recovered.

FREQUENCY DOMAIN COMPRESSIVE SENSING

This section considers the problem of reconstructing a signal from the partial knowledge of its Fourier coefficients. Suppose that we have a 2D discrete signal $f(i,j)$ and a set of frequencies Ω from 2D DFT. The question is "Is it possible to reconstruct signal from its incomplete DFT samples?".

Problem Formulation of Frequency Domain Undersampling

The first question is how many frequency samples we need to observe for exact reconstruction. Again, the minimum number of observations directly

Figure 7. (a) Original camera man image; (b) Reconstructed by the primal-dual algorithm and the wavelet inversion

depends on the sparsity property of the signal's DFT. The typical result is that any signal can be exactly reconstructed by convex programming from almost any set of frequency of size $0(L\log(N))$, where we assume the signal is a superposition of L spikes in the frequency domain and $N = X \times Y$ is the total number of pixels (Candès, Romberg & Tao, 2006). Given $0(L\log(N))$ samples of Fourier coefficients, the reconstruction problem is to find the best candidate by estimating the missing Fourier coefficients and minimising the l_1 or the l_2 norm.

Minimising the l_1 norm in the frequency domain (or minimal energy reconstruction) is as simple as making all of the missing coefficients zero. This strategy does not perform very well as the DFT coefficients are rarely zero. A good reconstruction algorithm proposed by Candès, Romberg and Tao (2006) is to estimate the missing DFT samples by minimising the total variance of the image in the spatial domain (in the l_2 sense). The standard form of the optimisation program is

$$\min \mathrm{TV}(g(i,j))$$

$$\text{subject to } \hat{g}\left(\omega\right) = \hat{f}\left(\omega\right) \text{ for all } \omega \in \Omega, \tag{9}$$

where ω is the observed partial frequency samples in Ω, and $\mathrm{TV}(g(i,j))$ is defined as:

$$\mathrm{TV}(g(i,j)) = \sum_{i,j} \sqrt{\left(g\left(i,j\right) - g\left(i-1,j\right)\right)^2 + \left(g\left(i,j\right) - g\left(i,j-1\right)\right)^2}. \tag{10}$$

In short, our problem is that given an incomplete frequency observation $\hat{f}\left(\omega\right)$, we seek a candidate image $g(i,j)$ with minimum complexity (total variance), and whose coefficients match those observations from the unknown image.

A powerful method to solve this second-order cone program (SOCP) is the log-barrier algorithm (Chong & Zak, 2008). A brief explanation on how the log-barrier can be used to solve the total variance minimisation convex problem follows.

Log-Barrier Algorithm

Candès and Romberg (2005) outline the log-barrier algorithm for second-order cone programs, which we used for the tests. For completeness, we briefly review their algorithms here. The standard form is:

$$\min c^{\mathrm{T}}x$$

$$\text{subject to } Ax=b,$$

$$f_i(x) \le 0,\ i=1,2,\ldots,m, \tag{11}$$

where each f_i describes either a linear or a second-order constraint. The standard log-barrier algorithm transforms the SOCP to several linear programs:

$$\min c^T x + \frac{1}{\tau^l} \times \sum_i -\log(-f_i(x))$$

$$\text{subject to } Ax=b, \tag{12}$$

where $\frac{1}{\tau^l}$ is the weighting factor for the lth step. The central idea of the log-barrier algorithm is to transform the non-linear constraints to a penalty function (log function), whose value increases exponentially when the interior points are close to the constraint boundary and becomes infinite when the constraint is violated. As τ^l becomes large, the solution approaches the solution of the original SOCP. At each step, we use Newton method to find the descent direction and we choose the step size so that the cost function decreases sufficiently while the next step solution point is still in the interior.

The following example shows a test of frequency domain compressive sampling. The original medical image is sampled in the frequency domain along 22 radius lines. The sampling rate is about 50 times lower than the Nyquist rate. Figure 8 (c) shows the effect of a missing coefficient (or the result of minimal energy reconstruction). The reconstruction using 17 log-barrier outer iterations and 58 Newton inner iterations is shown in Figure 8 (d). The reconstruction is nearly perfect (0.13% difference in the Euclidean sense).

Frequency domain compressive sensing yields extremely high reconstruction quality, since natural signals are more sparse in frequency expression. However, the cost is that image transformation (e.g., DCT) requires more computation, which may not be affordable in a typically energy constrained wireless sensor network. Nevertheless, compressed sampling in the frequency domain can be used as an extra stage for traditionally JPEG-based sensor systems to further compress data without introducing severe image distortions.

REDUCING THE EFFECT OF MISSING PACKETS

There is a significant difference between the effect of missing network packets in a CS-base network and that in a conventional compression technique-based network. In the network using a traditional image compression technique, lost packets will affect the reconstruction quality in some particular bandwidth or in some particular pixels. The lost information can normally be estimated from its neighbourhood values due to the coherent property of natural signals. However, the case of an image compressed by sparse sampling is totally different. In this section, we consider the reduction of image distortion caused by network packets lost in a CS-based WVSN. The focus is on the sampling scheme of a K-dimensional subspace projection (random measurement) that was introduced earlier in the "Spatial domain compressive sensing" section.

In projecting the N-dimensional image onto an K-dimensional subspace ($K<<N$), we randomly choose a measurement matrix of rank K. The number of measurements is also K for minimum redundancy, i.e., all of the measurements are linearly independent. Therefore, it is not possible to estimate the missing measurements from any of its neighbourhood. Furthermore, since both the measurement subspace and the missing packets are random, during the reconstruction process, the effect of lost information will spread over the whole image or the whole bandwidth. A trivial method to compensate for missing packets is to get more samples (measurements) in the front-end. However, this would require more energy and increase the network traffic. Below, we examine two efficient methods to compensate for a lost network packet.

Figure 8. (a) The test medical image. (b) Compressed sampling pattern in the frequency domain. (c) Minimal energy reconstruction. (d) Minimal total variance reconstruction.

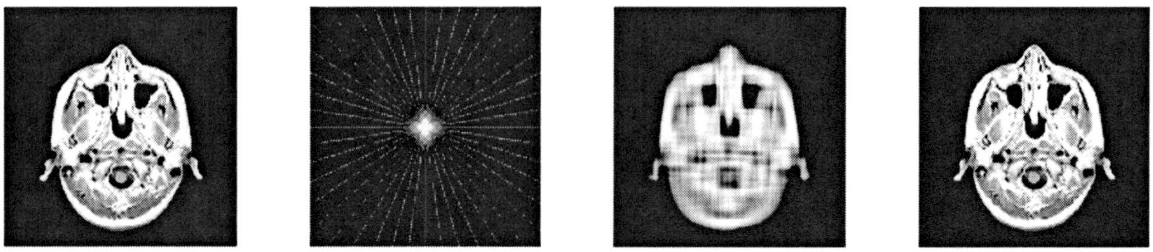

(a) (b) (c) (d)

Sampling Block Size Control

As the effect of lost measurements will spread all over the image after reconstruction, using a larger block during the sampling can effectively reduce "visible" image artifacts. Taking the 256 × 256 camera man photo as an example, we randomly compressed sample the image in each 8 × 8 and 16 × 16 blocks. Each block has one-pixel overlapping with its neighbourhood to avoid discontinuity after reconstruction. In total, we observe 5,000 measurements of this 65,536 pixel image. Every pixel is represented by 8 bits, and the total number of bits for transmission is 4.88KB. Then, to simulate the effect of packet loss, a set of randomly chosen 10% of the measurements are removed. The reconstructed image is given in Figure 9.

Although the quality of reconstructed images is nearly the same in the Euclidean sense (1.23% and 1.21% difference from original test image), there are more visually perceptible image artifacts in the 8×8 sampled image than those in the 16×16 sampled image. It is clear that this method can reduce the perceptible image distortions by spreading the effect of lost information to a larger area. However, the signal accuracy is not improved by much. The difference compared with the original test image is reduced from 1.23% to 1.21%. The next method can greatly improve the recovered signal accuracy by controlling the priority in the network flow.

Network Flow Control

We consider the adjustment of the network flow priority to reduce the effects of packet lost during the network transmission. The focus is on exploiting the special feature of the image Haar wavelet transform. That is, the transform not only reduces the number of non-zero points in the image, but also relocates the information distribution in the image. To explain, consider Figure 10 (a). Essential information is more concentrated on the left top corner. Now suppose that each part of the image suffers a different proportion of random packets missing, as illustrated in Figure 10 (b). With less lost on the top left and more missing in the bottom right area, the total number of missing packets is still 10%. Figure 10 (c) shows the reconstructed image from 10% of random missing packets and Figure 10 (d) shows the reconstruction from the uneven distribution of missing packets. Image quality evaluation shows that the Euclidean distances (normalised) between the recovered images and test image are 1.23% and 0.14%. The

Figure 9. (a) Test image. (b) Primal-dual reconstructed from 8×8 sampling with 10% of random packets missing. (c) Primal-dual reconstructed from 16×16 sampling with 10% of random packets missing.

(a) (b) (c)

Figure 10. (a) Two-step Haar transform on test image. (b) Redistribution of packets missing. (c) Recovered from randomly selected 10% of missing packets. (d) Recovered from redistributed missing percentile by network flow control.

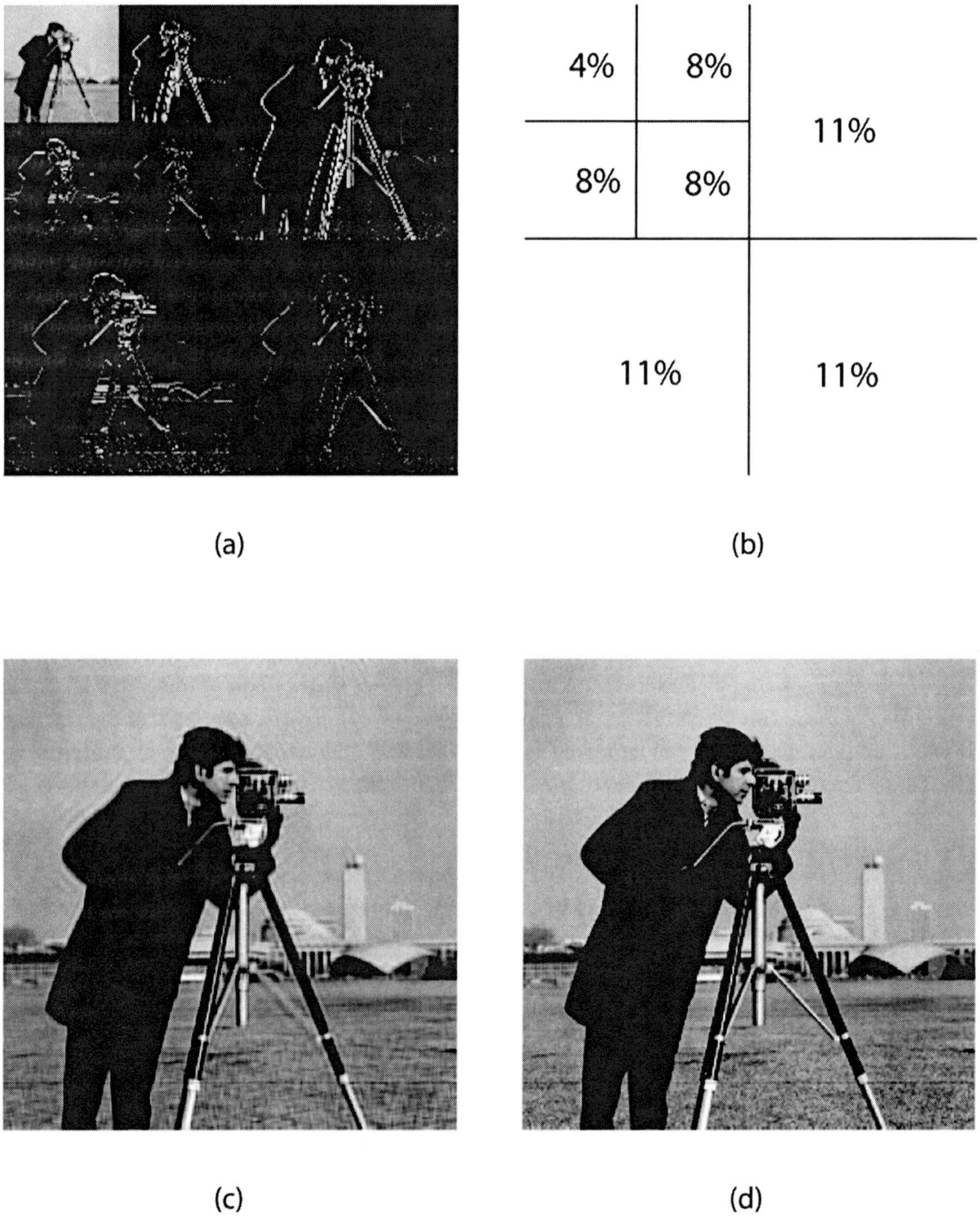

accuracy of the reconstructed image has been greatly improved.

Therefore, by setting different priority according to the importance of different parts of the image and incorporating appropriate network flow control, the effect of lost packets can be greatly reduced.

CONCLUSION

This chapter has proposed a compressive sensing based energy efficient design for visual signal compression in wireless visual sensor networks. The design consists of three parts: sparsity control, compressed undersampling and reconstruction.

The sparsity nature of a video signal is the basic prior condition for compressive sensing. The linear relationship between sparsity and compression ratio implies that appropriate image preconditioning is necessary for compressive sensing. The energy efficient 2D Haar wavelet transform uses a minimum number of computations to concentrate visual information and greatly increase the number of zero pixels. This transform can be implemented in sensor nodes without significantly draining the battery life to reduce the sparsity level for lower compression ratio. Compressive sensing explores the sparsity nature to compress visual signals during the data acquisition process. A lower sampling rate implies less energy consumption and requires less memory and micro-processing in sensor nodes. Time domain compressed sampling can be performed by randomly choosing sampling points or projecting the fully sampled signal to a randomly selected subspace. The reconstruction process aims at minimising the l_1 norm of the image and can be solved by the primal-dual algorithm. Time domain compressive sampling can greatly reduce the energy consumption in the sensor nodes, and the primal-dual reconstruction gives a high image recovery quality. Frequency domain compressed sampling requires a polar coordinate based transform as a prior stage. This sampling scheme is useful for conventional JPEG based wireless networks to further compress a video signal with a very small amount of added computation in the sensor nodes. The convex optimisation problem in the reconstruction can be solved by minimising the total variance of the image using the log-barrier algorithm.

The ability to maintain message quality, even when the message transmission is subject to packet loss, is important in wireless networks. This chapter has analysed several methods to reduce the effect of missing packets on network flow. Since in a CS-based WVSN, the effect of lost information will not be constrained to a particular spatial region of a particular bandwidth, but will spread to the whole sampling block, the block size control is helpful in missing packets compensation. Simulation results show that a larger sampling block size is more resistant to a certain percentage of missed packets. The collaboration of Haar wavelet transform and network flow control gives an alternative way to deal with the problem of missing packets. Not only can the Haar transform enhance the sparsity of a visual signal, but it also divides an image into several parts of different importance. By appropriate network flow control according to the importance of each packet, the effect of lost packets can be greatly reduced.

Compressive sensing is a novel technique for signal compression. The proposed energy efficient CS-based network is a potential network model for improving the performance of a wireless visual sensor network and prolonging the battery life of sensor nodes.

REFERENCES

Ash, R. (1965). *Information theory*. New York, NY: Inter Science Publishers.

Avcibas, İ., Sankar, B., & Sayood, K. (2002). Statistical evaluation of image quality measures. *Journal of Electronic Imaging*, *11*(2), 206–223. doi:10.1117/1.1455011

Blu, T., Dragotti, P. L., Vetterli, M., Marziliano, P., & Coulot, L. (2008). Sparse sampling of signal innovations: Theory, algorithms, and performance bounds. *IEEE Signal Processing Magazine*, *25*(2), 31–40. doi:10.1109/MSP.2007.914998

Boyd, S., & Vandenberghe, L. (2004). *Convex optimization*. Cambridge, UK: Cambridge University Press.

Candès, E., & Romberg, J. (2005). *l_1-MAGIC: Recovery of sparse signals via convex programming*. Retrieved on 10 June, 2010, from http://www.acm.caltech.edu /l1magic

Candès, E., Romberg, J., & Tao, T. (2006). Robust uncertainty principles: Exact signal reconstruction from highly incomplete frequency information. *IEEE Transactions on Information Theory*, *52*(2), 489–509. doi:10.1109/TIT.2005.862083

Candès, E., & Wakin, M. B. (2008). An introduction to compressive sensing: A sensing/sampling paradigm that goes against the common knowledge in data acquisition. *IEEE Signal Processing Magazine*, *25*(2), 21–30. doi:10.1109/MSP.2007.914731

Chartrand, R., Baraniuk, R. G., Eldar, Y. C., Figueiredo, M. A. T., & Tanner, J. (Eds.). (2010). Special issue on compressive sensing, *IEEE Journal of Selected Topics in Signal Processing, 4*(2).

Chen, S., Zhong, X., Chen, L., & Liao, X. (2006). Temperature measurement through wireless sensor networks, In *Proc. SPIE - The International Society for Optical Engineering* (vol. 6280 I).

Chi, T., Chen, M., & Gao, Q. (2008). Implementation and study of a greenhouse environment surveillance system based on wireless sensor network, In *Proc. International Conference on Embedded Software and Systems Symposia (IC-ESS2008)* (pp. 287- 291).

Choi, H., Choi, S., & Cha, H. (2008). Structural health monitoring system based on strain gauge enabled wireless sensor nodes, In *Proc. 5th International Conference on Networked Sensing Systems (INSS 2008)* (pp. 211-214).

Chong, E. P. K., & Zak, S. H. (2008). *An introduction to optimization* (3rd ed.). Hoboken, New Jersey: Wiley Interscience.

Donoho, D. L. (2006). Compressive sensing. *IEEE Transactions on Information Theory*, *52*(4), 1289–1306. doi:10.1109/TIT.2006.871582

Dufaux, F., & Nicholson, D. (2004). JPWL: JPEG 2000 for wireless applications, In *Proc. of SPIE - The International Society for Optical Engineering, 5558*(37), 309-318.

Han, B., Wu, F., & Wu, D. (2010). Image representation by compressive sensing for visual sensor networks. *Journal of Visual Communication and Image Representation*, *21*(4), 325–333. doi:10.1016/j.jvcir.2010.02.007

Lustig, M., Donoho, D. L., Santos, J. M., & Pauly, J. M. (2008). Compressed sensing MRI: A look at how CS can improve on current imaging techniques. *IEEE Signal Processing Magazine*, *25*(2), 72–82. doi:10.1109/MSP.2007.914728

Ma, Y., Baraniuk, R. G., Candes, E., & Elad, M. (Eds.). (2010). Special issue on applications of sparse representation and compressive sensing. *Proceedings of the IEEE*, *98*(6).

Nill, N. B. (1985). A visual model weighted cosine transform for image compression and quality assessment. *IEEE Transactions on Communications*, *33*(6), 551–557. doi:10.1109/TCOM.1985.1096337

Xie, W., Zhang, X., & Chen, H. (2007). Wireless sensor network topology used for road traffic, In *Proc. IET Conference on Wireless, Mobile and Sensor Networks (CCWMSN'07)* (pp. 285- 288).

Chapter 3
Wireless Video Sensor Networks:
Advances in Distributed Video Coding

Abdelrahman Elamin
Universiti Teknologi PETRONAS, Malaysia

Varun Jeoti
Universiti Teknologi PETRONAS, Malaysia

Samir Belhouari
Universiti Teknologi PETRONAS, Malaysia

ABSTRACT

Wireless Video Sensors Networks (WVSNs) generally suffer from the constraint that their sensor nodes must consume very little power. In this rapidly emerging video application, the traditional video coding architecture cannot be used due to its high encoding complexity. Thankfully, some theorems from Information Theory suggest that this problem can be solved by shifting the encoder tasks, partially or totally, to the decoder. These theorems are employed in the design of so-called Distributed Video Coding (DVC) solutions, the subject matter of this chapter. The chapter not only introduces the DVC but also reviews some important developments of the popular Stanford Wyner-Ziv coding architecture and caps it with latest research trends highlighting a Region-Based-Wyner-Ziv video codec that enables low-complexity encoding while achieving high compression efficiency.

INTRODUCTION: WVSNS AND DVC

Wireless video sensors network (WVSN) are fast finding many applications, chief among them being surveillance application in ad-hoc deployment scenarios. Examples of such surveillance applications are aplenty e.g. surveillance in a soccer-stadium, along certain traffic route, of a gathering of people in certain places etc. The video sensor nodes are deployed in an ad-hoc manner and many a times disposable. Such video sensors cannot afford the traditional encoding complexity of a video compression but still require good video compression efficiency.

DOI: 10.4018/978-1-61350-153-5.ch003

Today's digital video coding paradigm represented by the ITU-T and MPEG standards mainly relies on a hybrid of block-based transform and inter-frame predictive coding approaches. In this coding framework, the encoder exploits both the temporal and spatial redundancies present in the video sequence, which is a complex process and it requires a noticeable amount of resources (power and memory). As a result, all standard video encoders have much higher computational complexity than the decoder (typically five to ten times more complex) (Girod et al., 2005), mainly due to the temporal correlation exploitation tools used in the motion estimation task. As a result, the traditional video coding is no longer applicable for these WVSN applications. Appropriate video coding paradigm for these applications must have low encoding complexity. Lower encoding complexity can be achieved by moving some of the encoder tasks to the decoder part, particularly the complex motion estimation process.

Two notable theorems from information theory have paved the way for new video coding paradigm known in the literature as Distributed Video Coding (DVC). It allows low encoding complexity and approaches the efficiency of traditional video coding schemes. These two theorems are known as Slepian-Wolf theorem and Wyner-Ziv theorem (Slepian & Wolf, 1973; Wyner & Ziv, 1976). They suggest that, foe sources *X* and *Y*, separate encoding - joint decoding system can approach the efficiency of joint encoding-decoding system when the information about the correlation between *X* and *Y* is available at the decoder. The practical application of DVC (Aaron & Girod, 2002; Aaron, Zhang, & Girod, 2002; C. Brites, 2005; Fowler, 2005; R. Puri & Ramchandran, 2003; Rohit Puri & Ranichandran, 2003) is referred to as Wyner-Ziv video coding (WZ) where an estimate of the original frame herein called the "side information" is available at the decoder. The compression is achieved by sending only that extra information that is needed to correct this estimate. An error correcting code is often used with the assumption that the estimate is a noisy version of the original frame and the correction can be made with few extra parity bits that determine the rate. For the purpose of modeling, a virtual channel is assumed to represent the estimation noise (Westerlakena, Gunnewiekb, & Lagendijka, 2005) in the estimate of the original frame.

This chapter mainly aims at introducing the WZ video codec as light distributed video compression scheme. Section 1 covers the theoretical foundation of the distributed source coding (DSC) principle and the two major Information Theory theorems, namely the Slepian-Wolf and Wyner-Ziv theorems. Section 2 explains how the source coding, channel coding and estimation interplay to construct a distributed video coding solution. Section 3 presents a detailed review of the state-of-the-art solution, known as feedback channel DVC developed by Girod's group at Stanford University (B. Girod, 2003), as this system represents a good solution for WVSNs due to its low-encoding complexity. In section 4, some of the architectural developments which are derived from the initial Stanford architecture are given. Section 5 starts first by highlighting some performance issues in the state-of-the-art solution. It is followed by the analysis of various research efforts in the literature and capped with a solution called Region-Based DVC. Finally, as the WVSNs produce multiview video sequences the chapter introduces how to tailor the DVC system to match the feature of such video sequences.

THEORETICAL FOUNDATIONS OF DVC

This section provides the theoretical background and covers the basics concepts of distributed source coding DSC leading to DVC.

Figure 1. (a) Joint encoding of X and Y. The encoders collaborate and a rate H(X; Y) is sufficient. (b) Separate encoding of X and Y. The encoders do not collaborate.

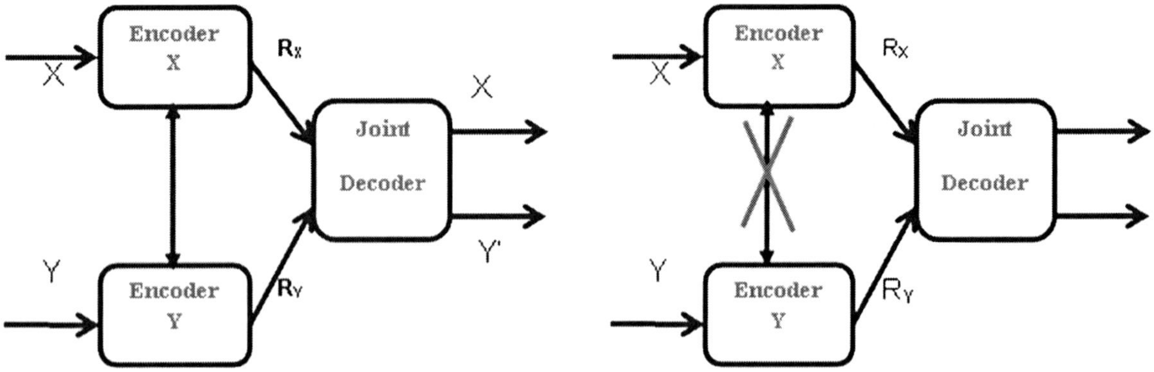

Slepian-Wolf Coding (SWC)

Let $\{(X_i,Y_i)\}$ be a sequence of independent and identically distributed (i.i.d.) drawings of a pair of correlated discrete random variables X and Y. For lossless compression with arbitrarily small error ($X = \hat{X}$ and $Y = \hat{Y}$) after decompression, it is known from Shannon's source coding theorem that a rate given by the joint entropy, $H(X,Y)$, of X and Y is sufficient if the two sources are jointly encoded (see Figure 1- (a)). For example, we can first compress Y into $H(Y)$ bits per sample, and based on the complete knowledge of Y at both the encoder and the decoder, then compress X into $H(X,Y)$ bits per sample. But if X and Y must be separately encoded for some user to reconstruct both of them the sufficient rate for lossless reconstruction is not known.

One can separately encode them with rate, $R=H(X)+H(Y)$, which is greater than $H(X,Y)$ if the two sources X and Y are correlated. In 1973, Slepian and Wolf (Slepian & Wolf, 1973) showed that $R=H(X,Y)$ is sufficient for lossless decompression, even for separate encoding of correlated sources (see Figure 1.(b)). In other words, for the rates

$$\begin{aligned} R_X &\geq H(X \mid Y) \\ R_Y &\geq H(Y \mid X) \end{aligned} \tag{1}$$

The combined rate is given by

$$R_Y + R_x + \geq H(X,Y) \tag{2}$$

The proof of the Slepian-Wolf theorem is based on random binning, which is non-constructive, i.e., it does not reveal how practical code design should be done (Xiong, Liveris, & Cheng, 2004). Binning is a key concept in DSC and refers to partitioning the space of all possible outcomes of a random source into disjoint subsets or bins. Wyner suggested the use of channel coding to approach the corner points A, B (Figure 2) of SW limits. Consider a binary sequence, X, to be encoded and a noisy version of it, Y, present at the decoder (Figure 4). To correct the errors between these two binary sequences, a channel code may be applied to the sequence X. The compression of X is achieved by transmitting only the parity bits. Hence, jointly with Y, the decoder uses the parity bits produced by the encoder to make error correction, achieving a perfect decoding of the sequence X.

Wyner-Ziv Coding

Wyner and Ziv (Wyner & Ziv, 1976) have studied a particular case of Slepian-Wolf coding corresponding to the rate point

Figure 2. The Slepian-Wolf Rate Region for Two Sources

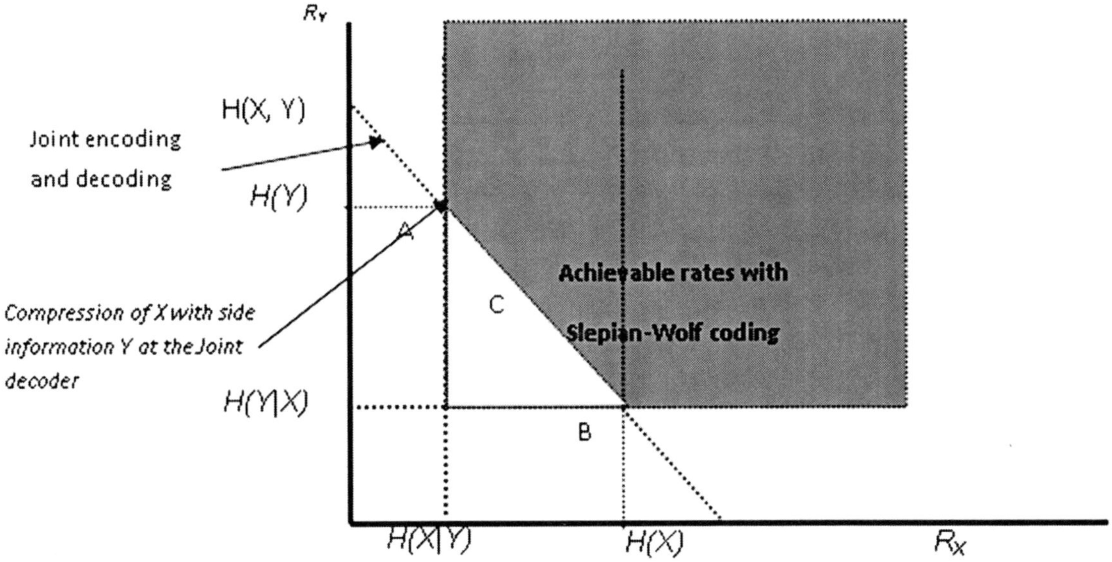

Figure 3. Relationship between channel coding and Slepian-Wolf coding

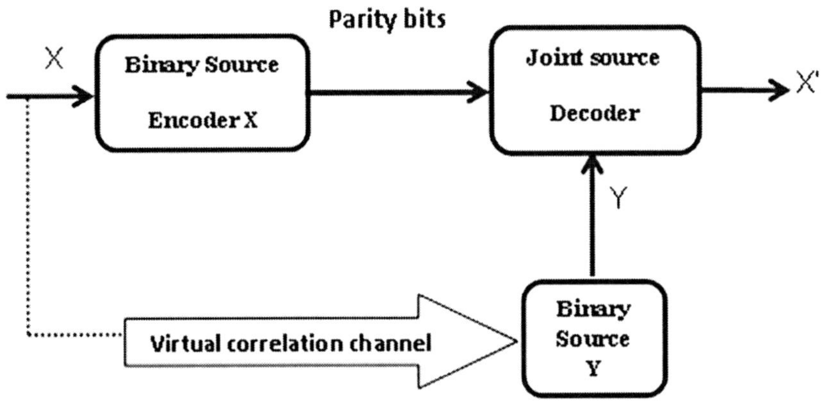

$(H(Y),H(X,Y))$ where X and Y are correlated sources. The work of Wyner-Ziv (Wyner & Ziv, 1976) has established the rate-distortion (RD), $R^*_{WZ}(D)$, necessary to encode X guaranteeing its reconstruction with an average distortion below D, assuming that only the decoder has access to Y. The results obtained by Wyner and Ziv indicate that when the statistical dependency between X and Y is exploited only at the decoder, the transmission rate increases compared to the case where the correlation is exploited both at the encoder and the decoder, for the same average distortion, D. Mathematically, the Wyner and Ziv theorem can be described by

$$R^*_{WZ} \geq R_{X|Y} \qquad D \geq 0 \tag{3}$$

where $R^*_{WZ}(D)$ represents the minimum encoding rate for X and $R_{X|Y}$ represents the minimum rate

Figure 4. Block diagram of the basic Wyner-Ziv codec

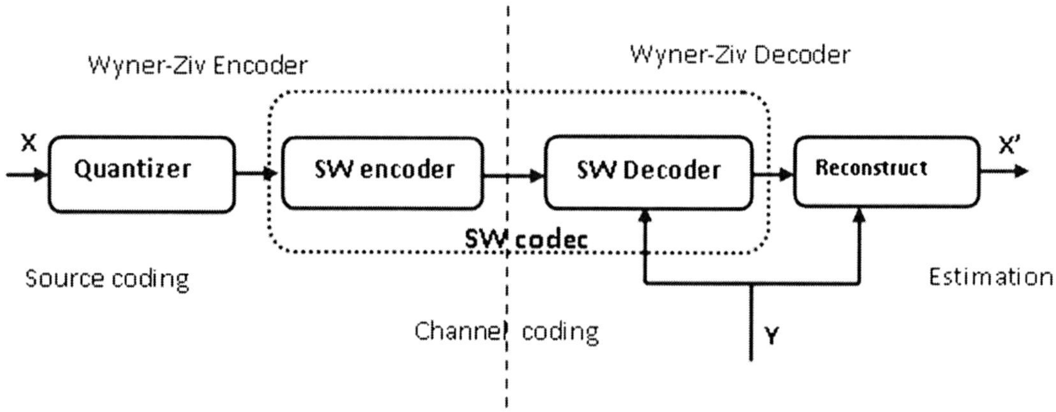

necessary to encode X when Y is available at the encoder and the decoder. Wyner and Ziv showed that there is no rate increase for all

$D \geq 0$, when X and Y are jointly Gaussian sequences and D, an MSE distortion measure. A major work on practical Wyner-Ziv code design called DISCUS (Pradhan & Ramchandran, 2003) recently extended the no rate loss condition for Wyner-Ziv coding even for case where only the innovation between X and Y is Gaussian, and X and Y could follow more general distributions. Because distortion is introduced to the source in WZC, source coding (or quantization) is needed to quantize X into X_q quantization indices. Typically, there is still correlation remaining in the quantized version of X and the side information Y and SW codec should be employed to exploit this correlation to reduce the rate. The SW decoder recovers the source codeword (or an estimate X'_q of the quantization index). The SW decoder is followed by a Minimum Mean Square Error (MMSE) estimation which searches for X', the reconstructed value of X, minimizing the expectation. $E[(X-X')^2|X_q, Y]$ The WZ coding involves three elements; the signal processing (source coding), communications (channel coding), and estimation theory.

In conclusion, the Slepian-Wolf and the Wyner-Ziv theorems suggest that it is possible to compress two statistically dependent signals in a distributed way (separate encoding, joint decoding) using a rate similar to that used in a system where the signals are jointly encoded and decoded like in traditional video coding schemes.

BASIC WYNER-ZIV VIDEO CODING ARCHITECTURE

Video coding scheme based on WZ coding system is known as Wyner-Ziv video coding. Such system architecture encloses three main modules; each one to perform a single function of the WZC elements. In traditional video coding schemes, the X and Y sources can be, for instance, the odd and even frames of a video sequence respectively. Following is a simplified walkthrough of the basic Wyner-Ziv coding architecture as shown in Figure 4.

Quantizer (Source Coding)

The source X is the pixel values of the even frames. The samples of the sequence X are first quantized using a uniform quantizer. The quantization compresses a range of values in the source into a single value and the space of a signal is divided into intervals called bins. Since the number of bins is less than the total number of the values a signal can take, the total bitrate is reduced. With the purpose of reducing the bitrate, the quantization stage is also applied to DVC. The conditions of a quantizer that suits the DVC were studied in (Rebollo-Monedero, Aaron, & B. Girod, 2003) where it studied the optimality conditions which such quantizers must satisfy, together with an extension of the Lloyd algorithm for a locally optimal design. Also the (Rebollo-Monedero et al., 2003) provided a theoretical characterization of optimal quantizers at high rates and their rate-distortion performance, and applied it to develop a theoretical analysis of orthonormal transforms for distributed coding. Bits are extracted from the quantized symbols (q) and grouped together to form bitplane starting from the most significant bit (MSB) toward the least significant bit (LSB).

Slepian-Wolf Codec (Channel Coding)

The goal of the Slepian-Wolf codec is to reduce the transmission rate of the main information so it can approach Slepian-Wolf limits. This process is performed by channel coding techniques. The Slepian-Wolf encoder module plays the role of generating redundant information (parity bits) as in channel codes. The Slepian-Wolf encoder is fed by the bitplanes extracted from the quantized symbols. The Slepian-Wolf encoder outputs the systematic bitplane and the parity bits. The systematic output is discarded and only the parity bits are transmitted. The idea is that, Y being the noisy version of X, the parity bits can correct this noisy estimate Y. At the basic Wyner-Ziv decoder, the quantized symbol stream is decoded through joint source-channel decoding with the aid of the sequence Y, hereafter also called the side information. Practical Slepian-Wolf coding is implemented by using near-Shannon-limit channel coding techniques such as Syndrome (Rohit Puri & Ranichandran, 2003), Turbo codes (Aaron et al., 2002; C. Brites, 2005) and Low-density parity check codes LDPC (X. Artigas et al., 2007).

Reconstruction (Estimation)

In this final stage of reconstruction (estimation), an estimate of the main information X is obtained using the decoded (q') quantized symbols and the side information Y. In DVC framework, the optimal reconstruction X' is the value that minimizes the distortion $d(X,X')$. In (Girod et al., 2005), the reconstruction is achieved by maximizing the conditional expectation (3) by choosing appropriate X' so as to reconstruct the source information X' (pixel value or DCT coefficient)

$$E(X|q',Y) \tag{4}$$

assuming a Laplacian distribution to model the residual between corresponding elements of X and Y (side information).In other works (Fowler, 2005; R. Puri & Ramchandran, 2003), a linear estimate algorithm is used to reconstruct the DCT coefficients as in (4)

$$x'=a_{0\cdot q}+a_1\cdot y \tag{5}$$

where a_0 and a_1 are obtained offline from certain training procedure.

Side Information (Y) Generation Process

Since the compression is achieved by sending the information that is needed to correct this side information Y, the reconstruction process of the decoded frame also incorporates the side information Y and the quantized symbol to reconstruct the original frame X. Therefore, if the Y is very similar to the original information X, few coding errors need to be corrected and less parity bits have to be sent to the decoder, thus allowing better overall compression. As the quality of the side information affects the distortion on the reconstructed frame, the performance of DVC system is fundamentally determined by it

(Aaron & Girod, 2002; Areia, C.Brites, Ascenso, & Pereira, 2007; Eduardo Peixtoto, L, Queiroz, & Mukherje, 2008; J. Ascenso & Pereira, 2007; Meyer, Westerlaken, Gunnewiek, & Lagendijk, 2005; Natrio, Brites, Ascenso, & F. Pereira, 2005; Yaacoub, Farah, & Pesquet-Popescu, 2009; Yangli, Jechang, & Chengke, 2008). To achieve better coding performance in terms of RD, one must make available at the decoder a better estimate of the side information. Although there exist temporal correlation between the odd frames and even frames, in cases where frame rates are low and the correlation is small between the odd and even frames, the performance of WZ video coding is degraded in terms of RD. Many studies have been conducted to obtain better estimate of the side information from temporally adjacent odd frames to improve the coding performance. Two major approaches found in the literature are:-

- The trajectory based motion compensation.
- The hash based motion estimation.

The *trajectory based motion estimation* approach completely adheres to the low complexity encoder constraints. The encoder does nothing to help the decoder estimate side information. This requires the decoder to fully exploit the temporal redundancy between adjacent odd frames. Research in the area of motion estimation/compensation done for H264 has provided many useful insights into the reconstruction of true motion information without the original video sequence. For example, linear motion model can be used, in which it is assumed that the motion in the current frame is a continuous extension of the previous frames' motion. While this is true for some video sequences, the motion of natural video sequences is not well defined and a simple model can be inadequate (Thoma & Bierling). Several implementations methods have been proposed for the trajectory based motion estimation approach and they can be divided into two classes, namely, the interpolation (Ascenso, Brites, & Pereira, 2005) based method and the extrapolation (Borchert, Westerlaken, Gunnewiek, & Lagendijk, ; Natrio et al., 2005) based methods.

The *hash based* approach, however, simply requires the encoder to participate and send some information to the decoder to help in producing the side information. The problem of extracting some information to help generate the side information is a tradeoff between extracting sufficient information and lowering the complexity of the encoder (Aaron, Rane, & Girod, 2004).

Virtual Channel (Estimation Noise) Modeling

Exact knowledge of the statistical dependence between *X* and *Y* is required (1) in order to make use of the side information to characterize the channel in the SW decoder; (2) to perform MMSE estimation in the inverse quantizer; and (3) to help control the SW code rate (Meyer et al., 2005). It is common in DVC literature (Aaron & Girod, 2002; Aaron et al., 2004; Baoguo & Hong, 2009; C. Brites & Pereira, 2007; Girod et al., 2005; Westerlakena et al., 2005) to use a Laplacian distribution to model the statistical correlation between the original frame and the side information. This distribution is used to convert the side information (pixel or DCT coefficient values) into soft-input information needed for channel decoding (Steinberg & Merhav, 2004). The correlation noise is interpreted as a virtual channel with an error pattern characterized by the said distribution (or model) since *Y* may be seen as a "corrupted" version of the original information. (*X-Y*) corresponds, in the Pixel domain scenario, to the residual between the *X* and *Y* frames while it corresponds, in the Transform domain scenario, to the residual between corresponding DCT bands of the *X* and *Y* frames. The virtual channel model is given by:

Figure 5. Stanford Wyner-Ziv video codec from ***(Girod, Aaron, Rane, & Rebollo-Monedero, 2005)***

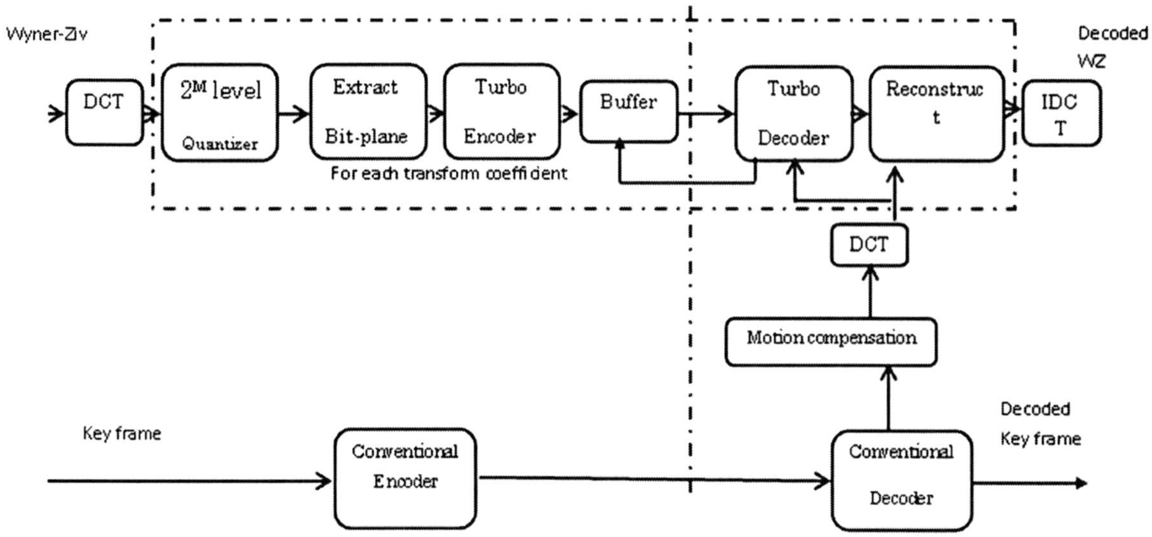

$$p[X-Y] = \frac{\alpha}{2} e^{-\alpha[X-Y]}$$
$$\alpha = \sqrt{\frac{2}{\sigma^2}} \qquad (6)$$
$$\sigma^2 = E([X-Y]^2) - (E[X-Y])^2$$

where $p(.)$ is the probability density function and α, the Laplacian distribution parameter. A bitplane-wise correlation model thus needs to be derived for controlling both the SW coder and decoder. A first model assumes the correlation channel between bitplanes of same significance of WZ and SI data to be binary-symmetric and characterized by a crossover probability, $p_{co,i} \equiv P_r(X_i|Y_i)$ which varies from bitplane to bitplane.

STANFORD WYNER-ZIV VIDEO CODING SOLUTION

One of the first practical WZ video coding solutions has been developed at Stanford University by Bernd Girod's group (Aaron & Girod, 2002; Aaron et al., 2004; Aaron et al., 2002; Ascenso et al., 2005; Girod et al., 2005). It is also referred to by Feedback Channel approach. This solution has become the most popular WZ video codec design in the literature, due to its simple encoder. The video frames are split into key frames and WZ frames. The key frames, KF, are encoded and decoded conventionally as in intra-frame codec. In between the key frames are Wyner-Ziv frames, which are intra-frame encoded but inter-frame decoded. In the following, we will present a more detailed description of the solution proposed by Bernd group [Aaron] illustrated in Figure 5. The encoding/decoding of the key frame follows the conventional intra-frame codec using H263+.

The WZ frames encoding and decoding are as follows:

Wyner-Ziv Encoding Procedure

Transform (DCT): The Wyner-Ziv frame is divided into 4-by-4 blocks and the resulting blocks are transformed using discrete cosine transform (DCT). The transformed coefficients of the entire frame are grouped together according to the position of the coefficient within the 4-by-4 blocks, forming transform coefficients bands.

Quantization: Uniform quantizer is designed for each transform coefficient bands with 2^{Mk}

levels. The quantization step is determined based on the number of the levels. This stage results in the quantized symbol bands.

Extract Bitplanes: The bits of quantized symbols band of the same index (importance, e.g. MSB) are grouped together forming a bitplane arrays.

Turbo Encoder: The bitplane arrays are fed individually to Turbo encoder. The generated parity information for each bitplane is stored in the buffer and transmitted in small amounts upon request from decoder via feedback channel.

The Slepian-Wolf codec is built based on Rate Compatible Punctured Turbo (RCPT) code structure. In a RCPT code structure, after turbo encoding an L-bit input sequence, the parity sequence generated by Turbo encoder is divided into P blocks of (L/P) bits each, where L is the interleaver length and P is often called the puncturing period. The encoder starts by sending one (L/P)-bit block of the parity sequence. If the decoder requests for more parity bits via the feedback channel, then the encoder transmits one more (L/P)-bit block of the parity sequence and so on, until no more requests for parity bits are made or all the parity bits have been transmitted.

The Wyner-Ziv Decoding Procedure

The key frames are decoded traditionally as in H.264 and then used in the next adjacent Wyner-Ziv frame decoding to generate estimate by means of motion estimation. The decoding of Wyner-Ziv is as following:

Motion-Compensation Extrapolation: The previously decoded key frames are used to generate an estimate of the Wyner-Ziv frame. DCT bands are then obtained with same procedure performed at the encoder side. To use side information at the Turbo decoding stage, a statistical dependence model between corresponding coefficients bands is considered. The solution uses Laplacian distribution to model the difference between corresponding bands in the original and the estimate of Wyner-Ziv frame.

Turbo Decoder: For each transform coefficient band, the Turbo decoder starts decoding the most significant bitplane followed by the decoding of the remaining bitplanes. The received parity bits and the side information together are used to decode the bitplanes. If the decoding fails, the decoder requests more parity bits via feedback channel. The usage of the feedback channel simplifies the rate control problem since the decoder can easily regulate the necessary bit rate as it knows the available side information. The request and decoding are repeated until the current bitplane error probability is lower than 10^{-3}. In general, due to the availability of the side information, the number of Wyner-Ziv bits required to determine in which quantization interval (level) a transform coefficient is mapped to, from the 2^{M_k} possible levels, is lower than M_k bits and thus compression efficiency achieved. By successfully decoding the last bitplane (LSB), the decoder has M_k bitplanes from which it can obtain the quantized symbol bands.

Reconstruction: The reconstruction of each transform coefficient band is performed based on the reconstruction method (reconstruct1) that uses the available quantized symbol and the side information.

Inverse Transform (IDCT): A block based 4-by-4 inverse discrete cosine transform is performed on the reconstructed transform coefficients.

A simpler system in pixel domain has also been developed [6] based on Stanford scheme, which generically makes use of a quantizer, a Turbo codec, a reconstruction module and an interpolation module.

ARCHITECTURAL DEVELOPMENTS

Due to its low encoding complexity the Stanford approach has become the most popular DVC system in the DVC research community. Although the architecture is same, different implementations have been made to improve its performance. Some have focused on the side information generation process (Ascenso et al., 2005), and some on the SW codec by employing the LDPC (X. Artigas et al., 2007), while others have focused on the feedback channel targeting its removal or its usage reduction. Among these works, DISCOVER project was the most significant contribution and is briefly described hereunder as follows:

DISCOVER uses the rate-compatible LDPC Accumulate (LDPCA) codes as Slepian-Wolf codec. An LDPCA encoder is composed from LDPC syndrome-former concatenated with an accumulator. For each bit plane, the created syndrome bits are accumulated modulo-2 to produce the accumulated syndrome. To establish if decoding is successful or not, the convergence is tested by computing the syndrome check error, i.e. the Hamming distance between the received syndrome and the one generated using the decoded bit plane, followed by a cyclic redundancy check (CRC). If the Hamming distance is different from zero, then the decoder proceeds to the next iteration. After a certain amount of iterations (experimentally it was found that 100 is enough), if the Hamming distance remains different from zero, then the bit plane is assumed to be erroneously decoded and the LDPCA decoder requests for more accumulated syndromes via the return channel.

In motion compensation interpolation (MCI), a motion field closer to the true motion is estimated between backward X_b (past) and forward X_f (future) reference frames. Then motion compensation between the two references is performed to obtain the side information. A block matching based on a modified MAD (mean absolute difference) criterion is used in order to regularize the motion vector field, which favors motion vectors closer to the origin. Then, bidirectional motion estimation is performed in order to find symmetric motion vectors from the current WZ frame to X_b and X_f. Spatial motion smoothing based on a weighted vector median filter is applied afterwards to the obtained motion field to remove outliers. Finally, motion compensation is performed between X_b and X_f along the obtained motion field, so as to generate the side information.

Minimum rate estimation module: In order to reduce the number of accumulated syndrome requests to be made by the decoder (which has a strong impact on decoding complexity), the encoder estimates a minimum number of accumulated syndromes to be sent per bit plane and per band.

ISSUES ON FRAME BASED

In this section, we review some issues that the frame based system presented in Figure 5 is still plagued with, namely, the accuracy of the correlation noise model, the poor rate control and the non-adaptive quantization.

Issue 1 – Inaccurate Correlation Noise Model: Although more sophisticated ME algorithms can be used to generate the side information, still the practical frame prediction is fundamentally faulty due to events like occlusions. Due to the limitation of side information prediction, the predicted frame is expected to have varying degrees of success along the predicted frame (Catarina Brites & Pereira, 2008; Meyer, 2004; Meyer et al., 2005; Westerlakena et al., 2005). These limitations result in location-specific non-stationary estimation noise, for example, when occlusion occurs. Although this last property is a problem but it can be accommodated within a SW codec without extra information about the location-specificity at the cost of a rate penalty. Distributed source coding relies heavily on efficient error correcting codes. The performance of these codes depends greatly on the choice of the noise model that characterizes the virtual dependency channel (Westerlaken, Borchert, Gunnewiek, & Lagendijk, 2006; Westerlakena et al., 2005). The lack of information about the unpredictable occlusion

Table 1. KS goodness-of-fit test

Video sequence	Resolution	Laplacian	Gaussian	Lap Mixture	Gauss Mixture
Carphone	144X176	0.3533	0.5423	**0.0248**	0.0477
Miss America	144X176	0.2504	0.7409	**0.0439**	0.0644
Foreman	144X176	0.3011	0.6840	**0.0348**	0.0641
Hall Monitor	144X176	0.3435	0.5856	**0.0335**	0.0428
Coast Guard	144X176	0.3257	0.5645	**0.0218**	0.0427
Akyio	144X176	0.3350	0.6340	**0.0391**	0.0656
Mother & daughter	144X176	0.3068	0.7530	**0.0506**	0.0655

noise can be defeated by the use of interleaver in the Turbo based systems because interleaving eliminates the location specificity of the noise at the cost of some rate penalty. As a result of the spatial variation in the side information, the noise process for the entire frame is represented by a group of uniform PDF models. This group is best described by a mixture of Gaussian or Laplacian to characterize the non-stationary nature of the virtual dependency channel. The hypothesis that the mixture models fit the dependency channel was validated by using the Kolmogorov-Smirnov goodness-of-fit test. Goodness-of-fit tests are used to examine hypothesis that a given data set comes from a model distribution with given parameters. The Kolmogorov-Smirnov (KS) test is a popular goodness-of-fit test. The KS statistic test measures the distance between the empirical CDF and the model CDF which measures the goodness-of-fit. If the empirical CDF is tested against several model CDFs, the model that gives the minimum KS statistic can be taken to be the best fit for the data. Kolmogorov-Smirnov values were obtained as shown in table1, for different video sequences as shown in table 1. For a typical significance level of 5%, those values imply accepting the hypothesis that it follows a Laplacian mixture model. The minimum statistics and the best fit distribution for dependency channel have been indicated in bold.

Issue 2 – Poor Rate Control: The current DVC solutions (Aaron & Girod, 2002, 2004; Ascenso et al., 2005; C. Brites, 2005; X. Artigas et al., 2007) work on frame level and assume that the quality of the side information is constant across the whole frame. Therefore, it uses one quantizer for the entire frame. Considering the varying motion characteristics and the amount of detail, the estimation of the original frame at the decoder is not the same for all regions of the frame. Based on this fact and utilizing the reconstruction function, one can conclude that the successfully estimated regions with coarser quantizer can achieve similar distortion reduction to the poorly estimated regions with finer quantizer. Therefore to achieve a certain operational characteristics in terms of distortion reduction, the quantizer must consider the quality of the side information. The intuition is that, in reconstructing the coded region, the joint decoder should rely more on side information when the rate is too low. On the other hand, when the rate is high, the decoded quantized symbol of the region becomes more reliable than side information, so the decoder should put more weight on the quantized symbol.

Issue 3 – Non-adaptive Quantization: The feedback channel (FC) plays the role of adapting the bit rate to the changing statistics between the side information and the frame to be encoded. The frame based DVC solutions assume that this changing statistics is constant across the whole frame. Each pixel of WZ frame is scanned row-by-row and then uniformly quantized. The quantized symbols form an input block which is binarized and fed into turbo encoder for coding and transmission. Therefore, the size of input block is assumed as arbitrarily large because it represents all bits from the whole frame or one bit-plane at once (Aaron & Girod, 2002; C. Brites, 2005). Too large input block will produce significant computation latency during the encoding and decoding process (Xue, 2009). The system will not be able to provide a timely WZ decoded output due to the vast Slepian-Wolf turbo coding and decoding delay (Xue, 2009). Frame based DVC systems make use of a rate-compatible-punctured codes (RCPT) codec as the module of the SW codec. The encoder "blindly" sends the parity bits according to the puncturing pattern determined by next lower coding rate. This causes the waste of transmission that some unnecessary parity bits are sent to help decoding some bits which are already correctly decoded (Aaron et al., 2002).

SOME RECENT ADVANCES

In recent years, a significant number of research groups around the world have adopted the Stanford WZ coding architecture and changed it in order to cope with above issues. Subsection 7.1 will briefly describe some related literature of work addressing these issues individually. Subsection 7.2 presents a solution proposed by us that addresses the above mentioned issues collectively.

GENERAL APPROACHES

Several works has been proposed for frame-based in order to cope with above drawbacks. In (Catarina Brites & Pereira, 2008), in order to exploit the spatial variability of the correlation noise, the models' parameters are calculated at the block level. The considered block size is equal to the one used in the frame interpolation process (8-by- 8) in order to match the frame interpolation errors more easily. The resulting RD performance is better than that where the noise channel modeling is frame level based. Non-stationary model to characterize the correlation noise is proposed in (Meyer, 2004), where two models are used, one to model the non-occluded regions and the other to model the occluded regions. The results show that the performance of the SISO decoder, and therefore of the overall system, can be improved greatly by classifying the decoder-generated side-information into two (or more) reliability classes. It also shows that the channel model should be an accurate representation of the real behavior of the channel, otherwise the decoding performance will heavily degrade. The lengthy block problem has also been addressed in (Chien, Karam, & Abousleman, 2007; Xue, 2009) by dividing the WZ frame, X, into M sub-images X_m, $m=\{0,1,2,3,\ldots,M\}$, and each sub-image is independently encoded using a Turbo-code based Wyner-Ziv encoder. Therefore, the block size of the Turbo encoder decreases to $1/M$ of the frame size. In the proposed system (Chien et al., 2007), Block-Adaptive wyner-Ziv coding (BAWZC)-based DVC system, the puncturing rate is independently adjusted for each block and the BAWZC decoder has to perform to inform the WZC encoder which blocks need more parity bits. The system proposed in (Xue, 2009), however, restricts the length of SW input to reduce the decoding complexity and the resulting delay. Both works (Chien et al., 2007; Xue, 2009) alleviate the problem of spreading the estimation errors all over the bitstream by keeping the localized errors close to each other in the generated symbol stream. The results from both works show that system performance is further enhanced. These works enhance the rate control performance marginally as they use the same correlation noise model to characterize the noise over all sub-images. More enhancement would, however, be possible if more accurate noise model is used.

REGION BASED ADAPTIVE DVC

Inspired from the traditional variable-size block video coding (Chan, Yu, & Constantinides, 1990; Silveira & Piedade, 1997; Vaisey & Gersho, 1992; Yu, Chan, & Constantinides, 1990), the region-based(A.Elamin, Jeoti, & Belhouari, 2010) approach proposes partitioning the WZ frame into many coding units, each unit having different DVC parameters (Channel model, scalar quantizer and rate). This partitioning process coincides with the spatial variation in the local characteristics of the video sequence. Typically, the frame partitioning is a process of classifying the side information into several reliability classes. Since the estimation of the DVC parameters at finer level enhances the overall performance, the DVC parameters are estimated per region. The scheme of partitioning frame into regions can be seen as allocating different channel and rate for each region conforming to the unequal error protection concept. The simplest method of unequal error protection is to allocate different channels for different types of data (Borade, Nakiboglu, & Zheng, 2008). Since the side information is unavailable at the encoder, the frame partitioning task cannot be performed at the encoder if its complexity is to be kept low. As the original WZ frame is not available at the decoder, the partitioning technique makes use of the estimated reliability information about the motion field in the SI generation process. The reliability is measured by quantifying the confidence on the motion vector estimation. It is common to use the two neighboring motion-compensated key frames to estimate the dependency model online (Catarina Brites & Pereira, 2008). The proposed frame partitioning is performed by using the two neighboring motion-compensated key frames at the decoder. The regions are constructed in two phases - an initial segmentation phase and grouping phase. In the first phase, the two neighboring motion-compensated key frames are divided into 8-by-8 blocks and the MSE between each block and its co-located block is computed. In the second phase, the part of the frame with MSE lower than certain threshold can be grouped into a region. For example, with this operation, the background part will possibly be grouped into one region. Straightforward coding of the region map is not very efficient. Furthermore, permutations of the region numbering do not affect the decoding process since each original region is decoded with its corresponding side information region and parity bits. We, therefore, consider coding the region map into quadtree. The use of quadtree is ubiquitous in image coding (Jiajun, Omair Ahmad, & Swamy, 1999; Lee, 1995; Strobach, 1991) because of the simplicity and efficiency in coding the block configuration, requiring only 1 bit per node of the tree.

Following is the simplified walkthrough of the proposed DVC coding architecture as shown in Figure 6. Note that the key frames (odd frames) are conventionally encoded and sent to the decoder.

The side information generation process performs the motion estimation and estimates the forward/backward motion vectors.

The two motion-compensated key frames are produced.

The partitioning process is performed and the side information is partitioned into N regions $\{Y_1,Y_2,\ldots,Y_N\}$ the corresponding regions map is created by using quadtree algorithm.

The regions DVC pararmeters (a_n,Q_n) are estimated at the decoder and quantizer set $Q=\{Q_1,Q_2,\ldots,Q_N\}$ is set transmitted to the encoder via the feedback channel.

The encoder decomposes the compressed regions map and partitions the original frame into corresponding N regions $\{X_1,X_2,\ldots,X_N\}$.

The samples of the first region X_1 are quantized with it corresponding quantizer, generating the quantized symbol stream.

The Slepian-Wolf encoder produces redundant information (e.g. parity bits) from the quantized symbol stream.

At the basic Wyner-Ziv decoder, the quantized symbol stream is decoded through joint source-

Figure 6. Architecture of the proposed solution

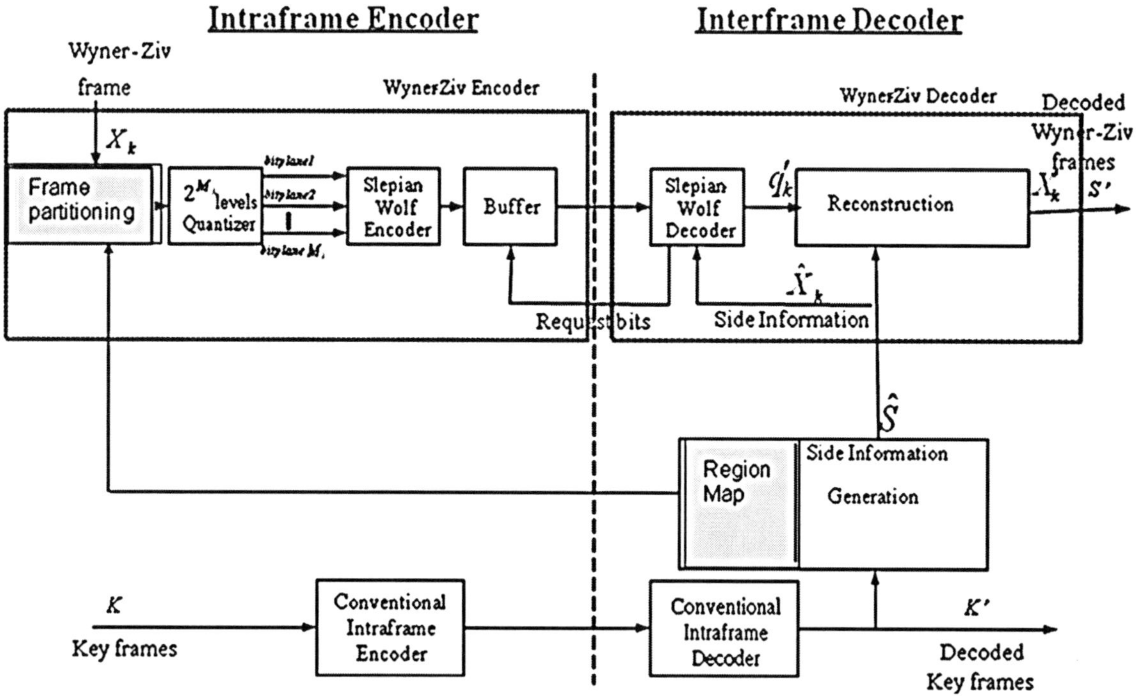

channel decoding with the aid of the region Y_l, the side information.

The decoded quantized symbol stream and the side information Y_l are then used together in a reconstruction module to reconstruct the pixel value X'_l.

The region Y_l (typically a decoder estimate of the region X_l to be Wyner-Ziv encoded) is considered to be available at the decoder.

The system repeats step 6-10 for all the remaining regions.

Figure 7 shows the comparison of rate-distortion performances of a typical frame based WZ codec and the adaptive region based WZ codec with MPEG codec. It can be observed that in terms of bitrate saving, the region-based solution outperforms the frame-based by 25-40%. This improvement can be justified by more accurate modeling of the dependency channel, the adaptive source coding for each region. Note that the region structure also tightens the control over the rate. As it can be observed from the results depicted in Figure 7, there is still a performance gap between DVC and MPEG+ interframe coding with I-P-I-P structure.

SINGLE VIEW DVC TO MULTIVIEW DVC

Video data in surveillance application consists of video sequences captured by multiple cameras from different angles and locations (Multiview). Generally, significant correlation exists among various views (Artigas, Angeli, & Torres, 2006; Xing, Hua, Jian-Guang, & Jiang, 2007). Thankfully, the different views exhibit a strong correlation which can be exploited during compression. High coding efficiency for these scenarios is obtained at the cost of a high computational complexity at the encoder to exploit the temporal correlation of single view and call for communication between the cameras to exploit the Multiview correlation. Since WVSNs have scarce resources

Figure 7. Comparison between RD performance for news and Carphone test sequence with frame based and region based codecs

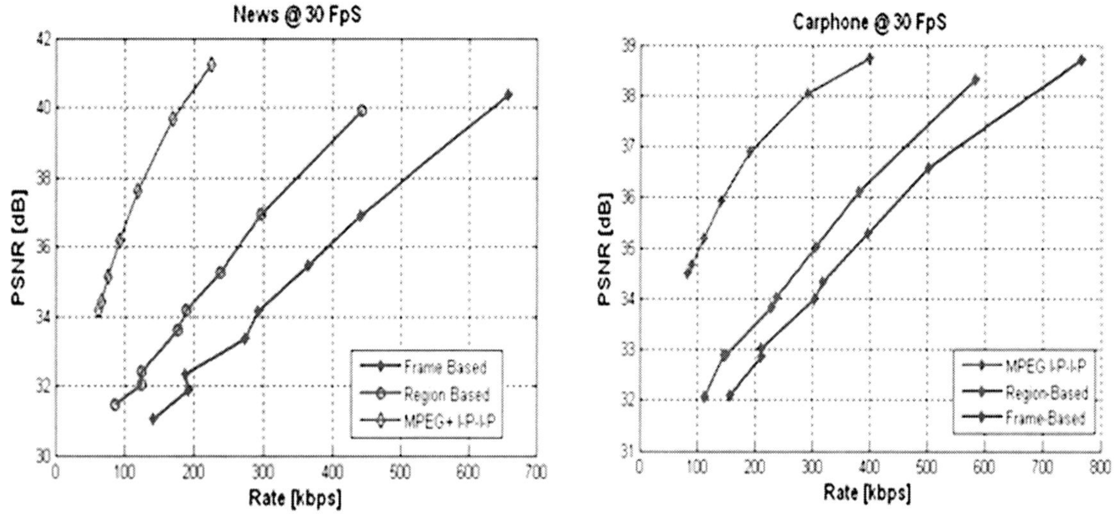

it is desirable to have a low complexity encoder and to avoid a complex inter-node communication system. Appropriate video coding scheme is obtained by promoting the single view DVC scheme into Multiview DVC. The Multiview DVC allows low encoding complexity and exploiting correlation between views without, or with limited, inter-sensor (that is inter-camera) communication. Multiview DVC incorporates all single video DVC components and differs from the single view case only at side information generation. More precisely, the side information is constructed not only using the frames within the same camera but using frames from the other cameras as well. Depending on cameras positions, spatial and temporal resolutions of the sequences and motion in the scene, temporal correlation may be higher than inter-view correlation or vice-versa (Ferre, Agrafiotis, & Bull, 2007; Shimizu, Tonomura, Kimata, & Ohtani, 2009). Techniques compensating the displacement of an object from one view to the other, called disparity are used to remove inter-view correlation (Adikari, Fernando, & Arachchi, 2006; Dong-Hoon & Yung-Lyul, 2008). Disparity vectors are function of depth, i.e., of the focal length and positions of the cameras. These techniques are referred to as pixel-based or block-based disparity-compensated view prediction techniques (Jong Dae & Rae-Hong, 2006) In a Multiview DVC scenario with no intercamera exchange, disparity estimation must be performed by the decoder. Capturing inter-view dependency turns out to be more difficult than for temporal dependencies, as, in general, multi-view images contain disparities much larger than displacements between successive frames.

CONCLUSION

Wireless Video Sensor Networks usually require the sensor nodes to be cheap and sometimes disposable. Video Coding in WVSNs must have least encoding complexity. In the light of this requirement, this chapter introduced the concept of distributed source coding where joint encoding-decoding compression performance can be attained by separate encoding-joint decoding approach based on Slepian and Wolf's and Wyner and Ziv's theorems of 1970s. It is shown that joint decoding results in considerable compression by exploiting the correlation at the decoder while keeping the encoding simple. This fulfils the requirements of real-world applications with tight operational constrains of WVSNs. The video coding paradigm, called distributed video coding (DVC) based on DSC concept is also developed for video applications in WVSNs where the interframe correlation is exploited at the decoder but not at the encoder, and, therefore, an intraframe low-complexity encoding can be combined with interframe high complexity decoding. The popular Stanford scheme based on the aforementioned approach is described. Recent advances beyond Stanford solution are also presented including a region based adaptive DVC scheme. However, the improvement in rate-distortion performance of DVC has still not breached the performance of a conventional interframe video coder. Future work in DVC is also likely to include multiview scenario. The spatial correlation between views can also be exploited at the decoder without, or with limited, inter-camera communication.

REFERENCES

Aaron, A., & Girod, B. (2002, April 2002). Compression with side information using turbo codes. Paper presented at the Proc. IEEE Data Compression Conference, Snowbird, UT.

Aaron, A., & Girod, B. (2004). Wyner-Ziv video coding with low-encoder complexity "Invited Paper". Paper presented at the Proc. Picture Coding Symposium.

Aaron, A., Rane, S., & Girod, B. (2004). Wyner-Ziv video coding with hash-based motion compensation at the receiver. Paper presented at the IEEE International Conference on Image Processing.

Aaron, A., Zhang, R., & Girod, B. (2002). Wyner-Ziv coding of motion video "Invited paper". Paper presented at the Proc. Asilomar Conference on Signals and Systems.

Adikari, B., Fernando, W. A. C., & Arachchi, H. K. (2006). A new motion and disparity vector prediction technique for H.264 based stereoscopic video coding. Paper presented at the Proceedings 2006 IEEE International Symposium on Circuits and Systems (ISCAS 2006).

Areia, J., Brites, C., Ascenso, J., & Pereira, F. (2007). Wyner-Ziv stereo video coding using a side information fusion approach. Paper presented at the IEEE International Workshop on Multimedia Signal Processing.

Artigas, X., Angeli, E., & Torres, L. (2006, 7-9 June 2006). Side information generation for multiview distributed video coding using a fusion approach. Paper presented at the Proceedings of the 7th Nordic Signal Processing Symposium (NORSIG 2006).

Artigas, X., Ascenso, J., Dalai, M., Klomp, S., Kubasov, D., & Ouaret, M. (2007). The DISCOVER codec: Architecture, techniques and evaluation. Paper presented at the Picture Coding Symposium.

Ascenso, J., Brites, C., & Pereira, F. (2005). Motion compensated refinement for low complexity pixel based distributed video coding. Paper presented at the IEEE International Conference on Advanced Video and Signal Based Surveillance.

Ascenso, J., & Pereira, F. (2007). Adaptive hash-based side information exploitation for efficient Wyner-Ziv video coding. Paper presented at the IEEE International Conference on Image Processing.

Baoguo, D., & Hong, S. (2009, 4-6 June). Encoder rate control for pixel-domain distributed video coding without feedback channel. Paper presented at the 3rd International Conference on Multimedia and Ubiquitous Engineering (MUE '09).

Borade, S., Nakiboglu, B., & Zheng, L. (2008, July 2008). Unequal error protection: Some fundamental limits and optimal strategies. Paper presented at the ISIT IEEE International Symposium Information Theory.

Borchert, S., Westerlaken, R. P., Gunnewiek, R. K., & Lagendijk, R. L. On extrapolating side information in distributed video coding. Paper presented at the 26th Picture Coding System.

Brites, C. (2005). *Advances on distributed video coding*. Lisbon, Portugal: Technical University of Lisbon.

Brites, C., & Pereira, F. (2007, Sept. 16 2007-Oct. 19 2007). Encoder rate control for transform domain Wyner-Ziv video coding. Paper presented at the IEEE International Conference on Image Processing (ICIP 2007).

Brites, C., & Pereira, F. (2008). Correlation noise modeling for efficient pixel and transform domain Wyner-Ziv video coding. IEEE transactions on Circuit and systems for video technology, 18(9).

Chan, M. H., Yu, Y. B., & Constantinides, A. G. (1990). Variable size block matching motion compensation with applications to video coding. Communications, Speech and Vision. *IEE Proceedings. Part I. Solid-state and Electron Devices*, *137*(4), 205–212.

Chien, W.-J., Karam, L. J., & Abousleman, G. P. (2007). Block-adaptive Wyner-Ziv coding for transform-domain distributed video coding. Paper presented at the 32nd IEEE International Conference on Acoustics, Speech, and Signal Processing.

Dong-Hoon, H., & Yung-Lyul, L. (2008, 13-15 Dec.). Fast mode decision using global disparity vector for multiview video coding. Paper presented at the 2nd International Conference on Future Generation Communication and Networking Symposia (FGCNS '08).

Eduardo Peixtoto, L. R. Queiroz, d., & Mukherje, D. (2008). On side information generation for Wyner-Ziv video coding. Paper presented at the XXVI SIMPOSIUM.

Elamin, A. (2010). Region-based adaptive distributed video coding. Unpublished doctoral dissertation, Universiti Teknologi PETRONAS, Malaysia.

Elamin, A., Jeoti, V., & Belhouari, S. (2010). Region-based distributed video coding. ETRI Jounal (2010 under-review).

Ferre, P., Agrafiotis, D., & Bull, D. (2007, Sept. 16 2007-Oct. 19 2007). Fusion methods for side information generation in multi-view distributed video coding systems. Paper presented at the IEEE International Conference on Image Processing (ICIP 2007).

Fowler, J. E. (2005). An Implementation of PRISM Using QccPack: Mississippi State University. (G. Institute o. Document Number)

Girod, B., Aaron, A., Rane, S., & Rebollo-Monedero, D. (2005, January). Distributed video coding. Paper presented at the Proc. IEEE Special Issue on Advances in Video Coding and Delivery.

Jiajun, Z., Omair Ahmad, M., & Swamy, M. N. S. (1999). Quadtree structured region-wise motion compensation for video compression. Circuits and Systems for Video Technology. *IEEE Transactions on*, *9*(5), 808–822.

Jong Dae, O., & Rae-Hong, P. (2006). Reconstruction of intermediate views from stereoscopic images using disparity vectors estimated by the geometrical constraint. Circuits and Systems for Video Technology. *IEEE Transactions on*, *16*(5), 638–641.

Lee, J. (1995). Optimal quadtree for variable block size motion estimation. Paper presented at the Proceedings IEEE Internatinal Conference Image Processing.

Meyer, P. F. A. (2004). *Slepian-Wolf turbo coding of video with non-stationary side-information*. Delf University of Technologoy.

Meyer, P. F. A., Westerlaken, R. P., Gunnewiek, R. K., & Lagendijk, R. L. (2005, July). Distributed source coding of video with non-stationary side-information. Paper presented at the Proc. Visual Communications and Image Processing.

Natrio, L., Brites, C., Ascenso, J., & Pereira, F. (2005). Extrapolating side information for low-delay pixel-domain distributed video coding. Paper presented at the Int. Workshop on Very Low Bitrate Video Coding.

Pradhan, S., & Ramchandran, K. (2003). Distributed source coding using syndromes (DISCUS): Design and construction. *IEEE Transactions on Information Theory*, *49*, 626–643. doi:10.1109/TIT.2002.808103

Puri, R., & Ramchandran, K. (2003, 6-10 April). PRISM: An uplink-friendly multimedia coding paradigm. Paper presented at the Proceedings of the 2003 IEEE International Conference on Acoustics, Speech, and Signal Processing (ICASSP '03).

Puri, R., & Ranichandran, K. (2003). PRISM: A "reversed" multimedia coding paradigm. Paper presented at the Image Processing International Conference, Sardinia, Italy.

Rebollo-Monedero, D., Aaron, A., & Girod, B. (2003). Transforms for high-rate distributed source Coding. Paper presented at the Proceedings of Asilomar Conference on Signals, Systems and Computers.

Shimizu, S., Tonomura, Y., Kimata, H., & Ohtani, Y. (2009, Aug. 30 -Sept. 2). Improved view interpolation for side information in multiview distributed video coding. Paper presented at the 3rd ACM/IEEE International Conference on Distributed Smart Cameras (ICDSC 2009).

Silveira, M., & Piedade, M. (1997, 9-12 Jun 1997). Variable block sized motion segmentation for video coding. Paper presented at the Proceedings of 1997 IEEE International Symposium on Circuits and Systems (ISCAS '97).

Slepian, D., & Wolf, J. K. (1973). Noiseless coding of correlated information sources. *IEEE Transactions on Information Theory*, *19*, 471–482. doi:10.1109/TIT.1973.1055037

Steinberg, Y., & Merhav, N. (2004). On successive refinement for the Wyner-Ziv problem. *IEEE Transactions on Information Theory*, *50*, 18. doi:10.1109/TIT.2004.831781

Strobach, P. (1991). Quadtree-structured recursive plane decomposition coding of images. *IEEE Transactions on Signal Processing*, *39*, 1380–1397. doi:10.1109/78.136544

Thoma, R., & Bierling, M. (1989). Motion compensating interpolation considering covered and uncovered background. Paper presented at the Signal Process., Image Commun.

Vaisey, D., & Gersho, A. (1992). Variable block-size image coding. *IEEE Transactions on Signal Processing*, *40*, 2040–2060. doi:10.1109/78.150005

Westerlaken, R. P., Borchert, S., Gunnewiek, R. K., & Lagendijk, R. I. L. (2006). Finding a near optimal dependency channel model for an ldpc-based wyner-ziv video compression scheme. Paper presented at the 12th Annual Conference of the Advanced School for Computing and Imaging.

Westerlakena, R. P., Gunnewiekb, R. K., & Lagendijka, R. (2005). The role of the virtual channel in distributed source coding of video. Paper presented at the ICIP IEEE International Conference image processing.

Wyner, A., & Ziv, J. (1976). The rate-distortion function for source coding with side information at the decoder. *IEEE Transactions on Information Theory*, *22*, 1–10. doi:10.1109/TIT.1976.1055508

Xing, S., Hua, C., Jian-Guang, L., & Jiang, L. (2007). Multiview image coding based on geometric prediction. Circuits and Systems for Video Technology. *IEEE Transactions on*, *17*(11), 1536–1548.

Xiong, Z., Liveris, A. D., & Cheng, S. (2004). Distributed source coding for sensor networks. *IEEE Signal Processing Magazine*, *2*, 421–440.

Xue, Z. (2009). *Research and developments of distributed video coding. Uxbridge*. London, UK: Brunel University.

Yaacoub, C., Farah, J., & Pesquet-Popescu, B. (2009). A genetic frame fusion algorithm for side information enhancement in Wyner-Ziv video coding. Paper presented at the 17th European Signal Processing Conference.

Yangli, W., Jechang, J., & Chengke, W. (2008, 27-30 May). An approach to side information estimation for Wyner-Ziv video coding. Paper presented at the Image and Signal Processing (CISP '08).

Yu, Y. B., Chan, M. H., & Constantinides, A. G. (1990, 3-6 Apr 1990). Low bit rate video coding using variable block size model. Paper presented at the International Conference on Acoustics, Speech, and Signal Processing (ICASSP-90).

KEY TERMS AND DEFINITIONS

Distributed Source Coding (DSC): Source coding (compression) paradigm for multiple correlated information sources in an independent way.

Distributed Video Coding (DVC): New video coding paradigm that uses channel coding techniques to achieve compression.

Multiview DVC: DVC technique for multiple cameras providing multiview.

Side Information: The estimate of the original source information by the decoder.

Slepian-Wolf Coding: The lossless version of distributed source coding.

Virtual Channel Noise: The resulting estimation "error".

Virtual Channel: It represents the ability of the decoder to estimate the source.

Wyner-Ziv (WZ) Coding: The lossy version of distributed source coding.

Chapter 4
Wavelet Filters Evaluation in Power Constrained Visual Sensor Networks

Mammeri Abdelhamid
University of Sherbrooke, Canada

Brahim Hadjou
University of Sherbrooke, Canada

Ahmed Khoumsi
University of Sherbrooke, Canada

Djemel Ziou
University of Sherbrooke, Canada

ABSTRACT

While designing a wavelet-based coder (WBC) in the context of Visual Sensor Network (VSN), engineers and designers must respect their strict constraint on power consumption. This makes the selection of the appropriate wavelet, among many existing competitors, not an easy task. A set of wavelet filters are evaluated and the best of them are selected. The comparison is performed in terms of the quality of the reconstructed image at the base station and the power consumption of a visual sensor (VS) processing the wavelet. Image quality is measured objectively using PSNR and SSIM, and subjectively using the mean opinion score of many viewers. For power consumption, we have developed a power model based on the number of times basic operations are performed by the filters. Moreover, we show and discuss some factors, such as decomposition level and filter length, influencing the power dissipation of a given VS while executing a given wavelet under evaluation. Our results provide a good reference for designers of WBC for power-constrained applications such as VSN.

DOI: 10.4018/978-1-61350-153-5.ch004

INTRODUCTION

Visual Sensor networks (VSNs) are gaining attention from researchers and engineers as they can be used efficiently in many domains. VSN is a distributed ad hoc network made up of small sensing devices equipped with low-power miniature cameras, called visual sensors. VSN are deployed around strategic areas of interest to collect and reconstruct, in multi-hop fashion, a monitored scene. Inside a VSN, each visual sensor (VS) takes, processes, and transmits wirelessly images to the base station (also called sink) through its neighbors located in its range of transmission. VSN can be used in a wide range of applications such as video-surveillance, environmental monitoring, and object detection/tracking. Unfortunately, VSN faces some challenges such as the limitation of their resources in battery and memory, and they handle a plethora of visual redundant flows, which may cause several problems.

One of the challenging problems facing VSN is the huge amount of visual information processed and sent by each power-limited VS. The processing and communication of such information shorten the lifetime of the visual sensors and shrink the connectivity of the whole VSN. This is due to the limited resources of small VS. This problem also concerns mobile devices such as PDA, cell phones, and laptop computers.

The crucial solution to manipulate these visual flows is through image compression, which by definition might reduce significantly an image's file size using lossy or lossless coding scheme. Very attractive and popular standards to achieve image compression are based on discrete cosine transform (DCT), such that JPEG, and discrete wavelet transform (DWT), such that JPEG2000. In past works, such as in (Mammeri et al., 2008a) or (Mammeri et al., 2008b), we have exploited the DCT in VSN, and have adapted JPEG to the requirements of such power-constrained networks. In this chapter, the DWT transforms evaluation is investigated in the context of VSN for the following reasons. DWT is a non-block-based transform, and thus, it allows to avoid the annoying blocking artifacts introduced by the DCT transform within the reconstructed image. Moreover, it has a good localization both in time and frequency domains. Furthermore, because of their inherent multiresolution nature, wavelet-based coders (WBCs) are especially suitable for applications where scalability and tolerable degradation are important (Mallat, 1989).

It is well known that the computational complexity of DWT is relatively higher than the DCT one. See (Grgic et al., 2001) for more details about the comparison between DCT and DWT. It is naturally that one could ask why DWT is preferred over DCT in VSN context. We believe that there exist at least two main reasons justifying this observation. Firstly, in VSN, it is highly recommended to reduce the size of the sent image for energy minimization purpose, thus the compression ratio CR should be high. In this case, DWT is favored over DCT (Grgic et al., 2001). Secondly, there exist a relatively new version of DWT implementation of low complexity (i.e., lifting scheme) than DCT (twice more efficient than convolutional-based wavelets). On the other it is reported in (Adams & Kossentini, 2000), that there is some correlation between the lifting version and the convolution version of the same wavelet filter.

A number of studies and reviews have evaluated the wavelet transforms in diverse domains, namely: medical image compression (Yao-Tien et al., 2005), audio compression (Srinivasan & Jamieson, 1999) and data compression (Grgic et al., 2001). While the aforementioned studies have considered the effectiveness of these transforms in terms of the perceived quality (in a non power-constrained application), none of them have focused on their use on special applications like VSN. Moreover, some research works in VSN have used some known wavelets within the compression stage of their WBC without any analysis or arguments, for instance the use of the Le-Gall 5taps/3taps in (Lecuire et al., 2007). In this work, an evaluation strategy has been conducted to investigate the impact of wavelets parameters (such as filter length and decomposition level) on both the power dissipated by a VS node while executing a wavelet, and the quality of the reconstructed image.

We note that our chapter is inspired by the work in (Grgic et al., 2001), but it has some major differences such as: our application domain, i.e., VSN, is totally different than the general imaging domain handled in (Grgic et al., 2001); the filter length equations used in (Grgic et al., 2001) are dedicated to both orthogonal and biorthogonal filters, while in this chapter, we have suggested different equations for each filter category; an extensive set of simulations, including objective (*PSNR* and *SSIM*) and subjective evaluations are made, on the contrary to (Grgic et al., 2001) where only few number of simulations were made; an in-depth explanation of various results are supplied. Moreover, in order to evaluate the power dissipated by a visual sensor executing the different wavelets under evaluation, we adopted the parameters of the microcontroller ATMEL ATmega128 (Atmel 8-bit Microcontroller, 2009) low-power CMOS 8-bit, which is one of the most widely used in sensor nodes. We note also that some results found in our simulations are different than those presented in (Grgic et al., 2001), for instance the best filter (*bior2.2*) in terms of power minimization and image quality maximization.

In VSN, the choice of wavelet filters is a critical issue that affects the battery state of a VS as well as the image quality at the base station. Thus, the use of an appropriate wavelet filter is mandatory. Our goal in this work is to provide a set of wavelet filters which can be used effectively within a WBC in power-constrained applications. Such selected wavelets have to respect the power constraint of VSN while having an acceptable quality of the reconstructed image at the base station. Note that the criterion of image quality is the resemblance of the original image with the image reconstructed at the base station.

In this work, a large set of wavelet filters are evaluated, and are selected those that make the best compromise between minimizing power consumption and maximizing image quality. We do not cover all wavelet filters or functions proposed in the literature, rather, we focus on those that are likely to be implemented in real-life design, i.e., Haar wavelet (*HW*), Daubechies wavelet (*DW*), Symlet wavelet (*SW*), Coiflet wavelet (*CW*), and biorthogonal (*bior*) wavelet (Daubechies, 1992). We evaluate and analyze both the power consumption behavior and visual quality results of the wavelet filters under evaluation. Image quality is measured objectively and subjectively. Two objective image quality assessments are used, the Peak Signal to Noise Ratio (*PSNR*) and the Structural SIMilarity index (*SSIM*). We use *PSNR* since it is the widely measure used in digital image processing. We supplement the objective measure by the relatively new tool called *SSIM*. To confirm or to correct objective evaluations, subjective assessments are introduced using the mean opinion score of 18 viewers.

To evaluate the power dissipation, we have developed a power model based on the number of times basic operations performed by a VS node while executing a 2-D DWT. Moreover, we show and discuss some factors influencing the power dissipation of a limited-power VS while executing a given wavelet transform, such as filter length and level of decomposition. From our analysis, we suggest which of DWT family can be used effectively in image compression in the context of VSN while obtaining an acceptable image quality at the sink.

In this work, we do not intend to describe and study the whole DWT-based compression chain; rather we focus on DWT phase since it has been shown that wavelet transform consumes 40% to 60% of the CPU time during compression process using JPEG2000 (Lee & Dey, 2002). After wavelet decompositions, a common scalar quantization followed by run length and Huffman encoding, is adopted in our work. More investigations about efficient compression algorithms are out of the scope of this paper. Our results provide a good reference for designers of WBC for power-constrained applications such as VSN. In order to avoid any confusion, along this chapter the term *energy* is used to mean the energy of the signal, while the term *power* is used to mean the consumed energy in Joule by VS battery.

This chapter is structured as follows. We review in Section *basic background* the necessary background related to DWT and its power consumption model. Some definitions related to VSN are introduced in Section *VSN requirements and characteristics*. In Section *factors influencing WBC in VSN context*, we introduce the wavelet filters used in our simulation, and we present some factors influencing DWT-based coders are discussed. In Section *simulation parameters setup*, we present the simulation parameters used in this chapter. The objective and subjective evaluations are presented in Sections *objective evaluation* and *subjective evaluation*, respectively. In Section *discussion*, we discuss in depth the simulations results. Finally, we summarize and present some future directions in Section *conclusion*.

BASIC BACKGROUND

In this section, the necessary DWT background to follow this chapter is reviewed. Specifically, Section *the DWT* briefly describes the 1-D and 2-D DWT decomposition, while Section *DWT power consumption model* presents the power consumption model used for the 2-D DWT phase. Basically, three elementary phases are involved in WBC. First, the original image is represented as a series of spatial frequency bands, also called sub-bands. The coefficients within each sub-band are then fed to the quantizer to reduce the encoding data. Quantization is a non-invertible process in which a set of sub-coefficients are approximated by discrete values. And finally, the reversible entropy-coding process (such as Huffman coding) is performed on the quantized data to generate fixed or variable length code words.

In this chapter, we focalize on DWT phase, instead of the whole DWT-based compression scheme, since DWT phase consumes 40% to 60% of the CPU time during the compression process using JPEG2000 (Lee & Dey, 2002). To accurately study the influence of DWT phase on the power consumption of a VS as well as the image quality, the quantizer and the entropy encoder effects are not considered.

The DWT

From a signal processing point of view, the 1-D DWT can be viewed as the multiresolution decomposition of an input signal *X[n]* having *N* samples. The 1-D DWT filter bank structure, realizing the one dimensional dyadic decomposition is illustrated in Figure 1. In this section, we denote the number of decomposition levels by *L*. The sequence *X[n]* is convolved with high-pass filter *h[n]* and low-pass filter *l[n]*, and each result is then downsampled by a factor of 2, yielding the transforms signal H_1 and L_1, at the first level. Let us consider T_H and T_L the number of taps of the high-pass and low-pass filters, respectively. The low and high frequency coefficients ($L_j[n]$ and $H_j[n]$) of the decomposition level *j* are computed as follows (Nguyen & Strang, 1996):

$$L_{j+1} = \sum_{i=0}^{T_L-1} l\left[i\right] \times L_j\left[2n - i\right] \qquad (1.1)$$

$$H_{j+1} = \sum_{i=0}^{T_H-1} h\left[i\right] \times L_j\left[2n - i\right] \qquad (1.2)$$

Where $j=0,\ldots, L-1$ and $n=0,1,\ldots, (N/2^{j+1}-1)$, and $L_0(n)=X[n]$.

Figure 1. One dimensional dyadic decomposition of DWT

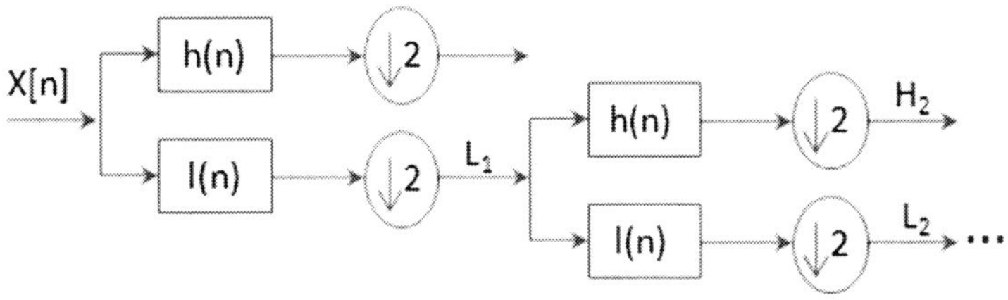

A straightforward implementation of the separable 2-D DWT can be performed by applying, at first, the 1-D DWT row-wise to produce *L* and *H* sub-bands, and then column-wise to produce four sub-bands *LL*, *LH*, *HL* and *HH* as illustrated in Figure 2, where the decomposition is performed on the *LL* sub-band. In this way, the 2-D decomposition of a two dimensional signal having *M* columns and *N* rows, is described as follows (Nguyen & Strang, 1996):

$$L_{j+1}\left[row\right]\left[m\right] = \sum_{i=0}^{T_L-1} l\left[i\right] \times LL_j\left[row\right]\left[2m-i\right] \tag{1.3}$$

$$H_{j+1}\left[row\right]\left[m\right] = \sum_{i=0}^{T_H-1} h\left[i\right] \times LL_j\left[row\right]\left[2m-i\right] \tag{1.4}$$

$$LL_{j+1}\left[n\right]\left[col\right] = \sum_{i=0}^{T_L-1} l\left[i\right] \times L_j\left[2n-i\right]\left[col\right] \tag{1.5}$$

$$LH_{j+1}\left[n\right]\left[col\right] = \sum_{i=0}^{T_H-1} h\left[i\right] \times L_j\left[2n-i\right]\left[col\right] \tag{1.6}$$

$$HL_{j+1}\left[n\right]\left[col\right] = \sum_{i=0}^{T_L-1} l\left[i\right] \times H_j\left[2n-i\right]\left[col\right] \tag{1.7}$$

$$HH_{j+1}\left[n\right]\left[col\right] = \sum_{i=0}^{T_H-1} h\left[i\right] \times H_j\left[2n-i\right]\left[col\right] \tag{1.8}$$

Where $j=0,\ldots,L-1$ and $row=0,1,\ldots,(N/2^{j+1}-1)$, $m=0,1,\ldots,(M/2^{j+1}-1)$, $col=0,1,\ldots,(M/2^{j+1}-1)$, $n=0,1,\ldots,(M/2^{j+1}-1)$, and $LL_0[n][m]=X[n][m]$.

Figure 2. Two dimensional dyadic decomposition of DWT

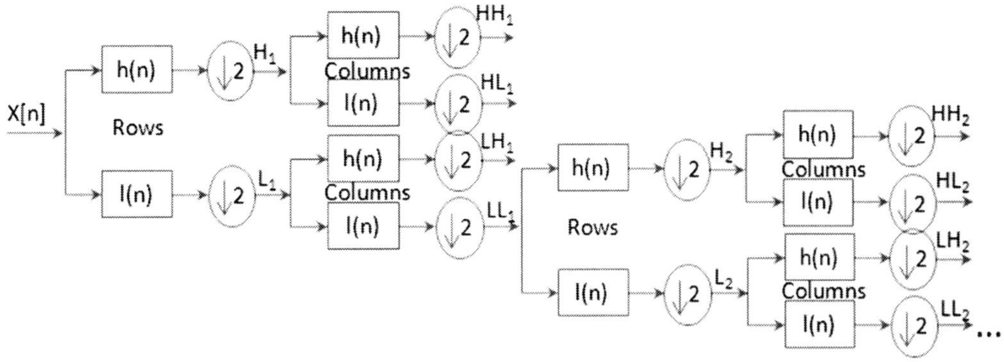

Figure 3. Example of wavelet decomposition of an image

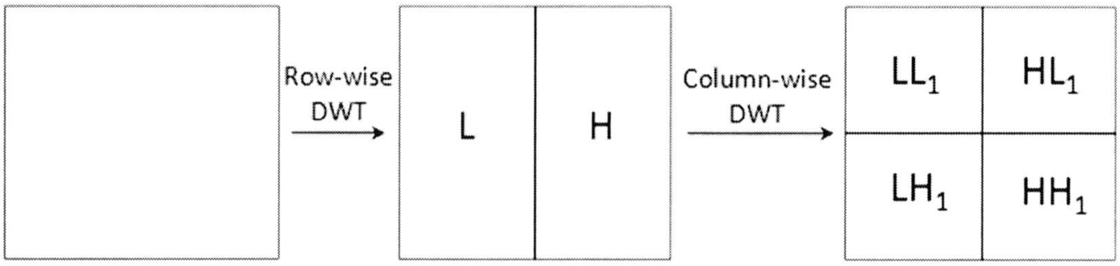

The HH_1, HL_1, LH_1 and LL_1 sub-bands are each of size $M/2 \times N/2$. The LL_1 outputs are sent to the second level of decomposition. The sub-bands outputs of the latter, namely, HH_2, HL_2, LH_2 and LL_2, are each of size $M/4 \times N/4$. Out of these, the LL_2 outputs are sent to third level, and so on.

The multiresolution nature of the wavelet decomposition allows to compact the image energy into small number of wavelet coefficients. For natural images, such as those handled by VSN, most of the image energy is concentrated in the *LL* band that corresponds to the coarsest scale. Moreover, the energy in the finer sub-bands is concentrated into a relatively small number of wavelet coefficients (Nguyen & Strang, 1996).

DWT Power Consumption Model

Let us now consider the power consumption of a given VS while performing a wavelet transform of a filter of length T (number of taps) and an image of size $N \times N$ pixels (instead of $M \times N$ for clarity purpose). In general, the power dissipation can be seen as the number of times basic operations, such as additions and multiplications are performed (Mammeri et al., 2008b). We model the power consumption of a wavelet filtering by counting the number of operations as explained bellow. On the contrary to the work of (Grgic et al., 2001), where the authors used the same model for both orthogonal and biorthogonal filters, our suggested model depends on whether the wavelet is orthogonal or biorthogonal. For orthogonal wavelet filters, both high-pass and low-pass filters (e.g., for analysis step) have the same lengths. While for biorthogonal filters, different lengths are used by low and high-pass filters.

Let us denote O_A and O_p the number of floating operations (additions and multiplications, respectively) of a 1-D signal per scale. In this study, we consider the dyadic signal decomposition with direct implementation (Nguyen & Strang, 1996). For orthogonal filters, O_A and O_p are given as functions of the filter length T as follows (Polyak & Pearlman, 2000):

$$O_A = 2\left\lfloor \frac{T+1}{2} \right\rfloor + 2\left\lfloor \frac{T-1}{2} \right\rfloor \tag{1.9}$$

$$O_P = 2T - 2 + 2\left\lfloor \frac{T-1}{2} \right\rfloor \tag{1.10}$$

Where $\lfloor a \rfloor$ means the greatest integer smaller than a.

In the case of biorthogonal filters with lengths T_r (for synthesis filter) and T_d (for analysis filter), O_A and O_p are given as follows (Polyak & Pearlman, 2000):

$$O_A = 2\left\lfloor \frac{T_r+1}{2} \right\rfloor + 2\left\lfloor \frac{T_d-1}{2} \right\rfloor \tag{1.11}$$

$$O_P = T_r + T_d - 2 \quad (1.12)$$

For images of size $N \times N$ with separable filtering and dyadic decomposition, we multiply the above formulas by a factor of two in order to account for horizontal and vertical filtering, respectively. Moreover, at each subsequent scale, the number of operations is divided by four, since the number of pixels on each scale is reduced by a factor of four comparatively to the previous one. Thus, the number of operations can be written as follows:

$$\begin{aligned} & 2(O_A + O_P)\left(N^2 + \left(\frac{N}{2}\right)^2 + \left(\frac{N}{4}\right)^2 + \ldots + \left(\frac{N}{2^{j-1}}\right)^2\right) \\ & = 2(O_A + O_P)N^2\left(1 + \frac{1}{2^2} + \frac{1}{4^2} + \ldots + \frac{1}{2^{2(j-1)}}\right) \\ & = 2(O_A + O_P)N^2\left(\frac{1 - \frac{1}{4^j}}{1 - \frac{1}{4}}\right) = \frac{8}{3}N^2(O_A + O_P)(1 - 4^{-j}) \\ & \sim O\left(N^2\right) \end{aligned} \quad (1.13)$$

$O(N^2)$means that the computation complexity is in order of N^2.

Let E_A and E_P be the power consumption of a given VS while executing an addition operation and multiplication operation, respectively. The power consumption E_{DWT} by a VS of an input image having $N \times N$ pixels processed with j levels of decomposition, while executing a 2-D DWT can be calculated from Eq. 1.13 as follows:

$$E_{DWT} = \frac{8}{3}N^2(O_A E_A + O_P E_P)(1 - 4^{-j}) \quad (1.14)$$

By substituting Eqs. 1.9-1.10 in Eq. 1.14, we obtain Eq. 1.15 that evaluates the power dissipated by orthogonal filters. And by substituting Eqs. 1.11-1.12 in Eq. 1.14, we obtain Eq. 1.16 that evaluates the power dissipated by biorthogonal filters. (Box 1)

This is a rough estimation of the power costs in relative terms. The real power consumption is implementation dependent. Our goal is to minimize the power consumption while satisfying an adequate quality at the reception, regardless of the exact absolute costs.

VSN REQUIREMENTS AND CHARACTERISTICS

This section provides some background information to follow this chapter. Recall that VSN are spatially distributed networks consisting of small sensing devices equipped with low-power CMOS

Box 1.

$$\begin{aligned} E_{DWT}^{O} &= \frac{8}{3}N^2\left(O_A E_A + O_P E_P\right)\left(1 - 4^{-j}\right) \\ E_{DWT}^{O} &= \frac{8}{3}N^2\left(2\left\lceil\frac{T+1}{2}\right\rceil + 2\left\lceil\frac{T-1}{2}\right\rceil\right)E_A\left(1 - 4^{-j}\right) + \frac{8}{3}N^2\left(2T - 2 + 2\left\lceil\frac{T-1}{2}\right\rceil\right)E_P\left(1 - 4^{-j}\right) \end{aligned} \quad (1.15)$$

$$\begin{aligned} E_{DWT}^{B} &= \frac{8}{3}N^2\left(O_A E_A + O_P E_P\right)\left(1 - 4^{-j}\right) \\ E_{DWT}^{B} &= \frac{8}{3}N^2\left(\left(\left\lceil\frac{T_r+1}{2}\right\rceil + \left\lceil\frac{T_d+1}{2}\right\rceil\right)E_A + \left(T_r + T_d - 2\right)E_P\right)\left(1 - 4^{-j}\right) \end{aligned} \quad (1.16)$$

Figure 4. Visual sensor network

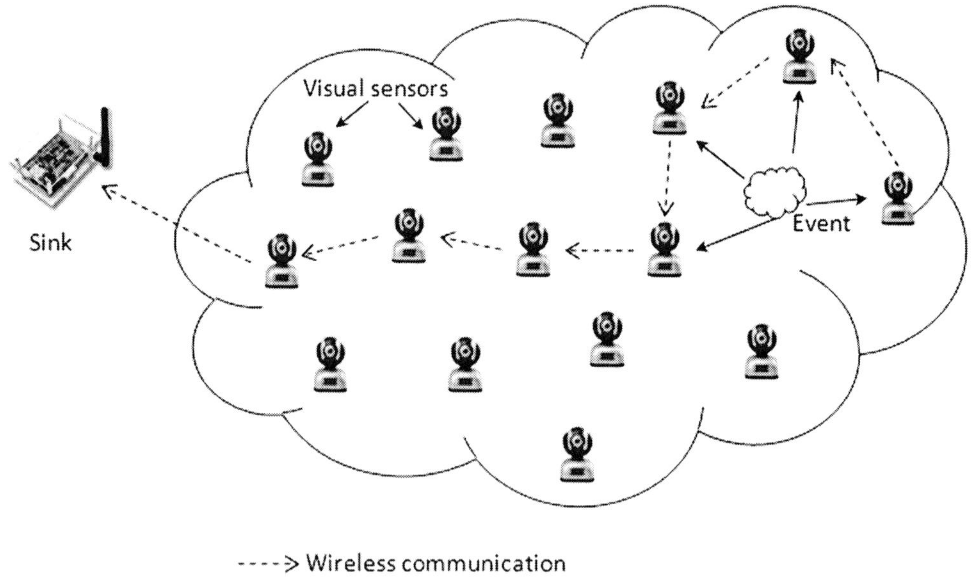

imaging sensors such as Cyclops. Ideally, VSNs are deployed in the region of interest to collect and fuse a captured data in multihop way. Inside a network, each VS captures, processes and transmits images to the sink through its neighbours located inside its range of transmission, see Figure 4. VSNs are involved in many domains such as, environmental monitoring, video-surveillance and object detection and tracking.

VSNs differ from their predecessor's scalar data wireless sensor networks (WSN) mainly in the following:

- Information volume and nature of VSN which is in general, pixel-based, is quite different than simple scalar data manipulated by WSN, such as temperature or humidity.
- Lost information in VSNs are tolerated since the redundancy nature of visual flows, unlike WSN where the lost of some packets may affect seriously the value of collected data (example, temperature value).
- VSN's camera has a restricted directional sensing field of view, which is not the case for the scalar data sensor.
- VS neighbors monitoring the same small local region of interest have multiple and different views of the same scene, comparatively to data sensor where the same value is collected by neighbors nodes (situated in the same region).
- Important resources in memory, processing, and communication power are required for VS node to manipulate the huge amount of visual flows.
- Efficient compression algorithms, in terms of power dissipation, are mandatory to handle information flows, compared to data scalar sensor where the compression is not very required.

In what follows, are described some factors influencing the WBC in the context of VSN.

FACTORS INFLUENCING WBC IN VSN

In general, the selection of the adequate wavelet filter for compression purpose is of high importance for designing WBC. This choice depends on few basic parameters such as: type of filter, length of filter, level of decomposition, and image parameters such as content and depth (Grgic et al., 2001).

In VSN, other parameters have to be taken into considerations while designing the DWT-based filters. These parameters are directly related to VSN nature, such as wireless ad hoc and distributed mode of deployment, and VS devices which are constrained in terms of battery, communication/processing capability, and achievable data rate. The wireless distributed mode of VSN deployment can be exploited by using the DWT-based distributed scheme. The latter can be performed for broadcasting and dispatching the workload of 2-D wavelet transform to several groups of nodes along the path between the VS source and the base station (Wu & Abouzeid, 2004), instead of processing the whole 2-D DWT by a single limited-power VS. In this chapter, we focus on studying and selecting the best DWT transforms suitable for VSN. Recall that we focalize on the application requiring a reduced power at the expense of receiving an image with lower (but acceptable) quality at the sink.

In the following subsections, we discuss the influence of the following parameters on WBC performance in the context of VSN:

- Image parameters.
- Perfect, near perfect or not perfect reconstruction.
- Filter length and filter order.
- Decomposition level.
- Choice of wavelet filter.

Image Parameters

As one might expect, image content is an important factor influencing filters effectiveness and compression results. Several types of images can be found; for example; medical images (MR, PET, Ultrasound, X-Ray images, etc.); synthetic images; natural images, etc. It is difficult to handle efficiently all of these kinds of images using the same wavelet filter. A particularly important observation is how to choose the test images set for evaluation. As almost all images handled by VSN are natural, we focus our analysis, simulations, and observations to natural images. Particularly, images with low and high spatial activity are considered. Other images such as #1 are used. Useful observation can be made for those natural images. Indeed, most of the image energy is concentrated in the *LL* band that corresponds to the coarsest scale (Nguyen & Strang, 1996). Moreover, the energy in the finer sub-bands is concentrated into a relatively small number of wavelet coefficients. The fact to consider only natural images, makes it easier to find efficient DWT transforms for image compression in VSN context.

Another factor related to image content is its bit-depth, which refers to the amount of bits per pixel. It is argued that bit-depth has some effects on compression performance (Adams & Kossentini, 2000). Note that the more bits per pixel an image contains, the larger is the image file size. Hence, this factor may affect the power dissipation by a VS and the image quality in the context of VSN.

The test data used in our study consist of the grayscale images, see Figure 5. These images cover a wide variety of contents (high and low spatial activities) as those handled by VSN. Particularly, images with high spatial activity (e.g., #1, #2 and #4) and low spatial activity (e.g., #3, #6 and #13) are considered.

Figure 5. Samples of images used in simulations

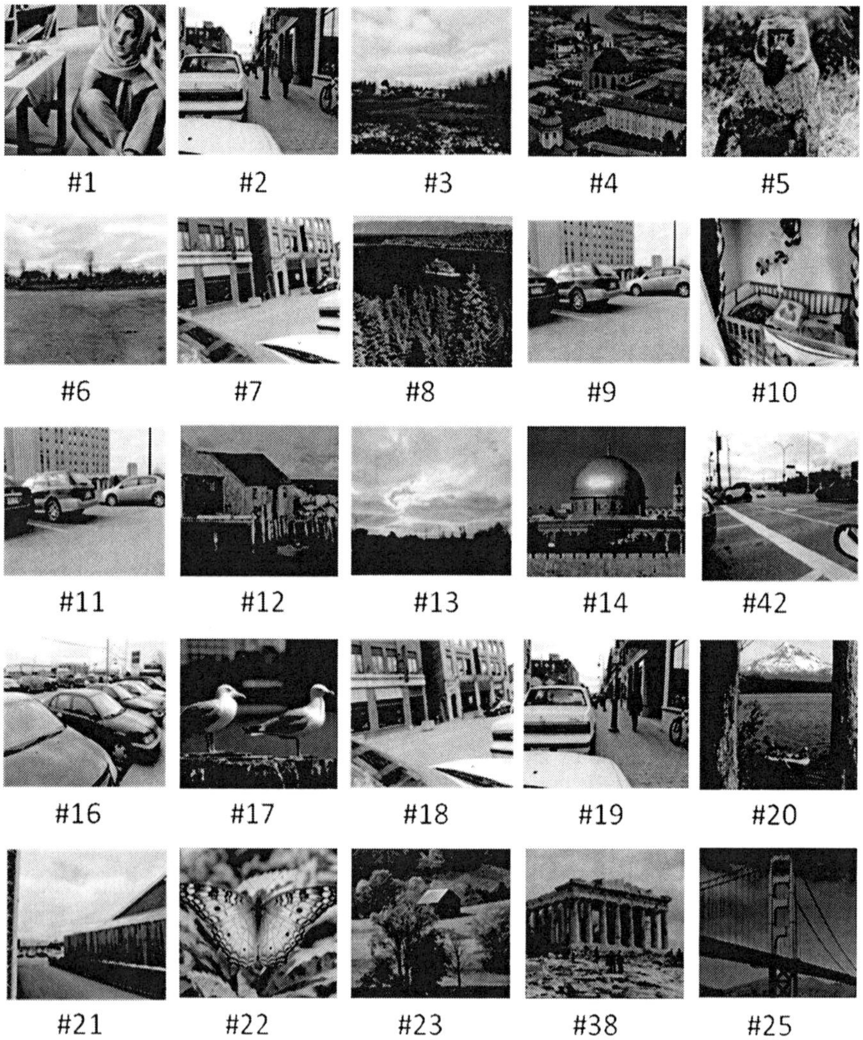

Perfect, Near Perfect or Not Perfect Reconstruction

We start this section by briefly defining the *perfect reconstruction* property, and after that we introduce its effect on image quality in the context of VSN.

Perfect reconstruction PR: Usually, a wavelet transform converts an original signal to a spatial-frequency domain and performs some operations on the transformed signal. The inverse wavelet transform accomplishes the inverse transformation back to the original domain in order to obtain a signal X that reconstructs *perfectly* the original signal X_o, possibly delayed or scaled by a constant C (i.e., $X_o[n]=C.X[n\text{-}delay]$), see Figure 6. Let $L(z)$ and $G(z)$ be the low pass analysis and synthesis filters, and $H(z)$ and $F(z)$ the high pass analysis and synthesis filters, respectively. The related question is, how to choose high pass and low pass analysis and synthesis filters to get perfect reconstruction: $X_o[n]=X[n\text{-}delay]$? The "gap" in Figure 6 shows where the signal might be sent or stored, after coding. At that point, we

Figure 6. Perfect reconstruction filter bank

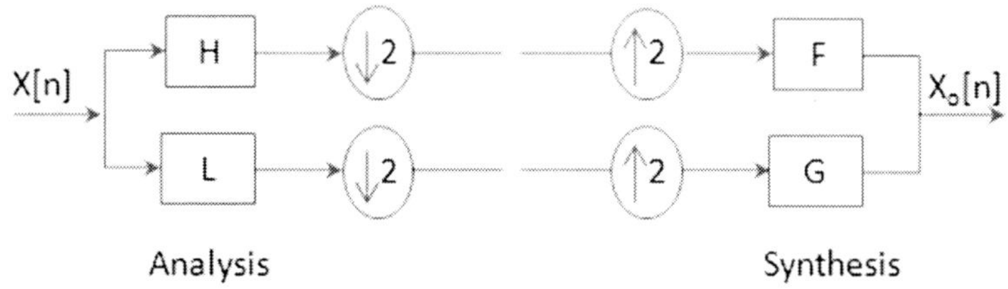

may compress and destroy information, in case of lossy compression. Perfect reconstruction assumes no compression or lost can be made, and the gap is closed.

Formally, to achieve perfect reconstruction, filters *L(z)*, *G(z)*, *H(z)* and *F(z)*, have to satisfy the following two conditions as shown in (Nguyen & Strang, 1996):

$$F(z)H(-z)+G(z)L(-z)=0 \qquad (1.17)$$

$$F(z)H(z)+G(z)L(z)=2z^{-delay} \qquad (1.18)$$

The first condition implies that the reconstruction is aliasing free and the second one means that the amplitude distortion of the reconstructed signal has value 1 (i.e., no distortion). Aliasing can be defined as spatial disturbing patterns in reconstructed images at the sink. More detail on this subject can be found in (Nguyen & Strang, 1996).

Often, DWT-based filters with PR property are desired in image coding systems (Nguyen & Strang, 1996). In VSN context, one can ask the following questions. Is this property always necessary? Is there some relationship between the dissipated power by a given VS and the PR property? If the PR is power consuming, is it acceptable to violate it when we deal with special applications like VSN (which require small power consumption even if we lose in the quality at the reception by the sink)? Is it possible to achieve both perfect reconstruction and very low power dissipation?

It is well known that, for lossless compression, PR is obviously mandatory (Nguyen & Strang, 1996). For low bit rate coding applications such as VSN, allowing small deviation from PR may be desirable if we gain in the dissipated power by a VS. If the PR condition is relaxed, complete aliasing at the reconstruction may not be avoided, and the cancelation of the aliasing will be required.

In application like VSN, such aliasing can be tolerated if the power gain is considerable and the image quality remains acceptable.

In this chapter, we do not investigate or discuss these theoretical aspects; rather, we focus in choosing the best wavelets having low dissipation at the price to loose in quality.

Decomposition Level

Another parameter influencing the compression performance is the decomposition level (*DL*), that is, the number of decompositions. It affects both the power dissipated by VS and the resolution of reconstructed image at the reception by the sink. After the first decomposition of the captured image, the decomposition process is performed iteratively on the low pass coefficients of the previous stage. For *DL* iterations, the target image is decomposed into *3DL+1* sub-bands. This iterative process yields better energy compaction due to the fact that low pass coefficients contain most of the original signal energy. Table 1 shows the energy distribution of the image #1 for 4-level decomposition. It is observed that most of the signal

Table 1. Energy distribution of image #1 for 4-level decomposition

Decomposition level	1	2	3	4
Energy(LL) (%)	99.04	97.40	95.85	93.97
Energy(HL) (%)	0.39	0.63	0.69	0.69
Energy(LH) (%)	0.42	0.72	0.51	0.47
Energy(HH) (%)	0.13	0.14	0.26	0.22

energy is concentrated in the *LL* sub-band, while other sub-bands contain very few information.

The quality of the compressed image depends on the level of decomposition *DL* used in the compression process. A larger *DL* increases the dissipation and decreases the compressed image file size. Moreover, the number of decompositions determines the resolution of the lowest level in wavelet domain. If a larger *DL* is used, we will be more successful in resolving important DWT coefficients from less important coefficients. The human visual system is less sensitive to removal of smaller details (Grgic et al., 2001). In simulation section, we show the effects of varying *DL* on the power dissipated by a VS node, and on a reconstructed image quality.

Increasing *DL* has two conflicting effects. On the one hand, it allows a better image quality while *DL* is smaller to a certain *DL*, and on the other hand, it augments the power dissipation. It is difficult to resolve this trade-off for all images. We show in simulations that, after certain number of iterations, the benefit gained in both energy compaction and reconstructed image quality becomes rather negligible compared to the computational effort performed by a VS node. The author in (Grigic, 2000) show that the sufficient number of decompositions is usually five, which is confirmed by our simulations. We highlight that smaller number of decompositions can also yield an acceptable image quality, at the price to gain in power consumption.

Filter Length and Filter Order

One of the parameters affecting the compression performance is filter length (*FL*). It is well known from wavelet theory that filters having high length yield wider functions with high degree of smoothness. On the contrary, filters with lower length have a better time localization and preserve important spatial information (Daubechies, 1992). Moreover, increasing *FL* increases the power dissipated by VS running a given wavelet. Equations 1.13 and 1.14 show how *FL* affects considerably the power dissipation. Hence, arbitrarily long filters cannot be used in the context of VSN. Further explanations about the influence of *FL* are found in simulation section. We will show by simulations how *FL* affects both the power dissipated by VS, and the degree of smoothness of the reconstructed image at the sink. More precisely, for each wavelet family, we show how *FL* affects the dissipation and the image quality for different kind of images.

The reader interested in the wavelet properties such as filter length, orthogonality, biorthogonality, symmetry, regularity and compact support is referred to (Daubechies, 1992) or (Nguyen & Strang, 1996). Table 2 summarizes some characteristics of the wavelet filters under evaluation.

It is important to clarify, in some words, the relationship between the filter length *T* and the filter order *I*. As aforementioned in Table 2, the filter length is directly proportional to the filter order in the case of orthogonal filters. For example, for *DW* and *SW* families the relationship is given by $T = 2.I$ and for *CW* family *T* is 6 times *I*. *HW* is the special case of *DW* with $I = 1$ and *T*

Table 2. Some characteristics of Wavelet filters under evaluation

	Haar	Daub	Coif	Sym	Spline
orthogonality	Ort.	Ort.	Ort.	Ort.	Bior.
Filter order(I)	1	2-20	1-5	2-10	
Compact support	Yes	Yes	Yes	Yes	Yes
Symmetry	yes	Far from	Close to	Close to	yes
Filter length	2	2I	6I	2I	$[2I_d, 2I_r]$

$= 2$. While for biorthogonal wavelets this relationship is given by $T = max(2.I_r, 2.I_d)+2$, but essentially for each low pass filter and high pass filter, different length T_r and T_d (corresponding to I_r and I_d, respectively) are used for each filter.

Choice of Wavelet Function

The selection of the appropriate wavelet filter for compression is of high importance, particularly in low-power applications such as VSN. It affects the quality of the reconstructed image at the sink and the battery state of VS.

Wavelets are generally characterized by their properties such as support, orthogonality or biothogonality, symmetry, regularity, the number of vanishing moments and the (explicit) existence of the analytical expression. In image compression domain, most desired properties are: compact support (FIR filters), symmetry (linear phase), orthogonality (signal energy conservation), regularity (smooth reconstruction) and the number of vanishing moments (more zero details coefficients). Among these properties, some of them are closely related, for instance regularity, number of vanishing moments and filter order (length). Unfortunately, some desirable properties such as, orthogonality, symmetry, and the number of vanishing moments are incompatible withitn each other. In what follows, let us expose very briefly some relations between wavelets properties used in this chapter. In fact, compactly supported orthogonal wavelets of Daubechies, which are built in order to maximize the number of vanishing moments, are not symmetrical, at the exeption of DW-1 (the Haar wavelet which has one zero moment). Note that Symlets wavelets which inherit the same properties as DW, are almost symmetrical. Coiflets wavelets are built by imposing a number of vanishing moments not only to the mother wavelet, as with DW, but also to the scaling function. On the contrary to the aforementioned orthogonal wavelets, Biorthogonal wavelets are more complicated and are defined based on a pair of scaling and wavelet functions. The use of biorthogonal wavelets has been found to be the most suitable for obtaining image wavelet transforms for compression applications. The biorthogonal filters have linear phase and symmetric properties that are useful in avoiding boundary artifacts in images.

This chapter does not describe or cover all wavelet functions proposed in the literature; rather, it focuses on those that are widely known and used in image compression. Indeed, orthogonal and biorthogonal wavelets families are considered. More precisely, we evaluate the biorthogonal wavelets (bior), and the following orthogonal wavelets: Haar wavelet (*HW*) (which is equivalent to Daubechies wavelet using *1* tap), Daubechies wavelet (*DW*), Coiflet wavelet (*CW*), and Symlet wavelet (*SW*). All of these wavelet families are one-dimensional (1-D) in nature and based on two-channel FIR filter banks. Each orthogonal wavelet family is parameterized by an integer *I*, which determines the order of the filter. For our simulation the following wavelets are used: *DW-I*, *I* varies from 1 to 20; *SW-I*, *I* varies from 1 to 10 and *CW-I*, *I* varies from 1 to 5. As biorthogonal wavelets family uses filters with similar or dissimilar orders for analysis and synthesis stages, two parameters I_d and I_r correspondent to the order of wavelet filter are used for analysis (decomposition) and synthesis (reconstruction), respectively, and they are denoted as in Matlab as follows: $biorI_r.I_d$. Thus the biorthogonal wavelets used in this work are: *bior1.1*, *bior1.3*, *bior1.5*, *bior2.2*, *bior2.4*, *bior2.6*, *bior2.8*, *bior3.1*, *bior3.3*, *bior3.5*, *bior3.7*, *bior3.9*, *bior4.4*, *bior5.5* and *bior6.8*.

Some reflections on filter length in the context of VSN can be drawn as follows. Relatively short analysis filters seem to be particularly effective for low rate image coding even they affects image smoothness. While designing filters in the context of power-constrained applications such as VSN, researchers and engineers have to be sure that the complexity of the analysis filter has to be minimized, at the cost of more complexity at the synthesis filter which is not assumed power-constrained (if implemented at the sink). For example in case of biorthogonal filters, T_d should be minimized at the cost of increasing T_r. This is due to the fact that the complexity of the filter increases with the increasing of the filter lengths. In this work, we do not design a new wavelet filter banks, rather, we compare existing wavelet filters.

SIMULATION PARAMETERS SETUP

Having briefly described the various wavelet families under evaluation and some factors influencing the DWT compression in VSN context, we present in this section the evaluation methodology used in this work as well as the simulation parameters. In Sections *objective evaluation* and *subjective evaluation*, we introduce the objective and the subjective simulations, respectively.

Factors influencing the performance of the DWT compression in VSN, described previously, are taken into account during our simulations. That is, several natural images are considered and compressed in a lossy way at several compression ratios (CRs) i.e., 100:1; 50:1; 25:1; 10:1; 2:1. Recall that the compression efficiency is defined by the parameter compression ratio CR, which is defined as the ratio between the original data and the compressed data, or also, as the ratio between the actual bpp and the reduced bpp. For instance, if the actual bpp = 8 and the reduced bpp = 0.5, then CR = 16:1.

In applications such as VSN, the perfect or near perfect reconstruction can be relaxed at the price to gain in the battery power. Hence, we consider the non perfect reconstruction of images under compression. For each test image, each CR, and each wavelet family described previously, we perform the following tests. For a given filter length (*FL*), we vary the decomposition level (*DL*) from 1 to 10 to study the effects of *DL* on both the VS battery state and the image quality at the base station. And complementary, for a given *DL*, we vary *FL* from 1 to its authorized value given in Table 2 to study the effects of *FL* on the same parameters, i.e., battery state and image quality.

To evaluate the power dissipated by a VS executing the different wavelets under evaluation, we adopted the parameters of the microcontroller ATMEL ATmega128 (Atmel 8-bit Microcontroller, 2009) low-power CMOS 8-bit, which is one of the most widely used in sensor nodes. The ATmega128 is able to perform an 8 bit multiplication in two cycles and an addition in one cycle. The power dissipated while executing a multiplication is $E_P = 6.6\ nJ$, while for an addition it is $E_A = 3.3\ nJ$ (Lecuire et al., 2007). To compute the power dissipated, Eqs. 1.15 and 1.16 are used for orthogonal and biorthogonal filters, respectively. In our simulations, it is supposed that enough memory in ATMEL Atmega128 is available to perform our tests, which is not guaranteed in the real testbed ATMEL Atmega128 (about 4KB for the internal SRAM). Anyway, our energy model is memory independent, and it takes into account only the dissipated energy.

The test data under evaluation consist of the grayscale natural images (Figure 5). To give more sense to our results, these images are natural and vary in size, and cover a wide variety of content (high and low spatial activity) as those handled by VSN. However, to be fair while comparing the power dissipated by different wavelets, the same image dimensions are used for the images under comparison.

The compression performance of the wavelet filters under consideration is assessed using both objective (quantitative) and subjective (qualitative) evaluations. Note that the criterion of image quality comparison is the resemblance between original and reconstructed images.

The goal of objective assessment (Section *objective evaluation*) is to predict the resemblance between original and reconstructed images accurately and automatically. Two objective image quality assessments are used, the most common Peak Signal to Noise Ratio (*PSNR*) (Wang and Bovik, 2006) and the Structural SIMilarity index (*SSIM*) (Wang et al., 2004). *PSNR* is the ratio between the maximum power of a signal (e.g., 255 for 8-bit image) and the power of signal's noise. A higher *PSNR* means a less noisy signal. In our simulation, the minimal acceptable quality of the reconstructed image at the base station corresponds to 24 dB. Formally, *PSNR* is expressed in decibels as follows, where a is the number of bits per pixel (x,y), f and f^r are the original and the reconstructed images having $N \times N$ pixels, respectively (Wang & Bovik, 2006):

$$PSNR = -10\log_{10} \frac{\frac{1}{N^2}\sum_{x=1}^{N}\sum_{y=1}^{N}\left|f(x,y) - f^r(x,y)\right|^2}{\left(2^a - 1\right)^2} \tag{1.19}$$

We supplement the objective measure by the relatively new tool called *SSIM*, which is defined as follows. *SSIM* is a method for measuring the similarity between two images. The reader interested in this subject, is referred to (Wang et al., 2004). Formally, *SSIM* is defined as follows (Wang et al., 2004):

$$SSIM\left(f,f^r\right) = \frac{\left(2\mu_f\mu_{f^r} + c_1\right)\left(2\operatorname{cov}_{ff^r} + c_2\right)}{\left(\mu_f^2 + \mu_{f^r}^2 + c_1\right)\left(\sigma_f^2 + \sigma_{f^r}^2 + c_2\right)} \tag{1.20}$$

Where μ_f and μ_{fr} are the mean intensities of f and f^r respectively; σ_f and σ_{fr} are the standard deviations of f and f^r respectively; cov_{ffr} is the covariance between f and f^r; and c_1 and c_2 are two small values used to avoid the division by 0 in Eq. 1.20. In our simulation, the following values of c_1 and c_2 are used, $c_1=(K_1L)^2$ and $c_2 = 0.02$, $L=255$ and $K_1 << 1$. The resultant *SSIM* index is a decimal value comprised between *-1* and *1*, the closer *SSIM* value to *1* the higher is the quality of the measured image.

We use *SSIM* since *PSNR* does not always provide a high quality measure (Wang et al., 2004). Moreover, it has been proven that *SSIM* exceeds *PSNR* in term of preciseness (Wang et al., 2004). However, as highlighted in (Wang & Bovik, 2006), the weakness of *PSNR* is less important in the case of multiresolution transforms as DWT. This observation will be confirmed in our simulation results, where the *PSNR* and the *SSIM* measures conduct to almost the same conclusions. For space limitation, only the *PSNR* measurements are presented and discussed.

In fact, the best way to evaluate the image quality is by *subjective* assessment of a human observer (Wang & Bovik, 2006). Indeed, the average score of the subjective assessment is used in our simulations (Table 3). This is performed as follows. Showing the original image captured by a VS source and the reconstructed image at the sink to a group of 18 viewers, and then averaging their evaluations. Each viewer was asked to assign one of the values presented in Table 3 to the image under evaluation. Due to its time consuming, the

Table 3. Rating scale of the television allocations study organization used in subjective measurement

Rate	Mention	Remark
1	Excellent	An image of extremely high quality.
2	Fine	An image of high quality (enjoyable viewing).
3	Passable	An image of acceptable quality.
4	Marginal	Poor quality; you wish you could improve it.
5	Inferior	Very poor image, but you could watch it.
6	Unusable	An image so that could not watch it.

subjective evaluation was restricted only to one of the best decomposition levels and some filter lengths. More explanations related to subjective assessments will be given in Section *subjective evaluation.*

Our objectives behind the simulations are summarized as follows:

1. For each wavelet family, the following tests are performed:
   - Study the effects of varying the decomposition level *DL* on the power dissipated by a given VS and the quality of the reconstructed image at the sink.
   - Study the effects of varying the filter length *FL* on both power dissipation and image quality
2. Determine which wavelet(s) perform(s) better for VSN.

In respect to the objectives cited above, the compression performance of the wavelet families under evaluation is assessed into two stages. In the first stage (Section *objective evaluation*), objective measurements using PSNR and *SSIM* are performed to determine the best wavelet filters under evaluation. In the second stage (Section *subjective evaluation*), subjective measurements are carried out and the results are compared to those obtained with objective measurement.

OBJECTIVE EVALUATION

Effects of Varying the Decomposition Level

We introduce some parameters that help the best understanding of the decomposition level. That is, we define $PSNR_l$ as a lowest quality measure of the reconstructed image required by the user at the base station (Thomos et al., 2006), and we set $PSNR_l$ to 24dB. $PSNR_l$ allows to determine the minimal number of decompositions DL_a required to obtain an acceptable image quality at the sink. The same definition can be applied for $SSIM_l$, and thus, $SSIM_l$ can be set to 0.65. Furthermore, we define DL_o and DL_m as follows. DL_o is the optimal number of decompositions required to obtain the highest quality, while DL_m represents the maximal number of decompositions which permits to obtain *PSNR* larger than or equal to $PSNR_l$, and is larger than DL_o and DL_a. With those parameters, we can define an interval of decompositions $[DL_a, DL_m]$ which represents the set in which we will select one *DL*. Note that this interval contains DL_o. Recall that since the same deductions are made for *PSNR* and *SSIM*, only the *PSNR* measurements are presented and discussed (at the exception of Figure 8).

To evaluate the effects of varying *DL* on both the power dissipated by the microcontroller ATmega128 and the quality of the reconstructed image at the sink, we perform the following tests. For each image, each CR, and for a given *FL* (measured by the number of taps) inside each wavelet family, we determine DL_o and DL_a. We select a small *FL* (e.g., *FL* = *2*), since it has not a strong influence on the quality of the reconstructed image, rather, it allows to dissipate small power compared to longer filter. More details about the effects of *FL* will be studied in Section *effects of varying filter length*. Recall that the image quality is assessed objectively using both *PSNR* and *SSIM*, and the power dissipated by a VS while executing a given wavelet is calculated using Eq. 1.15 for orthogonal filters and Eq. 1.16 for biorthogonal filters. In the remaining part of this subsection, we will first present the effects of *DL* on the image quality, and then the effects of *DL* on the power dissipation.

For all the wavelet families under evaluation the following results are true. As *DL* increases from 1 to 10 for a given *FL*, the *PSNR* (or *SSIM*) increases with *DL* until *DL* is equal to a value denoted DL_o, and after that, the *PSNR* decreases with *DL*. For very low CR (e.g., 2:1), DL_o is equal to the minimal *DL*, i.e., *PSNR* decreases with *DL* from the beginning. See Figure 7 for an example of biorthogonal wavelet, and Table 4 for an example of Daubechies wavelet. While for the dissipated power, it always increases with *DL* following Eq. 1.15 for orthogonal filters and Eq. 1.16 for biorthogonal wavelet. The dissipated power will be explained below. It is observed that DL_o differs slightly from one image to another, and from one wavelet family to another, for a given CR and *FL*. In Table 4, we see that as CR varies from 100:1 to 2:1, the *DL* required to obtain the highest *PSNR* decreases slowly until *DL=1* for CR = 2:1. In other words, increasing CR necessitates more decompositions to obtain an adequate image quality. For instance, for image #1 compressed using Daubechies family, with CR = 100:1 and *FL* = *2*, *5* decompositions (i.e., *DL=5*) are needed to obtain the maximum *PSNR*, i.e., 24.16 dB. As CR decreases until 2:1, only one decomposition (i.e., *DL=1*) is needed to obtain 42 dB. We note also that, as the CR decreases, the image quality measure (in terms of *PSNR* or *SSIM*) becomes better, while DL_o decreases until *1*. For instance for image #1, *PSNR* decreases from 42 dB till 24 dB with the increasing number of decompositions,

Figure 7. PSNR as a function of DL and bior2.2 for different CR values: 2:1, 10:1, 25:1, 100:1

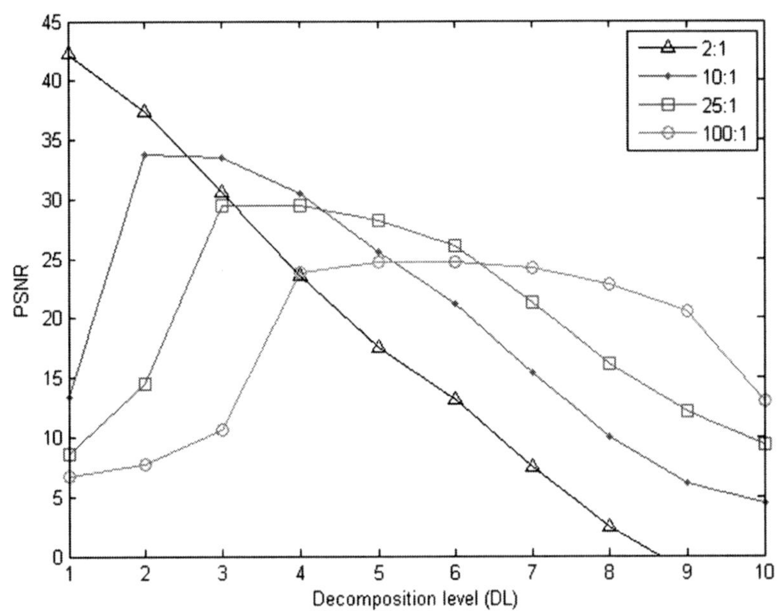

Figure 8. SSIM as a function of Decomposition level DL for bior2.2 for different CR values

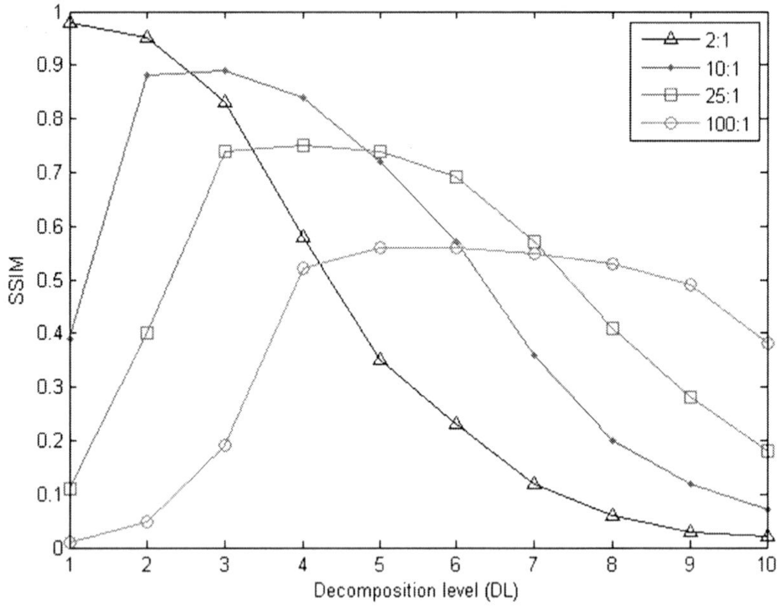

from *1* until *5*. Considering the aforementioned results, we conclude that DL_o is an increasing function of CR.

In Table 4, for *DL* = 4, 5 and 6, it is shown that the PSNR value increases from CR = 100:1 to 50:1. However, there is a drop in PSNR value at CR = 2:1 for high *DL*. It is anticipated that as CR decreases, the number of bits sent to the decoder will increase, this giving a better reconstructed image. An alternative explanation is related to the optimal DL_o (1 or 1 decompositions) reached when CR is low (e.g., CR = 2:1). After DL_o, the PSNR will naturally decrease with the increasing number of *DL*.

We note also that these values may change slightly with image content. Indeed, with images having low spatial activity such as #25, DL_a is slightly larger than those with high spatial activity images, such as #1. For instance, for Daubechies wavelet and CR=50:1, the *DL* permitting to obtain *PSNR* equal to or larger than $PSNR_l=24$ dB are: 3 to 7 for image #25, and 4 to 6 for image #1.

Almost the same concluding remarks can be drawn for all wavelets. See also Table 5, where DL_a is the minimum *DL* required to obtain an acceptable image quality (i.e., 24 dB), and DL_o is the *DL* yielding the best *PSNR*. In Table 5, for a given CR, we determine for each wavelet filter the optimal number of decompositions as well as the correspondent filter length and the PSNR.

The results given by *PSNR* are confirmed by those obtained by *SSIM* in almost all the cases (example Figure 7 for *PSNR* and Figure 8 for *SSIM*). For all wavelet families under consideration and for some CRs, the *DLs* giving the highest *PSNR* for natural images handled by VSN are listed in Table 6. For instance, for compression ratio 25:1, it is shown that 3 to 4 decompositions are required to obtain the highest *PSNR*. Also, it can be observed that for very small number of decompositions, an annoying blurring artifacts was introduced specially for high CR, for instance CR=100:1.

For both orthogonal and biorthogonal wavelets, we conclude for the decomposition level effects what follows. The *DL* required to obtain an acceptable image quality depends on the following parameters: CR, wavelet type, and image content.

Table 4. PSNR results for three CRs (100:1, 50:1, 2:1), and for different DL and FL, for Daubechies wavelet family for images #25 and #1

CR	*100:1*					
FL →	*DL →*	*1*	*3*	*4*	*5*	*6*
2	*# 25*	*10.14*	*12.70*	*28.40*	***29.32***	*29.22*
	# 1	*6.69*	*10.48*	*23.42*	***24.16***	*24.10*
4	*# 25*	*10.14*	*12.71*	*28.60*	***29.53***	*29.41*
	# 1	*6.69*	*10.52*	*23.96*	***24.70***	*24.56*
6	*# 25*	*10.14*	*12.71*	*28.48*	***29.33***	*29.24*
	# 1	*6.69*	*10.53*	*23.71*	***24.45***	*24.35*
8	*# 25*	*10.15*	*12.71*	*28.52*	***29.40***	*29.25*
	# 1	*6.96*	*10.51*	*23.77*	***24.44***	*24.35*
10	*# 25*	*10.15*	*12.72*	*28.40*	***29.18***	*29.06*
	# 1	*6.96*	*10.53*	*23.76*	***24.40***	*24.28*
CR	***50:1***					
FL →	*DL →*	*1*	*3*	*4*	*5*	*6*
2	*# 25*	*10.45*	*28.25*	*31.58*	***31.61***	*30.73*
	# 1	*7.35*	*23.31*	***26.63***	*26.45*	*25.70*
4	*# 25*	*10.46*	*28.50*	***31.80***	*31.78*	*30.72*
	# 1	*7.35*	*23.78*	***27.4***	*26.90*	*25.92*
6	*# 25*	*10.46*	*28.53*	***31.67***	*31.61*	*30.70*
	# 1	*7.36*	*23.87*	***26.84***	*26.63*	*25.76*
8	*# 25*	*10.47*	*28.50*	***31.60***	*31.57*	*30.45*
	# 1	*7.36*	*23.80*	***26.74***	*26.59*	*25.81*
10	*# 25*	*10.47*	*28.52*	***31.48***	*31.40*	*30.40*
	# 1	*7.36*	*23.94*	***26.73***	*26.53*	*25.62*
CR	***2:1***					
FL →	*DL →*	*1*	*3*	*4*	*5*	*6*
2	*# 25*	***47.74***	*37.59*	*31.01*	*24.24*	*17.91*
	# 1	***42.08***	*31.20*	*24.34*	*17.63*	*12.94*
4	*# 25*	***47.74***	*37.47*	*30.28*	*23.67*	*17.32*
	# 1	***42.18***	*31.10*	*24.16*	*18.68*	*12.69*
6	*# 25*	***47.94***	*36.73*	*30.50*	*23.50*	*17.61*
	# 1	***42.21***	*31.21*	*24.30*	*17.71*	*12.71*
8	*# 25*	***47.63***	*37.07*	*29.88*	*23.48*	*17.04*
	# 1	***42.11***	*30.85*	*24.18*	*18.39*	*13.31*
10	*# 25*	***47.82***	*36.82*	*30.09*	*23.16*	*17.14*
	# 1	***42.02***	*30.93*	*24.20*	*17.85*	*12.61*

Table 5. Number of decompositions and Filter length relationship for image #1. DL_a is the number of decompositions required to obtain an acceptable image quality, DL_o is the number of decompositions yielding the best PSNR, and FL is the filter length, while DL_m represents the maximal number of decompositions which is larger than or equal to $PSNR_l$, and larger than DL_o and DL_a.

	CR →	100:1		25:1		10:1		2:1	
DW	$[DL_a, DL_m]$	[4,8]		[3,6]		[2,5]		[1,4]	
	DL_o	5	Corresp. FL = 4	4	Corresp. FL=4	2	Corresp. FL = 2	1	Corresp. FL = 3
			PSNR = 24.7		PSNR = 29.12		PSNR = 33.39		PSNR = 42.21
SW	$[DL_a, DL_m]$	[4,7]		[3,6]		[2,5]		[1,4]	
	DL_o	5	Corresp. FL = 5	4	Corresp. FL=6	2	Corresp. FL = 6	1	Corresp. FL = 3
			PSNR = 24.85		PSNR = 29.25		PSNR = 33.57		PSNR = 42.23
CW	$[DL_a, DL_m]$	[4,7]		[3,6]		[2,5]		[1,4]	
	DL_o	5	Corresp. FL = 4	4	Corresp. FL=3	2	Corresp. FL = 3	1	Corresp. FL = 3
			PSNR = 24.72		PSNR =29.3		PSNR = 33.61		PSNR = 42.16
bior	$[DL_a, DL_m]$	[4,7]		[3,6]		[2,5]		[1,4]	
	DL_o	5	T_d=17, T_r=11	4	T_d=17, T_r=11	3	T_d=9, T_r=3	1	T_d=17, T_r=11
			PSNR =27.15		PSNR = 30.78		PSNR = 33.53		PSNR = 42.27

Table 6. DL values required to obtain the highest PSNR for different CR values, for filters under evaluations

CR	100:1	50:1	25:1	10:1	2:1
DW	5 to 6	4 to 5	4	2 to 3	1
CW	6	5	4	2 to 3	1
SW	5 to 6	4 to 5	3	2	1
bior	5	4	4	2	1

It is difficult to find a unique *DL* for all wavelets, CRs, or images. In general, for low CR, it is observed that only small number of decompositions is needed to obtain an adequate image quality, while for high CR more than 4 decompositions are required.

This deep analysis allows us to select an interval of *DL* giving an acceptable image quality, and thus, selecting the smallest number of *DL* permits to minimize the power dissipated by a VS for obtaining an acceptable image at the sink. For example for image #38, the number of decompositions for each CR are given as follows. For CR = 10:1; 25:1 and 100:1, the intervals of *DL* guaranteeing an acceptable-to-maximal quality (i.e., $PSNR \geq 24$ dB) are, [2,5], [3,6] and [4,7], respectively. If we need the smallest power, we select the smallest *DL* (denoted DL_a) of each interval. These intervals may change as functions of $PSNR_l$. We also note that DL_a increases with $PSNR_l$. From our simulations, we remark that 4 to 5 decompositions yield in general the best *PSNR* for almost all wavelets.

The power dissipated by VS increases with *DL*, following Eq. 1.15 for orthogonal filters (Figure 9), and Eq. 1.16 for biorthogonal filetrs (Figure 10). Figures 9 and 10 show how *DL* affects the power dissipated by a given VS while executing an orthogonal (for example *DW*) and biorthogonal wavelets, respectively. It is observed that filters used with lowest *DL* consume less power com-

Figure 9. DL effects on the power dissipated by a VS while executing orthogonal Daubechies wavelets DW-2, DW-4, DW-6 and DW-8 for an image of size 512 × 512 and 8 bpp.

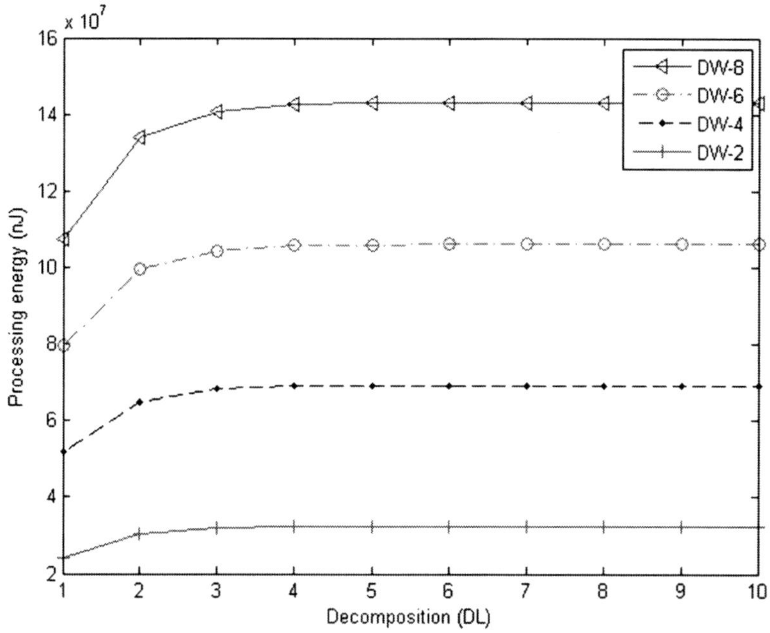

Figure 10. DL effects on the power dissipated by a VS while executing biorthogonal wavelets bior1.1, bior2.2, bior2.4, bior4.4 and bior6.8 for an image of size 512 × 512 and 8 bpp.

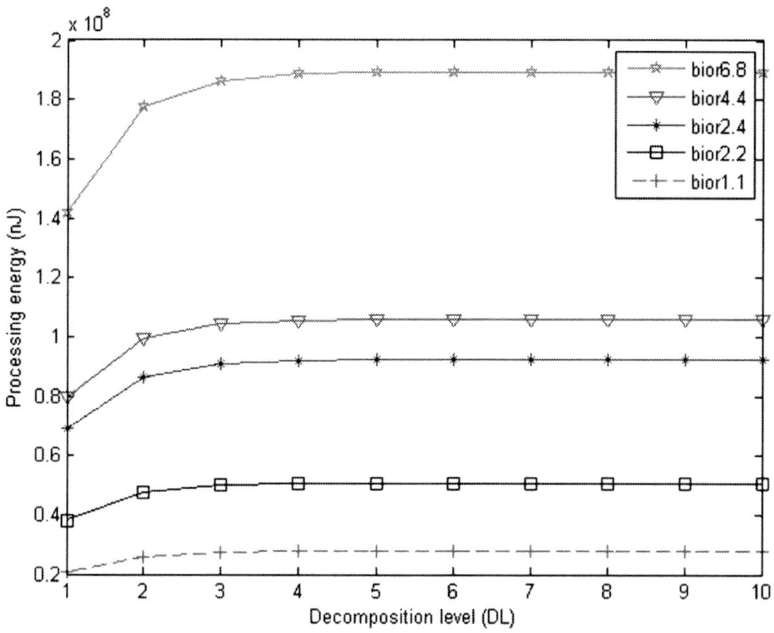

Figure 11. PSNR as a function of DL for different filter length values

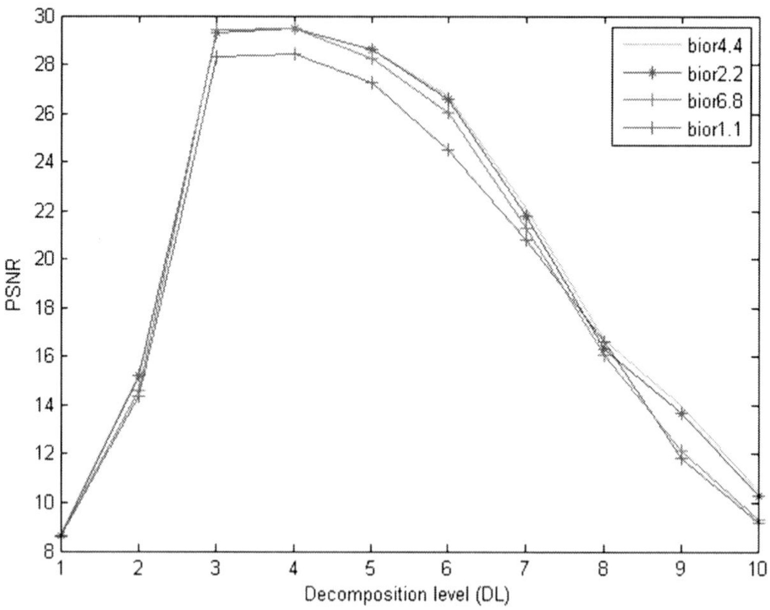

pared to those with important *DL*. Those filters, used with small *DL*, are highly important in VSN. It is also observed that the power dissipated increases rapidly with *DL*, until *DL*= *4* to *5*, and after that, the dissipated power becomes almost constant. This is due to the fact that the number of wavelet coefficients at each level of decomposition decreases with *DL*. Small *DL* implies less power dissipation, but, on the other hand, we obtain low to moderate quality of the reconstructed image specially with high CR. If the user at the base station wants to save the VS power, he/she can select small $DL \geq DL_a$ (i.e., permitting to obtain at least $PSNR_l$) at the expense to lose in the quality of the reconstructed image.

It is also shown, from Figures 9 and 10, that the number of taps of a given filter affects considerably the power dissipated by a VS. On the other hand, the effect of *FL* on the image quality seems negligible. The effects of *FL* on the power will be treated in the next Section.

Again, it is confirmed in Figure 11 that the quality of the reconstructed image depends on *DL*. It is observed that as the number of decompositions increases, PSNR is increased up to DL_o. Beyond that value, increasing the number of decompositions decreases the quality of the image. Hence, higher *DL*, doesn't imply better quality.

Effects of Varying the Filter Length and Filter Order

After evaluating the effects of *DL* on the power dissipation and the quality of the reconstructed image, let us now study the consequences of varying *FL* on the same factors. In the following sections the term filter length (*FL*) will be used more frequently than filter order (*FO*), since it reflects the number of taps of a filter under evaluation, and also it is used as an evaluation factor in the power estimated by Eqs. 1.9 to 1.16. Recall that *FL* is directly related to *FO* as explained in Sections *filter length and filter order* and *choice of wavelet filter*.

For all wavelet families, each of the test images was compressed using a selected *DL*, at several CRs, i.e., 100:1; 50:1; 25:1; 10:1; 2:1. We set *DL* to 5 since it yields good results in terms of image quality for a medium-to-high CRs (i.e., CR≥50:1), for almost all wavelet families under evaluation. In addition, we vary *FL* of each wavelet family from 2 to its authorized value given in Table 2. The image quality is assessed using both *PSNR* and *SSIM*, and the power dissipation is calculated using Eq. 1.15 for orthogonal filters and Eq. 1.16 for biorthogonal filters. Since the same deductions are made for *PSNR* and *SSIM*, we consider only the *PSNR* measurements. In the remaining part of this subsection, we will first present the effects of varying *FL* on the image quality and then its effects on the power dissipation.

In general, for all simulations the image quality measure (*PSNR* or *SSIM*) increases very slightly with *FL* until it reaches a maximum for a *FL* value denoted FL_o, and after that, the image quality fluctuates slightly around this maximum when *FL* continues to increase (Figure 12). Let us consider the example of the following images #1, #42 and #25 which represent images of low-to-high spatial activity. For each image (#1, #42, #25), each CR (100:1; 50:1; 25:1; 10:1; 2:1) and *DL=5*, we show in Table 7 the *PSNR*s obtained for different *FLs* and we highlight the best of them, for Daubechies wavelets, Symlet wavelets, Coiflet wavelets and biorthogonal wavelets. Clearly, no single *FL* performs best for all images. From these tables, the following observation can be drawn. *FL* has not a strong influence on the quality of the reconstructed image. Generally, the results presented in (Grgic et al., 2001) fit well with our finding. This is confirmed in Figure 12, where we see that the influence of *FL* on *PSNR* is insignificant.

As with DL_o, the value FL_o is not the same for all tests. It depends on wavelet type, *DL*, CR and image content. We conclude for each wavelet family what follows. For high CR (e.g., 100:1), the transforms *DW-4* and *DW-3* within Daubechies family, perform heavily better for almost all images (for example, Table 7 (a)). While for low CR (e.g., 2:1), *DW-1* (*HW*) and *DW-2* are better for medium-to-low spatial activity images such as #25, and *DW-4* for high spatial activity images such as #1. For Coiflet family, *CW-3*, followed by *CW-2*, yields the best results for high CR for almost all test images. While for low CR, *CW-1*

Figure 12. PSNR as a function of FL for DW and SW using DL=5. FLo correspond to 8.

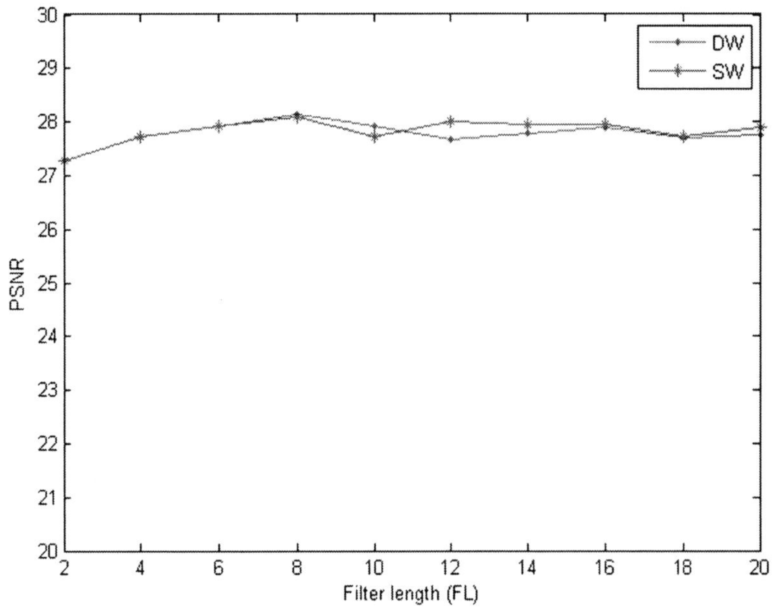

Table 7. PSNR results for different FL for (a) Daubechies, (b) Coiflet, (c) Symlet and (d) Biorthogonal families for different CRs and a fixed DL=5 (Images #1, #42 and #25). FL(k/p) means k taps for low pass filter, and p taps for high pass filter

(a)

-	Daubechies family (DL=5)															
	FO →	db1	db2	db3	db4	db5	db6	db7	db8	db10	db11	db12	db14	db16	db18	db20
CR ↓	FL →	2	4	6	8	10	12	14	16	20	22	24	28	32	36	40
100:1	#1	24.19	24.16	24.53	**24.70**	24.52	24.45	24.48	24.44	24.40	24.34	24.41	24.29	24.32	24.30	24.22
	#42	21.42	21.63	21.76	**21.86**	21.77	21.70	21.70	21.64	21.60	21.52	21.55	21.49	21.46	21.46	21.37
	#25	29.20	29.32	29.44	**29.53**	29.46	29.33	29.37	29.40	29.40	29.22	29.17	29.06	29.03	29.00	28.93
50:1	#1	27.27	27.72	27.92	**28.13**	27.91	27.68	27.79	27.88	27.76	27.55	27.72	27.53	27.47	28.66	27.44
	#42	26.45	26.92	**27.15**	27.03	27.13	27.06	26.95	26.82	26.91	26.68	26.65	26.55	26.47	26.42	26.49
	#25	33.15	33.14	**33.30**	33.24	33.13	33.09	33.08	32.98	32.85	32.74	32.79	32.57	32.58	32.42	32.39
10:1	#1	25.18	25.22	25.59	**27.91**	26.18	25.30	25.43	25.91	25.43	25.40	25.62	25.64	25.53	25.97	25.49
	#42	28.07	**28.47**	27.80	27.54	28.20	26.92	27.89	27.58	28.23	28.07	27.68	27.88	27.78	27.75	28.08
	#25	32.14	**33.14**	31.50	31.25	31.12	31.09	31.01	31.01	30.76	30.70	30.87	30.59	30.70	30.35	30.60
2:1	#1	17.77	17.63	18.01	**18.68**	17.99	17.71	17.84	18.39	17.85	17.84	18.12	18.15	18.01	18.51	18.00
	#42	**21.40**	21.34	20.53	20.15	21.06	21.53	20.65	20.30	21.21	21.12	20.50	20.90	20.82	20.79	20.87
	#25	**24.78**	24.24	23.94	23.67	23.57	23.50	23.43	23.48	23.16	23.15	23.29	23.02	23.17	22.76	23.04

(b) (c)

-	Coiflet family (DL=5)					Symlet family (DL=5)										
	FO →	CW-1	CW-2	CW-3	CW-4	CW-5	-	SW-2	SW-3	SW-4	SW-5	SW-6	SW-7	SW-8	SW-9	SW-10
CR ↓	FL →	6	12	18	24	30	-	4	6	8	10	12	14	16	18	20
100:1	#1	24.19	24.69	**24.77**	24.72	24.64	-	24.16	24.53	24.43	**24.85**	24.68	24.63	24.51	24.60	24.53
	#42	21.70	21.91	**21.93**	21.85	21.90	-	21.63	21.76	21.81	**21.93**	21.82	21.80	21.82	21.79	21.82
	#25	29.29	**29.62**	29.60	29.60	29.53	-	29.32	29.44	29.45	**29.69**	29.50	29.51	29.52	29.51	29.52
50:1	#1	27.87	**28.13**	28.09	28.09	28.04	-	27.72	27.92	**28.10**	27.72	28.01	28.54	27.94	28.01	27.89
	#42	26.96	27.36	**27.39**	27.22	27.36	-	26.92	27.13	27.22	**27.40**	27.27	27.29	27.28	27.26	27.27
	#25	33.32	**33.55**	**33.55**	33.54	33.50	-	33.14	33.30	33.40	**33.52**	33.42	33.41	33.39	33.40	33.38
10:1	#1	25.71	25.65	25.71	25.78	**25.87**	-	25.22	25.59	25.83	**24.94**	25.59	25.50	25.46	25.51	25.36
	#42	**27.88**	27.75	27.39	27.66	27.61	-	**28.37**	27.80	27.62	28.20	27.76	27.79	27.89	27.98	28.01
	#25	**31.99**	31.83	31.73	31.69	31.66	-	**31.81**	31.50	31.65	31.47	31.56	31.46	31.43	31.46	31.35
2:1	#1	18.17	18.08	18.11	18.20	**18.30**	-	17.63	18.01	**18.27**	17.34	17.98	17.88	17.83	17.78	17.73
	#42	**20.67**	20.41	20.31	20.27	20.22	-	**21.40**	20.53	20.25	21.00	20.41	20.61	20.60	20.71	20.76
	#25	**24.43**	24.27	24.19	24.15	24.13	-	**24.24**	23.94	24.09	23.95	23.99	23.89	23.86	23.91	23.78

(d)

-	biorthogonal family: biorx.y (DL=5)															
	x.y	1.1	1.3	1.5	2.2	2.4	2.6	2.8	3.1	3.3	3.5	3.7	3.9	4.4	5.5	6.8
CR	FL(k/p) →	2/2	6/2	10/2	5/3	9/3	13/3	17/3	4/4	8/4	12/4	16/4	20/4	9/7	9/11	17/11
100:1	#1	24.19	23.94	23.79	24.54	24.56	24.52	24.51	21.09	23.58	24.02	24.11	24.13	**24.68**	24.17	24.61
	#42	21.42	21.15	20.95	21.78	21.80	21.74	21.70	17.17	19.99	20.55	20.79	20.84	**21.97**	21.60	21.97
	#25	29.20	28.99	28.85	29.51	29.51	29.49	29.47	25.76	28.25	28.74	28.90	28.95	**29.60**	29.13	29.57
50:1	#1	27.27	26.86	26.62	**28.60**	28.60	28.52	28.44	23.52	26.87	27.52	28.71	27.81	28.26	26.75	28.16
	#42	26.45	25.92	25.60	27.41	27.37	27.30	27.21	23.18	26.02	26.58	26.64	26.68	27.65	26.85	**27.68**
	#25	33.15	33.60	32.43	33.80	**33.83**	33.77	33.67	29.86	32.34	32.66	32.80	32.88	33.74	32.57	33.69
10:1	#1	25.18	**26.86**	24.81	26.15	26.12	26.08	26.03	19.18	23.74	24.84	25.27	25.52	25.47	24.51	25.54
	#42	**28.07**	27.63	27.39	27.49	27.47	27.46	27.41	19.53	25.37	26.92	27.16	27.31	27.85	27.31	27.84
	#25	**32.14**	31.72	24.38	22.99	23.05	31.52	31.48	26.41	29.32	29.50	29.67	29.81	31.80	31.23	31.73
2:1	#1	17.77	17.79	17.75	16.78	16.74	16.70	16.66	8.70	12.92	13.95	14.39	14.58	17.84	**18.40**	17.57
	#42	21.34	20.97	20.79	18.82	18.60	18.60	18.59	10.62	15.79	17.17	17.32	17.39	20.65	**21.46**	20.18
	#25	**24.78**	24.55	24.38	22.99	23.05	23.00	22.97	16.86	19.68	19.76	19.92	20.05	24.26	24.58	23.98

is the best for medium-to-low frequency images (#25), and *CW-5* for high spatial activity image such as #1 (see Table 7 (b)). For Symlet family, *SW-5* followed by *SW-6* and *SW-4*, respectively yield the best results for high CR for almost all test images. For low CR, *SW-4* seems to be the best for high spatial activity images such as #1, and *SW-2* for low to medium spatial activity images such as #25, respectively (see Table 7 (c)).

For biorthogonal family, the best results are given by *bior4.4* and *bior6.8* followed by *bior2.4* and *bior2.2* for high CR. For low CR, it is observed that *bior5.5* performs better. Surprisingly, *bior1.1* yields also good result for low CR and low-to-medium spatial activity images. In general, *bior2.2* is found to get a good result (see Table 7 (d)), which is different than the results given in (Grgic et al., 2001).

We highlight that even with small *FL* (<3) we obtain an acceptable image quality (for example *bior1.1*), since *FL* has a very slight influence on the image quality. This result seems to be attractive in power-constrained application such as VSN. However, for very small *FL*, i.e, $FL = 1$, an annoying blurring artifacts was introduced specially for high CR, for exampe 100:1.

As said previously, image content is a primal factor affecting compression performance and the value FL_o. For that reason, the same simulations are performed for a large set of images having different contents, and almost the same observations and conclusions are made. Again, *FL* has not a strong influence on the quality of the reconstructed image at the sink.

For power dissipation considerations, the dissipation increases with *FL* following Eq. 1.15 for orthogonal filters (Figure 13 for *DW* and *SW*, and Figure 14 for *CW*), and Eq. 1.16 for biorthogonal filters (Figure 16). For orthogonal filters, such as Daubechies, Symlet or Coiflet, the power dissipated by a VS increases with FL, and the smallest power is dissipated by filters using 2 taps (i.e., *DW-1*). However, wavelets using 2 taps introduce an annoying blocking artifacts. For example, if we consider the image #1 of size 512 × 512 pixels and 8bpp, the power dissipated by an ATmega128 while executing *DW-1* using $DL=1$ and $DL=5$ is $0.347 \times 10^7 nJ$ and $0.46 \times 10^7 nJ$, respectively. This power increases with *FL* and is equal to $27.34 \times 10^7\ nJ$ and $36.41 \times 10^7 nJ$ while executing *DW-20* using $DL=1$ and $DL=5$, respectively. From Eq. 1.15, orthogonal filters having the same length consume the same power. For instance, *DW-x* and *SW-x* using *x* taps dissipate the same power. However and *CW-x* dissipates 3 times the power dissipated by *DW-x* and *SW-x*, since *T* (of *CW*) is 6 times the order *I*.

Let us now analyze the power dissipated by VS node executing biorthogonal filters which is represented by Eq. 1.16 and Figure 16. For simplification considerations, we refer to the biorthogonal filters by numbers 1 to 15, as follows: 1: *bior1.1*; 2: *bior1.3*; 3: *bior1.5*; 4: *bior2.2*; 5: *bior2.4*; 6: *bior2.6*; 7: *bior2.8*; 8: *bior3.1*; 9: *bior3.3*; 10: *bior3.5*; 11: *bior3.7*; 12: *bior3.9*; 13: *bior4.4*; 14: *bior5.5*; 15: *bior6.8*. It is observed that *bior1.1(2/2)* consumes the smallest power within biorthogonal family filter, due to its *FL*. Recall that numbers between brackets represent the number of taps for low pass and high pass analysis filters, successively. Thus, (*k/p*) means *k* taps for low pass analysis filter, and *p* taps for high pass analysis filter. The power dissipated by the microcontoler ATmega128, while executing the wavelet *bior1.1* using $DL= 1$ and $DL=5$ for an image of 512 × 512 pixels and 8bpp, is $2.07 \times 10^7\ nJ$ and $2.76 \times 10^7\ nJ$, respectively. It is clear from Figure 16 that *bior1.1(2/2)* dissipates the lowest power followed by *bior2.2(5/3)*, *bior1.3(6/2)* and *bior3.1(4/4)*. On the other hand, the biorthogonal wavelets consuming the highest power are *bior6.8(17/11)* followed by *bior3.9(20/4)*, *bior3.7(16/4)*, *bior2.8(17/3)* and *bior5.5(12/4)*. As with orthogonal filters, the power dissipation by biorthogonal filters is directly related to the number of taps.

Having briefly described the power dissipation of orthogonal and biorthogonal filters, let

Figure 13. FL effects on the power dissipated by a VS while executing orthogonal wavelets DW and SW for DL=1, 3, 5, and 7, for an image of size 512 × 512.

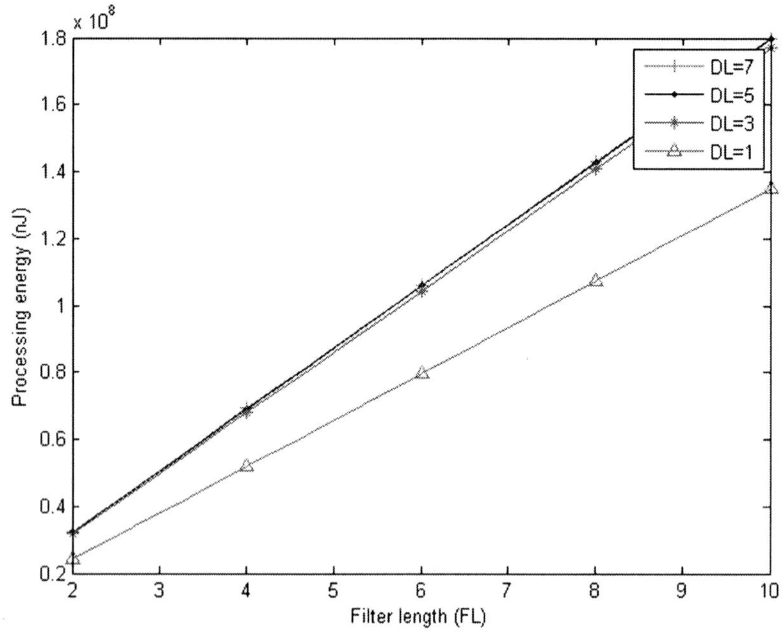

Figure 14. FL effects on the power dissipated by a VS while executing orthogonal wavelet CW for DL=1, 3, 5, and 7, for an image of size 512 × 512.

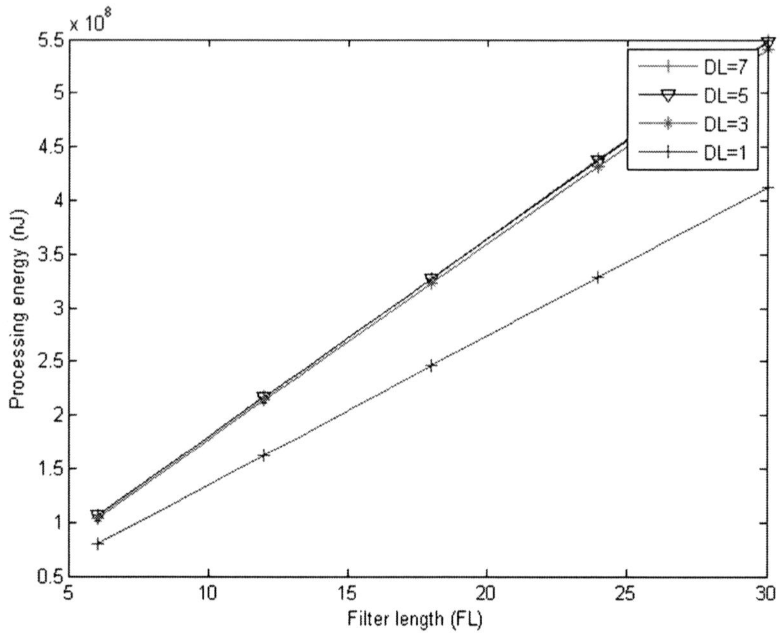

Figure 15. Classification of orthogonal and biorthogonal filters processing an image of 512×512 pixels (DL=5), using the microcontroller ATmega128. Filter categories **a)** *to* **s)** *in the x-axis refer to orthogonal and biorthogonal filters:* **a)** *Ort1;* **b)** *bior1.1;* **c)** *Ort2;* **d)** *Ort3 ;* **e)** *bior2.2;* **f)** *bior1.3 and bior3.1;* **g)** *Ort4;* **h)** *Ort5 and bior2.4;* **i)** *bior1.5 and bior3.3;* **j)** *Ort6, bior2.6 and bior4.4;* **k)** *bior3.5;* **l)** *Ort7;* **m)** *bior2.8 and bior5.5;* **n)** *bior3.7;* **o)** *Ort8;* **p)** *Ort9;* **q)** *bior3.9;* **r)** *Ort10;* **s)** *bior6.8.*

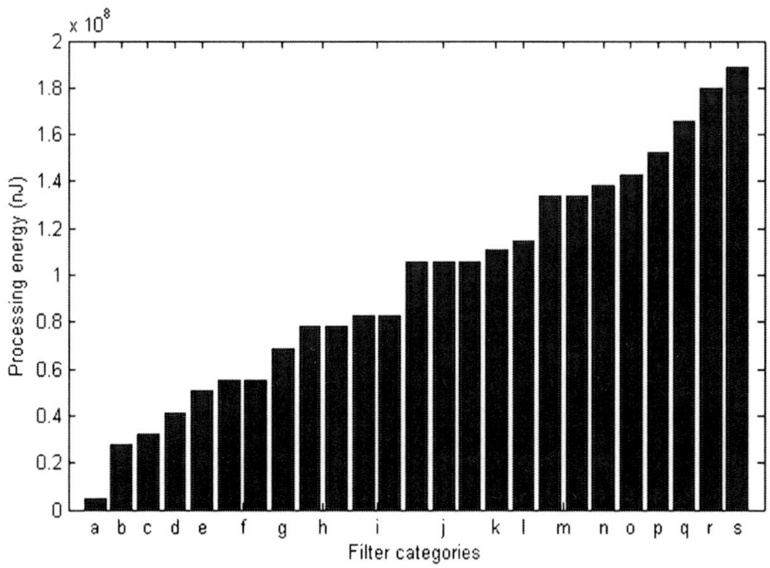

Figure 16. FL effects on the power dissipated by a VS while executing Biorthogonal wavelet for DL = 1,3, 5 and 7, for an image of size 512 × 512 and 8 bpp. Numbers 1 to 15 in the x-axis refer to the biorthogonal filters.

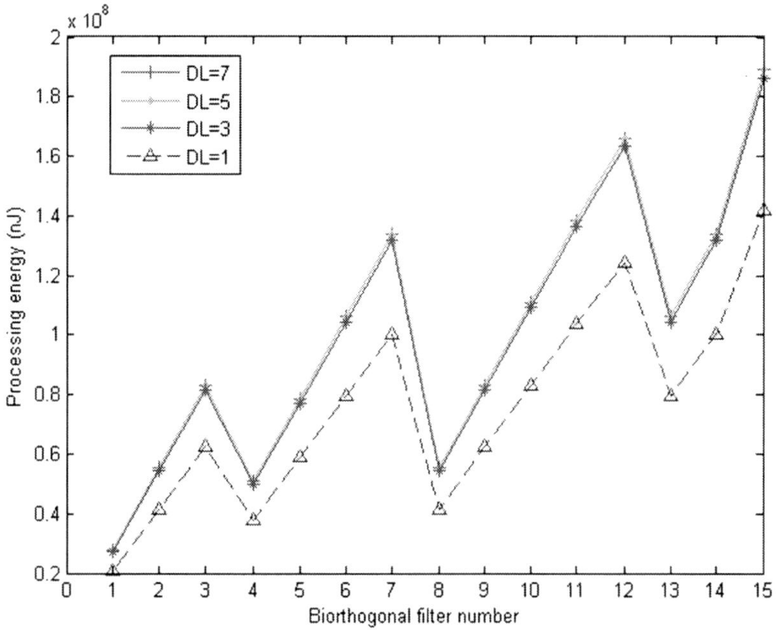

us classify those filters in terms of their dissipations, when processing the same image. For a given *DL*, orthogonal and biorthogonal filters under evaluation can be classified in ascending order, in terms of their dissipated power, as follows, where Ort-*x* means one of the orthogonal wavelets (*SW* or *DW*) studied in this work, using *x* taps: *Ort1*; *bior1.1*; *Ort2*; *Ort3*; *bior2.2*; *bior1.3* and *bior3.1*; *Ort4*; *Ort5* and *bior2.4*; *bior1.5* and *bior3.3*; *Ort6*, *bior2.6* and *bior4.4*; *bior3.5*; *Ort7*; *bior2.8* and *bior5.5*; *bior3.7*; *Ort8*; *Ort9*; *bior3.9*; *Ort10*; *bior6.8*.

Let us consider the following example of the microcontroller ATmega128 processing an image of 512×512 pixels. For a given *DL* (for example *DL=5*), orthogonal and biorthogonal filters under evaluation can be classified in ascending order, in terms of their dissipated power, as follows (Figure 15), where letters from *a* to *s* represent the filter categories as follows: a) *Ort1* (0.46×10^7nJ); b) *bior1.1* (2.76×10^7nJ); c) *Ort2* (3.22×10^7nJ); d) *Ort3* (4.14×10^7nJ), e) *bior2.2* (5.07×10^7nJ); f) *bior1.3 and bior3.1* (5.53×10^7nJ), g) *Ort4* (6.9×10^7nJ), h) *Ort5 and bior2.4* (7.83×10^7nJ), i) *bior1.5 and bior3.3* (8.29×10^7nJ), j) *Ort6, bior2.6 and bior4.4* (10.6×10^7nJ), k) *bior3.5* (11.06×10^7nJ), l) *Ort7* (11.5×10^7nJ), m) *bior2.8,bior5.5* (13.36×10^7nJ), n) *bior3.7* (13.82×10^7nJ), o) *Ort8* (14.3×10^7nJ), p) *Ort9* (15.21×10^7nJ), q) *bior3.9* (16.59×10^7nJ), r) *Ort10* (17.97×10^7nJ), s) *bior6.8* (18.9×10^7nJ).

For filter length effects, we conclude again that small *FL* are very attractive for VSN, since they offer an adequate image quality while minimizing the power dissipated by a given VS.

Considering the results presented above, we summarize in Table 5, some statistics related to the number of decompositions DL_a required to obtain an acceptable image quality, the number of decompositions DL_o which yield the best *PSNR* and its correspondent filter length *FL* for all wavelets under evaluation for low and high CR. Furthermore, in Table 6 we present the different *DL* values required to obtain the highest *PSNR* for different CR values. For power dissipation considerations, small filter length and low *DL* values are highly recommended to minimize the power dissipated by VSs.

In order to provide additional insight, subjective measures are introduced in the following section.

SUBJECTIVE EVALUATION

In applications such as VSN where images are viewed and evaluated by humans (e.g., parking monitoring), the objective tool such as *PSNR* is not always perfect for measuring the quality of the reconstructed image at the base station. Thus, the subjective measure might be necessary to supplement our objective measure, and then is used to select the most effective filter. That is, subjective evaluation is performed to confirm or to correct the objective measurement results performed in the previous section. Our evaluation strategy with subjective testing can be performed into two phases. In the first phase, the best wavelet within each family is selected. To do that, both the original image captured by a VS node, which serves for reference purpose, and the reconstructed image at the sink are shown to a group of 18 viewers. The averages of their evaluations are represented in Tables 8 and 9. Recall that each viewer was asked to attribute one value from Table 3 to the reconstructed image under consideration while comparing it to the original image. Due to its time consuming, the subjective evaluation was restricted only to one of the best decomposition levels *DL* found previously (for example *DL* = 5), and the filter length *FL* was varied from 1 to 10 for Daubechies and Symlet, and from 1 to 5 for Coiflet. All the biorthogonal wavelets used in objective evaluation are considered also for subjective evaluation. In addition, these tests are used only with two representative CRs: high CR (e.g., 80:1) and low CR (e.g., 10:1). Table 8 shows the average ranking of 20 images evaluated subjectively using Daubechies and Coiflet with *DL*=5, while Table 9 concerns the average ranking of the same 20 images evaluated subjectively using biorthogonal wavelet with 5 decompositions. These results confirm again our finding with objective evaluation.

In the second phase, the most effective wavelet inside each family found in the first phase, was selected and compared to each other to get the best filters. We highlight that in this test phase, the general consensus amongst the evaluation participants is that all of the best wavelets considered are very close in terms of visual quality especially for high CR. The possible explanation to this observation is that the human eye often cannot distinguish between the original image and a high CR reconstruction of an image. Table 10 summarizes the best wavelet filters inside each wavelet family considered in this work.

Considering the results from both phases cited above and given in Table 10, we conclude for high CR that *bior2.2* have the best performance, followed by *bior2.4*, *DW-4*, and *CW-2*, respectively. For low CR, *DW-4* seems perform heavily better, followed by *CW-2*, *bior2.4*, *bior2.2*, and *DW-5*, successively.

DISCUSSION

Having simulated the effects of varying *DL* and *FL* on both the quality of the reconstructed image at the sink and the power dissipated by VS, let us now discuss some theoretical aspects behind these results.

Clearly, the number of decompositions affects both the power dissipated by a VS node running the wavelet under evaluation, and the quality of the reconstructed image. We start this section

Table 8. Average ranking for Daubechies and Coiflet (DL=5)

-	*Average ranking for Daubechies and Coiffet (DL=5)*											
-	*Average ranking for Coiffet*				*Average ranking for Daubechies*							
CR	*CW-2*	*CW-3*	*CW-4*	*CW-5*	*db1*	*db2*	*db3*	*db4*	*db5*	*db6*	*db8*	*db10*
80:1	*1.9*	*2.1*	*3.4*	*3.2*	*3.1*	*3.3*	*2.1*	*1.9*	*2*	*2.4*	*2.7*	*2.5*
10:1	*1.6*	*1.8*	*3.3*	*2.5*	*3*	*2.8*	*2.7*	*1.5*	*1.7*	*2.1*	*2.9*	*2.8*

Table 9. Average ranking for biorthogonal family biorx.y (DL=5)

class	*Average ranking for biorthogonal family biorx.y (DL=5)*														
CR	*1.1*	*1.3*	*1.5*	*2.2*	*2,4*	*2.6*	*2.8*	*3.1*	*3.3*	*3.5*	*3.7*	*3.9*	*4.4*	*5.5*	*6.8*
80:1	*2.7*	*2.8*	*2.7*	*2.3*	*2.1*	*2.2*	*2.2*	*3.6*	*2.9*	*2.6*	*2.6*	*2.5*	*1.9*	*2.3*	*1.9*
10:1	*2.2*	*2.1*	*2.2*	*2.3*	*2.2*	*2.4*	*2.3*	*4.9*	*5*	*4.9*	*4.8*	*4.7*	*4.8*	*2.2*	*2.3*

Table 10. Average ranking for db, SW, CW and bior wavelets

-	*Average ranking for db, SW, CW and Bior wavelets*										
CR	*db3*	*db4*	*db5*	*db6*	*SW-4*	*SW-5*	*CW-2*	*CW-3*	*bior2.2*	*bior2.4*	*bior2.6*
80:1	*2.1*	*1.9*	*2*	*2.4*	*2.2*	*2.1*	*1.9*	*2.1*	*1.8*	*1.9*	*2.1*
10:1	*2.7*	*1.5*	*1.7*	*2.1*	*2.3*	*2*	*1.6*	*1.8*	*1.7*	*1.6*	*2*

by the explanation of some theoretical aspects behind the effects of varying *DL* on the quality of the reconstructed image at the sink. After that, we present some justifications of the same effects on the power dissipation. We have shown that as *DL* increases, the *PSNR* increases up to $PSNR_l$ corresponding to DL_o. Our argumentation behind this observation, is that in order to obtain the highest *PSNR*, a larger *DL* (corresponding to DL_o) is required to separate low frequency coefficients from high frequency coefficients. Beyond DL_o, the quality will decrease as *DL* augments. A possible explanation for this observation is given as follows. It is well known that for the perfect reconstruction of a 1-D signal $L_0[n]=X[n]$, the following equation has to be iterated:

$$L_j\left[n\right] = \sum_i L_{j+1}\left[i\right] l\left[n-2i\right] + \sum_i H_{j+1}\left[i\right] h\left[n-2i\right] \quad (1.21)$$

To reconstruct $X[n]$ after *DL* iterations, the following coefficients are involved: $L_{DL}[n]$, $H_{DL}[n]$, $H_{DL-1}[n]$,..., $H_1[n]$. It is well known that during compression process, many high frequency coefficients are neglected. As *DL* increases, more high frequency coefficients are set to 0, which justify that beyond DL_o, the quality measured by *PSNR* will decrease.

The power dissipation increases with *DL* until certain value, where its increasing becomes very slow. The explanation for this finding is related to the number of coefficients processed at each level *DL*. Since the passage from each level *DL* to a subsequent level $DL+1$ requires some additional operations, which are equal to $(N/2^{DL})^2$ where N^2 is the size of the original image, the consumed power represented by Eqs. 13 and 14 will increase, even if the computation at each level is reduced by a factor of $(1/2^{DL})$, which may justify the slow increasing in the energy. As *DL* increases, the increasing in the power becomes negligible. Let us analyze mathematically Eqs. 13 and 14. It is observed that at level *DL*, the power dissipated by VS is proportional to $(1-4^{-DL})$. The analytical study of the mathematical function:

$f(x) = (1-4^{-x})$, $x \geq 0$ may explain the simulations results. Indeed $f(x)$ increases rapidly at the

Figure 17. Analytical study of the effects of DL on the power dissipation by the function f(x)=1-4-x

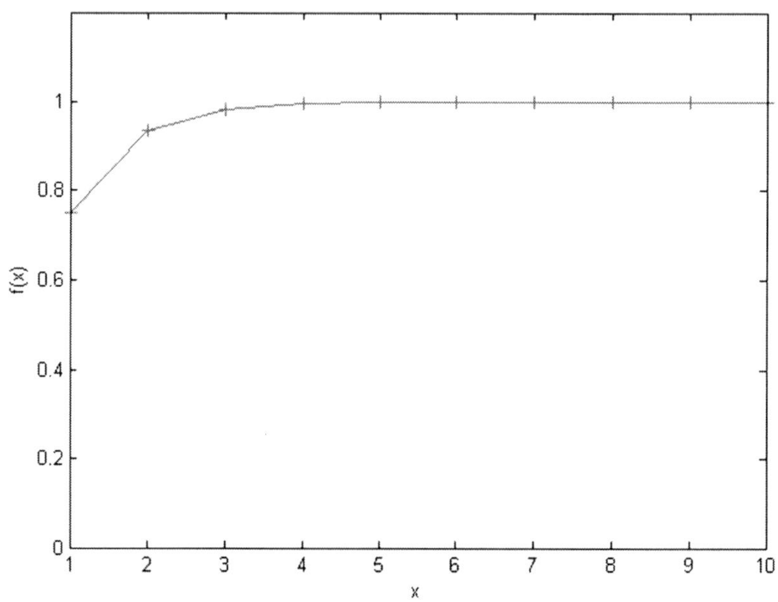

beginning, and after $x=4$, its increasing becomes rather negligible (Figure 17).

By analogy to capacitor (C) charge/discharge over a resistor (R) of analog circuit domain, we can supplement our argumentation by the fact that after $t = 5\tau$, where ($\tau = RC$), the steady state is reached. In our case, $f(t)$ can be written as follows:

$$f\left(t\right) = 1 - e^{-\left(\ln 4\right)t} \tag{1.22}$$

If we put $\tau = 1/\ln 4$, we obtain $t = 5\tau = 5/\ln 4 = 3.6$. Which means that we reach the steady state after 3.6. In our case, as DL is an integer, its value may be surrounded to 4, which fits well with Figures 9 and 10.

Another important relation is between DL_o and CR. It is observed that at very low CR (e.g., 2:1), only one decomposition (i.e., $DL_o = 1$) is needed to obtain the highest *PSNR*. As CR is low, we do not need large number of decompositions to represent our image, this is why we need only low DL_o. As CR increases, larger number of decompositions are required to resolve important DWT coefficients from less important coefficients. Moreover, it has been reported in the literature that the number of decompositions determines the resolution of the lowest level in wavelet domain.

At the end, let us consider the filter length and its effects on the compression performance in the context of VSN. Clearly, from the above simulations, filter length has not a strong influence on the quality of the reconstructed image at the sink. That is, increasing the filter length does not imply automatically better image quality. However, the increasing the filter length affects considerably the power consumption of a VS as shown by Eqs. 13 and 14. Generally the results of (Grgic et al., 2001) concerning the relation between filter length and image quality coincide with our findings.

After examining the results of the influence of *DL* and *FL* on the performance of the wavelets under evaluation, we provide some guidelines for users designing WBC for power-constrained applications such as VSN. Of the wavelet filters considered above, several were found to behave particularly well depending on the evaluation criterion, i.e., power dissipation or image quality. Obviously, the best choice among them, is the one that maximize the image quality and minimize the power dissipation. Such a choice depends also on the CR and image content as explained previously.

For low CR, 1 to 2 decompositions are sufficient to obtain high *PSNR* while using a relatively low number of taps, as explained in Table 4. That is, wavelets such as *SW-2*, *SW-3*, *DW-2*, *DW-3*, *bior1.1*, *bior2.2* are examples of wavelets dissipating less power while having an acceptable image quality, and can be used efficiently in the context of VSN. While for high CR, it is observed that 4 to 5 decompositions are sufficient to obtain high *PSNR*.

FUTURE RESEARCH DIRECTIONS

The design of a wavelet-based coder in the context of VSN is mainly based on the selection of the appropriate wavelet filter, since the latter consumes the biggest part of the whole compression scheme. An eventual open research work should be the conception of a new wavelet filter dedicated to VSN incorporating the energy as a design parameter. After that, the remaining task is to incorporate the selected filters within a real testbed, and show its effectiveness for minimizing the power consumption by a VSN.

Wavelet filters based on lifting theory is another promising subject. Due the fact that the lifting-based filters are less energy consuming than their predecessor, their integration in the WBC is very mandatory. An open research issue is the evaluation of lifting-based filters parameters in the context of VSN. For instance, study the effects of the number of lifting steps on both energy and image quality.

Another related point is the selection of an appropriate quantization matrix, the entropy encoder, such Huffman or Golomb-Rice encoder, and study the interconnection between the different stages of a compression chain, i.e., transform (DCT or DWT), quantization matrix and the entropy encoder.

CONCLUSION

In this work, a large set of wavelet filters are evaluated in the context of VSN, and the best of them which minimize the power consumption, while having an adequate image quality are selected. Moreover, factors influencing the power dissipation of a limited-power VS while executing a wavelet filter, are discussed. We evaluated both the power consumption behaviour and visual quality results of the wavelet filters under evaluation. Image quality is measured objectively using *PSNR* and *SSIM* and subjectively using the mean opinion score of 18 viewers. To evaluate the power dissipation of a VS node while executing a 2-D DWT, we have developed a power dissipation model based on the number of times basic operations are performed.

Orthogonal and biorthogonal wavelets are evaluated and compared in function of the decomposition level *DL* and the filter length *FL*. These evaluations are performed for several compression ratios and different image contents. It is observed that *DL* affects both the power dissipation and the quality of the reconstructed image at the sink. It is observed that the *DL* required to obtain an acceptable image quality depends on the following parameters: CR, wavelet type, and image parameters such as their size and content. It is difficult to find a unique *DL* for all wavelets, CRs, or images. In general, for low CR, it is showed that only small number of decompositions are needed to obtain an adequate image quality, while for high CR at least 4 decompositions are required. For power dissipation, it is observed that filters having lowest number of decompositions consume less power compared to those with important number of decomposition. Those filters are highly important in VSN.

Of the transforms considered above, filters having small length in each family are appreciated. For instance in Daubechies family the following filters are recommended: DW-2, DW-3 and DW-4. In Coiflet wavelet family, transforms such as CW-2, CW-3 and CW-4 are suggested. The same conclusion may be applicable for Symlet wavelet. For biorthogonal wavelet, filters such as bior2.2, bior2.4 are recommended. We highlight also that very small filters, such as DW-1, dissipate the smallest power, at the price to lose in the quality. The second parameter affecting both quality and power dissipation is related to the number of decompositions required for each wavelet. We have found that optimal number of decompositions is usually 5 in high CR. With low CR, less than 3 decompositions are recommended.

For as the 5/3 and 2/6 transforms have the lowest computational complexity while the 9/7-F transform has the highest. Unfortunately, the 9/7-F transform, although very good for lossy compression, is significantly higher in complexity than all of the other transforms, and performs poorly for lossless compression. Obviously, no one transform has superior lossy and lossless compression performance for all classes of images in addition to low computational complexity. Therefore, tradeoffs are involved, and the most appropriate choice of transform depends on the needs of the application at hand. For example, in the case of applications requiring low computational complexity, the 5/3 transform is particularly attractive. It performs reasonably well for both lossy and lossless compression, and has the lowest computational complexity of all of the transforms evaluated.

ACKNOWLEDGMENT

We would like to acknowledge and thank the anonymous reviewers for their constructive and insightful comments.

REFERENCES

Adams, M. D., & Kossentini, F. (2000). Reversible integer-to-integer wavelet transforms for image compression: Performance evaluation and analysis. *IEEE Transactions on Image Processing*, *9*(6), 2279–2288.

Atmel 8-bit Microcontroller (2009). *Atmel Corporation*. Retrieved from http://www.atmel.com.

Chen, Y.-T., Tseng, D.-C., & Chang, P.-C. (2007). Wavelet-based medical image compression with adaptive prediction. *Computerized Medical Imaging and Graphics*, *31*(1), 1–8. doi:10.1016/j.compmedimag.2006.08.003

Daubechies, I. (1992). *Ten lectures on wavelets*. Philadelphia, PA, USA: Society for Industrial and Applied Mathematics.

Grgic, S., Grgic, M., & Zovko-Cihlar, B. (2001). Performance analysis of image compression using wavelets. *IEEE Transactions on Industrial Electronics*, *48*(3), 682–695. doi:10.1109/41.925596

Grigic, S. (June, 2000). *Optimal decomposition for wavelet image compression*. Paper presented at the 1st International Workshop on Image and Signal Processing and Analysis, Pula, Croatia.

Lecuire, V., Duran-Faundez, C., & Krommenacker, N. (2007). Energy-efficient transmission of wavelet-based images in wireless sensor networks. *EURASIP Journal on Image and Video Processing*, *2007*(1).

Lee, D.-G., & Dey, S. (April-May 2002). *Adaptive and energy efficient wavelet image compression for mobile multimedia data services*. Paper presented at the IEEE International Conference on Communications, New York, NY, USA.

Mallat, S. (1989). A theory for multi-resolution signal decomposition: The wavelet representation. *IEEE Transactions on Pattern Analysis and Machine Intelligence*, *11*(7), 674–693. doi:10.1109/34.192463

Mammeri, A. Khoumsi. A., Ziou, D., & Hadjou, B. (2008, April). *Energy-aware JPEG for visual sensor networks*. Paper presented at the MCSEAI conference, Oran, Algeria.

Mammeri, A. Khoumsi. A., Ziou, D. & Hadjou, B. (2008, August). *Modeling and adapting JPEG to the energy requirements of visual sensor networks*. Paper presented at the ICCCN Conference, Virgin Islands, USA.

Nguyen, T. Strang, G. (1996). *Wavelets and Filter Banks*. Wellesley-Cambridge: Wellesley.

Polyak, N., & Pearlman, W. (2000). A new flexible bi-orthogonal filter design for multiresolution filter banks with application to image compression. *IEEE Transactions on Signal Processing*, *48*(8), 2279–2288. doi:10.1109/78.852009

Srinivasan, P., & Jamieson, L. H. (1999). High quality audio compression using an adaptive wavelet packet decomposition and psychoacoustic modeling. *IEEE Transactions on Signal Processing*, *46*(4), 1085–1093. doi:10.1109/78.668558

Thomos, N., Boulgouris, N. V., & Strintzis, M. G. (2006). Optimized transmission of JPEG2000 streams over wireless channels. *IEEE Transactions on Image Processing*, *15*(1), 54–67. doi:10.1109/TIP.2005.860338

Wang, Z., & Bovik, A. C. (2006). *Modern image quality assessment*. USA: Morgan and Claypool.

Wang, Z., Bovik, A. C., Sheikh, H. R., & Simoncelli, E. P. (2004). Image quality assessment: From error visibility to structural similarity. *IEEE Transactions on Image Processing*, *13*(4), 600–612. doi:10.1109/TIP.2003.819861

Wu, H., & Abouzeid, A. (August, 2004). *Energy efficient distributed JPEG2000 image compression in multi-hop wireless networks*. Presented at the 4th Workshop on Applications and Services in Wireless Networks, Boston, Massachusetts, USA.

ADDITIONAL READING

Akyildiz, I. F., Melodia, T., & Chowdhury, K. R. (2007). A survey on wireless multimedia sensor networks. *Journal of Computer Networking*, *4*(51), 921–960. doi:10.1016/j.comnet.2006.10.002

Bracamonte, J., Ansorge, M., & Pellandini, F. (1996, October). *VLSI systems for image compression. A power-consumption/image resolution trade-off approach.* Paper presented at the International Society for Optical Engineering, Annapolis, Maryland, USA.

Chew, L. W., Chia, W. C., Ang, L., & Seng, K. P. (2009). *Very low-memory wavelet compression architecture using strip-based processing for implementation in wireless sensor networks.* EURASIP Journal of Embedded Systems.

Chow, K.-Y., Lui, K.-S., & Lam, E. Y. (2006, January). *Balancing image quality and energy consumption in visual sensor networks.* Paper presented at the 1st International Symposium on Wireless Pervasive Computing, Phuket, Thailand.

Chow, K. Y., Lui, K. S., & Lam, E. Y. (2007). Efficient on-demand image transmission in visual sensor networks. *EURASIP Journal on Applied Signal Processing*, *1*, 225–225.

Ferrigno, L., Marano, S., Paciello, V., & Pietrosanto, A. (July, 2005). *Balancing computational and transmission power consumption in wireless image sensor networks*, Giardini Naxos, Italy.

Lu, Q., Luo, W., Wang, J., & Chen, B. (2008). Low-complexity and energy efficient image compression scheme for wireless sensor networks. *Journal of Computer Networking*, *52*(13), 2594–2603. doi:10.1016/j.comnet.2008.05.006

Magli, E., Mancin, M., & Merello, L. (July, 2003) *Low-complexity video compression for wireless sensor networks.* Paper presented at the 2003 International Conference on Multimedia and Expo, Washington, DC, USA.

Mammeri, A. Khoumsi. A., Ziou, D. & Hadjou, B. (2008, October). *Energy-efficient transmission scheme of JPEG images over visual sensor networks.* Paper presented at the 4th IEEE International Workshop on Performance and Management of Wireless and Mobile Networks (P2MNET), in Conjunction with 33rd IEEE Conference on Local Computer Networks (LCN), Montreal, Canada.

Mammeri, A., Khoumsi, A., Ziou, D., & Hadjou, B. (2008, April). *Energy-aware optimized JPEG for visual sensor networks.* Paper presented at the Maghrebian Conference on Software Engineering and Artificial Intelligence, Oran, Algeria.

Mammeri, A., Khoumsi, A., Ziou, D., & Hadjou, B. (2009). Semi-reliable encoding and transmitting scheme for visual sensor networks driven by energy considerations. *International Journal of Computer Science and Information Security*, *6*(2), 249–258.

Margi, C. B., Roberto, M., & Obraczka, K. (2006, May). *Energy consumption tradeoffs in visual sensor networks.* Paper presented at the 24th Brazilian Symposium on Computer Networks, Curitiba-PR, Brazil.

Namgoong, W., & Meng, T. H. (1997). A low-power encoder for pyramid vector quantization of Subband coefficients. *Journal of VLSI Signal Processing Systems*, *16*(1), 9–23. doi:10.1023/A:1007905725622

Obraczka, K., Manduchi, R., & Garcia-Luna-Aveces, J. J. (October, 2002). *Managing the information flow in visual sensor networks.* Paper presented at the 5th International Symposium on Wireless Personal Multimedia Communication, October 2002, Honolulu, Hawaii.

Sayood, K. (2005). *Introduction to data compression.* San Francisco, CA, USA: Morgan Kaufmann Publishers Inc.

Shapiro, J. M. (2001). Embedded image coding using zerotrees of wavelet coefficients. In *Readings in multimedia computing and networking* (pp. 124–141). San Francisco, CA, USA: Morgan Kaufmann Publishers Inc.

Swami, A., Zhao, Q., & Hong, Y. W. (2007). *Wireless sensor networks: Signal processing and communications*. USA: Wiley.

Taubman, D. S., & Marcellin, M. W. (2001). *JPEG 2000: Image compression fundamentals, standards and practice*. Norwell, MA, USA: Kluwer Academic Publishers.

Wang, W., Peng, D., Wang, H., & Sharif, H. (2007). A novel image component transmission approach to improve image quality and energy efficiency in wireless sensor networks. *Journal of Computer Science*, *3*(5), 353–360. doi:10.3844/jcssp.2007.353.360

William, A. P. (2004). Asad-Islam & Nithin, N. & Amir, S. (2004). Efficient, low-complexity image coding with a set-partitioning embedded block coder. *IEEE Transactions on Circuits and Systems for Video Technology*, (14): 1219–1235.

Wu, H., & Abouzeid, A. (2005). Energy efficient distributed image compression in resource-constrained multihop wireless networks. *Journal of Computer Communication*, *28*(14), 1658–1668. doi:10.1016/j.comcom.2005.02.018

Yong, S., Hui, Z., & Guangshu, H. (2002). Real-time implementation of a new low-memory SPIHT image coding algorithm using DSP chip. *IEEE Transactions on Image Processing*, *11*(9), 1112–1116. doi:10.1109/TIP.2002.802533

KEY TERMS AND DEFINITIONS

Decomposition Level: Related to the number of decompositions. It affects both the power dissipated by a visual sensor and the resolution of reconstructed image at the reception by the sink. After the first decomposition of the captured image, the decomposition process is performed iteratively on the low pass coefficients of the previous stage. For DL iterations, the target image is decomposed into 3DL+1 sub-bands.

Energy Minimization: In our context is a task of reducing the energy while processing and sending images through wireless networks.

Filter Length: Represents the number of taps of a given filter. It is well known from wavelet theory that filters having high length yield wider functions with high degree of smoothness. On the contrary, filters with lower length have a better time localization and preserve important spatial information. It is considered in this study, since it affects both image quality and energy consumption.

Image Quality: Reflects the satisfaction of the user while monitoring a given image. It is measured objectively using PSNR or SSIM values.

Wavelet-Based Coder: A coding engine based on wavelet instead of other transforms such as discrete cosine transforms.

Wireless Sensor Network: A set wirelessly connected small sensors. It has almost the same application fields of VSN.

Wireless Visual Sensor Network: An ad-hoc wireless sensor network made of visual sensors (VS) incorporating image retrieving, processing and communication functionalities. VSN is gaining attention in a wide range of applications such as video-surveillance, object detection/tracking and environmental monitoring.

Chapter 5
Low Level Representation of Data for Visual Sensor Network

Ruth Aguilar-Ponce
Universidad Autonoma de San Luis Potosi, Mexico

J. Luis Tecpanecatl-Xihuitl
Universidad Autonoma de San Luis Potosi, Mexico

Alfonso Alba-Cadena
Universidad Autonoma de San Luis Potosi, Mexico

ABSTRACT

Wireless Sensor Network future direction is going towards more complex sensor such as camera sensor. Therefore, a very active research field is Visual Sensor Network. This type of network brings new challenges such as processing and transmitting a massive amount of data generated by the camera sensor. The efforts into decreasing the amount of data to be transmitted are going towards two directions: data encoding and data filtering. This chapter introduces an algorithm for each direction. Visual data encoding is performed by means of Predictive Video Encoding using Phase-Only Correlation function to achieve motion estimation. Visual data filtering is done at the lowest level of abstraction and is performed in three phases: pixel classification, background update and detection. The algorithms involved in each phase are light in terms of complexity and memory resources.

INTRODUCTION

Wireless Sensor Network (WSN) have been in study for the last 20 years; therefore, there are already some network deployed such as ZebraNet, a network to monitor wild life in Africa (Juang, 2002) or Great Duck Island Network (Mainwaring, 2002). However, most of the research has been focused on sensors that collect scalar data such as temperature, humidity, vibration and acoustic. However, these types of sensors do not provide enough information for some applications such as distributed surveillance systems, traffic monitoring or health care for elderly people. Therefore, a new research direction is been focused in Visual Sensor Network. Visual Sensor Network (VSN) is a large set of camera nodes, where each node is capable of processing images locally and

DOI: 10.4018/978-1-61350-153-5.ch005

extracting important information. Moreover, VSN collaborates with other camera nodes to achieve and specific application task as well as provide high-level descriptors of the capture images. Imaging technology has advance notably, producing small, low power, and low cost image sensor devices making VSN attainable in the near future. VSN provide a more complete set of information since a camera sensor is involved (Charfi, 2009). VSN have several applications such as distributed video surveillance, elderly health care assistance, traffic control, environmental and structural monitoring within others. Distributed video surveillance uses networked smart camera nodes that cooperatively process the information to prevent crime, locate missing persons, identify criminals or prevent potentially harmful activities. The smart camera nodes must use computer vision techniques as well as image processing algorithm to achieve its mission. Traffic control and monitoring network may provide rerouting traffic to avoid congestion points as well as assist efficient parking detection. While telemedicine sensor network integrates body sensing lectures with images of the elderly people inside their own environments to allows remote health care assistance. The structural stress may be detected with a network of camera sensor distributed along the building or bridge to provide a 24 hrs monitoring (Akyildiz, 2002).

VSN represents new challenges with respect to scalar WSN, such as how to manage the amount of information generated by the cameras, how to integrate highly complex vision algorithms into a limited embedded system, what are the best way to represent visual data, how can the data be fused to obtained only the important information out, how to efficiently scale VSN with respect of bandwidth management, transmission power, etc. Let´s take a look into some of these challenges.

The amount of information generated by this type of sensor is large because represents the scene with a two-dimensional set of data points (Soro, 2009). Consider a low resolution 352 × 288 pixels camera using a RGB format produces 72 Mbps.

$$\left(\frac{352 \times 288}{frames}\right)\left(\frac{30\ frames}{\sec}\right)\left(\frac{3\ colors}{pixel}\right)\left(\frac{8bits}{colors}\right) = \frac{72Mb}{s}$$

However, WSN are bandwidth limited in nature. WSN are densely populated; consequently bandwidth is a scarce resource. Moreover, a node is energy-constrained and transmission is one of the most expensive operations in terms of energy. For that reason, data must be process within the node to decrease the amount of data before its send to the base station. VSN bring new challenges to the WSN research such as camera coverage, network architecture to guarantee the ability to communicate a massive amount of data over a limited bandwidth network, and data processing to accomplish a data representation suitable for VSN environment.

A camera sensor provides information of distant objects and scenes from certain direction. Therefore, these sensors have directional sensing mode. In a scalar WSN the sensing mode is omnidirectional; therefore the fusion algorithms developed for WSN consider the vicinity of the nodes to fuse data. VSN must consider the area that is cover by the camera; also it must developed algorithm to establish the overlap of such areas to be considered within the fusion algorithm. The representation of the data obtained by the camera node must also be considered in developing the fusion algorithm.

The focus of this work is data processing to efficiently communicate visual nodes.VSN as well as WSN collect the data and processed it to stream only the relevant data to the Base Station (BS). Since a low-resolution camera may produce a 72 Mbps data stream is impractical to stream all data due to bandwidth constraint and energy limited nodes. Since processing cost is more economical than communication cost in terms of energy, it makes more sense to process the data and send only the relevant information to BS. There are two approaches to eliminate redundant

Figure 1. Levels of Representation of Data for VSN

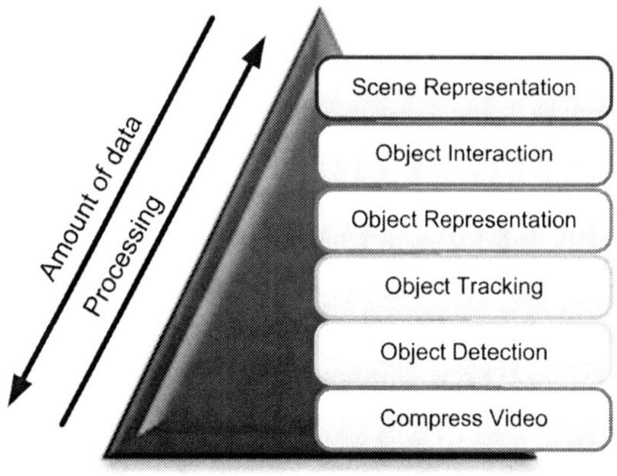

data, visual data filtering and visual data coding (Charfi, 2009).

Visual Data Filtering

Visual data filtering is a technique that processes the visual data to extract the relevant information for the application as well as compress data for transmission. There are several levels of abstraction that can be implemented in a VSN. The level of abstraction dictates the amount of data that will flow through the network (Obraczka, 2002). These levels are represented in Figure 1.

The first level of representation is the 2D data set compressed by any method. This level do not perform any abstraction rather shows the total information gathered by the camera sensor in a suitable manner for transmission. The second level of abstraction refers to object detection. The node determines if the gathered visual data contains a moving object and then this information is send to BS. Only the frame containing the object is compressed and sent to BS. The next level obtains the route of the moving objects. Automated tracking algorithms are required and their processing is heavier. However, the information is reduced to routing data only. The next level describes the objects based on the detection and tracking information. Subsequently, the routing information along with object classification is used to infer the object interaction. Finally, the data gathered from object detection up to object interaction allows a high level description of the scene. For example, two individuals chat in the middle of the room and disperse after 10 minutes. The lowest level of abstraction requires lower processing complexity and small or null cooperation with the neighboring nodes. The upper abstraction levels involve complex algorithms and cooperation between nodes to achieve a global view of the scene. As the processing gets more complex towards the upper level of the triangle, the amount of data is decreasing but the cooperation is increasing.

Applications of VSN require real-time execution of the algorithm for data representation, and then tight boundaries of maximum execution time are imposed. Moreover, algorithms must execute in a resource-limited platform using lightweight embedded processors. Therefore, architecture node must change to incorporate powerful processors and only simple vision algorithm can be considered.

Considering these restrictions, medium to low levels of abstraction are suitable for the bandwidth and energy limited nodes due to low complexity of the algorithm involved and also the amount of cooperation between nodes is limited. BS is the best place to perform the medium to upper levels of abstraction due to the complexity of the algorithm involved. Moreover, the view of the BS is more general because received the information of all nodes.

Visual Data Coding

Visual Data Coding (VDC) is the lowest level of abstraction of the above model. VDC is considered as the second approach to eliminate redundant information. Video signal has temporal, spatial and color space redundancy. Moving objects change from 1 to 15 pixels between frames; consequently, there is high redundancy between frames know as temporal redundancy. Moreover, the background image is constant for each frame. Also, the images contain a high spatial redundancy due to objects spanning over several pixels. The color space redundancy is due to the image representation. For example, RGB format represents the same scene but in red, green and blue color. Therefore, RGB components are correlated among themselves. Visual data coding explodes the temporal, spatial and color redundancy to represent the video in a more compact format.

Predictive Video Encoding (PVE) uses the information of the previous coded frames to predict the movements of the current frame (Coga, 1981). Therefore, PVE encodes the visual data into three types of coded frames, I-frame or Intra-coded frame; it is coded independently of all other frames (using image compression techniques). P-frame or Predict coded frame, which is coded based on previously coded frame. Finally, the B-frame or Bi-directionally predicted frame, which is coded based on both previous and future coded frames. Source coding is a mature field; however, there are several things to consider before adapting this scheme to VSN.

Source Video Encoding is a technique designed for TV broadcasting, where the transmission scheme is one-to-many. Therefore, the source encoder is complex while the sink decoder is simpler. This scheme works for TV broadcasting where the source code can perform complex task but the receptors are simpler. In VSN, the transmission scheme is many-to-one. Moreover, the source is resource limited. Therefore, the complexity of the encoder must decrease. There have been two strategies, Distributed Source Coding or decreasing complexity on the source encoder (Charfi, 2009).

Distributed Source Coding

Distributed Source Coding (DSC) considers the compression of multiple correlated sensor outputs with limited or no cooperation from the source nodes and a joint decoding at central decoder. DSC changes the traditional coding paradigm to many-to-one. This paradigm applies to VSN because there are several camera nodes looking at the same scene and sending all their information to a single BS. This paradigm considers a light-weight coding algorithm but a highly complex decoder. Therefore, camera nodes only have to support simple algorithm and can send the information directly to BS without cooperating with the neighbor nodes.

The first algorithm to establish DSC was proposed by Slepian and Wolf in 1973. The algorithm considers that there are several sources acquiring correlated data (Slepian, 1973). Then only one of the sources must send data compress in a lossless format known as side information, all other sources only send compress partial information known as *syndrome*. Then, sink decoder can reconstruct data using the syndrome information and the side information send by only one of the sources. Then, the bit rate sent by the sources is reduced as the number of sources sending correlated information to the decoder increases. Figure 2 shows the diagram of the lossless Slepian-Wolf encoder-decoder. The Slepian-Wolf algorithm works for lossless channels. However, in sensor network the transmission channels are lossy. In 1976, Wyner and Ziv generalized the DSC with side information to lossy environments. The Wyner-Ziv encoder separates the signal into mutually exclusive set producing a signal X_Q as shown in Figure 3. Then Slepian-Wolf encoding is applied to this quantized signal. In the decoder, the received signal is decoded with side information until it achieves minimum distortion (Wyner 1975).

In VSN, Wagner proposed a DSC for images capture by camera nodes with overlapping camera coverage. The scheme identifies overlap images from neighbor camera nodes by employing stereo-image techniques. Then, low resolution version of the overlap image is sent to the BS. The BS uses super-resolution technique to reconstruct it (Misra, 2008).

Figure 2. Diagram of the Slepian-Wolf Encoder-Decoder

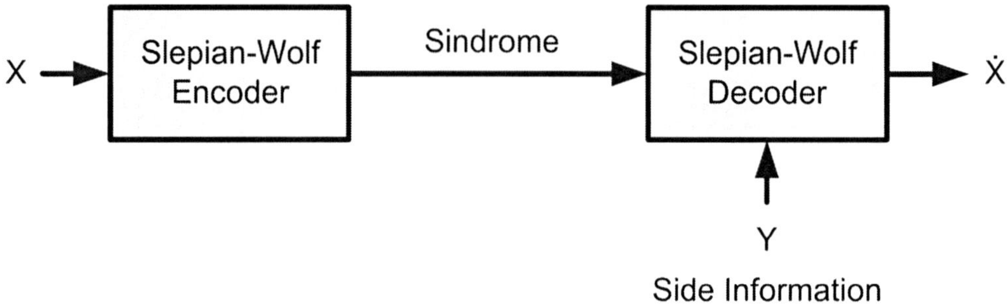

Figure 3. Diagram of the Wyner-Ziv Encoder-Decoder

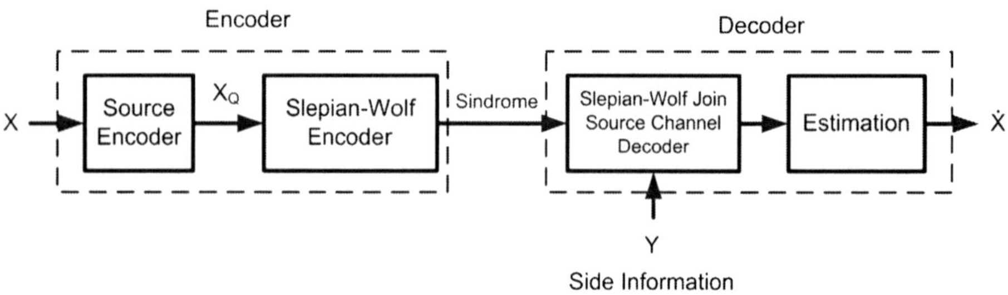

Majumdar and Ramachandran proposed a Power-efficient, Robust, high compression, Syndrome-based Multimedia coding (PRISM). In PRISM, motion compensation is done at the BS allowing coding effort of camera nodes to be low. Motion estimation is done based on the correlation of the preceding frame instead of the frame itself (Majumdar, 2004).

DSC is an energy-efficient technique for VSN that are highly populated. The success of the technique depends on the overlapping fields of view from the camera nodes. However, the energy savings depends on the density of the network and the size of the overlapped fields of view. One of the major drawbacks of DSC is that requires tight synchronization between camera nodes (senders) and the BS (decoder). Furthermore, it is necessary to know the correlation model of the data from the multiple sources to establish overlap regions, and this is a hard problem to solve. Additionally, nodes must have some degree of cooperation to establish which node is going to send the whole information and which nodes are going to send only syndrome information.

Despite these disadvantages of DSC, it is a promising technique for video coding in VSN to encode video efficiently in terms of energy and saving effort in the highly-restricted camera nodes.

Independent Source Coding

Independent Source Coding (ISC) only uses the information of one source camera node to encode the video. ISC is dominant in every multimedia scheme because its simplicity. Additionally, only the video of the node is used for the encoder, then it does not require cooperation between nodes. In dense VSN, the amount of overlapped regions is large and this scheme result in redundant information at BS as well as useless transmission of the same data. However, if the overlapped regions are small, then this scheme is a good option since no cooperation is required and the amount of redundant data is small.

Distributed Source Coding (DSC) separates encoding at a number of nodes and joint decoding at BS. This scheme shifts the complexity towards the BS in order to get a simpler source encoder. Then, the coder is simpler and it's able to operate with low power consumption. However, most of the research has been focused on analyzing DSC under several asymptotic assumptions (Charfi, 2009). Although, DSC is not ready to by efficiently used on VSN due to the lack of accurate representation of the spatial correlation in the real-world images. ISC can be used in dense VSN by using aggregation nodes scheme where the redundant information is discarded before it is send to BS. VSN is going to need DSC or ISC with fusion schemes or a combination of both (Masri, 2008).

The rest of the chapter is organized as follows: section 2 introduces our approach to data filtering by means of object detection as well as our simulation results and comparison with the most representative algorithm previously proposed. Section 3 introduces video encoding algorithm using phase only correlation for motion estimation. Future research directions are discussed in section 4. The conclusions are drawn in section 5.

VISUAL DATA FILTERING FOR VISUAL SENSOR NETWORK

Visual data filtering using the lowest level of representation is object detection. Object detection algorithms are well suited for Visual Sensor Network (VSN) due to the low complexity. Accurate detection save unnecessary effort in later processing block such as tracking, classification and scene representation; consequently, change detection plays a key role in real-time image analysis. VSN requires reliable object detection to decide when to send visual data.

There are two techniques used for object detection, background subtraction and optical flow. However, optical flow is resource demanding and highly complex making it hard to effectively implement in VSN. Background subtraction is a commonly used technique for motion detection in static scenes and attempts to detect moving regions by subtracting the current image from a reference background image created by analyzing the statistic properties of previous frames over an initialization period. The challenges involving background subtraction are robustness against changes in illumination, avoiding detection of non-stationary background objects such as swinging leaves, rain, snow, or shadows cast by moving objects. Additionally, the internal background model should react quickly to changes in background such as the starting and stopping of vehicles.

There are four major steps in background subtraction algorithms, preprocessing, background modeling, foreground detection, and data validation. Preprocessing is a simple image-processing task to transform the raw input video into a format that can be processed. In order to reduce camera noise a temporal or spatial smoothing filter is used. Most background subtraction algorithms use luminance intensity for the background model because there is only one scalar per pixel. However, color spaces such as RGB (Red Green Blue) or HSV (Hue Saturation Value) are becoming more popular in background subtraction algorithms.

Background modeling uses the incoming video frame to calculate and update the model. The background model is a statistical description of the entire background scene. Foreground detection is performed after the model has been updated. The detection of moving objects is achieved by checking the significance of the difference between the input pixels and the corresponding background estimate. This action is done by the following equation

$$\left|I_t\left(x,y\right) - B_t\left(x,y\right)\right| > T \tag{1}$$

Figure 4. Flow diagram of a generic background subtraction algorithm

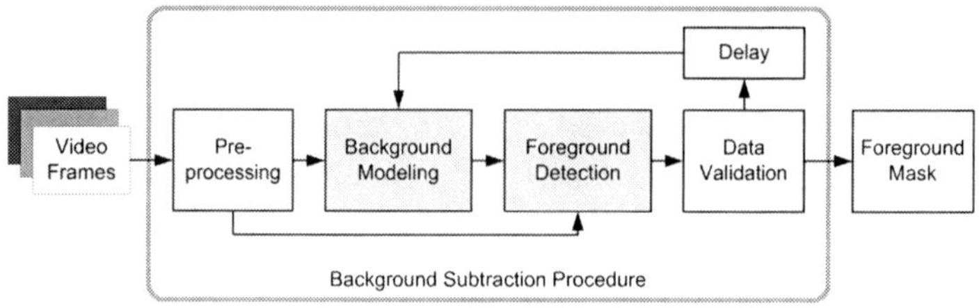

where $I_t(x,y)$ corresponds to luminance pixel intensity at time t and $B_t(x,y)$ is its background estimate. The threshold T is determined experimentally.

Finally, data validation examines the candidate mask, eliminates those pixels that do not correspond to the actual moving object, and outputs the final foreground mask. Figure 4 shows the generic processing flow of background subtraction techniques. Some of the most representative background subtraction algorithms are introduce in the following section.

Background Subtraction Algorithms

The most instinctive technique is adjacent frame differencing followed by thresholding. Change is detected if the difference of the corresponding pixels in two or three consecutive frames in a video exceeds a preset threshold. The method is highly adaptive to dynamic scene changes and has the lowest computational complexity of all the background subtraction algorithms. However, the foreground mask found by this method does not contain interior pixels of an object with uniform intensity. Additionally, this method fails in detecting some types of moving objects such as objects with similar background colors. This technique is very susceptible to noise and illumination changes. The complexity of the algorithm is $O(1)$ per pixel.

Median filter is one of the most popular background subtraction techniques (Cheung, 2004). The median of each pixel of all the frames in the buffer constitutes the background model. Background pixels are considered those that stay on more than half of the frames on the buffer. Median filter can be easily extended to color images by replacing the median by the medoid (Cheung, 2004).

$$B_n(x) = Median\left(I_{n-1}(x), I_{n-2}(x), \ldots, I_{n-L}(x)\right) \tag{2}$$

The main disadvantage of this technique is that it requires a buffer large enough to hold the L last frames of the video sequence. Additionally, the median filter does not provide a rigorous description of the background and does not offer information to adapt the subtraction threshold. The complexity of the algorithm is $O(L \log L)$ per pixel and the memory requirement is an L+1 frame buffer.

Recursive background techniques do not require a buffer of previous frames. Instead, they recursively update the background model based on each input frame. Any error in the background estimation can remain for a long period due to their recursive nature. The most popular recursive technique is Mixture of Gaussians (MoG) (Cuchiara, 2003). MoG can robustly deal with lighting changes, repetitive motions, clutter, the introduction or removal of objects from the scene, and slowly moving objects.

Stauffer and Grimson determine that a unimodal background model could not handle image acquisition noise, light change, and multiple surfaces for a particular pixel at the same time (Stauffer). Therefore, they proposed a mixture of

Gaussians distributions to represent each pixel in the model. In this model, the values of an individual pixel over time is considered as a "pixel process" and the recent history of each pixel, $\{X_1, \ldots, X_t\}$, is modeled by a mixture of K Gaussian distributions. The probability of observing current pixel value then becomes:

$$P_n(x) = \sum_{i=1}^{K} \omega_{i,n} \eta\left(x, \mu_{i,n}, \Sigma_{i,n}\right) \tag{3}$$

where $\eta\left(x, \mu_{i,n}, \Sigma_{i,n}\right)$ is the Gaussian distribution. Each K Gaussian distribution estimates the background and foreground objects. In practice, K is set to a number between 3 and 5. Each Gaussian describes red, green, and blue values. If these values are independent, then the covariance matrix $\Sigma_{i,n}$ simplifies to a diagonal. Additionally, if the standard deviation for the three colors is assumed to be the same, then the covariance matrix is reduced to $\Sigma_{i,n} = \sigma^2 I$ where I is a diagonal matrix.

MoG offers a model for background and foreground objects, but then a criterion to discriminate between them is required. First, all distributions are ranked based on the ratio between their peak amplitude and the standard deviation. The background scene has a higher and compact distribution. Therefore, the likelihood of a pixel belonging to background is high if its distribution is high and compact. The following equation illustrates this criterion:

$$B_n(x) = \arg\min_K \left(\sum_{i=1}^{K} \omega_{i,n} > T \right). \tag{4}$$

The technique is computationally intensive; its parameters require careful tuning and it is very sensitive to sudden changes in global illumination. The memory requirement is a 5-frames buffer. The complexity per pixel is $O(K) + \beta\, O(2K)$ where β represent the update rate of the background model.

The approach proposed by Oliver et al. (Oliver, 2000) is based on eigenvalue decomposition of the whole image. The idea is to obtain the eigenvectors that described the background and applied principal component decomposition. Then only the most important eigenvectors are selected to describe the background model. When a new frame arrives, a projection into the space form by the eigenvectors is performed and the foreground mask is estimated.

During the learning phase, the method acquires N images and computes the mean image I_μ, and all the images are mean-subtracted. Then the covariance matrix is computed and the best M eigenvectors are stored in a matrix Φ_M of size M per the number of pixels in each frame. The classification phase begins after selecting the eigenvectors. The incoming frame is projected onto the eigenspace and back projected onto the image space as follows:

$$\begin{aligned} I'_n(x) &= \Phi_M\left(I_n(x) - I_\mu(x)\right), \\ B_n(x) &= \Phi_M^T I'_n(x) + I_\mu(x). \end{aligned} \tag{5}$$

Then, foreground points are detected by subtraction of the background model from the current frame followed by thresholding.

The original presentation of the work by Oliver did not explicitly specify what images should be part of the initial sample or how the model should update over time or update frequency. However, the best results are obtained when motionless images are used in the learning procedure. If such images are not available, the results are not as accurate. The main disadvantage of this method is that the learning phase is computationally intensive and requires N frames and M frames for the classification procedure (Piccardi, 2004). Therefore the memory requirements are N+M+2 frames and the complexity per pixel is $O(M) + \beta\, O(2N^2+N)$, where β is the updating rate of the model.

A more reliable technique is Wronskian Change Detector (WCD) that detects changes based on

Figure 5. Region of support for pixel x_5

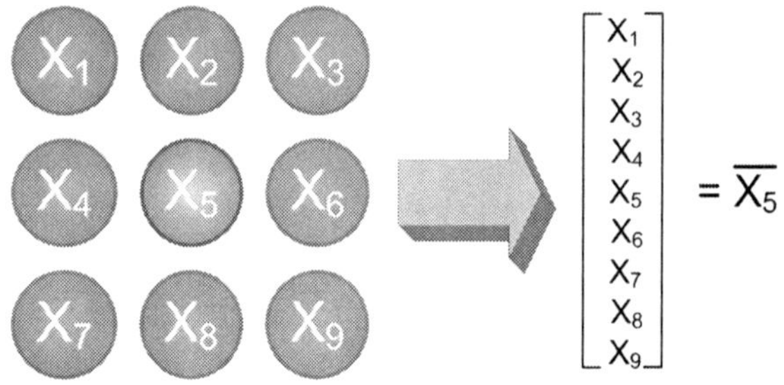

the variance of intensity ratios. The WCD takes advantages of the spatial correlation by creating a region of support of each pixel and analyzes the variance of the intensity ratios of this region. However, the algorithm makes no use of the temporal correlation existing in the video stream. WCD takes advantage of the spatial correlation by using a region of support for each pixel. However, neglects temporal correlation of the visual data.

Wronskian Change Detector

Wronskian Change Detector (WCD) is a non-recursive background subtraction technique that distinguishes changes based on intensity values (Durucan, 2001). Consequently, WCD requires conversion of images into luminance values. Changes are detected based on the variance of intensity ratios. In order to determine if a change has occurred, a region of support is assigned to each pixel as shown in Figure 5. The size of the region of support can be 3×3, 5×5, or 7×7 pixels.

WCD exploits the fact that the luminance ratios from two sources help to quantify the difference between the light sources. In order to measure the change, the Wronskian value of a pixel is calculated by subtracting the sample mean of the intensity ratios inside the region of support from the squared ratios. A large mean or large variance of the intensity ratios increases the Wronskian value. Wronskian calculations are followed by thresholding where a change is detected if the values exceed a preset threshold. Detection is calculated by equation 8, where x and y are the intensity values of the current and the previous frame, respectively. The previous frame is taken as the background model. The dimension of the region of support is n.

$$W\left(\frac{x}{y}\right) = \frac{1}{n}\left[\sum_{i=0}^{n}\frac{x^2}{y^2}\sum_{i=0}^{n}\frac{x}{y}\right] = 0 \qquad (6)$$

$W(x/y)$ detects changes corresponding to dark zones, while its inverse ratio $W^*(x/y)$ finds if a change has occurred in bright zones. Therefore, computing both values allows robust detection against global illumination changes.

The algorithm requires a 2-frame buffer and the complexity per pixel is $O(n^2)$ where n is the size of the region of support. The algorithm exploits the spatial correlation achieving a 21% of errors and is robust against illumination changes due to the use of intensity ratios as a measurement of change.

Temporal Correlation in Wronskian Change Detector

WCD employs a previous frame as a background model. Then, Wronskian determinant is used as a change measure. However, visual data exhibits a high temporal correlation that can be exploited by using a statistical background model. Average filter is the simpler statistical background model, where the model is constructed based on an average of n-previous frames. The model improves

Figure 6. Phases of the Proposed Algorithm

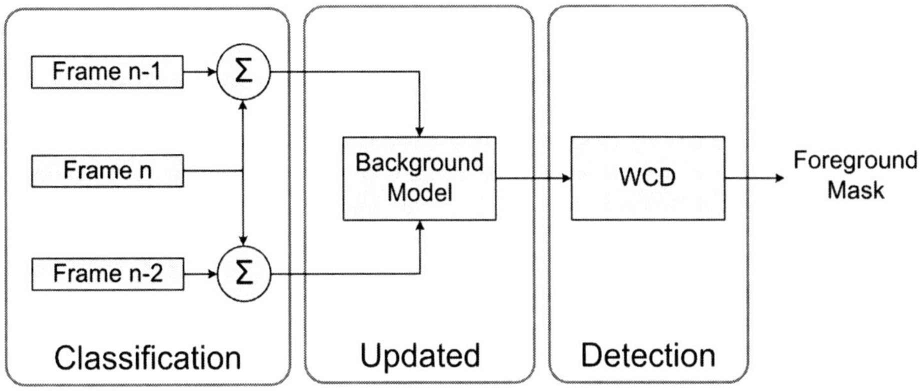

relatively with the amount of previous frames used to obtain the model. However, the complexity as well as the amount of memory increases with the number of previous frames. A technique that avoids the increment on memory and complexity is Recursive Running Average (RRA). The background model obtains by employing RRA is described in the following equations:

$$B_{n+1}(x) = \begin{cases} aB_n(x) + (1-a)I_n(x) & \textit{if } x \textit{ is moving} \\ B_n(x) & \text{otherwise} \end{cases} \tag{7}$$

$$T_{n+1}(x) = \begin{cases} aI_n(x) + 5(1-a)\,|\,I_n(x) - B_n(x)\,| & \textit{if } x \textit{ is moving} \\ T_n(x) & \text{otherwise} \end{cases} \tag{8}$$

where α is a time constant that specifies the changing rate of the information, B_n is the background model at time n, T_n is the threshold at time n used to determine if a pixel has change. The pixels are classified into moving and non-moving by employing a three-frames differencing. The addition of the background model includes 2-frame buffer to the WCD and some extra complexity. The algorithm includes three phases, classification of pixels by three-frames differencing, upgrading the background model and detection by WCD. These phases are illustrated in the Figure 6.

Simulation Results

The proposed algorithm was implemented in Matlab. In order to establish the improvement in the precision of the algorithm a standard benchmark was used. The Wallflower benchmark was introduced in (Toyama, 1999). The benchmark typifies seven canonical situations of characteristic problems encountered in object detection, such as an object has been moved, it should be part of the foreground only as it moved. After the object has settled, it should be part of the background. The rest of the situation are gradual illumination changes, sudden changes in illumination, foreground object with characteristic similar to the background, elimination of uninteresting motion as well as training without movement.

The Wallflower benchmark consists of seven sequences taken with a 3-CCD camera recording to digital tape, sampled at 4Hz and stored at size 160×112 pixels. On every output image, a speckle removal was employed to eliminate islands of four pixels. The threshold in each set was unchanged for the whole sequence and the threshold was not changed between sequences.

The results for frame differencing, running average, WCD, eigenbackground subtraction and

Table 1. Percentage of Error for background techniques using the Wallflower Benchmark

Algorithm	Error Type	Moved Object	Time of Day	Light Switch	Waving Trees	Camou-flage	Boot-strap	Foreground Aperture	Total
Frame Differencing	False Neg	0.00	80.04	76.75	53.49	93.75	68.41	78.02	64.35
	False Pos	0.00	0.87	0.48	25.93	0.38	1.73	3.07	4.64
	Total	0.00	40.46	38.61	39.71	47.07	35.07	40.54	34.49
Running Average	False Neg	0.00	78.55	88.25	37.66	99.82	85.81	46.22	62.33
	False Pos	0.00	0.18	0.00	8.57	0.05	0.90	37.52	6.75
	Total	0.00	39.37	44.13	23.12	49.93	43.35	41.87	34.54
Wronskian Change Detector (RS = 3)	False Neg	0.00	35.53	45.05	13.47	53.05	31.04	39.47	31.09
	False Pos	0.00	1.73	2.33	48.32	5.31	19.61	4.94	11.75
	Total	0.00	18.63	23.69	30.89	29.18	25.33	22.20	21.42
Eigenbackground	False Neg	0.00	56.78	29.58	17.46	3.31	10.39	48.61	23.73
	False Pos	5.55	0.09	2.27	15.44	14.65	37.66	3.79	11.35
	Total	5.55	28.44	15.93	16.45	8.98	24.03	26.20	17.94
Proposed	False Neg	4.00	32.94	14.15	4.85	26.47	29.85	22.18	19.21
	False Pos	0.00	2.93	14.31	36.85	9.09	18.85	33.20	16.46
	Total	0.00	17.94	14.23	20.85	17.78	24.35	27.69	17.55

the proposed technique are reported in Table 1. Table 1 shows the number of false positive and negative. A false positive is a pixel that is classified as foreground but in reality is a background pixel, while a false negative is a pixel classified as a background pixel but is a foreground pixel. However, the number of false positive or negative pixel is not giving a good idea of the performance of the algorithm. Then a percentage of false negative and positive is presented in the total row.

Eigenbackground subtraction is a nonrecursive technique that achieves the highest precision. The background model is performed by extracting the principal component of the correlation matrix formed by the last n previous frames. Then, the background model is the projection of the present frame into the space form by the m-most important components of the matrix (Brand, 1997).

The propose scheme achieves similar performance than the eigenbackground subtraction technique with only 4-frames buffer instead of 50-frames buffers used in the latter technique. The small memory requirements make this scheme a better fit for embedded applications, where memory resources are highly constrained.

VIDEO PREDICTIVE ENCODING FOR VISUAL SENSOR NETWORK

The lowest level of representation of data in a VSN is 2D data set. However, this data set can be highly reduced by employing video compression. The information contained in a video sequence is highly redundant due to the inter-frame correlation. The goal of video compression is to eliminate redundant information so that the video sequence can be efficiently stored or transmitted.

Predictive Video Encoding (PVE) takes advantage of the temporal correlation by using a previous frame to predict the movements of the blocks in the following frame. Then, the frames are divided in a lattice of macroblocks. Each macroblock is subject to Motion Estimation (ME) to determine the vertical and horizontal displacement of the block. PVE is based on two premises: the

Figure 7. Video Compression Data Flow

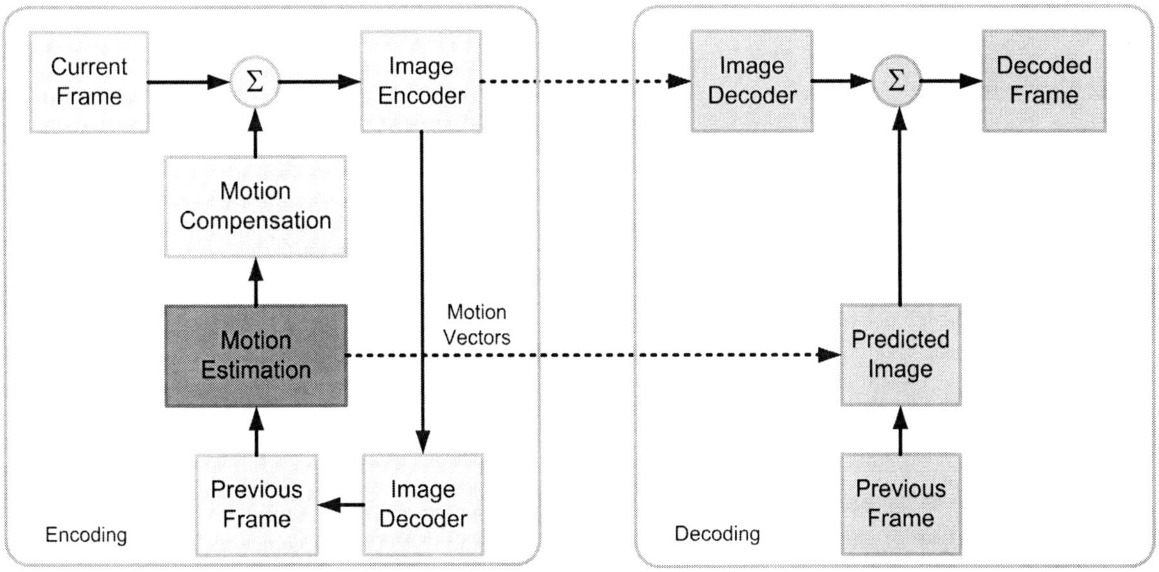

whole block moves in a single direction and the blocks only has horizontal and vertical displacements, known as motion vector. Once the motion vector for each macroblock is determined, then motion compensation is performed to estimate the residual error between the original frame and the compensated frame. The MV along with the residual error is stored or transmitted. The compression data flow is illustrated in Figure 7. The key elements in the compression dataflow are motion estimation and compensation.

The most popular technique for ME is Block Matching (BM) due its low complexity. The bottleneck in BM is finding the best match for each of the macroblocks. Therefore, BM involves search algorithm and matching criteria. These two elements determine the quality of the compress video as well as the complexity of the algorithm.

The matching criteria include Normalized Cross-Correlation Function (NCCF), Mean Square Difference (MSD), Mean Absolute Differences (MAD), and Sum of Absolute Differences (SAD) between others. The widely used criterion is SAD due to its low complexity and simple hardware implementation as can be seen in equation 9.

$$\mathrm{SAD}(x,y,m,n) = \sum_{i=0}^{w}\sum_{j=0}^{w} \left| F_{ref}(x+i, y+j) - F_c(x+i+m, y+j+n) \right| \tag{9}$$

where F_{ref} is the reference frame, F_c is the candidate frame, and w is the macroblock size.

Most of the computation in a ME algorithm is dedicated to the search method. The method that can give us the best quality is Full Search (FS). FS explore a search area around the block of ±p pixels. For each location, a SAD value is calculated and the one with the lowest SAD is chosen as the motion vector. However, FS is costly because every candidate block within the search window is evaluated. Therefore, FS considers $(2p+1)^2$ candidate blocks. SAD is calculated for each block, leading to w^2 operations per block. Then, FS performs $(2p+1)^2w^2$ operations per block. Considering typical values such as p equal to seven and w equal to 16, then FS performs 57600 operations per block. Most of the cases, the MV is equal to zero due to the amount of blocks belonging to background scene, which does not move. Therefore, most of these operations are useless.

In order to eliminated useless operations most of the research has been focusing into fast search techniques that can located the best match with a minimum candidate block evaluation.

Several fast search strategies have been proposed such as Three Step Search (TSS), New TSS (Li, 1994), Simple and Efficient TSS (SETSS) (Lu, 1997), Four Step Search (4SS) (Po, 1996), Diamond Search (DS) (Zhu, 2000), Adaptive Rood Pattern Search (ARPS) (Nie, 2002).

FAST SEARCH TECHNIQUES

TSS algorithm makes the assumption that the error surface due to motion in each block is unimodal. Therefore, the search points are distributed S points away from the center of the block, where S is the step size. Then, each search point is defined as ±S points from the origin. The SAD is calculated at each location and the point with lowest SAD is chosen as the new origin. Then, S/2 is chosen as the new step size. The process is repeated until the step size is one. The technique achieves a reduction of search points by a factor of 9. However, the technique fails to capture small movements (Koga, 1981).

In order to improve detection of small movements, a new algorithm was proposed know as New TSS using the pattern defined by the TSS and adding 8 extra point around the origin. The first 8 points are defined at a step size of 4, and the 8 extra points are set at a step size of 1. If the inner points give a lower SAD, then the origin is chosen as the motion vector. However, if the outer points give a lower SAD, then the algorithm follows the TSS pattern. The algorithm must check a minimum of 17 points for each block and a maximum of 33 in the worst case scenario (Li, 1994).

Simple and Efficient Search algorithm divide the area into four quadrants, next the algorithm selects the origin A = (0, 0) and points B = (0, S) and C = (-S,0) where S is the step size. The matching function is calculated at each point. If the matching function is better for B and C then the search area moves between them. If B and A are best match, then the search area is defined between them. These patterns are illustrated in Figure 8c. SETSS gives lower PSNR than TSS; therefore it was not widely accepted.

Four-Step Search is another modification over TSS. This algorithm selects the 9 points with a fixed step size of S = 2. Next, the search origin is moved to the point with the lowest cost. And again 9 points are selected with the same step size. This process is repeated two more times, but the last iteration chooses a step size S = 1.

Diamond Search has two predefined patterns known as Large Diamond Search Pattern (LDSP) and Small Diamond Search Pattern (SDSP). The first step defines the LDSP centered in the origin, as shown in Figure 8e. The cost function is calculated for each point and the point with lower cost is selected as the new origin for the LDSP. The procedure repeats until the minimum cost is centered at the diamond. Then, the algorithm uses de SDSP center at the last point and the vector motion is chosen. The algorithm does not have a limit in the number of steps; therefore can get a global maximum.

Adaptive Rood Pattern Search exploits the fact that the movement of the object in the video sequence is coherent. Therefore, there is a high probability that the current block moves in the same direction that the block to its immediate left. The ARPS searches in a rood pattern of step size equal to the maximum value of the coordinates of the previous motion vector $s=max(|x|,|y|)$. Then the point with the lowest SAD is taken as the center for a Small Diamond Search Pattern (SDSP). The advantage of this method over DS is that the first rood pattern puts the SDSP in the immediate neighborhood of the true motion vector.

These methods make the assumption that the error surface is unimodal. However, this assumption not always holds. Therefore, they achieve a sub-optimal solution that is reflected in loss of quality indicated by the Peak Signal to Noise

Figure 8. Pattern Search for Fast Search Techniques, a) TSS, b)NTSS c)SES d)FSS e)DS,f)ARPS

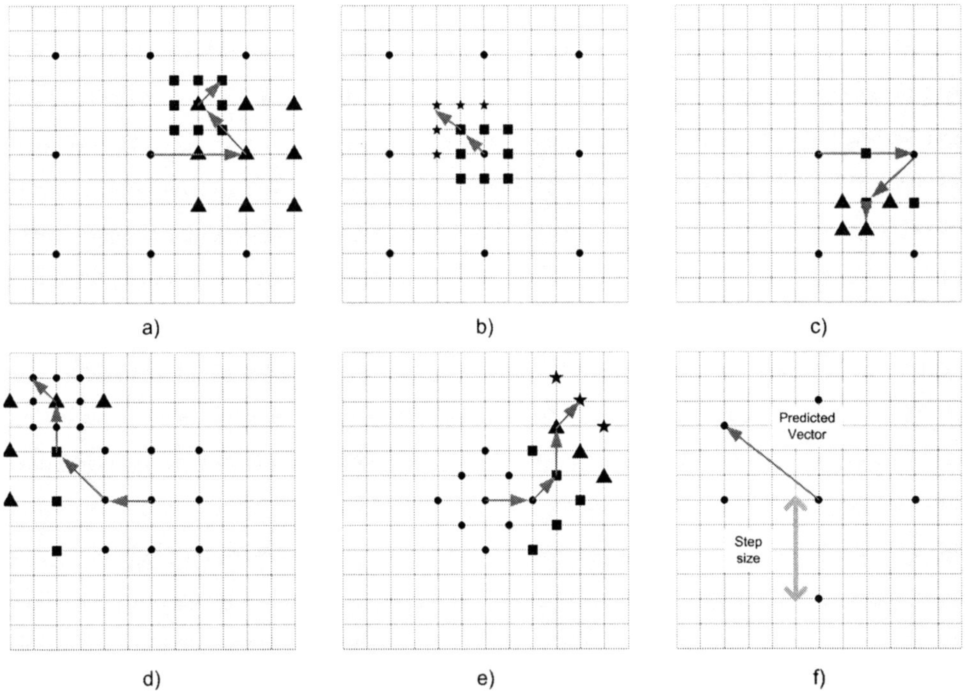

Ratio (PSNR) value. The PSNR is a measure of the quality of the reconstructed image and is defined in equation 10.

$$PSNR = 20\log_{10}\left[\frac{MAX}{\sqrt{MSE}}\right] \quad (10)$$

where

$$MSE = \sum_{i=0}^{m-1}\sum_{j=0}^{n-1}[F_{ref}(i,j) - F_{com}(i,j)]^2$$

and F_{ref} is the reference frame and F_{com} is the compensated frame, both of size m × n and MAX is the maximum value per pixel that has been set to 256.

The proposed technique uses a deterministic approach to find the motion vectors by employing the Phase-Only Correlation (POC) function. POC uses the normalized cross-correlation to determine the displacement between two images. Our method employs this information to create a table of motion vectors. The estimated motion vectors are tested for each block. Then, the index on the table is saved to indicate the motion vector for that specific block. The number of estimated motion vectors determines the PSNR values achieve by the technique.

Phase-Only Correlation Function

The estimation of the motion vectors is done by matching objects between two frames. The matching procedure is done in the spatial domain. Therefore, the estimation is done by comparing the block with a set of candidate blocks. Another approach to this problem is taking into consideration that a displacement in the spatial domain is mapped as a change of phase in the frequency domain. The shift property of the Fourier transform states that if $F(k)$ is the Fourier transform of a real-value scalar signal $f(x)$, and $g(x)=f(x-d)$ is a delayed version of $f(x)$ by d samples. Then the Fourier transform of $g(x)$ is given by:

$$G(k) = e^{-j^{2\pi kd}/N} \quad (11)$$

This result can be extended to two-dimensional signals. Consider two $N \times M$ images $f(x,y)$, and $g(x,y)$. Then, the Fourier transform of the images are defined as,

$$F(k_1,k_2) = \sum_x \sum_y f(x,y) e^{-j\frac{2\pi}{N}k_1 x} e^{-j\frac{2\pi}{N}k_2 y} \quad (12)$$

$$G(k_1,k_2) = \sum_x \sum_y g(x,y) e^{-j\frac{2\pi}{N}k_1 x} e^{-j\frac{2\pi}{N}k_2 y} \quad (13)$$

If $g(x,y)$ is a displace version of the image $f(x,y)$. Then the normalized cross-spectrum of the images is compute as,

$$R(k_1,k_2) = \frac{F(k_1,k_2)G^*(k_1,k_2)}{|F(k_1,k_2)G^*(k_1,k_2)|} = e^{-j^{2\pi k_1 \delta_1}/N} e^{-j^{2\pi k_2 \delta_2}/N} \quad (14)$$

where $G^*(k_1,k_2)$ is the complex conjugate of $G(k_1,k_2)$ POC function is obtained by taking the inverse Fourier of the normalized cross-spectrum,

$$r(x,y) = \sum_{k_1} \sum_{k_2} R(k_1,k_2) e^{j\frac{2\pi}{N}k_1 x} e^{j\frac{2\pi}{N}k_2 y} \quad (15)$$

POC function $r(x,y)$, approximates to an impulse centered at (δ_1,δ_2). Therefore, the displacement required to realign the images so that $f(x,y)=g(x-\delta_1,y-\delta_2)$ can be found as

$$(\delta_1,\delta_2)=\text{argmax}\{r(x,y)\} \quad (16)$$

If the image $g(x,y)$ is only a displace version of $f(x,y)$, then POC is only an impulse centered in the displace amount. In a video sequence, moving objects are represented by one impulse. The size of the peak is related to the size in pixels of the moving object. Figure 9 shows a synthetic example of the POC function. In this example there are two moving areas of different sizes. POC gives a peak in each of the artificially-produce displacements, and additional peak is produce due to the non-moving background.

POC function has been successfully used for many applications such as rigid image registration (Keller, 2005), optical flow estimation (Chien, 2004), and high-accuracy stereo disparity estimation (Muquit, 2006).

POC-Guided Area Matching Method

Block matching for Motion Estimation (ME) divide the image in non-overlapping block. A motion vector for each block is calculated based on search method and a matching criterion. For this method, the Sum of Absolute Difference (SAD) has been employed as a matching criterion.

The proposed search method is based on the POC function. POC function gives us the displacement information of each of the moving objects in the frame. POC uses the frequency information of the whole frame. Therefore, there is no restriction on the size of displacement that may occur. In the other hand, FS considers a restriction on the displacement because it considers search windows between ±6 to ±15 pixels around the block. Then, POC function is calculated for the whole frame. Only the N most relevant peaks are selected and stored in a lookup table. The selected motion vectors are tested for each of the blocks. Therefore, the amount of search points is restricted to a set of N points. The motion vector with the lowest SAD within our pool of motion vectors is selected for that specific block. Only the index on the table is saved for the selected motion vector.

The number of chosen peaks is selected as a power of two to achieve a good compression ratio. This selection is to use efficiently the number of bits. Therefore, for $N=2^M$ motion vectors, only M bits are used to store the index of the lookup table.

Figure 9. Artificially Created Images (a) Reference image, (b) Image with Moving Objects (c) Resulting POC Function

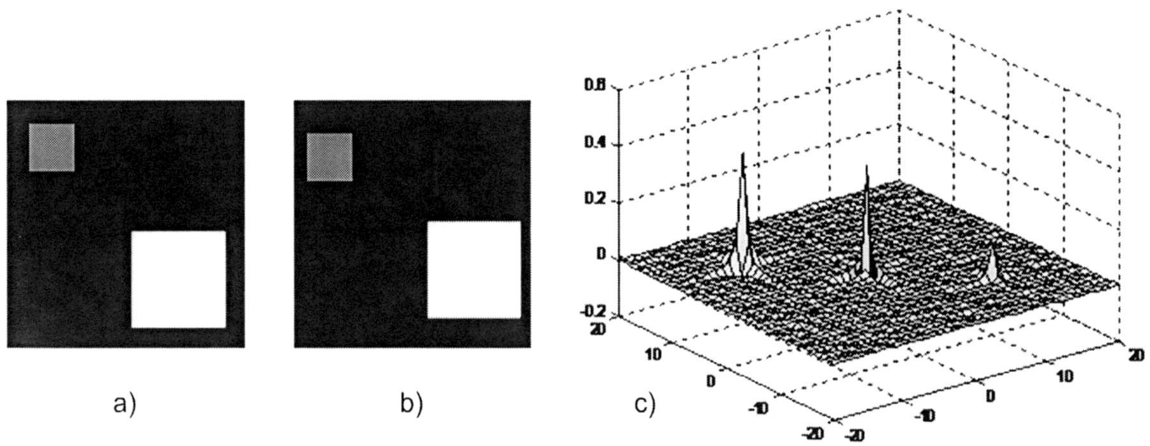

Our proposed method called POC-Guided Area Matching (PGAM) is described as follows:

- Obtained the POC function $r(x,y)$ for images $f(x,y)$ and $g(x,y)$.
- Select $N=2^M$ motion vectors representing the N biggest peaks of the POC function $\{(\delta_1,\delta_2)\}_N=\text{argmax}\{r(x,y)\}$
- For each block, the selected motion vectors are tested. The pool of motion vectors represents our search points. The one with the lowest SAD value is selected for that specific block.

The number of selected motion vector has been varied from 4, 8, 16, 32, 64 and 128 peaks.

Result and Comparison

The proposed method was tested with Wallflower Benchmark (Toyama, 1999) consisting on seven video sequences of 160 × 112 pixels. The movement on the sequences shows low and medium speed. The comparison is performed with FS, TSS, NTSS, SETSS, 4SS, DS and ARPS. Table 2 shows our results considering 4 to 128 peaks.

PGAM with 4 peaks only test 4 points for each block and achieves the greatest compression rate, but loses some accuracy as can be seen from the PSNR values. The lost of quality is between 0.01 to 2 dB. However, the complexity drops dramatically because only 4 testing points are considered. The number of testing points for each of the reviewed method is illustrated in Figure 10 compared with 8 peaks for PGAM. PGAM with 8 motion vectors achieved the same or better performance than FS with only 8 search points. The results for *waving tree* sequence give us lower PSNR values because there is a tree moving in the scene. Therefore the peaks correspond to the waving leaves. For this sequence, the size of the peaks is less representative.

Our method achieves the same PSNR value that Fast search techniques with lowest number of tested points as can be seen from Table 1. However, using a comparable number of search points as can be seen from Figure 10, our technique improves PSNR values. Compared with FS using a window search of ± p pixels, with p=7. FS computes SAD for $(2p+1)^2$=225 search locations, while our technique only uses 8 search locations to outperform it.

Table 2. Average PSNR for the Wallflower Benchmark

Method	Moved Object	Bootstrap	Camou-flage	Foreground Aperture	Light Switch	Waving Trees	Time of Day
FS	39.6605	21.8118	19.9492	42.6538	38.4107	18.5446	44.0678
TSS	39.6507	21.6427	19.8168	42.6498	38.4054	18.438	44.054
SETSS	39.6341	21.3842	19.6353	42.628	38.392	18.1634	44.0384
NTSS	39.6489	21.6135	19.8485	42.6502	38.4051	18.402	44.0534
4SS	39.6391	21.4631	19.6298	42.644	38.4017	18.3475	44.0506
DS	39.6397	21.4454	19.6889	42.6453	38.402	18.3014	44.0498
ARPS	39.6309	21.4011	19.6576	42.6449	38.4005	18.2657	44.0408
4 peaks	39.5395	20.8832	17.6759	42.627	38.3439	17.1103	43.9676
8 peaks	39.6245	21.5296	20.4614	42.679	38.3944	17.4808	44.0198
16 peaks	39.6715	22.0337	22.2746	42.7204	38.4397	17.6982	44.0551
32 peaks	39.7024	22.3915	23.2754	42.7377	38.4701	17.8603	44.0745
64 peaks	39.7237	22.628	24.051	42.7502	38.4987	18.0015	44.0889
128 pks	39.7416	22.8052	24.5995	42.7605	38.5146	18.1257	44.0994

Figure 11 shows the PSNR value for each of the frames for the Camouflage sequence. As we can see, our method achieves better PSNR values for that sequence. Figure 12 shows compensated images for the Camouflage sequence for the frame number 290.

Higher values of N improve the PSNR but also increase the amount of computations. A good tradeoff between amount of computation and quality is 16 motion vectors. In most cases our method achieves better PSNR for the tested sequences. However, the method should consider the overhead of calculating POC for each image.

Figure 10. Average number of seach points for each of the tested sequences

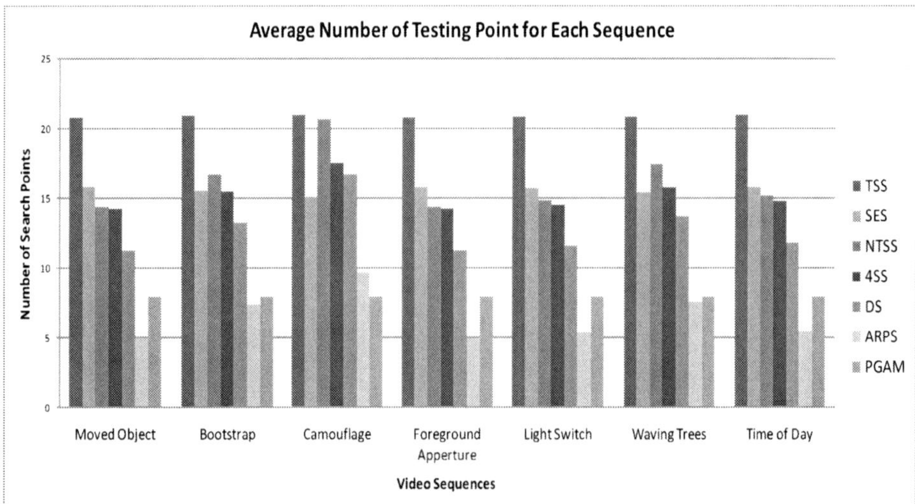

Figure 11. PSNR values for the Camouflage Sequence

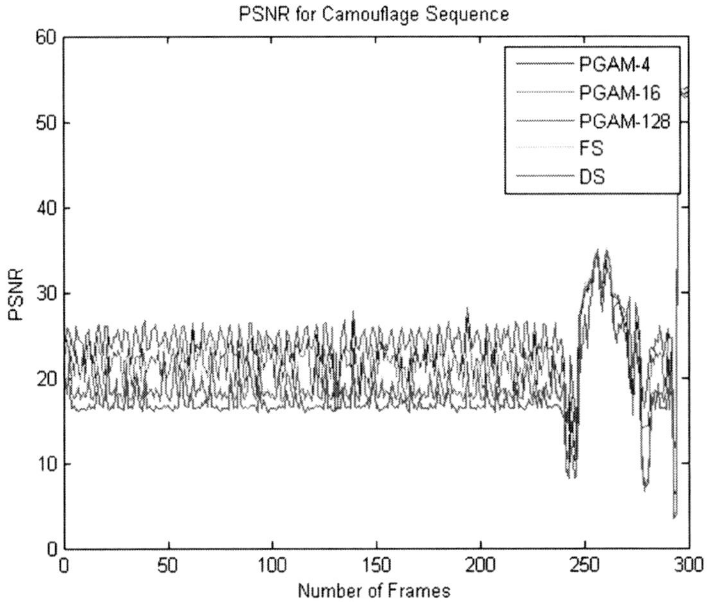

In order to calculate POC, a 2D-FFT must be used. It is well known that the FFT requires $N \cdot log(N)$ operations where N is number of points. However, a hardware accelerator to perform the FFT can be used, decreasing the computation time for POC. The major advantage of using this method is that larger movements can be capture; while FS is incapable to track such movements.

Each of these peaks represents moving objects on the scene. However, when there are several moving objects or objects that have a very dynamic

Figure 12. Compensated Images for frame 290 of Camouflage sequence, (a) Original Frame, (b) FS, (c)PGAM-4 peaks, (d)PGAM-8 peaks, (e)PGAM-16 peaks, (f)PGAM-32 peaks, (g)PGAM-64 peaks, (f) PGAM-128 peaks

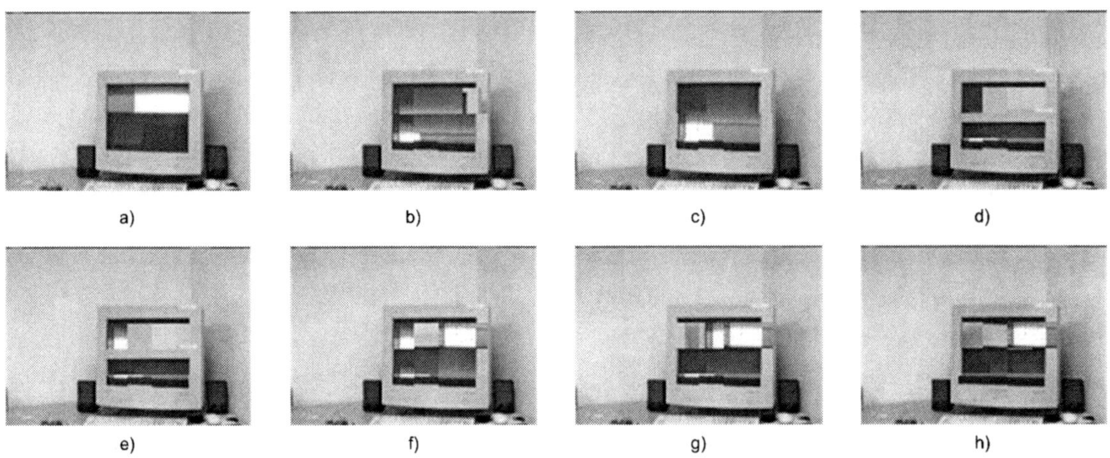

movement such as waving trees, our method will give a lower performance than FS.

In order to solve this problem, information about the size of the peaks can be taking into consideration to divide the image into equally divide regions. POC is calculated for each region and N_i motion vectors for region R_i are selected. Preliminary results show that PSNR is improving but the complexity of the method increases due to multiple calculation of POC function. An adaptive algorithm that considers the size of the peaks and decides when to divide the image into regions tries to compromise the complexity with the quality of the video compression.

FUTURE RESEARCH DIRECTIONS

In order to prolong network lifetime, node must change data rate transmission. Different operation modes based on power-aware techniques can prolong the life of the node (Zhang, 2009). If the node has a full battery, then may uses the highest data rate, while the battery is dying should use the lowest data rate. Also, the conditions of the channel dictate the data rate that can be used to ensure effective transmission. Visual Data Coding for VSN require adapting to the changes in data rate. Scalable Video Coding is capable of adapt to different data rate.

Scalable Video Coding (SVC) encodes the signal once at highest resolution, but partial streams may be decoded based on the specific rate and resolution that fit the data rate transmission at hand. SVC offers a simple and flexible solution for transmission and display of video over a variety of devices and proving adaptability for bandwidth variations and error conditions.

The scalability of the video coding is performed in any of the following dimensions: spatial, temporal or quality-level resolution also known as Signal-to-Noise Ratio (SNR) scalability. The coding can be done in any of the dimensions or a combination of two or more dimensions.

Motion-compensated wavelet filtering has enabled the efficient implementation of scalable video codecs that achieves a high compression performance. Wavelet video coding schemes provide flexible spatial, temporal and SNR with fine granularity over a large range of bit rates (Ohm, 2005).

The general processing of scalable predictive codec is illustrated in Figure 13. The main idea is to encode by motion estimation a base resolution image. Then, this image is used to generated an enhance version of it for higher resolution.

The base image is sent for low data rate transmission. However, if the channel allows an enhance version of the image, it is sent. The enhance image is achieve by upsampling of a reconstructed version of the base image. The base image is generated by decimation (preprocessing), the appropriate quantizer leads to the desired quality level of the spatio-temporal signal resolution at this level. The reconstructed base signal is an approximation of all higher layer resolution levels. The higher layers perform up-sampling of the lower layers signals. The interpolation and decimation is performed to go from a base signal to an enhance signal or vice versa. The information is propagated from lower into higher resolution layers during encoding and decoding. However, each layer should be self-contained, i.e. the prediction should not use any information from other layers. If this condition does not hold, then different estimates may be used at the encoder and decoder sides, resulting into a drift effect.

Future research direction is to employ POC-Guided Area Matching (PGAM) technique in a scalable video encoder to allow adaptation of the encoder to variable data rate transmission.

CONCLUSION

WSN is a highly populated network of smart nodes that are composed by sensing unit, processing unit, communication, and power management unit.

Figure 13. Scalable Predictive Video Coding Scheme

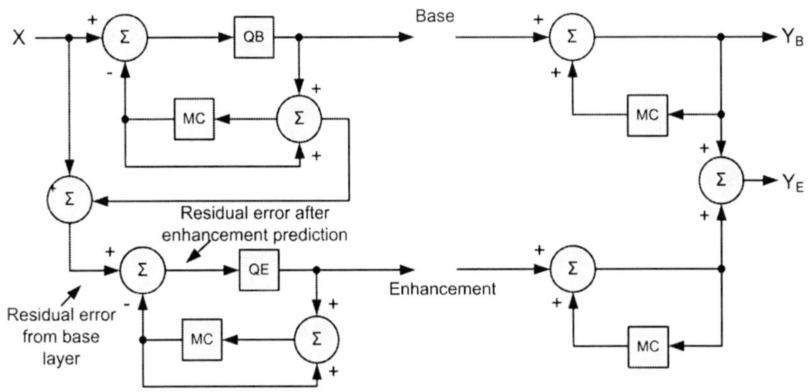

Sensor nodes are design to process the acquired data before transmission. This feature is unique in WSN and gives them the possibility of process the data inside the network. Therefore, this paradigm allows the adoption of preventing events to occur instead of sending only the information of the event that just occurs. Then, WSN changes our perspective from containing the damages cause by any event to prevent the event from happening. This fact has motivated the wide study of this type of network.

VSN gathered 2D data set representing the scene under observation by the camera sensor. However, nodes are highly restricted in terms of energy. Therefore, VSN requires techniques to reduce the amount of data. There are two strategies to reduce the amount of data, visual data filtering and visual data coding. Two algorithms have been introduced, PGAM for visual data coding and Temporal Correlated WCD for object detection.

The proposed method called POC-Guided Area Matching (PGAM) takes the information provided by the POC function to select a number of motion vectors that are selected as the search points for each block. The number of search location is comparable with ARPS. However, the PSNR value is similar than FS for almost all the tested sequences. While 16-peaks PGAM achieves better results than PGAM with only 16 search locations. POC captures better the movements of blocks because is not restricted to a search window. When there are objects with a more dynamic movement on each frame, the method should divide the image into regions and chose a number of motion vectors for each region as a search location for the block within the region. The number of search vector can be dynamically select based on the size of the peaks.

The temporal correlation property of the video stream is exploited in the proposed scheme by introducing a recursive background model to the Wronskian Change Detector. The WCD already takes advantage of the spatial correlation existing on the images. Using the temporal and spatial correlation results in improvement over the WCD of 20% and 50% over the frame differencing and running average alone. The performance of the proposed scheme is comparable to the eigenbackground subtraction technique without requiring a large buffer for the previous frames due to the recursive nature of the background model.

ACKNOWLEDGMENT

This work was supported by grant PROMEP/UASLP/10/CA06. A. Alba was partially supported by grant PROMEP/103.5/-09/573. R. Aguilar-Ponce was partially supported by grant PROMEP/103-5/09/4244.

REFERENCES

Akyildiz, F., Melodia, T., & Chowdhury, K. R. (2007). A survey on wireless multimedia sensor networks. *Elsevier Computer Network*, *51*(4), 921–960. doi:10.1016/j.comnet.2006.10.002

Brand, M., Oliver, N., & Pentland, A. (1997). *Coupled hidden Markov Models for complex action recognition*. In *Proceedings of IEEE Computer Society Conference on Computer Vision and Pattern Recognition* (pp. 994-999).

Charfi, Y., Canada, B., Wakamiya, N., & Murata, M. (2009). Challenging issues in visual sensor network. *IEEE Wireless Communication*, *16*(2), 44–49. doi:10.1109/MWC.2009.4907559

Cheung, S.-C. S., & Kamath, C. (2004). *Robust techniques for background subtraction in urban traffic video*. In *Proceedings of Video Communication and Image Processing Conference*. (UCRL-JC-153846-ABS, UCRL-CONF-200706)

Cucchiara, R., Piccardi, M., & Prati, A. (2003). Detecting moving objects, ghost and shadows in video stream. *IEEE Transactions on Pattern Analysis and Machine Intelligence*, *25*(10), 1337–1342. doi:10.1109/TPAMI.2003.1233909

Durucan, E., & Ebrahimi, T. (2001). Change detection and background extraction by linear algebra. *Proceedings of the IEEE*, *89*(10), 1368–1381. doi:10.1109/5.959336

Juang, P., Oki, H., Wang, Y., Martonosi, M., Peh, L. S., & Rubenstein, D. (2002). *Energy-efficient computing for wildlife tracking: Design tradeoffs and early experiences with ZebraNet*. Paper presented at Architectural Support for Programming Languages and Operating Systems Conference, San Jose, CA.

Keller, Y., Shkolnisky, Y., & Averbuch, A. (2005). The angular difference function and its application to image registration. *IEEE Transactions on Pattern Analysis and Machine Intelligence*, *27*(6), 969–976. doi:10.1109/TPAMI.2005.128

Koga, T., Iinuma, K., Hirano, A., & Iijima, Y. (1981). Motion compensated interframe coding for video conferencing. In *Proceedings of NTC'81* (G5.3.1-G5.3.5).

Li, R., Zeng, B., & Liou, M. L. (1994). A new 3-step search algorithm for block motion estimation. *IEEE Transactions on Circuits and Systems for Video Technology*, *4*(4), 438–442. doi:10.1109/76.313138

Lu, J., & Liou, M. L. (1997). A simple and efficient search algorithm for block matching motion estimation. *IEEE Transactions on Circuits and Systems for Video Technology*, *7*(2), 429–433. doi:10.1109/76.564122

Mainwaring, A., Polastre, J., Szewczyk, R., Culler, D., & Anderson, J. (2002). *Wireless sensor networks for habitat monitoring*. Paper presented at 2002 ACM International Workshop on Wireless Sensor Networks and Applications (WSNA '02), Atlanta, GA.

Majumdar, A., & Ramchandran, K. (2004). PRISM: An error-resilient video coding paradigm for wireless networks. In *Proceeding of 1st International Conference on Broadband Networks* (pp. 478-485).

Misra, S., Reisslein, M., & Xue, G. (2008). A survey of multimedia streaming in wireless sensor networks. *IEEE Communications Surveys & Tutorials*, *10*(4), 18–39. doi:10.1109/SURV.2008.080404

Muquit, M. A., Shibahara, T., & Aoki, T. (2006). A high-accuracy passive 3D measurement system using phase-based image matching. *IEICE Transaction on Fundamental Electronics, Communication, and Computer Science. E (Norwalk, Conn.)*, *89-A*(3), 686–697.

Nie, Y., & Ma, K. K. (2002). Adaptive rood pattern search for fast block-matching motion estimation. *IEEE Transactions on Image Processing*, *11*(12), 1442–1448. doi:10.1109/TIP.2002.806251

Obraczka, K., Manduchi, R., & Garcia-Luna-Aceves, J. J. (2002). *Managing the information flow in visual sensor networks*. Paper presented at the 5th International Symposium on Wireless Personal Multimedia Communications.

Ohm, J. R. (2005). Advance in scalable video coding. *Proceedings of the IEEE*, *93*(1), 42–56. doi:10.1109/JPROC.2004.839611

Oliver, N. M., Rosario, B., & Pentland, A. P. (2000). A Bayesian Computer vision system for modeling human interactions. *IEEE Transactions on Pattern Analysis and Machine Intelligence*, *8*(22), 831–843. doi:10.1109/34.868684

Piccardi, M. (2004). Background subtraction techniques: A review. In *Proceedings of IEEE International Conference on System, Man, and Cybernetics* (pp. 3099-3104).

Po, L. M., & Ma, W. C. (1996). A novel 4-step search algorithm for fast block motion estimation. *IEEE Transactions on Circuits and Systems for Video Technology*, *6*(3), 313–317. doi:10.1109/76.499840

Slepian, D., & Wolf, J. (1973). Noiseless coding of correlated information sources. *IEEE Transactions on Information Theory*, *19*(4), 471–480. doi:10.1109/TIT.1973.1055037

Soro, S., & Heinzelman, W. (2009). A survey of visual sensor networks. *Hindawi advances in multimedia*. Article ID 640386 (21 pages).

Stauffer, C., & Grimson, W. E. L. (1999). *Adaptive background mixture model for real-time tracking*. Paper presented at IEEE Conference on Computer Vision and Pattern Recognition.

Toyama, K., Krumm, J., Brumitt, B., & Meyers, B. (1999). Wallflower: Principles and practice of background maintenance. In *Proceedings of 7th International Conference on Computer Vision,* Kerkyra, Greece (pp. 255-261). Washington, DC., USA: IEEE Computer Society Press.

Wyner, A., & Ziv, J. (1976). The rate-distortion function for source coding with side information at the decoder. *IEEE Transactions on Information Theory*, *22*(1), 1–10. doi:10.1109/TIT.1976.1055508

Zhang, J., Wu, D., & Ci, S. (2009). Power-aware mobile multimedia: A survey. *The Journal of Communication*, *4*(9), 600–613.

Zhu, S., & Ma, K. K. (2000). A new diamond search algorithm for fast block-matching motion estimation. *IEEE Transactions on Image Processing*, *9*(2), 287–290. doi:10.1109/83.821744

Chapter 6
Multimedia Transmission over Wireless Sensor Networks

Nalin Sharda
Victoria University, Australia

ABSTRACT

This chapter presents an overview of multimedia information transmission over Wireless Sensor Networks (WSNs). These WSNs have evolved since the 1980s and their evolution can be divided into three generations. Wireless Multimedia Sensor Networks (WMSNs) have become viable in recent years with the availability of inexpensive video cameras, increase in procession power and memory capacity of nodes, and better power sources and their management. Multimedia information requires higher bandwidth and lower delay and delay jitter to provide the required Quality of Service (QoS) for multimedia transmission. Further research is being conducted, and can be taken even further, in the areas of advanced algorithms for content capture, compression and communication of multimedia information over WSNs.

INTRODUCTION

A Wireless Sensor Network (WSN) is a system of distributed nodes connected by wireless communication technologies. Each node performs the function of sensing some physical parameter, such as temperature and pressure, and passing on the captured information to one or more destination nodes. Each node in the WSN can also be involved in routing the information from the source to the destination.

While early WSNs tended to capture and transmit only low volume data, newer systems are being developed to capture and transmit multimedia information as well. In general, multimedia information requires much higher bandwidth for timely transmission. Furthermore, multimedia content, in general, and video in particular, is adversely affected by transmission impairments such as delay and jitter.

DOI: 10.4018/978-1-61350-153-5.ch006

The objectives of this chapter are to:

- Present an overview of the evolution of Wireless Sensor Networks, and multimedia transmission in such networks.
- Discuss factors such as bandwidth, delay and delay jitter and how these influence the Quality of Service (QoS) in the delivery of multimedia information.
- Explore the special issues and limitations related to Wireless Sensor Networks, such as limited energy, computational power, and bandwidth.
- Discuss current and further research directions in coding, compression and communication of multimedia information.

WIRELESS SENSOR NETWORKS OVERVIEW

Wireless Sensor Networks (WSNs) are essentially wirelessly connected sensors that can monitor the physical environment in which these are deployed. The parameters monitored by these sensors can include temperature, sound volume, vibration, pressure, motion, machine wear, and even the concentration of pollutants in the environment.

Originally WSNs were developed for military applications such as the surveillance of a battlefield. As often happens, these have now found many industrial as well as civilian application, such as: industrial process monitoring, machine health monitoring, environment and habitat monitoring, healthcare, home automation, and traffic control (Kay & Mattern, 2004).

Each node in a WSN contains one or more sensors. Some nodes are designed such that different sensors can be plugged in a motherboard, as required. Each note also comprises a radio transceiver linked to other wireless communications components, a low power microcontroller, and some power source, such as a battery or a solar cell (Kay & Mattern, 2004).

These sensor nodes can vary considerably in terms of size and complexity. Often the size and the cost of the node are linked to the node's performance and properties such as: power source, its capacity, memory size, computational speed and communication bandwidth (Kay & Mattern, 2004).

A complete WSN if built using a wireless ad-hoc network, i.e. a network in which the routing of data from a source to a destination node is based on a multi-hop routing algorithm, in which the same data packet may be forwarded via multiple routes to the destination nodes. Such protocols are not the best for transmitting multimedia information. Therefore, many new communication protocols have been developed to address issues related to the environment and applications where WSNs are deployed.

History of Wireless Sensor Networks

The evolution of WSNs started in the 1980s with the advent of the microcontroller chips, and has progressed incessantly since then to the present day, and, of course, will continue in the future. Sarigiannidis (2006) divided the evolution of WSNs into three generations, based on the information originally presented by Chong and Kumar (2003). The classification given in Table 1 is based on the same.

While the WSNs for the first two generations have been deployed widely, the proposed third generation WSNs are yet to become widespread. Furthermore, it will be even more challenging to create dust particle size multimedia WSNs, as multimedia information places much higher demands on all aspects of the node: i.e. cameras, processors, power sources, memory capacity and communication bandwidth; all of these need to be of higher specifications. Nonetheless, Nikoletseas, et.al (2004) have presented protocols for data communication in dust size WSNs.

Table 1. Three generations of Wireless Sensor Networks (Chong & Kumar, 2003)

Parameter	**Generation-1**	**Generation-2**	**Generation-3**
Period	1980's to1990's	Early 2000s	Mid 2000s onwards
Size	Large shoe box and above	Pack of cards to small shoe box	Dust particle to coin
Weight	Kilograms	Grams to Kilograms	Milligrams to grams
Node architecture	Separate sensing, processing and communication	Integrated sensing, processing and communication	Integrated sensing, processing and communication
Topology	Point-to-point, & star	Client server & Peer-to-peer	Peer-to-peer
Power supply	Large batteries	AA batteries	Solar, ambient temperature/ pressure
Lifetime	hours, days and longer	days to weeks	months to years
Deployment	Vehicle-placed or air-dropped individual sensors	Hand placed	Embedded or "sprinkled", and left behind.

Wireless Multimedia Sensor Networks (WMSNs)

Nodes used in WSNs inherently have low power, limited buffer memory and restricted processing capabilities, as compared to standard workstations. Therefore, creating WSNs capable of carrying multimedia information poses many challenges. Some of the limitations are being overcome with time; nonetheless, as we demand more of multimedia information to be transmitted through WSNs, new communication techniques and protocols need to be devised to tackle these challenges.

The development of Wireless Multimedia Sensor Networks (WMSNs) has been boosted in recent years with the availability of inexpensive CMOS still and video cameras, along with associated hardware such as microphones (Akyildiz, Melodia, & Chowdhury, 2007).

Furthermore, after capturing multimedia information, these WMSNs are able to store and process (e.g. compress) this information in real-time, and transmit the same using protocols targeted towards efficient transmission of multimedia information.

Architecture of a Typical Multimedia Wireless Sensor Network

In general, a WMSN allows users to connect to it via the Internet, using the Internet Protocol (IP) protocol to be able to send queries to a IP-less WMSN (Akyildiz, Melodia, & Chowdhury, 2008). The various components used to build the WMSN can include the following:

- **Video and Audio Sensors** to capture sound, still or moving images. These sensors can be arranged in a single-tier network, or in a hierarchical configuration, using a multi-tier network.
- **Scalar Sensors** to capture physical attributes such as temperature, pressure, and humidity. They are typically resource-constrained devices in terms of energy supply, storage capacity, and processing capability.
- **Multimedia Processing Hubs** are devices with relatively large computational resources making these suitable for collating multimedia streams from the individual sensor nodes. These can integrate and reduce the volume of data transmitted to the sink node.

- **Storage Hubs** allow further processing of multimedia data, such as feature extraction to identify the important characteristics of the content, before sending it to the end user.
- **Sinks** are responsible for converting high-level user queries to network-specific directives and return the required parts of the multimedia stream back to the user.
- **Gateways** connect the sinks to the Internet; often these are the only IP-addressable component in a WMSN.
- **Users** are the devices (e.g. computers / mobile phones) that –often through a human user– issue monitoring commands to the WMSN. These are generally identified by using an IP address. Application-level software running on such devices issue queries and display results obtained from the WMSN.

Applications of Multimedia Wireless Sensor Networks

The proposed Wireless Multimedia Sensor Networks will provide opportunities for developing many new applications such as tracking individuals, home automation, and environmental monitoring.

Some of these new multimedia-based applications can include the following (Akyildiz, Melodia, & Chowdhury, 2007):

- **Multimedia surveillance sensor networks:** These Wireless Multimedia Sensor Networks (WMSNs) use miniature video cameras, connected to low-power processors and wireless transceiver. Such video sensors (in conjunction with microphones) can be used to create advanced surveillance systems that can be used for the detection of crime and terrorist attacks.
- **Storage of potentially relevant activities:** These WMSNs would use multimedia sensors to record and analyze day-to-day hazardous activities such as thefts, car accidents, and traffic violations. The stored video/audio information could then be used to create reports in response to future queries.
- **Traffic avoidance, enforcement and control systems:** These WMSNs could be used to monitor traffic on highways and provide routing advice to avoid congested areas. Also, smart parking systems based on such WMSNs could monitor available parking spaces and provide parking advice to drivers.
- **Advanced health care delivery:** By integrating third generation (3G) wireless networks with a WMSN, telemedicine services can be provided anywhere anytime, even at the scene of an accident on a highway. For this, patients will need to carry with them sensors that can monitor body temperature, blood pressure, pulse rate, ECG, and breathing etc.
- **Automated assistance for the elderly and family monitors:** WMSNs can be used to monitor and capture the state of health of elderly people. By using wearable video and audio sensors, intelligent agents can be used to detect and warn medical services and family member of emergency situations, thereby providing assistance sooner than possible by the traditional means.
- **Habitat and Environmental monitoring:** Developing sustainable habitats in built and natural environments is becoming very important. To achieve this, it is important to be able to monitor environmental parameters and their current status. An array of video cameras can be used to keep constant watch on a given habitat, and record the changes that occur over time, or in response to specific catastrophic events such as storm or earthquake.

- **Person locator services:** This type of service will require image processing software that can scan a large number of video streams and still images, to locate missing persons, or identify criminals or terrorists.
- **Industrial process control:** Industrial process control requires monitoring of a myriad of parameters, e.g. temperature and pressure. If these exceed some preset values, timely reporting is essential to avoid a disaster. This process monitoring can be enhanced by providing live audio and video streams. WMSNs can be deployed to improve manufacturing process used in manufacturing semiconductor chips, automobiles, food or pharmaceutical products, to name but a few (Akyildiz, Melodia, & Chowdhury, 2007).

While the complete set of applications is limited only by one's imagination, the above set of applications provide a good representation of the way in which multimedia information might be used in WMSNs. This information can be used to build an understanding of what features are required of WMSNs to be able to carry multimedia information effectively.

MULTIMEDIA INFORMATION

Multimedia information, i.e. a combination of various types of media, such as text, audio, still images and moving images is now widely used for a variety of applications (Sharda, 1999). While different types of analog media content have been in use individually for many decades, it is only over the last couple of decades that their combined use has become widespread, leading to the concept of multimedia. This widespread use of multimedia can be attributed to advances in digital technology. This digital technology includes coding of media content as digital information, processing of the same using digital signal processing techniques and miniaturization of digital hardware, such as digital cameras.

Importance of Multimedia Information

Multimedia information has become widely used because of its expressive nature. For some applications it provides substantial competitive advantage, whereas for some applications it is almost mandatory. For example, when monitoring a natural habitat, its temperature, pressure and humidity provide a wealth of information as a record of its ecological health; nonetheless, after a severe storm, only pictures can tell if the trees / houses are still standing or have been flattened.

"However, this does not mean that text is unimportant. Under some situations text is more important than images. For example, it will be better to SMS the flight number to a friend picking you at the airport, rather than sending your plane's photos or video" (Sharda, 2007, p. 526).

Applications of Mobile Multimedia

In the past many applications that use multimedia content were restricted to wired networks and desktop computers. However, over the last decade many of these applications have migrated to wireless networking environments. Some of the novel mobile multimedia applications that have been developed include: e-marketing, e-learning, and e-tourism (Sharda, 2009). And now we can add the applications related to Wireless Sensor Networks to the gambit of applications that use multimedia information.

Improvements in the ability to transmit multimedia information over mobile systems came about with the advent of the 3rd generation (3G) mobile systems. Multimedia information transmission capability will improve further with the development of the 4th generation (4G) mobile technologies in the next few years; though it is taking longer than originally envisaged.

Multimedia Coding

Digital coding of multimedia content has made it much easier to store, transmit and process. However, this ease comes with some drawbacks; and some of these drawbacks relate particularly to the transmission of multimedia information. Originally all multimedia content types are analog quantities. These have to be digitized and processed for their effective use in digital systems.

The coding and compression techniques used for transforming analog multimedia information into digital content have a significant bearing on the artifacts and impairments introduced in the content as it is finally displayed. Each of the multimedia content types can be classified further into sub-categories (Sharda, 2006). Still images can be bi-tonal, grayscale, or full-color continuous tone (as in a photograph) or with sudden variation in the intensity (as in a printed page).

Many text representation techniques and associated digital coding techniques are in use. Similarly, different techniques are in use for coding still images; the choice depends upon the image type and its application. Audio and video are time varying entities and require continuous sampling over a period of time. "Errors and delays introduced at any stage of sampling, encoding, transmitting and decoding of audio and video can lead to reduction in the quality of the final presentation" (Sharda, 2008, p. 46).

For most multimedia content, compression becomes essential to reduce the storage space and transmission time. However, the advantage with uncompressed multimedia content is that it has in-built redundancy; consequently, a few corrupted bits do not change the look and feel of the contents dramatically. On the other hand, compressed media is compacted by eliminating redundancy. Therefore, any errors introduced during transmission or processing show up in the final content, and at times, as severe degradation. This degradation can manifests itself as missing blocks and frames in a digital video, and / or intermittent audio. This can reduce the user's ability to recognize objects in the video; whereas, similar errors in an analog video would lead to 'forgiving' degradation of the image.

Most mobile transmission systems are more error-prone than wired transmission systems, due to noise introduced by other electromagnetic signals present in the environment. How successfully a particular media type can be transmitted over a network depends upon coding and compression techniques; and the applications in which the media is being used. As multimedia content is ultimately for human consumption, what is acceptable quality finally depends upon human perception. "For example, streamed video can accept a few seconds of delay, but live video conferencing becomes rather ineffective if the round-trip delay exceeds even a tenth of a second" (Sharda, 2008, p. 46).

Coding Text

The American Standard Code for Information Interchange (ASCII) has been one of the most widely used text codes. Over the years Extended ASCII codes have been developed to handle special characters and languages other than English. Unicode was developed as code that is independent of platform, program or language. In Unicode a unique 16-bit number is reserved for every character.

A Short Message Service (SMS) 7-bit code was designed to send and receive text messages on mobile phones. The SMS code set was originally developed for the second generation (2G) mobile technology, and it works well with 2G as well as third generation (3G) systems. For text messages the main Quality of Service (QoS) problem can be errors; as bandwidth, delay and delay jitter do not significantly affect the transmission of text messages.

Table 2. Required data volume for various multimedia types (Sharda, 2007)

Information	**Encoded as**	**Size and bandwidth**
TEXT	ASCII	**2 KB** (Kilo Bytes) / page
GRAPHICS	Bit mapped, Still images	**64 KB** / image (B&W) **7.5 MB** (Mega Bytes) / image (color)
AUDIO	8-bit mono 8K samples/s	**64 Kbps** (Kilo bits per second)
AUDIO	16-bit stereo, 44.1K samples/s	**1.4 Mbps** (Mega bits per second)
VIDEO ***Uncompressed*** ***Compressed***	640x560 pixels 24 bits / pixel 30 frames / second MPEG 2 MPEG 4	**258 Mbps** **4-10 Mbps** **9-40 Kbps**

Coding Audio

Traditionally, analog audio signals are digitizing using Pulse Code Modulation (PCM), in which an analog audio signal is sampled at double the rate of the highest frequency in the signal, and each sample is stored using n-bits; generally n=8 for phone quality audio, and n=16 for CD quality audio.

Coding Still Images

The fundamental technique for the coding still digital images is based on dividing an image into a grid of picture elements, called Pixels. Each pixel can be coded with one bit for pure black and white image, or with more than one bits for a gray scale or color image. With n-bits used for coding, each pixel can take 2^n shades of gray or 2^n color hues.

Coding Moving Images

Moving image coding is based on creating upto 30 still image frames in every second, and displaying these at the same rate at which these were captured. Therefore, moving image coding is, to some extent, based on the fundamentals used in still image coding; however, with the additional requirement of building timing relationship between the consecutive frames. Any loss of correct timing relationship between the consecutive frames leads to serious impairments being introduced in the final output.

Advanced Coding Techniques

A number of advancements to the traditional coding techniques have been developed in recent years; some of these are aimed particularly at transmitting multimedia information over wireless network and WSNs. Some of these advanced techniques will be introduced in later sections.

Volume of Multimedia Information

The volume of data required to code and store multimedia content increases as we move from plain text to more expressive image-based content. Table 2 lists the size for some commonly used multimedia content.

A single page of plain text needs a few kilo bytes (KB) of data, however, this increases as pages are formatted in a specific system, e.g. Microsoft Word or Adobe PDF format. The additional formatting information can increase the size to as much as double to that of the original plain text. Plain black and white (B&W) images need as much as tens of KB for data; however, color images need many more mega bytes (MB) of data storage.

Continuously varying multimedia information produces streaming data, i.e. a data stream is generated that produces a certain volume of bits every second. What is called the telephone quality audio, is sampled at 8 kilo samples per second and stored with 8-bit resolution; this generates 64 kilo bits per second (Kbps) of data stream. Whereas the CD quality stereo audio, sampled at 44.1 kilo samples per second and stored with 16-bit resolution generates streaming data of 1.4 mega bits per second (Mbps).

Video is nothing but a sequence of still images concatenated in time. The number of frames required to produce smooth moving images is linked to the persistence of vision of the human eye; i.e. any image projected on the human eye persists for about 40 mille seconds (ms). Therefore, if consecutive images are presented at the rate of 25 frames per second, or higher, the human visual system perceives this as a smooth moving image. Lower frame rates produce images that look like a sequence of still images.

Uncompressed video is nothing but a sequence of digitized still images, and therefore, it generates streaming data at the rate of hundreds of Mbps. To reduce the data volume for video it is compressed in time and space. Compression in space implies that each frame is compressed as if it were a still image. Compression in time finds the correlation between consecutive frames and reduces the data volume by removing redundant information. When video is compressed using the MPEG-2 standard, the data rate reduces to under 10 Mbps. With the MPEG-4 compression standard –designed for video conferencing type of applications– the data rate comes down to less than 40 Kbps, as it uses smaller image size and lower frame rate.

Multimedia Compression

As seen in the preceding section, the total volume of data produced by multimedia information can be rather large, and thus, difficult for the WMSNs to handle effectively. Therefore, compressing this data is often desirable, and at time mandatory. However, compressed data streams become more venerable to impairments caused particularly in WSNs. The compression technique used has a clear bearing on the way multimedia information is affected by factors such as noise, delay and delay jitter.

Lossy and Lossless Compression

Data compression is considered lossless if the original information can be fully regenerated from the compressed information. Generally, lossless compression gives lower compression ratios. Lossy compression reduces the total volume of data stored by removing information that is not significant to the application or for the human perception. Lossy compression generally provides higher compression rations; however, it is likely to be affected more severely by impairments encountered during data transmission on a wireless network.

Audio Compression

One of the most widely used standards for storing and transmitting high quality stereo audio is the MPEG audio Layer 3 compression format called MP3. It is one of three audio coding schemes associated with the MPEG video compression standard, and provides the highest level of compression, achieved with the use of perceptual audio coding. This psychoacoustic compression method removes all redundant and irrelevant parts of a sound signal that the human ear does not hear. MP3 files being small, they are easily transferred across wireless networks (Sharda, 2006), including WMSNs.

For high quality stereo audio, compression becomes highly desirable for transmission over any type of wireless networks. However, this makes the audio signal highly susceptible to transmission impairment. When audio is transmitted as an integral component of a video stream, one has to bear in mind that the human ear is more sensitive to errors than the human eyes. Hearing acts like a differentiator, thereby accentuating any variations, while human eye behave like an integrator, smoothening out variations (Sharda, 1999). Therefore, an audio stream must be given higher priority as compared to text or an image data stream, in particular when transmitted over a wireless network (Sharda, 2006).

Still Image Compression

The compression techniques used for still images depend upon the image type and the application. Some of the commonly used compression standards are JPEG (Joint Photographic Experts Group), GIF (Graphics Interchange Format), and PNG (Portable Network Graphics) (Sharda, 2006).

The JPEG codec (coder–decoder) was designed to be suitable for lossy compression of continuous tone images, such as photographs. GIF codec was designed for use over the Internet for bitmapped pictures with only 256 colors. PNG codec was designed as an enhancement to the GIF codec, while using lossless compression. The PNG codec is an improvement upon the GIF codec, and is available freely.

JPEG 2000 standard was developed by using wavelet transform instead of the Discrete Cosine Transform (DCT), which is used for the JPEG codec. As a result, JPEG 2000 gives higher compression ratios without introducing the blocky and blurry effects introduced by the original JPEG standard. Furthermore, JPEG 2000 allows progressive downloading of image with different resolution, quality, components, or spatial regions, eliminating the need to decompress the entire image before it can be displayed. This feature is particularly useful for WMSNs.

In general, wireless networks have higher errors rates, and lower bandwidth; consequently, these are more severely tested when transmitting digital images. Because JPEG 2000 provides higher compression ratios, it is highly suitable for the low bandwidth wireless networks. However, issues such as error resilience over wireless networks need to be addressed, these have been tackled by the JPEG 2000 Wireless (JPWL) effort (Dufaux & Nicholson, 2004). "Their aim is to standardize tools and methods for efficient transmission of JPEG 2000 images over error-prone wireless networks. One of the techniques developed by JPWL makes the code stream more error resilient by adding redundancy, or by interleaving data" (Sharda, 2008, p.50).

In JPWL, the decoder detects errors, and corrects some, wherever possible. Another technique changes the sensitivity of different parts of the code stream to errors. By adding such techniques to JPWL, JPEG 2000 is being made more errors resilient, making it a good choice for the transmission of digital images and video over wireless media, including WMSNs.

Moving Image Compression

For the transmission of video over wireless networks, codecs have been developed that work with smaller image size and lower frame rate. These codecs fall under the category of video conferencing standards; for the foreseeable future WMSNs more likely to use these videoconferencing standards for transmitting video.

Video conferencing standards such as H.261 have provided the means of transmitting low resolution and low frame rate videos over networks since the 1990s, with a data rate that is a multiple of 64 Kbps. Its enhanced version H.263 (1995) provided a replacement that could work at all bit rates. It was then enhanced as H.263v2 (1998) and H.263v3 (2000). The H.263 codec is similar to

H.261 codec, but with improved performance and error recovery. As these standards were designed to work with data rates in multiples of 64Kbps, these are sometimes called px64 (where p can be 1 to 30) (Sharda, 2008).

The H.263 standard is also the baseline standard for the newer 3G-324M standard that works with 3G wireless networks (Smith & Jabri, 2004). An option in the 3G-324M specification is MPEG-4 AVC, the next generation video coding standard. This was approved in 2003 and has been called by various names: MPEG-4 AVC or ITU-T H.264, or simply Advanced Video Coding (AVC). The MPEG-4 AVC codec doubles the compression ratio, but with the same picture quality (Navakitkanok & Aramvith, 2004). Therefore, it is far better than the earlier standards for WMSNs. It also offers better resilience to transport errors, improved bit rate scalability, and stream switching for transmission over a less reliable network infrastructure, which is inherent in WMSNs.

The Motion JPEG 2000 standard performs video compression by applying intra-frame compression only. This makes Motion JPEG 2000 very well suited for video transmission over wireless networks, including WMSNs. Motion JPEG 2000 outperforms MPEG-4 vis-à-vis coding efficiency, error resilience, complexity, scalability and coding delay (Tabesh, Bilgin, Krishnan, & Marcellin, 2005).

Quality of Service (QoS) Parameters

Quality of Service (QoS) for multimedia transmission can be categorized under three aspects, namely Temporal Facet, Quality Resolution, and Cost Factor (Georgievski & Sharda 2003). For each of these categories, some parameters are independent, i.e. these can be selected independent of each other. The dependent parameters then can be calculated from the values chosen for the independent parameters.

Temporal Facet

The Temporal Facet includes parameters that affect the timing relationships between some related objects in a multimedia presentations, including parameters related to time for a particular session. To maintain good quality for a presentation one needs to maintain an accurate timing relationship between the related objects in a media stream, e.g. consecutive frames in a video, and between related objects in two connected streams, e.g. the video and the audio. Some of the important independent and dependent temporal parameters are (Georgievski & Sharda 2003):

Independent Parameters

- **Jitter:** The acceptable instantaneous difference between the desired presentation times and the actual presentation times.
- **Skew:** The acceptable average difference between the desired presentation rate and the actual presentation rate.
- **Packet Error Rate (PER):** The acceptable number of corrupted packets per unit time.
- **Bit Error Rate (BER):** The acceptable number of corrupted bits per unit time.
- **Session Time (hh:mm:ss):** The run length time for a particular session taking place.

Dependent Parameters

- **Delay:** There are two types of delays:
    - **Response Time Delay** is the roundtrip delay encountered in an interactive activity.
    - **End-to-End Delay** is the sum of all delays encountered in the processing performed on the end systems and the network.

Quality Resolution

The Quality Resolution includes parameters related to the specification of the quality for the multimedia presentation. The independent and dependent parameters related to the quality are the following (Georgievski & Sharda, 2003):

Independent Parameters

- **Multimedia Standard:** On selecting a Multimedia Standard, the following parameters will be set according the standard's specification.
- **Picture Resolution:** The horizontal and vertical pixels per frame.
- **Frame Rate:** Defines frames per second (fps).
- **Color Resolution:** Bits per pixel (bpp).
- **Audio Channels:** The number of channels to be used for the audio stream.
- **Audio Sampling Resolution:** The number of bits used for storing the amplitude of the audio samples.
- **Audio Sampling Rate:** The number of samples taken per second for the audio stream.
- **Compression Ratio:** The compression ratio for corresponding multimedia objects. Different values may be used for audio, video and still images.

Dependent Parameters

- **Application Bit Rate (ABR):** Specifies the rate at which the application generates information bits. ABR can either be constant, or variable.
    - **For Constant Bit Rate (CBR)**, the Independent Parameters within the QoS Resolution aspect can be used to calculate the ABR.
    - **For Variable Bit Rate (VBR)** (which is generated by compressed media), the calculated value for the ABR is dependent on the Multimedia Standard and the compression ratios.
- **Burstiness Ratio:** The traffic Burstiness generated by a compressed media stream. This value is calculated by dividing Mean Bit Rate (MBR) by the Peak Bit Rate (PBR).
- **Throughput:** The effective rate of transmission of information bits. The throughput is calculated by dividing the number of information bits transferred by time taken to transfer bits.
- **Required Bandwidth:** The bandwidth required for a particular multimedia transmission/session.
- **File Size:** The size, in Megabytes (MB), for the stored multimedia objects.

Cost Factor

The Cost Factor includes parameters related to the cost of transmitting multimedia information over a communication network. There are two main costs: 1. Bandwidth, 2. Monetary. Independent and dependent parameters included in this aspect of QoS are as follows (Georgievski & Sharda 2003):

Independent Parameters

- **Bandwidth provided (bps):** The Bandwidth that the service provider delivers.
- **Time Charge ($):** The cost per unit of session time based on a particular bandwidth provided.
- **Volume Charge ($):** The cost per transmitted/received megabyte.

Dependent Parameters

- **Total Cost ($):** The Total Cost depends upon the particular cost model that the service provider employs.

Quality of Service Issues in Transmitting Mobile Multimedia

The Quality of Service (QoS) issues related to the transmission of multimedia information over WMSNs have much in common with the similar issues for the other wireless networks. Therefore, in the following we first discuss issues related to multimedia information transmission over wireless networks and then consider the WMSNs. Though, WMSNs present greater challenges due their resource restrictions and ad-hoc networking protocols.

Transmission of digital multimedia content over packet switching networks remains a fundamental challenge. This situation has improved over the recent years, as more data transmission bandwidth becomes available; however as the data volume increases a more systematic approach is necessary to use multimedia information more effectively (Sharda, 2006).

When transmitting multimedia information, one needs to transmit the core multimedia content as well as associated meta-information, such as: information arrival intimation, e.g. bell, ring, beep, vibrate; abbreviated information such as subject and caller ID. Which type of meta-information is required and how it is used, depends upon the core information type, the application, user preferences, user devices and the required QoS (Sharda, 2006).

Digitized text, audio and video can be transmitted over digital networks with much greater ease using packet switching networking technology. However, with packet switching technology it is possible for different packets to encounter different delays during transmission. This creates various artifacts in the media content that can be detrimental to the audio or the video as it is presented.

Text codes do not need high bandwidth, and can tolerate delays and delay variance (i.e. delay jitter and skew) for transmission. Hence, maintaining the required QoS when transmitting text is generally not a problem.

The QoS requirements for bi-directional audio transmission, as required for a conference call, are more stringent than those for unidirectional audio transmission. For unidirectional data transmission, buffering can be used to remove any delay jitter. However, for conversational audio applications excessive buffering leads to unacceptable delay.

Still images can be transmitted easily over wired networks. However, over wireless networks, especially over WMSNs, the delay involved in transmitting still images will depend upon the bandwidth of the network. Nonetheless, delay jitter will not have any substantial effect on still image transmission.

Delivery of the required QoS is most important when transmitting video. Video conferencing is the most difficult application vis-à-vis data transmission over mobile systems. Any delay, and in particular delay jitter, will introduce significant artifacts in the video images being displayed.

Challenges in Using Multimedia Information

As observed earlier, transmission of multimedia information becomes challenging when it has to be transmitted over mobile networks. As WMSNs have even less bandwidth, memory, processing speed and battery power than the standard infrastructure based mobile networks, and use ad-hoc routing protocols, transmission of multimedia information becomes even more challenging.

Multimedia content is ultimately to be viewed by human beings. Communication channels and protocols consider factors such as bandwidth, delay, jitter, skew, and error rate. However, the application users view these factors as these manifest into shaky picture, interrupted transmission, and blocking artifacts (Sharda, 2006).

Multimedia applications require more elaborate communication protocols than those used for simple data transmission. As multimedia information is often compressed, it becomes more venerable to the introduction of undesirable artifacts as it is transmitted over mobile networks. Even a few bits in error can destroy complete blocks of information that has been compressed with JPEG or MPEG codecs. Newer codecs such as JPEG 2000 can overcome this problem to some extent, by using techniques that lead to graceful degradation of the content, i.e. few error bits lead to artifacts that are barely perceptible.

CURRENT AND FURTHER RESEARCH DIRECTIONS

Having explored the nature of multimedia content, its coding techniques and the impact of factors such as delay and jitter on its QoS, we can now explore some of the special requirements for creating WMSNs (Akyildiz, Melodia, & Chowdhury, 2007). These requirements point to the opportunities for future research.

In-Network Distributed Processing

In the past, the task of processing multimedia content, and the task of transmitting it over the networking infrastructure have been tackled as two independent problems, by and large. Some efforts in combining these are based on joint source-channel coding and channel-adaptive streaming (Akyildiz, Melodia, & Chowdhury, 2007).

Hence, new research that addressed the content delivery aspects along with data transmission aspects has received some attention, and we need even more in the future. This research needs to consider the characteristics of the source, as well as cross-layer interactions, particularly among lower layers of the communications protocol stack.

These new protocols need to recognize that the capture, processing and delivery of multimedia content are not independent tasks, and their interaction has a major impact on the level of QoS that can be delivered. This can be achieved more readily in the WMSNs that allow in-network processing of multimedia data, i.e. the processing is done within the network and not just on the end nodes. This is akin to distributed processing. Furthermore, it would allow distribution of the buffer required for removing delay jitter and skew from the data stream.

Higher Bandwidth

Multimedia content, especially bi-directional streaming video requires bandwidth that is an order of magnitude higher than that is available on the current WSNs. Hence, high data rate and low-power consumption transmission techniques need to be leveraged. In this respect, the ultra wide band (UWB) transmission technique seems particularly promising for WMSNs (Akyildiz, Melodia & Chowdhury, 2007).

Coding Techniques

Coding techniques used for video compression are based the assumption that to reduce the bit rate generated by the codec needs to exploit statistics embodied in the source information. This is done at intra-frame level and inter-frame level. Intra-frame compression techniques first break images into blocks and then reduce redundancy within each frame block-by-block. Inter-frame compression uses predictive coding or motion estimation techniques to locate redundancy among consecutive frames.

Block-based intra-frame coding and inter-frame predictive coding require complex encoders, and therefore powerful processors; requiring high energy consumption. This is not the most suitable methodology for WMSNs. A new possibility is to find the balance between a complex and a simple decoder by using distributed source coding. The aim is to use source statistics in the decoder, and shift the complexity from the source end, allowing the use of simple encoders in the sensor. This methodology holds much promise for WMSNs. This is especially true for a network of video sensors, where it is not possible to use current video encoders due to processing and energy constraints (Akyildiz, Melodia, & Chowdhury, 2007).

Advanced Video Coding, Compression and Communication Techniques

In recent years some advanced video coding, compression and communication (CCC) techniques have been developed. Some of these techniques exploit the synergy between two or more aspects of the CCC aspects related to video capture and transmission.

Distributed Video Coding

"Distributed coding is a new paradigm for video compression, based on Slepian and Wolf 's and Wyner and Ziv's information- theoretic results from the 1970s" (Girod et al., 2005). In the currently used video coding techniques, such as those developed by MPEG and ITU-T, the encoder exploits the source signal statistics. The Distributed Video Coding (DVC) algorithms also exploit the source statistics, however, at the decoder end. This reduces the processing load on the video capturing node in a WMSN.

Pereira et al. (2008) list the main advantages and the potential disadvantages of Distributed Video Coding (DVC) as:

Advantages of DVC

- **Lower encoding complexity:** DVC can shift computational complexity from the encoder to the decoder. Some DVC codecs provide better rate-distortion (R-D) performance and encoder complexity tradeoffs, as compared to H.264 or AVC Intra coding techniques.
- **Lower size and weight devices:** With the reduced complexity of the encoder, the size and the weight of the devices that capture the video can also be reduced.
- **Lower encoding power consumption:** The lower encoder complexity can also reduce the power consumption, leading to longer battery life and / or reduced battery size, and / or enhanced transmission range.
- **Improved error resilience:** The traditional predictive video coding approach is significantly affected by the errors encountered in the transmission channel. The DVC approach can be more resilient to such errors.

Disadvantages of DVC

- **Higher decoding complexity:** DVC shifts the complexity from the encoder to the decoder. However, the required decoding complexity is rather high; which can be a significant drawback in applications requiring decoding in real-time.
- **Lower compression efficiency:** DVC has not achieved the same level of compression efficiency as the predictive coding schemes such as the H.264/MPEG-4 Advanced Video Coding standard –H.264/AVC, (Wiegand & Sullivan, 2007).

There is a growing interest in conducting further research on DVC based codecs and their applications. It is expected that with further research into the DVC methodology its performance will improve, making it a viable video compression technique for WMSNs (Pereira et al., 2008).

Compressive Sampling

Image compression algorithms compress high-resolution images to substantially reduce the raw data set. Compressive sampling (CS) is technique that avoids creating a large data set in the first place. It aims to incorporate data compression into the data acquisition process itself (Romberg, 2008).

The basic premise underpinning CS is that one can recover certain signals and images with far fewer samples than dictated by the traditional sampling theorems. To achieve this, CS uses two principles: Sparsity and Incoherence (Candès & Wakin, 2008). Using these two concepts one can design efficient sampling algorithm to capture the useful information embodied in a sparse signal, directly in a compressed format. Therefore, such sampling algorithms are useful for capturing multimedia information that needs to be transmitted over WMSNs.

Network Coding

Network coding implies that each node in a network can perform some computation (Chou & Wu, 2007). Traditional routers merely forward messages from the source to the destination. However, with network coding as messages travel over the network these can be encoded and / or recoded within the network. These messages are then decoded at the final destination. With further research, network coding is likely to find efficacious application in WMSNs, thereby improving their throughput and robustness (Thomos & Frossard, 2009).

Power Consumption

One of the main challenges in developing viable WMSNs is the reduction of power consumption. The imperative to make WSNs small and light-weight contradicts the need for more power required for multimedia sensors, higher processing power, and high transmission rates. In traditional sensor nodes energy consumption is often dominated by the communication functionality; however, in WMSNs more power is required for content capture and processing as well.

He, Cheng, and Chen (2008) have used power-rate-distortion (P-R-D) optimization to develop a new approach for energy minimization in WMSNs. They exploit energy tradeoff between video encoding and wireless communication, by using non-stationary characteristics of input video data. They demonstrate, analytically and experimentally, how to achieve significant energy savings. Their adaptive scheme can estimate P-R-D parameters and perform online resource allocation and energy optimization for real-time video encoding. By using the proposed P-R-D optimization methodology, they can reduce energy consumption of video encoding by up to 50%; this is especially possible in delay-tolerant video communication applications.

Integration with Other Networking Technologies

The TCP/IP protocol has become the backbone of all networking systems. Therefore, it becomes a commercial imperative to development WMSNs communication protocols that can interoperate with the Internet using the TCP/IP architecture. However, Akyildi et al. (2007) suggest the "the characteristics of WSNs rule out the possibility of all-IP sensor networks and recommend the use of application level gateways or overlay IP networks as the best approach for integration between WSNs and the Internet".

CONCLUSION

The first generation WSNs began appearing in the 1980s, and now the third generation WSNs are becoming capable of handling multimedia information, morphing into Wireless Multimedia Sensor Networks (WMSNs). Factors such as bandwidth, delay and delay jitter have a significant effect on the Quality of Service (QoS) that can be delivered in transmitting multimedia information over Wireless Sensor Networks (WSNs). For successfully transmitting multimedia information WSNs require high bandwidth, lower delay and even lower delay jitter. Further research needs to focus on reducing the power consumption for media capture, compression, and information transmission in WMSNs.

REFERENCES

Akyildiz, I. F., Melodia, T., & Chowdhury, K. R. (2007). A survey on wireless multimedia sensor networks. *Computer Networks*, *51*(4), 921–960. doi:10.1016/j.comnet.2006.10.002

Akyildiz, I. F., Melodia, T., & Chowdhury, K. R. (2008). Wireless multimedia sensor networks: Applications and testbeds. *Proceedings of the IEEE*, *96*(10), 1588–1605. doi:10.1109/JPROC.2008.928756

Candès, E., & Wakin, M. (2008). An introduction to compressive sampling. *IEEE Signal Processing Magazine*, *25*(2), 21–30. doi:10.1109/MSP.2007.914731

Chong, C. Y., & Kumar, S. P. (2003). Sensor networks: Evolution, opportunities, and challenges. *Proceedings of the IEEE*, *91*, 1247–1256. doi:10.1109/JPROC.2003.814918

Chou, P. A., & Wu, Y. (2007). Network coding for the Internet and wireless networks. *IEEE Signal Processing Magazine*, *24*(5), 77–85. doi:10.1109/MSP.2007.904818

Dufaux, F., & Nicholson, D. (2004). JPWL: JPEG 2000 for wireless applications. *Proceedings of the Society for Photo-Instrumentation Engineers*, *5558*, 309–318.

Georgievski, M., & Sharda, N. (2003). *A taxonomy of QoS parameters and applications for multimedia communications*, International Conference on Internet and Multimedia Systems and Applications (IMSA2003), Aug 13-15, Hawaii, USA.

Girod, B., Aaron, A. M., Rane, S., & Rebollo-Monedero, D. (2005). Distributed video coding. *Proceedings of the IEEE*, *93*(1), 71–83. doi:10.1109/JPROC.2004.839619

He, Z., Cheng, W., & Chen, X. (2008). Energy minimization of portable video communication devices based on power-rate-distortion optimization. *IEEE Trans. on Circuits and Systems for Video Technology*, *18*(5), 596–608. doi:10.1109/TCSVT.2008.918802

Kay, R., & Mattern, F. (2004). The design space of wireless sensor networks. *IEEE Wireless Communications*, *11*(6), 54–56. doi:10.1109/MWC.2004.1368897

Navakitkanok, P., & Aramvith, S. (2004). *Improved rate control for advanced video coding (AVC) standard under low delay constraint*, International Conference on Information Technology: Coding and Computing (ITCC'04) (vol. 2).

Nikoletseas, S., Chatzigiannakis, I., Euthimiou, H., Kinalis, A., Antoniou, A., & Mylonas, G. (2004). Energy efficient protocols for sensing multiple events in smart dust networks, *Proceedings of the 37th Annual Simulation Symposium (ANSS'04)*.

Pereira, F., Torres, L., Guillemot, C., Ebrahimi, T., Leonardi, R., & Klomp, S. (2008). Distributed video coding selecting the most promising application scenarios. *Signal Processing Image Communication*, *23*(5), 339–352. doi:10.1016/j.image.2008.04.002

Romberg, J. (2008). Imaging via compressive sampling. *IEEE Signal Processing Magazine*, *25*(2), 14–20. doi:10.1109/MSP.2007.914729

Sarigiannidis, G. (2006). *Localization for Ad-Hoc wireless sensor networks*, Master of Science Thesis, The Technical University of Delf, The Netherlands.

Sharda, N. (1999). *Multimedia information networking*. Upper Saddle River, NJ: Prentice Hall.

Sharda, N. (2007). The role of mobile multimedia systems in digital business ecosystems, *IEEE-DEST 2007: IEEE International Digital EcoSystems Technologies Conference,* Cairns, Australia (pp. 523-528).

Sharda, N. (2008). Quality of service issues in mobile multimedia transmission. In Karmakar, G., & Dooley, L. (Eds.), *Mobile Multimedia Communications: Concepts, Applications and Challenges* (pp. 45–63). Hershey, PA: Idea Group Inc.

Sharda, N. (2009). *Multimedia and communication technologies in digital ecosystems*, 7th International Conference on Information, Communications and Signal Processing, Macau.

Smith, J. R., & Jabri, M. A. (2004). The 3G-324M protocol for conversational video telephony. *IEEE MultiMedia, 11*(3), 102–105. doi:10.1109/MMUL.2004.18

Tabesh, A., Bilgin, A., Krishnan, K., & Marcellin, M. W. (2005). JPEG2000 and motion JPEG2000 content analysis using codestream length information, In *Proceedings of the Data Compression Conference (DCC'05)* (pp. 329-337).

Thomos, N., & Frossard, P. (2009). Network coding: From theory to media streaming. *Journal of Communications, Special Issue on Multimedia Communications, Networking, and Application, 4*(9), 628–639.

Wiegand, T., & Sullivan, G. J. (2007). The H.264/AVC video coding standard. *IEEE Signal Processing Magazine, 24*(2), 148–153. doi:10.1109/MSP.2007.323282

KEY TERMS AND DEFINITIONS

Compressive Sampling (CS): A sampling technique that avoids creating a large data set in the first place by incorporating data compression into the data acquisition process.

Distributed Video Coding (DVC): A video coding algorithms in which the source statistics is exploited at the decoder end.

In-Network Distributed Processing: A methodology for multimedia information capture and transmission in which processing is done also within the network and not just on the end nodes.

Lossless Compression: Compression technique that allows original information to be regenerated; it gives lower compression ratios.

Lossy Compression: Compression technique that reduces the total volume of data stored by removing information that is not significant to the application or for the human perception; it gives higher compression ratios.

Multimedia Surveillance Sensor Networks: Wireless Multimedia Sensor Networks (WMSNs) that use miniature video cameras to create surveillance systems.

Network Coding: A coding technique in which each node in a network performs some computation.

Quality of Service (QoS): The quality with which multimedia information can be transmitted; it can be categorized under three aspects: Temporal Facet, Quality Resolution, and Cost Factor.

Wireless Multimedia Sensor Networks (WMSNs): A WSN capable of capturing multimedia information, as well as storing and processing this information.

Wireless Sensor Network (WSN): A system of distributed nodes connected by wireless communication technologies.

Chapter 7
Energy-Efficient Backoff Algorithms for Wireless Sensor Networks

S. Mehta
Inha University, Korea

K.S. Kwak
Inha University, Korea

ABSTRACT

A prime goal of wireless sensor networks (WSNs) is to minimize energy consumption than high channel efficiency and low channel access delay. Existing MAC protocols for WSNs reduce energy consumptions by introducing variation in an active/sleep mechanism, but they cannot save energy during the execution of backoff algorithm. In this chapter, we study and compare the different backoff algorithms for wireless sensor networks. We also use the concept of a geometrically increasing probability distribution for contention process. This allows us to introduce improved backoff (IB) algorithm for energy efficient MAC protocol in WSNs, where binary exponential backoff (BEB) based algorithm is widely used. With the help of numerical results we show that the IB gives edge over BEB in throughput, channel access delay, and energy efficiency under varying traffic conditions.

INTRODUCTION

Communication in wireless sensor networks is divided into several layers. Medium Access Control (MAC) is one of those layers, which enables the successful operation of the network. MAC protocol tries to avoid collisions by not allowing two interfering nodes to transmit at the same time. The main design goal of a typical MAC protocols is to provide high throughput and QoS. On the other hand, wireless sensor MAC protocol gives higher priority to minimize energy consumption than QoS requirements. Energy gets wasted in traditional MAC layer protocols due to idle listening,

DOI: 10.4018/978-1-61350-153-5.ch007

collision, protocol overhead, and over-hearing (I.F Akayildiz et.al, 2002; W. Heidemann et. al., 2002). There are some MAC protocols that have been especially developed for wireless sensor networks. However, we discussed a very few MAC protocols here just to present some examples of Carrier Sense Multiple Access/Collision Avoidance (CSMA/CA) based MAC protocols for WSNs. Typical examples include S-MAC, T-MAC, and H-MAC (W. Heidemann et. al., 2002; T.V.Dam et.al., 2003, S.Mehta et.al.,2009). In large amount of literature authors proposed new MAC protocols for WSNs but in this chapter our approach is to concentrate on backoff algorithm than a complete MAC protocol. In this chapter we introduce an energy efficient backoff algorithm, which is easy to integrate with existing energy efficient MAC protocols for WSNs.

Before going into more details about the backoff algorithms, it is useful for readers to take a glimpse at some important sensor networks MAC protocols.

S-MAC is proposed to improve energy efficiency in wireless sensor networks. It divides the time into large frames. Every frame has two parts: an active part (on time) and a sleeping part. A node turns off its radio during the sleep time to preserve the energy. During the active time, a node can communicate with its neighbors and transmits the queued packets during the sleeping time. Hence, S-MAC saves the unnecessary waste of energy on idle listing. Periodic sleep may result in high latency (for mutihop sensor networks) and low throughput problems (Heidemann et. al., 2002). Timeout-MAC protocol is proposed to enhance the performance of S-MAC under variable traffic conditions. It defines the minimum time out timer TA. T-MAC works similar to S-MAC with one difference of time out timer. In T-MAC, active time ends when no activation has occurred for a time out timer TA. Time-out timer may result in early sleeping and low throughput problems (T.V.Dam et. al., 2003). Data gathering MAC falls into duty cycle based MAC protocols category. The main aim of D-MAC is to achieve low latency as well as energy-efficiency. D-MAC is an improved slotted aloha protocol where slots are assigned to the sets of nodes based on data gathering tree (Parent-child topology). In D-MAC, low latency is achieved by assigning subsequent slots to the nodes that are successive in the data transmission path. However, D-MAC works well only for tree based structure and may cause a collision problem (G. Lu et. al., 2004). Pattern-MAC is a 'time slotted' protocol. It adaptively determines the sleep-wake up schedules for a node based on its own traffic and the traffic patterns of its neighbors. In P-MAC, a node gets information about the activity in its neighborhood before hand through patterns. Based on these patterns, a node can put itself into a long sleep for several time frames when there is no traffic in the network. If there is any activity in the neighborhood, a node will know this through the patterns and will wake-up when required. Hence, P-MAC saves more energy compared to S-MAC without compromising on the throughput. But it is worthwhile to mention about complexity, collision, and overhead problems in P-MAC (T. Zheng et. al, 2005). In (S.Mehta et.al., 2009) authors proposed a hybrid MAC to achieve the energy efficiency along with good QoS. H-MAC is based on IEEE 802.11's PSM mode and slotted aloha. In H-MAC, time is divided into large frames, every frame has two parts: an active part (on time) and a sleeping part. Active part is further divided into N slots and can dynamically assign to nodes. However, H-MAC protocol requires a good synchronization method to work efficiently. In most of the sensor networks MAC protocols, to maximize the battery lifetime, implement the variation of active/sleep mechanism. With this mechanism most of the MAC protocols trades networks QoS for energy savings while H-MAC protocol reduces the comparable amount of energy consumption along with maintaining good network QoS such as latency, throughput, and channel utilization. However, most of the sensor networks MACs' backoff algorithm is similar to that of the IEEE 802.11 Distributed Coordinated Function (DCF), which is based on Carrier Sense Multiple Access with binary exponential backoff (BEB) based algorithm as Collision Avoidance (CSMA/CA) Mechanism. Generally, the energy consumption of CSMA/CA based backoff scheme is high when nodes are in backoff procedure and in idle mode. Moreover, a node that successfully transmits resets it Contention Window (CW) to a small, fixed minimum value of CW. Therefore, the node has to rediscover the correct CW, wasting channel capacity and increase the access delay as well. From (W. Heidemann et. al., 2002; T.V.Dam et.al., 2003, S.Mehta et.al.,2009; G.Lu et. al., 2004; T. Zheng et. al, 2005), we compare some of the popular sensor networks MAC protocols with respect to different parameters, and summarize them as shown in Table 1.

BACKOFF ALGORITHMS

As mentioned earlier, mobile ad hoc networks have two major problems, the shared wireless channel and power saving. When designing the network, these two factors raise the need for an optimum usage of the medium via implementing a suitable Backoff algorithm as a part of the MAC protocol. The choice of the backoff technique affects the throughput and the delay over the network. For an easier understanding of the general form of Backoff functionality, this study divides Backoff algorithms into two main categories; static and dynamic backoff algorithms.

Static Backoff Algorithms: Some researchers (J.Deng et. al., 2003) have proposed using an optimal fixed value as backoff period suggesting a backoff period of the form:

```
Backoff Timer =I, where I is an in-
teger
```

In spite of the fact that the value of I can be carefully chosen depending on many factors; such as the number of nodes in the network, having a fixed value can work under a certain scenario for a specific network topology. In the case of Ad-hoc networks, the major challenges would be mobility and dynamic topology, i.e. positions of nodes within the network area.

Dynamic Backoff Algorithms: In the second type of backoff algorithms, backoff periods are changed depending on many factors. The most common factor used is the result of last attempt of transmission by the node requesting channel access. The input of the formula is the current size of Contention Window (CW) and the result of this formula is the new size of Contention Window (CW_{new}) CW_{new} is limited between a maximum value and a minimum value referred to as CW_{max} and CW_{min} respectively. CW_{new} is used then to randomly choosing the value of Backoff timer

Table 1. Comparison of different MAC protocols for sensor networks

Protocol	S-MAC	T-MAC	D-MAC	P-MAC	H-MAC
Time-Sync.	Yes	Yes	Yes	Yes	Yes
Point to Point	Suitable	Suitable	Not-Suitable	Suitable	Suitable
Broadcast	Not-Suitable	Not-Suitable	Not-Suitable	Suitable	Suitable
Convergecast	Not-Suitable	Not-Suitable	Suitable	Suitable	Suitable
Mobility	Not-Suitable	Not-Suitable	Not-Suitable	May be	May Be
Type	CSMA	CSMA	TDMA/S. Aloha	Slotted Aloha	CSMA/S. Aloha
Adaptive to change	Ok	Good	Weak	Good	Good
Half/Full Duplex	Half	Half	Full	Full	Full

(BackoffTimer) according to respective backoff algorithm.

Some Important Backoff Algorithms: As we discussed above Backoff algorithm has a notable impact on the overall performance of MAC layer, so many researchers have proposed their backoff algorithms for Ad-hoc/Sensor networks. Here we present some of the important backoff algorithms in brief, for more information interested readers are referred to respective algorithm at mentioned references.

Binary Exponential Backoff (BEB): This is the basic and widely accepted backoff algorithm for MAC protocols. In the BEB algorithm, the contention window is doubled every time a node experiences a packet collision. If a node is successful in its packet transmission, the contention window is reset to the minimum value. In order to avoid the contention window from growing too large or shrinking too small, two bounds CWmin and CWmax are used (IEEE LAN Standard, 1997).

Double Increment Double Decrement (DIDD): The authors in (P. Chatzimisios et.al., 2005), points out that when the number of contending stations increases the sudden reduction of the contention window when success in BEB can lead to significant performance degradation. So DIDD suggests that if a packet collides, DIDD will double the contention window to reduce the probability of a packet collision as BEB does. However, in the case of a successful packet transmission, DIDD will halve the CW (will not go back to CW_{min}) in order to avoid potential future packet collisions.

The Fast Collision Resolution (FCR): The authors in (Y. Kwon et. al., 2003) is proposed to use a much smaller initial (minimum) contention window size CWmin and a much larger maximum contention window size CWmax in order to improve the channel busy, the FCR will increase the contention window size of a station when it is in either collision state or deferring state; And when the channel is idle, FCR will reduce the back-off timers exponentially faster after a prefix number of consecutive idle slots. The prefix number is set to [(CWmin+1)×2-1].

Multiple Increase Linear Decrease (MILD): In (V. Bharghavan et. al., 1994) authors proposed the idea of MILD. If a packet collides, the contention window is increased by a multiplicative factor (1.5); and if a packet transmission is successful, the contention window is decreased by one. In this way, MILD can provide a better fairness for different contending flows in a dense network. But the linear decrease of contention window also brings the throughput down in a sparse network.

Exponential Increase Exponential Decrease (EIED): In (N. Song et. al., 2005) authors proposed the idea of EIED. The goal of a backoff algorithm is to minimize the probability of transmission collision by randomizing the transmission timings: a node chooses a random transmission opportunity within a contention window. The CW is dynamically adjusted depending on the collision history. In EIED, the contention window size is exponentially increased by a backoff factor r_I>1 whenever a packet is involved in a collision, and is exponentially decreased by a backoff factor r_D>1 if a packet is transmitted successfully.

From the above discussion, we compare these backoff algorithms with respect to different parameters, and summarize them as shown in the Table 2[2].

In this chapter we aim to study the influence of backoff algorithms on performance of wireless sensor networks. In the following section we present the performance of six backoff algorithms (BEB, DIDD, FCR, MILD, EIED, and IB) through simulator. We use the MATLAB, NS-2 and OPNET[3] simulators for evaluation of different backoff algorithms. All nodes communicate over identical half-duplex wireless bandwidth of 2 Mb/s. we assume an ideal channel condition and transmission errors are only due to packet collision The transmission range is 250 meters and rest of the parameters are listed in Table 3

Figure 1 presents the comparative performance of six backoff algorithms in single-hop topology

Table 2. Comparison of different backoff algorithms

Backoff Algorithm	BEB	DIDD	FCR	MILD	EIED	IB
CW Size	Variable	Variable	Variable	Variable	Variable	Fixed
Transmission probability	Uniform Distribution	Uniform Distribution	Uniform Distribution	Uniform Distribution	Uniform Distribution	Geometrical Distribution
Adaptability	Not for all	Not for all	Not for all	Not for all	Not for all	Possible for all
Energy Efficiency	Not-Suitable	Not-Suitable	Not-Suitable	Not-Suitable	Not-Suitable	Suitable
Fairness	Poor	Poor	Need Extra Mechanism	Moderate	Poor	Need Extra Mechanism
QoS Support	Moderate	Moderate	Moderate	Poor	Moderate	Good
Delay	Poor	Poor	Good	Poor	Good	Good
Complexity	Simple	Difficult	Difficult	Difficult	Difficult	Simple
Scalability	Medium	Medium	Small	Medium	Medium	Large

Table 3. Simulation Parameters

Parameters	Values
CW_{min}	8~32
ACK Packet Size	24 Bytes
Data Packet Size	1024 Bytes
Nodes	5~50
Data Rate	2 Mbps
Transmitting Energy	50x10^{-6} J/Bit
Idle/listening Energy	75x10^{-6} J/Bit

with saturated traffic condition. It shows that the backoff algorithm affects the throughput performance evidently. The throughput is improved when IB, FCR, and DIDD algorithms are used. IB improves the channel capacity by nearly 40%. Because IB uses non-uniform distribution for the contention process which reduce the chance of collision to minimum level consequently achieves higher throughput. More details on IB algorithm will be discussed later in this chapter. FCR and DIDD provide relatively small contention windows in rest of the back off algorithms. So it cost less for backoff time and the throughput is improved. BEB, EIED, and MILD algorithms show almost similar throughput performance. We can also see that as the number of nodes increases, the throughput of all six algorithms decreases. It's because when the contending nodes are increasing, collision probability increases to highest. As the collision probability increases, backoff algorithms take long time and number of retries to resolve it. So when number of contenders is high, throughput goes down to lower value.

Figure 2 shows the access delay performance of all six back off algorithm. Here, we define the access delay as the time from the moment a packet is ready to be transmitted to the moment the packet starts its successful transmission. As IB has fixed and smallest backoff interval time along with minimum collision property, it shows smallest access delay time. Next to IB, FCR shows smaller access delay, because it always chooses small back off time which includes two more collisions in dense networks. EIED provides the fair access delay but the throughput performance is worse than DIDD. Because DIDD uses a slow decrease in contention counter compared to EIED. BEB, MILD, and DIDD algorithms show almost similar access delay performance due to their backoff policy.

Figure 3 illustrates the energy efficiency performance of all six backoff algorithms. Here we define the energy efficiency as energy required to successfully transmit one bit of data packet. As

Figure 1. Throughput Performance of different back off algorithms

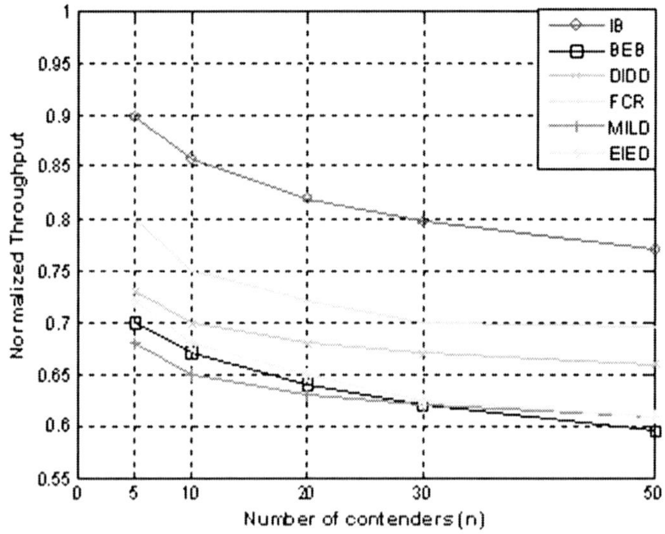

Figure 2. Average access delay Performance of different back off algorithms

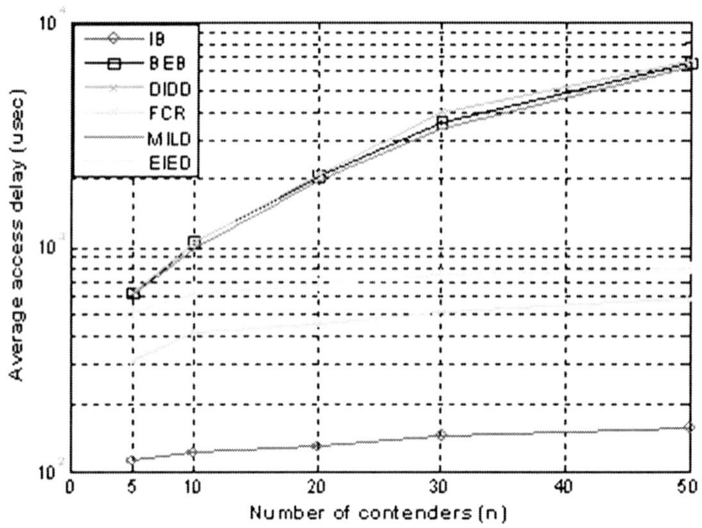

collision and number of retransmissions increases, more energy gets waste. IB shows best energy efficiency due to its fixed and short contention cycle. Moreover, its unique characteristic to avoid collision amplifies the result. FCR and DIDD algorithms maintain fairly better energy efficiency than BEB, MILD, and EIED algorithms because of their collision avoidance process and small contention window size.

Here we compared six different backoff algorithms such as BEB, MILD, DIDD, FCR, EIED, and IB with different performance metrics. These results showed the impact of backoff algorithms on sensor networks performance and the advantage

Figure 3. Energy Efficiency of different backoff algorithms

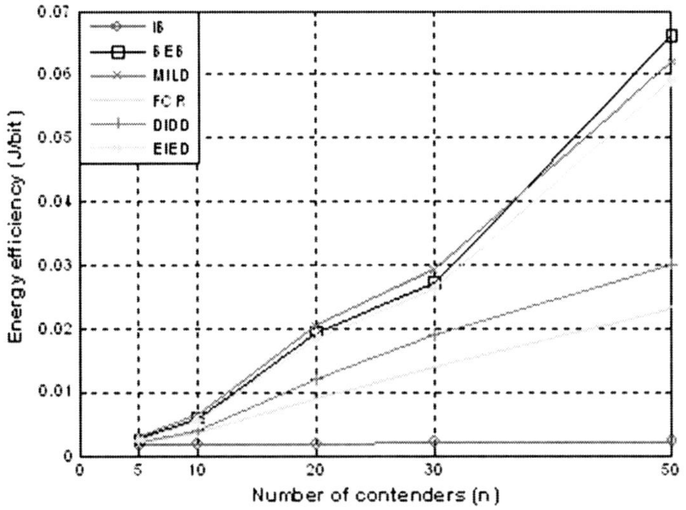

to use IB as backoff algorithm in sensor MAC protocol.

WSN MAC Protocol

In this section, we 'consider a MAC protocol' with active/sleep duty cycle4 to minimize the energy consumption of a node. In this MAC protocol time is divided into super-frames, and every super frame into two basic parts: active part and sleep part, as shown in Figure 4. During the active part a node tries to contend the channel if there is any data in buffer, and turn down its radio during the sleeping part to save energy. With this frame work of 'consider MAC Protocol' we can easily cover the major MAC protocols available for the sensor networks. In our opinion, it is better to consider a general frame work of a MAC protocol which can represent a complete branch of sensor MAC protocols than any individual sensor MAC protocol. So now onwards we will compare this general frame structure of a MAC protocol for any comparison with IB protocol. It is also worth to mention that we choose BEB algorithm with carrier sense multiple access with collision avoidance mechanism in 'consider MAC protocol' for medium contention due to it simplicity and widely acceptance in research literature as well as in the different standard documents.

Figure 4. Active/Sleep Mechanism

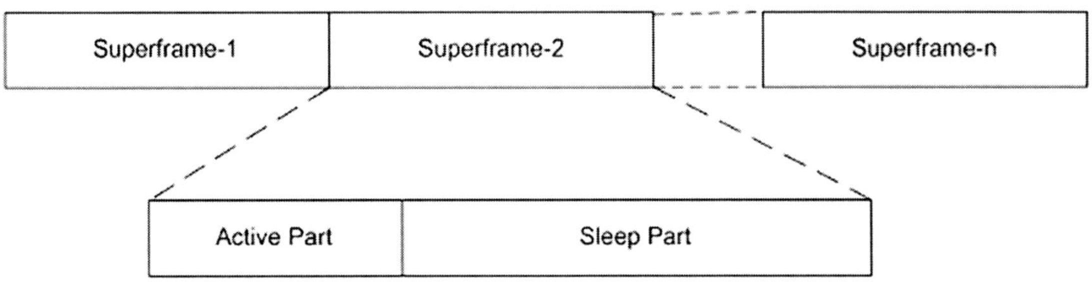

In WSN MAC, if the channel is idle for back-off inter frame space (BIFS) or short inter frame space (SIFS) the node start executing the BEB scheme. At each packet transmission, the back off counter is uniformly selected from the given range of [0, CW-1]. Here, CW is known as contention window and its value depends on the number of failed transmissions for the packet. In this chapter for "Consider MAC Protocol", at the first transmission attempt, CW is set to minimum value of 8 and if transmission attempt fails then its value gets double, and again set to minimum value on successful transmission. Furthermore, the maximum contention window value is set to 645 slots (7, 15, 31, and 64) with the maximum retry limit up to 3 counts. After selecting CW node decrement its value by 1 as long as channel is sensed idle and freeze its value when channel is sensed busy. It is worth to note that traditional6 backoff method has the following disadvantages:

- Traditional back-off takes time for CW (Contention Window) to correctly adapt to the right value when the active population (of nodes) is large.
- If CW is already large then the latency performance is high.
- A station that successfully transmits resets its CW to a small, fixed minimum value of CW. Consequently, the node has to rediscover the correct CW, wasting some precious bandwidth.
- Traditional backoff algorithm could not take the advantage of shared learning when number of nodes is increasing.

Traditional back-off's success probability degrades with increasing number of nodes.

Traditional back-off is also not suitable for power saving networks application.

In this chapter, we adopt the concept of a geometrically increasing probability distribution for contention process from (Y.C. Tay et. al., 2004) and introduce improved backoff (IB) for WSNs. In IB, we use a fixed size contention window, but a non-uniform, geometrically-increasing probability distribution for picking a transmission slot (i.e transmitting probability) in the contention window interval instead of traditional backoff procedure. This method does not require any complex or hard method to estimate the number of nodes. Furthermore, IB can easily accommodate the changing dynamics of WSNs, especially, saturation traffic and non-saturation traffic conditions which are most likely in sensor networks.

Motivation for Improved Backoff Algorithm

Now based on (S. Mehta et. al., 2010), and using the parameters listed in Table 1.5. In BEB based algorithm, we show the relation between throughput and contention window in Figure 5 with different number of nodes.

As shown in Figure 5, the value of throughput firstly increases and then decreases for given number of nodes (n) as the value of CW increases from 1 to 1000. For the small number of nodes first throughput is increasing and then decreasing while for large number of nodes throughput is increasing slowly before its maximum point. The reason behind this is very obvious, at the lower number of nodes, less waiting time in backoff procedure during low contention window size. At the higher number of nodes, at first CW is too small to adjust with the number of nodes hence high collision, but later it is adjusted with the number of nodes so less collision and less waiting time in backoff procedure. However, all the nodes in the network achieved nearly same maximum throughput as shown in Table 4.

Similarly, in Figure 6 we show the relation between average access time and contention window with different number of nodes. From Figure 6 we can observe that the average access delay time for different number of nodes is different. However, for given number of nodes, after certain length of contention window, the access

Figure 5. Relation between throughput and contention window

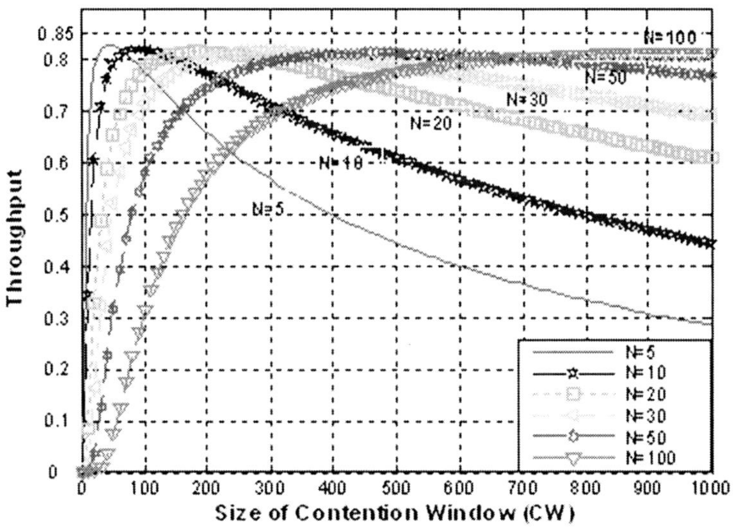

Table 4. Number of nodes vs. maximum throughput

Number of Nodes (n)	Maximum Throughput	CW Size
5	0.825	45
10	0.820	97
20	0.819	195
30	0.818	300
50	0.817	450
100	0.816	800

delay time does not jitter and it is almost constant for rest of the contention window size. It is worth to note that the size of the superframe was kept fixed in order to obtain the results presented in Figure 5 and 6.

As we mentioned earlier energy efficiency of MAC protocol in WSN is very sensitive to number of nodes competing for the access channel. It will be very difficult for a MAC protocol to accurately estimate the different parameters like collision probability, transmission probability, etc., by detecting channel. Because dynamics of WSN keep on changing due to various reasons like mobility of nodes, joining of some new nodes, and dying out of some exhausted nodes. Also, estimating about the other neighboring nodes information is too complex, as every node takes a distributed approach to estimate the current state of networks.

So, from the aforementioned explanation and results presented in Figure 5, 6 and Table 4, we can observe that if we can adjust the size of the window or transmitting probability according to the number of competing nodes the maximum throughput can be achieved. This gives us an intuition to use Improved Backoff (IB) scheme as an optimal and simple solution for designing an energy efficient MAC protocol for WSN.

Improved Backoff (IB) Algorithm

In this section we introduce the improved backoff (IB) algorithm, and it is very simple scheme to integrate with any energy efficient MAC protocols for WSNs. This method does not require any complex or hard method to estimate the number of nodes, nor does it need any extra overhead in the algorithm. Furthermore, IB can easily accommodate the changing dynamics of WSNs.

Figure 6. Relation between average access delay and contention window

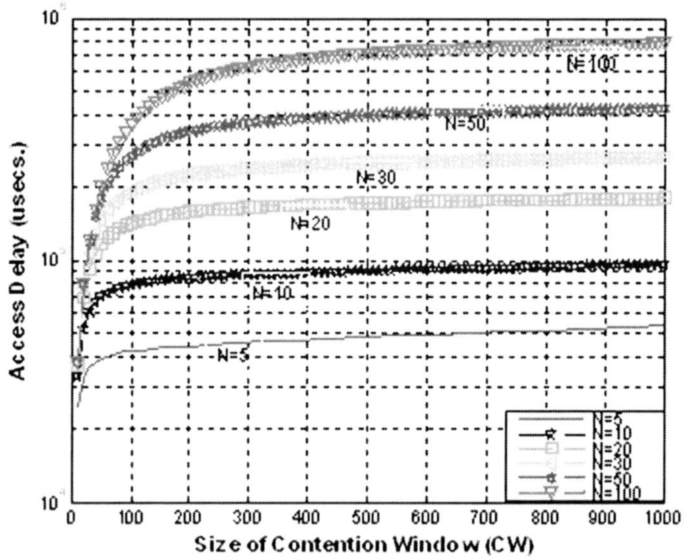

IB Mechanism

In contrast to traditional backoff scheme, IB scheme uses a small and fixed CW. In IB scheme, nodes choose non-uniform geometrically increasing probability distribution (P_{t_r}) for picking a transmission slot in the contention window. Nodes which are executing IB scheme pick a slot in the range of (1, CW) with the probability distribution P_{t_r} . Here, CW is contention window and its value is fixed. Figure 7 shows the probability distribution P_{t_r} . The higher slot numbers have higher probability to get selected by nodes compared to lower slot numbers.

In physical meaning we can explain this as: at the start node select a higher slot number for its CW by estimating large population of active nodes (n) and keep sensing the channel status. If no nodes select the first or starting slots to transmit in then each node adjust its estimation of competing nodes by multiplicatively increasing its transmission probability for the next slot selection cycle. Every node keeps repeating the process of estimation of active nodes in every slot selection cycle and allows the competition to happen at geometrically–decreasing values of n all within the fixed contention window (CW).

In contrast to the probability distribution P_{t_r} , in uniform distribution, as shown in Figure 7, all the contending nodes have the same probability of transmitting in a randomly chosen time slot. From Figure 7 we can depict that when the population of competing nodes (n) is large, most of the nodes will choose medium to high slot numbers as their CW and a very few nodes will choose low slot numbers, hence, a collision-free transmission will take place in a low slot number. When n is medium, most nodes will choose higher slot numbers and a collision- free transmission will take place in a medium slot number. Similarly, when n is small, a collision free transmission will take place in a high slot number. Thus for any value of n, and for any fast change in n, a collision free transmission can take place. If only one node gets the chance to select the contention slot within the fixed CW, it will transmit in that slot. While other nodes will select new random contention slots for next competition to

Figure 7. Difference between uniform and truncated geometric distributions

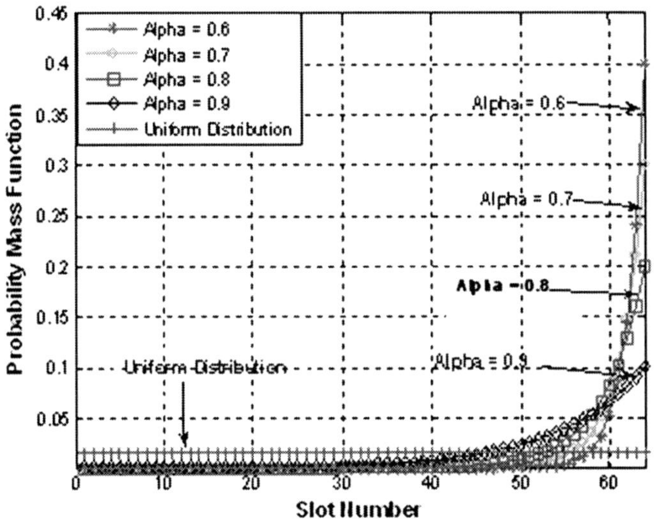

win channel medium, regardless of success or failure of transmission of winner node.

Let's say each node picks a slot $t_r \epsilon [1,CW]$ using the probability distribution P_{t_r}. Then slot t_r defines as an idle slot if no nodes choose it, and also as a collided slot if more than one node chooses it. A node can get the chance to transmit if it is the only node to choose slot t_r. As we mentioned earlier IB uses a truncated, increasing geometric distribution, as presented in (S. Mehta et. al., 2010), and is given by

$$p_{t_r} = \frac{(1-\alpha)\alpha^{CW}}{1-\alpha^{CW}} \alpha^{-t_r} \; for \; t_r = 1,\ldots,CW, \tag{1.1}$$

where $0<\alpha<1$ is a distribution parameter. In this range of α, p_{t_r} increases exponentially with t_r, so the later slots have higher probability. Here, it is worth to note that IB scheme doesn't use timer suspension like in IEEE 802.11 to save energy and reduce latency in case of a collision. The only problem with the IB is fairness, however, for WSNs, fairness is not a problem due to two main reasons. First, overall network performance is more important rather than an individual node. Second, all nodes don't have data to send all the time (i.e. unsaturated traffic condition). Using IB may give us the optimum network performance as it reduces the collision to minimum.

Analytical Modeling

In this section we present the general frame work to model the backoff algorithms 7. This frame work basically consists of three steps: finding an attempting probability for a node in backoff, finding the transition probability for a given channel state, and modeling the stationary probabilities of the channel state for required protocol details. Here we derive channel efficiency, channel access delay and energy efficiency with these three basic steps. As we mentioned earlier BEB is widely accepted and studied in the research literature, we choose only BEB out of previously discussed backoff algorithms for detailed analytical modeling to compare with IB.

Approach and Assumptions

Most of the studies on backoff algorithm (BA) are focused on the stability issue rather than performance analysis of backoff algorithm. In this paper our main focus is to analyze and compare the performance of both IB and BEB with respect to network load in steady state condition. Here, we define network load in terms of the number of nodes that are contending for the access medium. Another approach is to consider total arrival packet rate to the network as an offered load. The main purpose of BA is to reduce the effect of contention among the nodes and try to adopt the population of nodes so the number of nodes contending for medium is a more favorable way to define as an offered load for analyzing BA. Here, we assume a fixed number of nodes in saturated and non-saturated conditions. Saturation conditions mean every node always has a packet to transmit and similarly in non-saturation case every node receives a packet with probability λ. More on saturation and non-saturation will be explained later. The channel is an ideal and introducing no errors to the reception of a packet other than collision. Also, capture effect is not considered. The BA performed in a time-slotted fashion. A node attempts to attain the access the channel only at the beginning of a slot. Furthermore, all nodes are well synchronized in time slots and propagation delay is negligible compared to the length of an idle slot.

BEB and IB Analytical Modeling

In BEB, at first transmission, a packet is transmitted after waiting for the number of slots randomly chosen from the given range of contention window (0,1, 2….CW_0-1), where CW_0 is the minimum contention window size. This contention window is used as time unit for a node to detect the transmission of a frame from any other node. This time unit is to be defined as 'slot time' and this is different from the data transmission slot. Generally data transmission slot is quite long compared to contention window slot. We model the operation of BEB at an individual node using the state diagram shown in Figure 8. This diagram is based on the model presented in (G. Binachi, 2000; M.Ergen et. al., 2005) including the freezing and retry limit parameters.

As shown in the Figure 8 let j denote the backoff stage, where j=0, 1, 2, 3 (assuming immediate acknowledgement for a data/request packet). So, we have CW_0=8, CW_1=16, CW_2=32, and CW_3=64. Let b(t) is defined as a random process representing the backoff counter of a node and s(t) is representing random process of the back stage j. b(t) is decremented at the start of every idle backoff slot. It is important to note that time scale for b(t) doesn't represent real time but it observes only backoff slots and its suspended for the duration of all transmissions and interframe spaces (i.e. SIFS). And whenever b(t) reaches zero the station transmits and regardless of the outcome of the transmission, uniformly chooses a new value for b(t) from (0,1,…,CW_j-1)(i.e. new backoff counter value). Here, we define p_c as the conditional collision probability and we also assume that it is independent and constant, regardless the number of retransmissions attempted. p_c also represents the probability of detecting the channel busy. Thus, the two dimension process, $\{s(t),b(t)\}$, is a discrete-time Markov Chain. Therefore, the state of each node is described by {j, k}, where j stands for the backoff stage, and k stands for the backoff timer value. In addition to normal state diagram we also add 2 extra states to model the non-saturation traffic condition. A node may now wait in the idle state for a packet from upper layers before going into backoff procedure. This corresponds to a delay in the idle state and it is represented by upper left two sates in the Figure 8. The delay in the idle state is modeled geometric with parameter λ. The state transition diagram of the Markov chain model shown in Figure 8 has the following transition probabilities:

Figure 8. Markov Chain Model for BEB algorithm of a "Consider MAC Protocol"

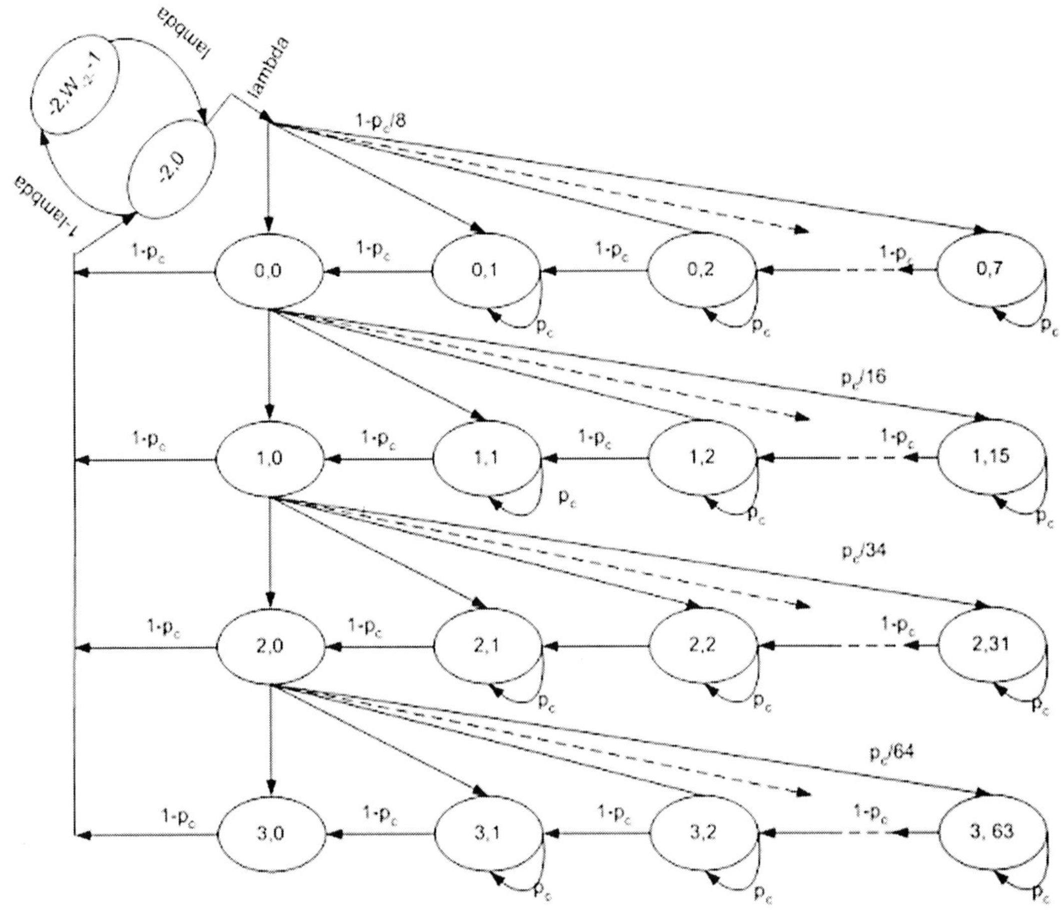

$$\begin{cases} P\left\{j,k \mid j,k+1\right\} = (1-p_c) & k \in (0, CW_j - 2) \;\; j \in (0,3) \\ P\left\{j,k \mid j,k\right\} = p_c & k \in (1, CW_j - 1) \;\; j \in (0,3) \\ P\left\{j,k \mid j-1,0\right\} = p_c / CW_j & k \in (1, CW_j - 1) \;\; j \in (1,3) \\ P\left\{m,k \mid m,0\right\} = p_c / CW_j & k \in (1, CW_j - 1) \\ P\{-2, CW_{-2} - 1 \mid -2, CW_{-2} - 1\} = (1-\lambda) & \\ P\{0,k \mid -2,0\} = \lambda(1-p_c)/CW_0 & k \in (1, CW_j - 1) \\ P\{-2,1 \mid -2,0\} = (1-\lambda) & \\ P\{-2,0 \mid -2,1\} = \lambda & \end{cases} \tag{1.2}$$

The first equation in (1.2) indicates that at the beginning of each slot time, the backoff counter is decremented if the channel is sensed idle. The second equation shows that the backoff counter is frozen if channel is sensed busy. The third and fourth equations, respectively, indicate that following an unsuccessful transmission, the node back off stage (j-1) selects a backoff interval uniformly in the range of (0,CW_j-1) and when the backoff stage reaches m, CW_m stays constant. Rest of the equations shows the transition probabilities for two extra sates we added. Here, we take CW_{-2}=2 to introduce 2 extra states. These transition probabilities are straight forward as shown in Figure 8.

Similar to BEB algorithm, Figure 9 shows the state diagram of IB algorithm at an individual node. As we explained in earlier, IB does not use contention counter suspension and there is only one stage (i.e, fixed backoff window). In IB, each node selects a contention slot with a geometrically increasing distribution as presented in (1) within the range of (1,..., CW), where *CW* is the fixed contention window size. This contention window is used as time unit for a node to detect the transmission of a frame from any other node. This time unit to be defined as 'slot time' and this is different from the data transmission slot. Generally data transmission slot is quite long compared to contention window slot. Using similar notation as in (G. Binachi, 2000) for IB, here the state of each node is described by {j, k}, where j stands for the backoff stage, and k stands for the backoff timer value (For IB, j=0 and max $\{k\}=CW$). Here, p_{CIB} represents as the collision probability and also p_{CIB} represents the probability of detecting the channel busy. Therefore, Figure 9 shows the one dimension discrete-time Markov Chain for IB at an individual node. In this Markov Chain, the non null one step transition probabilities are as follows:

$$\begin{cases} P\{0,k \mid 0,k+1\} = 1 - p_{CIB} & k \in (1,CW) \\ P\{0,1 \mid 0,k\} = p_{CIB} & k \in (1,CW) \\ P\{0,k \mid 0,1\} = p'_{k} & k \in (1,CW) \\ P\{-2,W_{-2}-1 \mid -2,W_{-2}-1\} = (1-\lambda) & \\ P\{0,k \mid -1,0\} = \lambda / p'_{k} & k \in (1,CW) \\ P\{-2,1 \mid -1,0\} = (1-\lambda) & \\ P\{-1,0 \mid -2,1\} = \lambda & \\ P\left\{-1,0 \mid 0,1\right\} = 1 - p_{CIB} & \end{cases} \tag{1.3}$$

The first equation in (1.3) indicates the backoff counter which is decremented if the channel is sensed idle. The second equation in (1.3) indicate the node defers the transmission of a new frame and enters stage 0 of the backoff procedure if it detects a successful transmission of its current frame or finds the channel busy or if it detects that a collision occurred to its current not successfully transmitted frame. The third equation in (1.3) indicates the node selects a backoff interval non-uniformly in the range of (1, CW) following an unsuccessful transmission. Rest of the equations shows the transition probabilities for two extra sates we added. Here, we take CW_{-2}=2 to introduce two extra states. The fourth equation in (1.3) represents the node waiting in the idle state for packet to arrive from the upper layer. The fifth equation in (1.3) shows the buffered packet enter to backoff procedure. The equations six and seven in (1.3) represent the transition between buffer to idle state and back to buffer state according to availability of a packet, respectively. The last equation in (1.3) represents transition of backoff procedure to buffer state in case of a successful packet transmission.

In IB Scheme, a node is randomly selecting a contention window from the (1, CW) and transmit with the probability p'_{k}, where p'_{k} is based on the

Figure 9. Markov Chain for IB Method

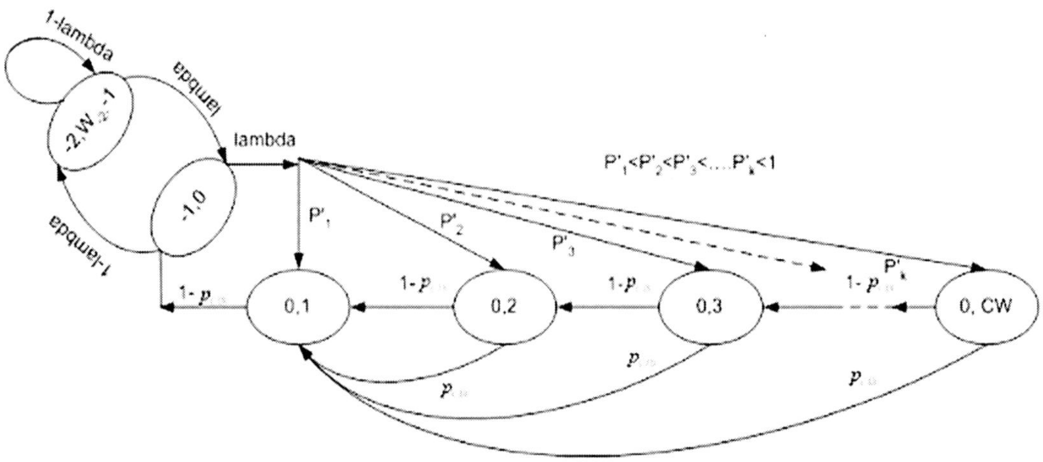

non-uniform increasing geometry distribution as given in (1.1) and define as

$$p'_1 < p'_2 <, \ldots, < p'_k < 1 \quad k \in (1, CW) \tag{1.4}$$

To understand (1.4) readers are advised to refer Figure 7 where the different values of p'_k with different values of α are plotted.

Throughput Analysis

In BEB, let $b_{j,k} = \lim_{t \to \infty} \Pr\{s(t) = j, b(t) = k\}$ be the stationary distribution of the Markov chain [8, 9]. Let p_{tr} be the probability that a node transmits during a generic slot time. A node transmits when its backoff counter reaches zero. Similar to (G. Binachi, 2000; B.J. Kwak et.al., 2005) p_{tr} is given by

$$p_{tr} = \frac{2(1-2p_c)(1-p_c)}{(1-2p_c)(CW_{\min}+1) + p_c CW_{\min}(1-(2p_c)^m)} \tag{1.5}$$

and as we mentioned earlier and shown in Figure 8, p_{tr} with non-saturation traffic condition is given by

$$p_{tr} = \frac{1}{\frac{(1-2p_c)(CW_{\min}+1) + p_c CW_{\min}(1-(2p_c)^m)}{2(1-2p_c)(1-p_c)} + (1-p_c)(1/\lambda^2 - 1)} \tag{1.6}$$

Here, equation (1.6) will reduce to (1.5) if we put λ=1 (i.e., saturated case). From (1.5) and (1.6) we see that $p_{tr} = p_{tr}(p_c, m, CW_{\min}, \lambda)$ and depend on p_c. The relation between p_{tr} and p_c is given by

$$p_c = 1 - (1 - p_{tr})^{n-1} \tag{1.7}$$

Here, n denote the number of nodes. Equations (1.5) and (1.7) form a nonlinear system with two unknowns p_{tr} and p_c. This nonlinear system can be solved using numerical methods. Figure 10 shows the plot of p_{tr} as a function of p_c with various values of CW and n. This graph is plotted with m, the retry limit, equals to zero. As shown in Figure 10 the unique intersection of equations (1.5) and (1.7) gives us the value of p_c and p_{tr} for given CW and n. The plot shows p_{tr} and p_c converging to 0.5 and zero, respectively, as the number of nodes increases [17]. Similarly, we can obtain the solution for p_{tr} and p_c when retry limit, m, is $m \neq 0$ and traffic condition is non-saturated (i.e., $\lambda \neq 1$).

Let p_b define as the probability that there is at least one transmission in a slot and is given by

$$p_b = 1 - (1 - p_{tr})^n \tag{1.8}$$

Let p_i define as the probability that there is not a single transmission in a slot and is given by

$$p_i = (1 - p_{tr})^n \tag{1.9}$$

A successful transmission occurs when there is only one transmitting node. Thus, the probability that there will be a successful transmission in a time slot and is given by

$$p_s = np_{tr}(1 - p_{tr})^{n-1} \tag{1.10}$$

It is important to note that a collision occurs if there are multiple nodes transmitting in the same time slot. Thus, the probability that a collision 8 will occur in a time slot and is given by

$$P_c = 1 - (1 - p_{tr})^n - np_{tr}(1 - p_{tr})^{n-1} \tag{1.11}$$

In IB Scheme, a node is randomly selecting a contention window from the (1, CW) and transmit with the probability p'_k, where p'_k is based on the non-uniform increasing geometry distribution as given in (1.1) (Y.C. Tay et. al., 2004) and define as

Figure 10. Probabilities of transmission and Collision

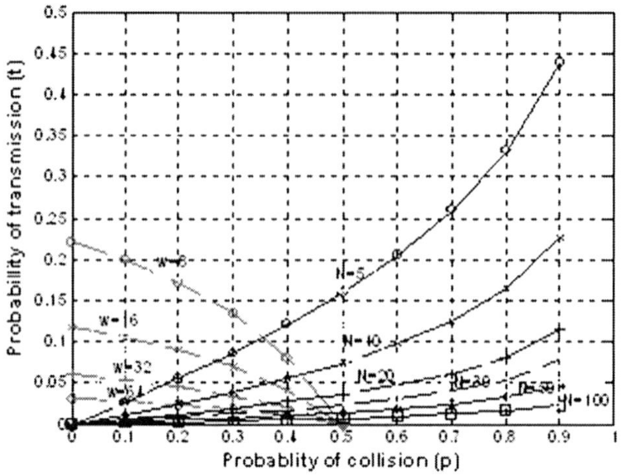

$$p'_1 < p'_2 <, \ldots, < p'_k < 1 \quad k \in (1, CW) \tag{1.12}$$

To understand (1.12) readers are advised to refer Figure 7 where the different values of p'_k with different values of α are plotted. Now similar to BEB scheme, we can define the probabilities of busy medium, idle medium and successful transmission in a time slot in IB scheme, respectively, as follow

$$p_{IBb} = 1 - (1 - (p'_1 + \ldots. + p'_k))^n \tag{1.13}$$

$$p_{IBi} = (1 - (p'_1 + \ldots. + p'_k))^n \tag{1.14}$$

$$p_{IBs} = np'_k(1 - (p'_1 + \ldots. + p'_k))^{n-1} \tag{1.15}$$

Now, the probability of collision in IB is given by

$$P_{IBc} = 1 - (1 - (p'_1 + \ldots. + p'_k))^n - np'_k(1 - (p'_1 + \ldots. + p'_k))^{n-1} \tag{1.16}$$

Using aforementioned equations we can define the throughput as the fraction of time that the channel is used for successful transmission. The time that the channel remains empty or busy with collision is wasted. Here, successful transmission includes data/request frame with an acknowledgement. The channel efficiency for BEB scheme is given by

$$\eta_{BEB} = \frac{T_S p_s}{T_i p_i + T_S p_s + T_c p_c} \tag{1.17}$$

Here, T_i is constant and defined in the standard (of duration δ), T_S and T_C are the average time duration of successful transmission and collision, respectively. The duration of successful and collision slots are similar; hence the duration of a collision can be approximated by the duration of a successful slot $T_C \approx T_S$. T_S and T_C are given by (both are same in case of BEB and IB).

$$T_S = H_{P+M} + T_{Data/Req} + SIFS + P + T_{ACK} + P + SIFS \tag{1.18}$$

$$T_C = H_{P+M} + T_{Data/Req} + SIFS + P \tag{1.19}$$

where H_{P+M} is the MAC and Physical layer frame headers and P is the propagation delay. $T_{Data/Req}$

is the duration of data packet or request packet. Using this approximation and equations (1.8, 1.9, and 1.10, we can simplify equation (1.17) as

$$\eta_{BEB} = \frac{np_{tr}(1-p_{tr})^{n-1}}{1-\frac{T_S-T_i}{T_S}\left(1-p_{tr}\right)^n} \tag{1.20}$$

In a same way, simplified throughput for IB is given by

$$\eta_{IB} = \frac{np'_k(1-(p'_1+....+p'_k))^{n-1}}{1-\frac{T_S-T_i}{T_S}(1-(p'_1+....+p'_k))^n} \tag{1.21}$$

Delay Analysis

Here, medium access delay is defined as the time elapsed between the generation of a frame and its successful reception. Let D be the medium access delay and E [D] indicate its mean value. Then, E [D] is given by

$$E[D] = E[N_c](E[B_d] + T_C + T_O) + (E[B_d] + T_S) \tag{1.22}$$

where E[N_c] represents average number of collisions of a frame until its successful reception. E[B_d] represents average backoff delay spent by a node before accessing the medium in busy channel condition and T_O represents the time that a node has to wait before sensing the channel again in case of its frame collision. Now E[B_d] is given by

$$E[B_d] = E[X] + E[N_F](p_s T_S + (1-p_s)T_c) \tag{1.23}$$

where E[X] is the average back off value and E[N_F] is the average number of times that a node freezes its counter before reaches state 0. Equations (1.20) is applicable for both IB and BEB, however in IB E[B_d] is given by

$$E[B_d] = E[X] \tag{1.24}$$

In IB, backoff counter is not suspended if a node detects transmissions from other nodes before its counter reaches state 0 but it takes new value of backoff counter for next attempt (IB has a fixed contention window range). In case of BEB and IB, E[N_c] is given by

$$E[N_c] = \frac{1}{p_s} - 1 \tag{1.25}$$

In BEB, E[X] and $E[N_F]$ are given by

$$E[X] = \frac{CW_0}{2}\left[2(1-p_c)\times\frac{1-(2p_c)^{m+1}}{1-2p_c}\right] + \frac{CW_0}{2}\times p_c^{\,m+1}\left[2^{m+1}+\frac{2^m}{1-p_c}\right] - \frac{1}{2(1-p_c)} \tag{1.26}$$

$$E[N_F] = \frac{E[X]}{\max(\mathrm{E}[\psi],1)} - 1 \tag{1.27}$$

where $E[\psi]$ is the mean number of consecutive idle slot times before a transmission starts. In IB, E[X] is calculated as simply mean of geometrically increasing distribution with fixed interval. Finally, the time T_O depends on the access method and is given by

$$T_O = InterframeSpace + ACK\_timout \tag{1.28}$$

Energy Efficiency Analysis

In this subsection we consider only two power levels, P_{tx} and P_{rx}. We assume that a node consumes power P_{tx} for transmitting and P_{rx} for being idle and listening. Total power consumption to transmit a data/request packet is given by

$$E_T = E_B + E_L + E_S + E_C \quad (1.29)$$

where E_B, E_L, E_S, and E_C represents energy consumption in backoff stage, over hearing in backoff stage, energy consumption for successful transmission and energy consumption in collision, respectively. We define the energy efficiency as energy required to successfully transmit one bit of data/request packet and is given by

$$\eta_e = \frac{E_T}{L} \quad (1.30)$$

where L is the length of data/request payload. The energy consumption when back off is decrementing, E_B, is given by

$$E_B = E[X] \bullet E[N_c] \bullet Slot \bullet P_{rx} \quad (1.31)$$

where slot is the duration of an idle slot (of duration δ). The energy consumption when overhearing other nodes' transmission during the backoff procedure is given by

$$E_L = E[N_c] \bullet E[X] \bullet P_{rx} \quad (1.32)$$

In case of IB method E[X] will be different from equation (1.26). The energy consumption for successful transmission is given by

$$E_S = (H_{P+M} + T_{Data/\mathrm{Req}}) \bullet P_{tx} + (SIFS + P + T_{ACK} + P + SIFS) \bullet P_{rx} \quad (1.33)$$

The energy consumption due to collision is given by

$$E_C = (H_{P+M} + T_{Data/\mathrm{Req}}) \bullet P_{tx} \bullet E[N_c] + (SIFS + P + T_0) \bullet P_{rx} \bullet E[N_c] \quad (1.34)$$

Energy efficiency model is valid for both BEB and IB methods.

Performance Analysis

In this subsection we present the performance comparison of our "consider" or "normal" MAC protocol, and IB based MAC protocol in terms of channel efficiency, medium access delay and energy-efficiency. Both the protocols are same in nature except for their backoff procedure. Here, we fixed the channel rate to 1 Mbps with an ideal channel condition. Since the channel conditions are assumed to be ideal, transmission errors are a result of packet collision only. For the "normal" MAC protocol maximum retry limit is set to 3 (m=3), minimum contention window is set to 8 for "normal" MAC and fixed contention window of 16 for IB based MAC, and traffic model is set to saturation and non-saturation modes, respectively. The backoff algorithm (BA) performed in a time-slotted fashion. A node attempts to attain the access the channel only at the beginning of a slot. Furthermore, all nodes are well synchronized in time slots and propagation delay is negligible compared to the length of an idle slot. For the performance evaluation we carried out simulation in Matlab.

Here, we define network load in terms of the number of nodes that are contending for the access medium. Another approach is to consider total arrival packet rate to the network as an offered load. The main parameters for our simulation are based on (G. Binachi, 2000) and listed in Table 5. For calculating the energy consumption in nodes we choose ratio of idle: listen: transmit as 1:1:1.5, as measured in (M. Stemm et. al., 1997). Here we considered the decentralized single hop topology scenario where all nodes are trying to send their traffic to one destination node or sink so we do not have any hidden terminal problem. With this topology we can represent the 'last hop' communication scenario in wireless sensor

Table 5. Simulation Parameters

Parameters	Values
CW_{min}	8
ACK Packet Size	24 Bytes
Data Packet Size	1024 Bytes
Nodes	5~100
Data Rate	1Mbps
Transmitting Energy	50x10^{-6} J/Bit
Idle/listening Energy	75x10^{-6} J/Bit

networks. Also, we assumed no capturing effect. For the simulation results we doesn't consider the technology adopted at the Physical layer, however the physical layer determines some network parameter values like inter-frame spaces. Whenever necessary we choose the values of the physical layer dependent parameters by referring to (G. Binachi, 2000). The results obtained here are the average values of our collected data.

The normalized throughput is mostly depends on number of active nodes and contention window size. We investigated various values of CW and n for IB and found the CW=16 is the most suitable value to get best delay and energy efficiency performances while maintaining the channel efficiency above 50% for the given number of nodes in given traffic conditions. Similarly, we engineered the value of α to α=100$^{-1/16}$≈0.74. As shown in Figure 11, for "consider" MAC, results with CW_{min}=8 gives high channel efficiency values at lower number of nodes. The reason is very obvious, less collision and low waiting time in backoff procedure, and as number of contenders increases channel efficiency start decreasing. At first graph shows lower channel efficiency, and then increases to higher values, and later maintains nearly constant values. This is because at first CW is too big for given number of nodes and later it adjusted with the number of nodes, hence, nearly constant collision and waiting time in backoff procedure. In contrast to "consider" MAC, IB based MAC maintains high throughput due to its unique quality of collision avoidance among the competing nodes. In IB most of the nodes choose higher contention slots while very few nodes selects lower contention slots, hence, less or no collision and low waiting time in backoff procedure. Now, we compare the "consider" MAC with IB based MAC under the non-saturation condition as shown in Figures 12 and 13.

In the non-saturation case, especially for low n, competition for channel access is low so the

Figure 11. Number of contenders vs. Channel efficiency

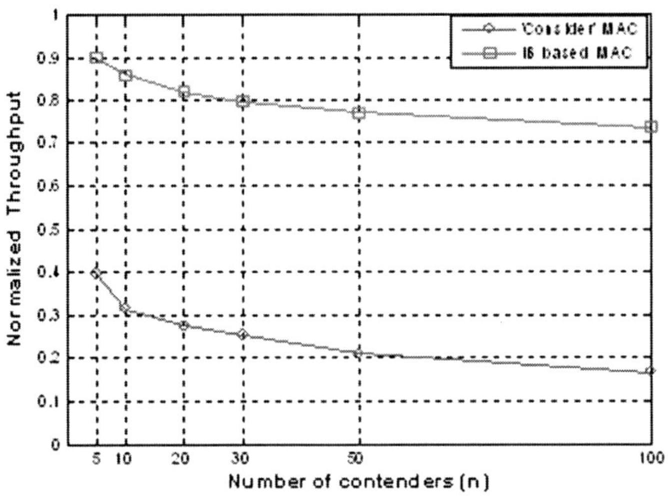

Figure 12. Number of contenders vs. Channel efficiency (Lambda=0.5)

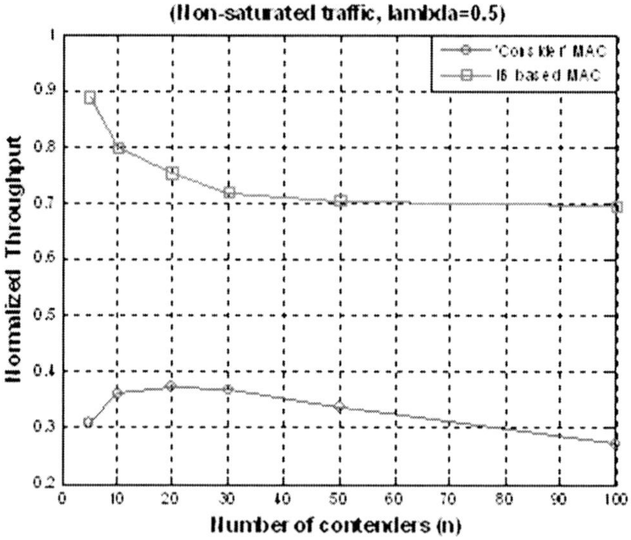

Figure 13. Number of contenders vs. Channel efficiency (Lambda=0.3)

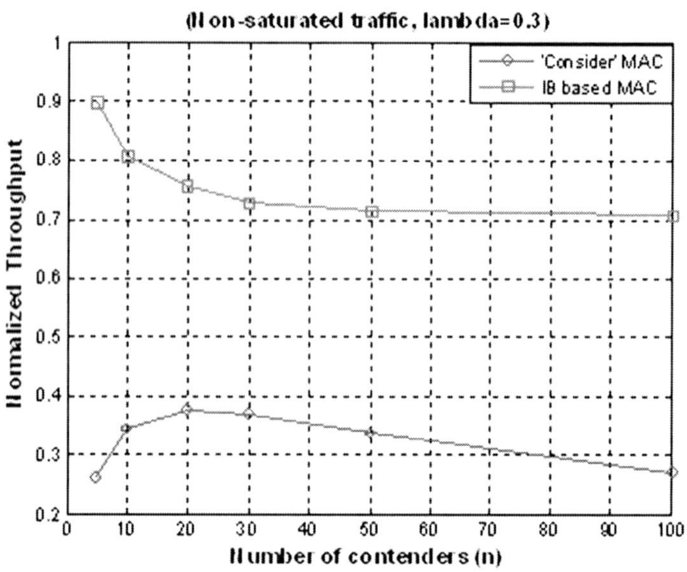

graph for CW_{min}=8 is initially starting from low value, reach to maximum and again start decreasing. From these figures we can also find the optimum value of CW for a given value of n.

Figure14, 15, and 16 shows the average medium access delay performances of "consider" MAC and IB based MAC for saturation and non-saturation traffics, respectively. In "consider" MAC, as a large number of stations attempt to access the medium, more collision occurs so the number of retransmissions increases and the nodes suffer longer delays. In IB based MAC, as we expected access delay is very low compared to "consider" MAC. This is because of low or

Figure 14. Number of contenders vs. Average access delay

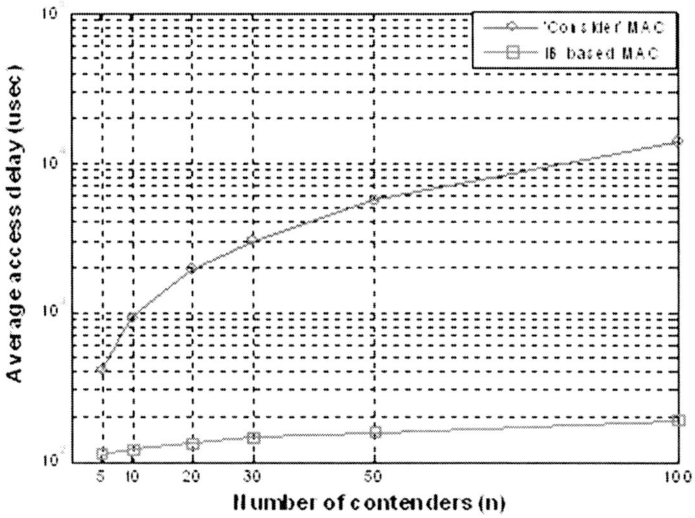

Figure 15. Number of contenders vs. Average access delay (Lambda=0.5)

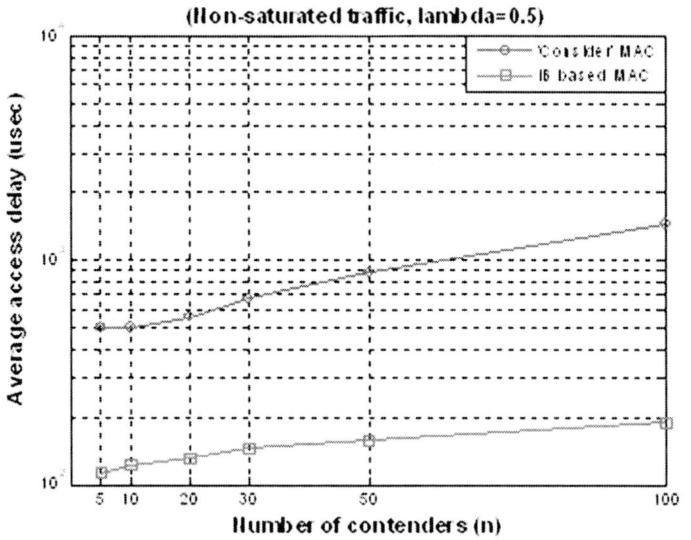

no collision and less idle waiting time in backoff procedure.

Figures 17, 18, and 19 illustrate the effect of bakeoff schemes on energy efficiency under the saturation and non-saturation conditions, respectively. From Figure 17 we can see that in "consider" MAC more energy waste, as collision and number of retransmission increases. As CW value increases the collision probability and the number of retransmissions reduces, hence less energy waste. In contrast IB wastes very less energy due to its unique characteristics of collision avoidance. From all aforementioned results we can see the superiority of IB based MAC over "consider" MAC. Accepting IB as backoff scheme can increase the overall performance of MAC

Figure 16. Number of contenders vs. Average access delay (Lambda=0.3)

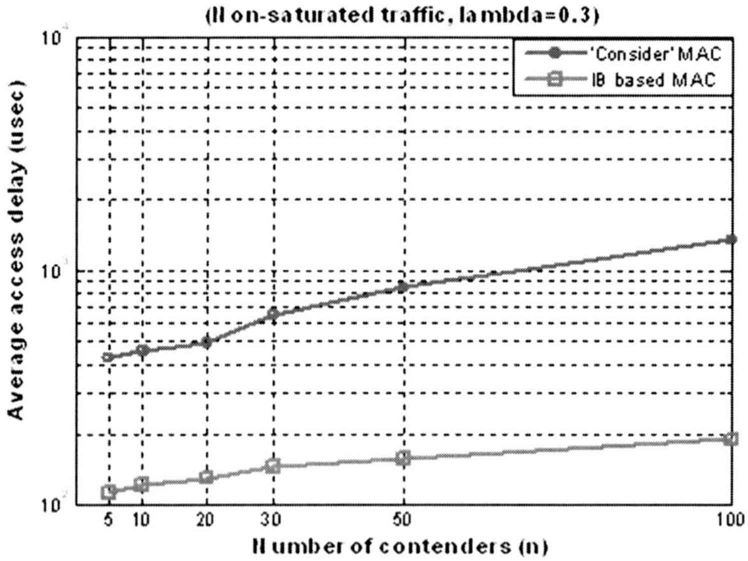

Figure 17. Number of contenders vs. Energy efficiency

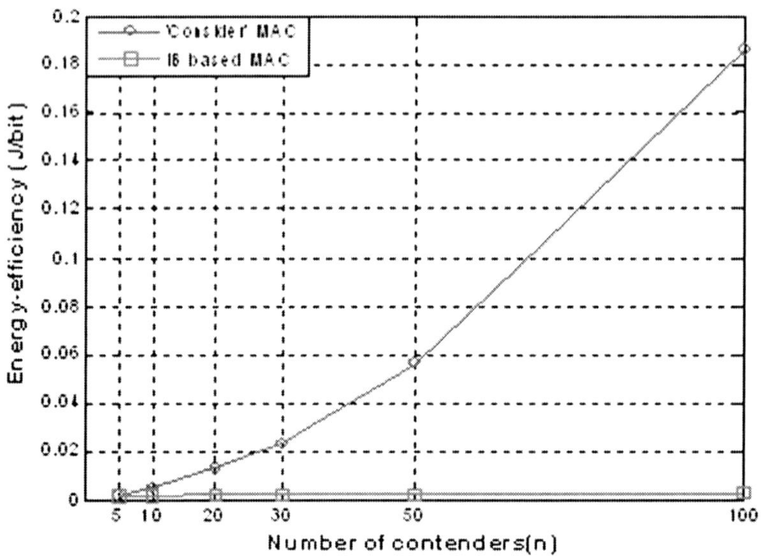

Figure 18. Number of contenders vs. Energy efficiency (Lambda=0.5)

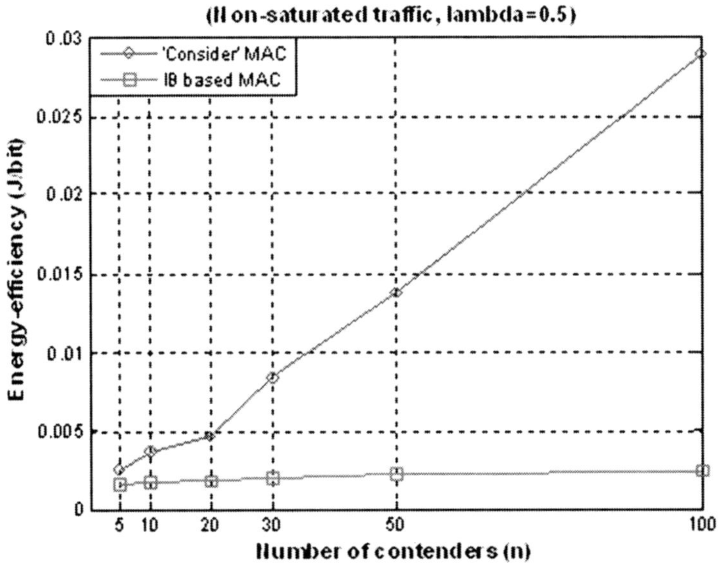

Figure 19. Number of contenders vs. Energy efficiency (Lambda=0.3)

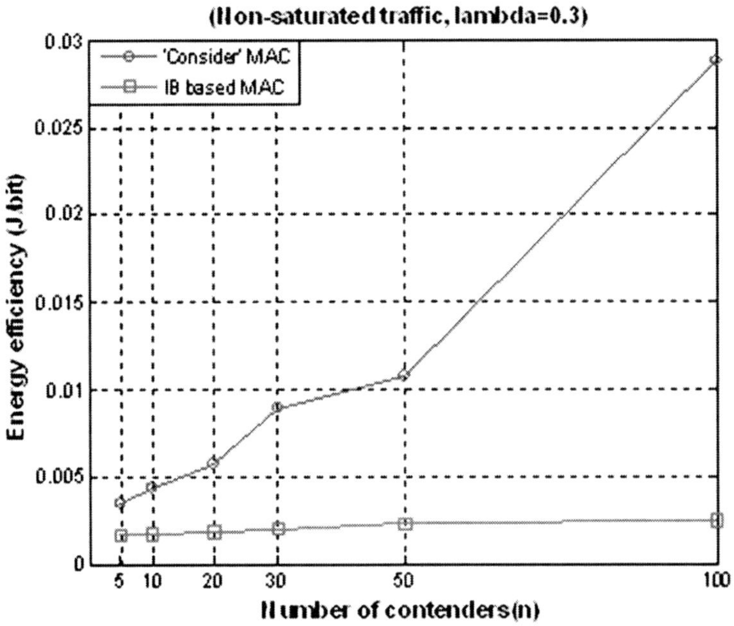

protocol to a large extend, and we can also get an optimal and simple solution to design an energy efficient MAC protocol for WSNs.

CONCLUSION

In this chapter we studied some of the well known back off algorithms and compared them with respect to different parameters. We also studied some of the popular sensor MAC protocols and based on their general structure we conceived an idea of 'consider MAC' protocol with BEB algorithm. Later, we used the concept of a geometrically increasing probability distribution for contention process. This allows us to introduce CSMA/CA based improved backoff (IB) algorithm for energy efficient MAC protocol, where binary exponential backoff (BEB) based algorithm is originally used. We applied IB scheme to WSN MAC to enhance its performance. We have also proposed the stochastic models to compute the performance of IB in WSN MAC. Using the proposed analytical models, we analyze the performance of BEB based algorithm and IB in terms of throughput, channel access delay, and energy efficiency under the saturation and non-saturation traffic conditions. Our results show that IB exhibits an edge over BEB based algorithm in WSN MAC. Furthermore, IB can achieve high energy efficiency while maintaining high throughput and low channel access delay. We hope this chapter will provide basic concepts of sensor MAC protocol designing and will serve as an initial starting point for students, researcher, and engineer in the field of wireless sensor networks.

ACKNOWLEDGMENT

This research was supported by the MKE (The Ministry of Knowledge Economy), Korea, under the ITRC (Information Technology Research Center) support program supervised by the NIPA (National IT Industry Promotion Agency) (NIPA-2010-C1090-1011-0007). This work was supported by the National Research Foundation of Korea (NRF) grant funded by the Korea government (MEST) (No. No.2010-0018116).

REFERENCES

Akayildiz, I. F., Su, W., & Yogesh, S. (2002). A survey on sensor networks. *IEEE Communications Magazine*, (August): 2002.

Bharghavan, V., Demaers, A., & Shenker, S. (1994). MACAW: A media access protocol for wireless LAN's. In *Proceedings of ACM SIGCOMM* (pp. 212-225).

Bianchi, G. (2002). Performance analysis of the IEEE 802.11 distributed coordination function. *IEEE Journal on Selected Areas in Communications, 18.*

Chatzimisios, P., Boucouvalas, A. C., Vitsas, V., Vafiadis, A., Oikonomidis, A., & Huang, P. (2005). A simple and effective backoff scheme for the IEEE 802.11 MAC protocol. *In Proceedings of the International Conference on Cybernetics and Information Technologies, Systems and Applications* (*CITSA 2005*), Orlando, USA.

Dam, T. V., & Langendone, K. (2003). An adaptive energy-efficient MAC protocol for wireless sensor networks. In *Proceeding of SenSys '03*, Los Angeles, CA (pp. 171-180).

Deng, J., & Hass, Z. J. (2003). On optimizing the backoff interval for random access schemes. *IEEE Transactions on Communications*, *51*(12), 2081–2090. doi:10.1109/TCOMM.2003.820754

Ergen, M., & Variya, P. (2005). Throughput analysis and admission control for IEEE 802.11a. In *Proceeding of Springer Mobile Networks and Applications*.

Heidemann, W., Ye, J., & Estrin, D. (2002). Energy-efficient MAC protocol for wireless sensor networks. In *Proceeding of INFOCOM 2002*, New York, NY (pp. 1567-1576).

IEEE Computer Society LAN MAN Standards Committee. (1997). *Wireless LAN medium access control (MAC) and physical (PHY) layer specifications. IEEE Standard 802.11-1997*. New York, NY: The Institute of Electrical and Electronics Engineers.

Kwak, B. J., Song, N. O., & Miller, L. E. (2005). Performance analysis of exponential backoff. In *IEEE/ACM Trasection on Networking, 13*(2)

Kwon, Y., Fang, Y., & Latchman, H. (2002). A novel medium access control protocol with fast collision resolution for wireless LANs. In *Proceeding of IEEE INFOCOM '03, 2*, 853-862, San Francisco, USA.

Lu, G., Krishnamachari, B. & Ragavendra, C. (2004). An adaptive energy-efficient and low-latency MAC for data gathering in sensor networks. *In proceeding of WMAN'04.*

Mehta, S. & Kwak, K. S. (2009). H-MAC: A hybrid MAC protocol for wireless sensor networks. Accepted for publication in *International Journal of Computer Networks & Communications,* Dec. 2009.

Mehta, S., & Kwak, K. S. (2010). *An energy efficient improved backoff scheme for sensor networks (Technical Report-7)*. Incheon, South Korea: Inha University.

Mehta, S. & Kwak, K. S. (2010). Performance analysis of binary exponential backoff and improved backoff for WAPN. Published in *EURASIP Journal on Wireless Communications and Networking.*

Song, N., Kwak, B. J., & Miller, L. E. (2005). Analysis of EIED backoff algorithm for the IEEE 802,11 DCF. In *Proceeding of IEEE VTC 2005-Fall, 4*, 2182-2186.

Stemm, M., & Katz, R. H. (1997). Measuring and reducing energy consumption of network interfaces in hand-held devices. *IEICE Transactions on Communications. E (Norwalk, Conn.), 80-B*(8), 1125–1131.

Tay, Y. C., Jamieson, K., & Krishnan, B. (2004). Collision-minimizing CSMA and its applications to wireless sensor networks. *IEEE Journal on Selected Areas in Communications*, *22*(6), 1048–1057. doi:10.1109/JSAC.2004.830898

Zheng, T., Radhakrishnan, S., & Sarangan, V. (2005). PMAC: An adaptive energy-efficient MAC protocol for wireless sensor networks. In *Proceeding of IPDPS'05* (pp. 65-72).

ENDNOTES

1. Some part of this chapter is taken from [S.Mehta et.al.,2010].
2. For the comparison purpose we include IB at this point, more details on IB has been discussed later in this chapter.
3. We used MATLAB for IB, NS-2 for BEB, and rest of algorithms are implemented in OPNET.
4. We can easily relate the "Consider MAC Protocol" with available MAC protocols and standards for wireless sensor networks, as most of the popular WSN MAC protocols are based on the active/sleep cycle mechanism.
5. This assumption is valid as most of the popular WSN MAC protocols have limited retry limit with maximum contention window set to some fixed value.
6. Here, traditional backoff procedure means CSMA scheme with binary exponential

backoff (BEB), unless and otherwise specified.

7 In this chapter, we use words 'algorithm', 'scheme', and 'method' interchangeably.

8 The probability P_c is different from the conditional collision probability (p_c). P_c denote the probability that a slot is busy with collision, where p_c denote the probability that a collision seen by a packet being transmitted on the channel.

Section 3
Collaborative Information Processing in VSN

Chapter 8
Visual Human Tracking in Wireless Cameras Networks:
A SURF-Based Approach

Yi Zhou
University of Technology of Troyes, France & Shanghai Jiao Tong University, China

Hichem Snoussi
University of Technology of Troyes, France

Shibao Zheng
Shanghai Jiao Tong University, China

Fethi Smach
University of Rennes I, France

ABSTRACT

In wireless camera networks, the communication load between cameras is a major concern for visual tracking. To save the bandwidth, traditional applications transfer the spatial coordinates under the precondition of camera calibration, which is computationally unreasonable for large and mobile camera networks. In this chapter, we exploit the use of distinctive and fast to compute local features to represent the non-rigid targets. Transmission of feature descriptors between cameras is done without any calibration. Combining the haar-like patterns and relative color information, our local features succeed to re-identify and relocate the target among the distributed cameras. Furthermore, efficient interest point detection and matching scheme are proposed for the visual tracking under real-time constraints.

INTRODUCTION

In this chapter, the camera network objective is to online detect and track a target moving in hostile environments. The ability of the system to detect the target state, in an automatic way, is very important for its efficient functioning or for security purposes. In fact, according to each state, the system should adopt a specific behavior. One can also mention the use of camera networks for the monitoring of production systems in order to face the industrial risks, the monitoring of the houses for safety or the house automation, the air and transport control in general, the agriculture of

DOI: 10.4018/978-1-61350-153-5.ch008

Figure 1. Distributed wireless embedded smart cameras

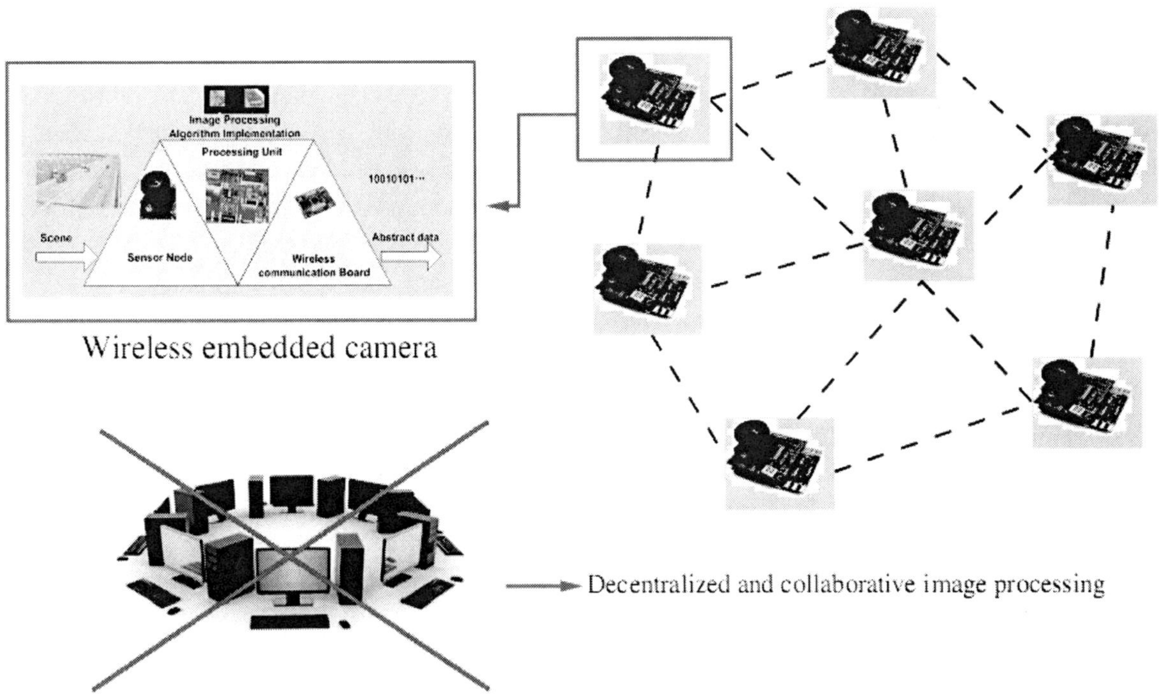

precision or intelligent alarms for the prevention of the natural disasters. With such systems, the automatic control of an event or an incident rests on the reliability of the network for an efficient and robust decision-making.

Collaborative information processing in the camera networks is becoming a very attractive field of research. In such a camera network, the imaging sensors role is not limited to detect and transmit the data to a central unit where they are processed. Individual cameras have the capability to process the raw data and transmit only abstracted information to a fusion unit. The sensors have the ability to collaborate, exchange information to ensure optimal decision. Such imaging sensors are called "smart" cameras. Contrary to the centralized approach, the system does not depend on a unique processing unit which damaging leads to the entire system failure (see Figure 1). Every smart sensor is able to play a central role and provide a suboptimal decision. The system is thus very robust against a probable foreign attack or a technical failure of the central unit. In addition, as collected data are locally processed, only relevant information is exchanged between smart nodes, limiting hence the required channel communication bandwidth. In fact, in a centralized network, all sensors transmit raw data to a unique processing unit, increasing the required communication bandwidth.

For visual tracking in wireless camera networks, the local image processing is highly affected by the limited processing unit. Therefore, the design of a wireless camera network should pay more attention to saving the overheads of local computation. Visual tracking usually consists of two steps: the first step is representing the target as a reference model in the form of feature descriptors; while the second step is inferring the best location by comparison between the reference model and the current frame. However, in the location procedure, it is challenging to prove whether a candidate box involves the true target or not. This can be simply explained by the unstable appearance of the target across time. Typically, in human tracking case, the non-rigid object model (size, shape, color, etc.), irregular movement and frequent occlusion require more robustness of the tracker. Probabilistic methods have been introduced to infer the location under uncertain and complex situations. Particle filtering (Doucet, Godsill, & Andrieu, 2000; Isard & Blake, 1998; Pérez, Vermaak, & Blake, 2004) is widely used in the literature. However, it has been shown that the particle filter might fail in the high dimension situation and also that the number of particles is critical to its performance. If the tracking environment is difficult, it would sacrifice more particles to track the target position. Variational inference (Vermaak, Lawrence, & Pérez, 2003) is an interesting alternative to the particle filter tracking. However, the variational inference still uses the Monte Carlo sampling to compute the likelihood function. The performance of the above-mentioned two Bayesian inference algorithms highly depends on the distinctive and consistent reference model. In fact, if the true target's feature model varies across the scene, no inference tools would function well by only using the initialized reference model.

On another hand, unstable model means that it is hard to find the consistent global reference features, especially in different camera views (see Figure 2). For example, the whole person color distribution is changing along with the illumination and camera optical parameters. The contour of the same person may be deformed with varying poses and different camera views. Recent research shows that the local features of the target are more robust to such variations. Here, the local aspect means dividing the whole target into smaller regions, such as sub-rectangles in (Adam, Rivlin, & Shimshoni, 2006), interest points around regions in scale invariant feature transform (SIFT) in (Lowe, 2004) and the recently proposed speed-up robust features (SURF) in (Bay, Tuytelaars, & Gool, 2006). In each sub-region, specific features are abstracted for different purposes. This kind of feature vector is also named "bag-of-features" (Mikolajczyk & Schmid, 2005; Marszaek & Schmid, 2006). This means that taking abundant local features enhances the robustness of the tracker, while global features are not robust to partial occlusions or non-stationary environments.

Furthermore, due to their distinctiveness, local features get the potential to recognize or re-identify targets in distributed camera networks. For continuous tracking in wireless camera networks, spatial information is often used to locate the target and initialize the tracking in another camera view. Many multiple cameras applications use homography (Lee, Romano, &Stein, 2000; Black, Ellis, & Rosin, 2002) to map the spatial points in different camera views. However, the measure load of computing the homography parameters is too heavy to carry out in a broad and unfixed wireless camera networks (WCN). In this work, in order to circumvent the spatial mapping, we explore the view independent local features in different views.

The chapter is organized as follows. We first give a background description for local features in visual tracking. Next, we focus on the SURF descriptor and we propose a robust and fast technique for its real-time computation in visual tracking framework. Then, a set of experiments is shown and commented. Finally, future trends and conclusions end the chapter.

BACKGROUND

In distributed wireless camera networks, the hand-off information between the cameras plays an important role in visual tracking. For saving the communication bandwidth and energy, simple spatial coordinates of the target are often sent between the networked cameras. Then, the correspondence coordinates are computed by geometric mapping for relay tracking (Medeiros, Park, & Kak, 2008). To compute the correspondence mapping matrix, careful calibration is required. However, calibra-

Figure 2. The layout of camera networks

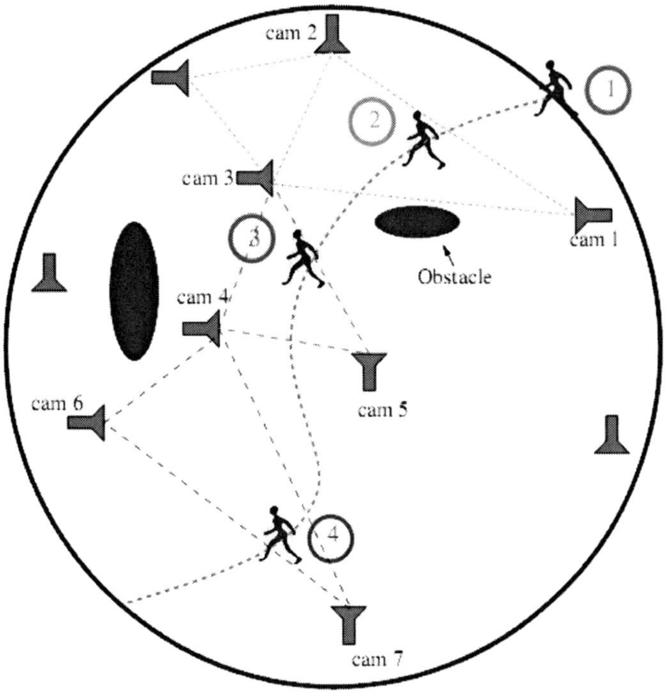

tion in a large camera networks is a heavy task. In (Tsai, 1992), by using the internal and external camera parameters, homography matrix can be computed in the same ground plane. This mapping requires the ground mark in the overlapped view of two cameras like in (PETS2006). To get less precise metric, under the assumption that the Regions of Interesting (ROI) in both camera share the same ground plane, homography matrix could be induced by the mass of correspondence points with local pixel coordinates (Black et al., 2002; Lee et al., 2000; PETS2001). To get the correspondences, lots of metric ground truth data need to be marked manually at first. For mobile cameras networks, the calibration method is even more complicated as manually setting the ground mark could be unfeasible.

In the distributed camera networks, Devarajan (2006) utilized the communication graph and vision graph to cooperate with two neighbor for defining the connected boundary and spatial relation. To get an automatically vision graph, local features are detected and matched in two overlapped views. However, the local feature-based vision graph is highly demanding, as it requires abundant correspondences in an overlapped view. Computing and matching the features in a whole scene between every two cameras are expensive for dense networks. Furthermore, the neighbor-based vision graph would not be easily reconfigurable in case of network topology variations (e.g. new sensors joining the network). In (Bai & Qi, 2010), to balance between the resource consumption and performance accuracy, light-weight corner detector based on discrete wavelet transform and compact feature descriptors are used for image comparison in the visual sensor networks.

In this chapter, we focus on tracking an object in the camera coordinates rather than estimating the object location in the world coordinates. As an alternative way to circumvent the calibration in a network, distinctive and informative features are exploited for target tracking and re-identifying across the camera networks. This method is inspired by the human visual system and is more flexible to be applied to dense (mobile) camera networks.

In image matching, registration, segmentation, object recognition and tracking, the first step consists of extracting features from the distinctive parts of the target. For regular objects, the global information is enough. For non-rigid, non-regular and texture-varied objects, dividing the whole target into sub-regions is a convenient technique to enhance the robustness of the tracker. In many cases, global information such as contour distribution (Isard & Blake, 1998a), dense HOG vector (Dalal & Triggs, 2005), Haar-like features (Viola & Jones, 2002) could be used to identify the inter-category, like the human or non-human classification or face detection. Nevertheless, because of the highly global normalization, they get little distinctiveness in intra-category applications, such as recognition or tracking. In (Yilmaz, Javed, & Shah 2006), object tracking could be categorized into 3 classes: (1) point-based techniques, (2) region-based techniques (Collins, 2003) and (3) contour-based techniques (Yilmaz, Li, & Shah, 2004; Zhao, Nevatia, & Wu, 2008). The latter two types of human tracking use global information to describe the target. For instance, global features like the color histogram (Comaniciu, Ramesh, & Meer, 2003) or contour distribution (Yilmaz, Li, & Shah, 2004), have been used to represent the targets. However, the global method performance is poor when occlusion, deformation or illumination changes occur during the tracking process. In (Adam et al., 2006), a local feature based representation was proposed. The target was divided into small sub-regions. The local specific feature is abstracted in each sub region. In practice, the smaller sub-regions are robust to outside variations and partial occlusions. Because the target gets abundant representative parts, when some parts of the sub-regions are affected by the environment changing, the unchanged sub-regions still ensure the estimation of the true target position. With only fixed patches gray histograms, the tracking succeeded to handle partial occlusions in a low-resolution monocular view.

To get a fine division into sub-regions, local feature detectors are widely used to find the informative regions. For different purposes, different detectors and variant features are used for interest regions division and object representation. In (Gabriel, Hayet, Piater, & Verly, 2005), a Harris corner detector located the interest points. Then, their local jet features and geometry information were exploited as feature descriptors. Harris matrix $\boldsymbol{H}$ indicated the weighted Sum of Squared Differences (SSD) around an image patch. In (Low 2004), Difference of Gaussian (DOG) operator is convoluted with the pyramid images. Local space-scale extrema points are detected in a multi-scale space. Around each detected points, the local feature in a 20x20 pixel rectangle is computed and concatenated into a 128 dimension-histogram of oriented gradients. Later, the SURF technique (Bay et al., 2006) outperformed the SIFT descriptor both in speed and robustness. Contrary to the required convolution in the pyramid images, SURF makes use of the integral image and size-variable box filters to simulate the multi-scale space. After the non-maximum suppression, local extreme points were selected. For the sake of fast computing, haar-like patterns were extracted as local features. Here, the haar-like patterns are similar to the signed rectangle in a 5x5 pixel region. They are convoluted on the integrated image which is already computed in the interested points detection step. Each 5x5 rectangle produces 4 dimensions feature about the sum value on the x and y directions. Therefore, in the 20x20 pixel region around the interested point, there is a 64 dimensions local feature vector. In (Strandmark & Gu, 2009), the SURF technique was used to extract interest points that are tracked in the video sequence.

Recently, in (Calonder, Lepetit, & Fua, 2008; Calonder et al., 2009), random tree classifier has been used to handle a virtually unlimited number of key points. The classifier is able to recognize a 3D point's feature under varied viewing conditions. In their proposal, the interest point detection step is the same as with the SIFT technique. By using a classifier in the matching step, the image comparison time is decreased especially when a big number of key points are involved. In (Vijay et al., 2010), local features such as HOG and SIFT are compressed and transmitted in the mobile networks. Their experiments showed that the compressed local features outperformed the MPEG-7 image signatures in the image matching applications.

The SURF descriptor is mainly used for image matching and registration, where the real-time constraint is not required. However, its implementation for visual tracking raises the problem of its computational load. In order to save the computation time and get more supporting interest points, in (Dreuw, Steingrube, Hanselmann, & Ney, 2009), a grid based SURF descriptor is used for face recognition. By exploring the coarse to fine feature distinctiveness, in (Alahi, Vandergheynst, Bierlaire, & Kunt, 2010), cascade descriptors around the SURF interest points are proposed. Recently, an alternative based on the covariance feature, fusing the color derivation and the spatial coordinates into low dimensional matrix, has been proposed in (Tuzel, Porikli, & Meer, 2006). It allows achieving a more discriminative performance, sacrificing more computing time during the matrix comparison procedure. Although more distinctive, the computation overhead makes the covariance technique far from respecting the real-time constraints.

TRACKING BASED ON THE IMPROVED SURF

Original SURF Properties

Before introducing the use of local features for visual tracking in a wireless camera networks, we recall two conditions to be taken into account. First, local features should be distinctive enough even in different cameras views, and second, the dimension of the feature descriptors should be as small as possible in order to reduce the wireless communication burden. As mentioned in the background section, the sparse distinctive point detection and region-based feature extraction perform well in the image matching literature. Also, the SURF descriptor is well adapted for the visual tracking context. For instance, the scale invariant feature allows to efficiently deal with an object size gradually changing in a scene. In the following, we give a brief review of the SURF technique and its use for visual tracking.

Prior to SURF, many salient points or region detectors have been investigated. For instance, Harris corner (Harris & Stephens 1988) was proposed for detecting the corner. Maximally stable extremal regions (MSER) (Matas, Chum, Urban, & Pajdla, 2002; Forssén & Lowe, 2007) was used for blob detection. Scalable difference of Gaussian (DOG) filters produced the local extremal interest points in (Lowe, 2004). In SURF (Bay et al., 2006), the Hessian matrix took the place of the DOG filters to detect the local extremal points and showed the advantage of repeatability compared to its counterparts. It relies on the determinant of the Hessian (Lindeberg, 1998). Given a 2D image *I*(x,y), where x, y are the pixel coordinates, the Hessian matrix *H*(x,y) at scale σ is computed as follows:

$$H(x,y,\sigma) = \Phi(x,y,\sigma) \otimes \begin{pmatrix} \frac{\partial^2 I}{\partial x^2} & \frac{\partial^2 I}{\partial xy} \\ \frac{\partial^2 I}{\partial xy} & \frac{\partial^2 I}{\partial y^2} \end{pmatrix} \quad (1)$$

where $\Phi(x,y,\sigma)$ is a Gaussian filter mask around point (x, y) with scale σ The convolution between the Gaussian filter and the second order derivative matrix produces the Hessian matrix. To get a faster detection, Bay et al. (2006) approximate

Figure 3. The output of interest points and local extreme regions in two scales

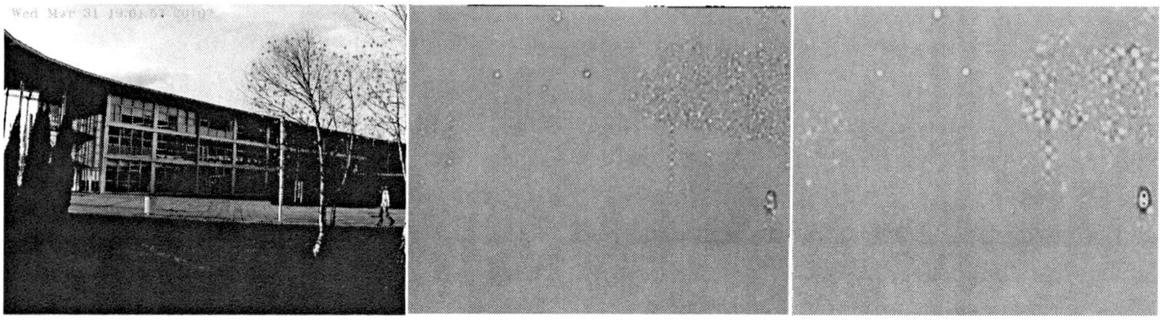

the Hessian matrix by box filters (Bxx, ${}_B$yy, ${}_B$xy)${}_i$n its three directions (xx, yy, xy). Then, by using integral image, the box filtering computation cost is sharply decreased comparing to the Gaussian convolution. Integral image was first introduced by (Viola and Jones 2002), for computing the haar-like patterns in face detection applications. In order to build an integral image, accumulating is operated in each pixel as follows:

$$II(x,y) = \sum_{i=0}^{i \le x} \sum_{j=0}^{j \le y} I(i,j) \tag{2}$$

where I means the intensity image and II(x, y) represents the integral image value at coordinates (x, y). Once building up the intensity image, only 4 memory accesses and 3 addition operators are needed for computing box region intensity. This technique speeds up the box filter-based image convolution. Interestingly, in SURF, the integral image is fully exploited for both Hessian matrix approximation and haar-like descriptors extraction. Therefore, the computation time was greatly saved, compared to its counterpart SIFT. The determinant of the approximated Hessian matrix

$$\det(H_{approx}) = B_{xx}B_{yy} - (wB_{xy})^2 \tag{3}$$

is used to validate an interest point.

In Figure 3, from left to right, the SURF interest points are marked with the red color in a testing scene foreground. The right two frames impressively give the extreme region response in the map of H_{approx} under scale 1.2 and 2 respectively. The hottest color means the highest value. As it is shown in the scene, each interest point is located in the local extreme region.

Taking the advantage of box filters, the SURF uses the scaled box filters other than the image pyramid to find the salient points in different scales. This makes the SURF faster than the SIFT technique which is based on the pyramid difference of Gaussian images for scale variable interest points detection.

After the region searching, the scale space extreme points are located by using non-maximum suppression in 3x3x3 neighborhoods. Through this method, the comparatively low value points but with similar features are omitted. Each interest point in SURF defines a scaled region. Every region is partitioned into a regular grid. In a sub-rectangle, haar-like box filters in both x and y directions (d_x,d_y) are used to compute the coarse gradient information based on the integral image. Therefore, in an interest point region, the feature descriptor is represented as a vector:

$$\vec{v} = \left(\sum d_x + \sum d_y + \sum |d_x| + \sum |d_y|\right) \tag{4}$$

The Drawbacks of SURF for Visual Tracking

SURF was firstly applied for image matching. Inspired by the robustness and the distinctiveness of SURF features, Strandmark et al (2009) investigated the use of the SURF descriptors for visual tracking in low-resolution videos. However, in practical situations, there are still some drawbacks for the use of SURF in visual tracking.

As we mentioned before, SURF adopts the non-maximum suppression in spatial and scale space. This scheme can eliminate the similar feature points, which can confuse the matching in the next step. However, in the visual tracking application, the target only covers a relatively small part of the whole image. Strictly selecting the interest points does not only increase the computation time, but also decreases the feature points in the target region. In worse conditions, there are no points at all for a human in a video sequence. Strandmark et al (2009) compute the mapping matrix and the consensus points in order to take into account the spatial information of the interest points distribution. But, if the detected interest points is lower than 4, one cannot compute the mapping matrix.

The use of SURF in tracking consists of matching the target interest points in successive frames. Matching is based on the Euclidean distance between the descriptors. A ratio-based procedure is classically used in order to ensure the similarity between the matched interest points. However, the ratio- based matching excludes to select the points distributed in a closer region. As a result, only one or two points are matched during the tracking process. This low matching rate would confuse the tracker with other objects in the scene.

Another drawback of the use of SURF in tracking is related to its low distinctiveness with intra-category objects. In fact, it uses Haar-like features which are broadly used for face or human detection (Monteiro, Peixoto, & Nunes, 2006; Chunhui, Yan, Yongmei, & Zhaoxia, 2008; Viola, Jones, & Snow, 2003). However, these features are less successful for recognition. In practice, the SURF features yield error-prone matching for two different persons in the same scene.

Improved SURF

The diagram, in the Figure 4, summarizes the proposed tracking procedure based on the SURF descriptor. By using background model, the detecting range of Interest Points (IP) is greatly shrunk. With lower threshold in the detecting step, more points would be taken from the small target region. The closely distributed interest points are clustered into a group. This group reflects the object distinctiveness in the feature space. After extracting features and matching with the reference, a cluster will be recognized as the target or not. If it is considered as a target, all the unmatched points in this cluster will be stored in the unmatched data sets. If the unmatched points keep on appearing for several times, then they are transformed to reference points.

Fast Detection of Interest Points

The detection of interest points and the extraction of high dimensional feature descriptors in SURF are time consuming. For example, to get the feature vector in a 640x480 frame, it takes more than 200 milliseconds to detect the extreme points and additional 200 milliseconds to compute the corresponding feature descriptors. For video surveillance applications, the region of interest is mainly focusing on the moving foreground in a limited region. Besides the static image content, the video sequence yields information about the time correlation. Based on the background model, the foreground region could be easily obtained by background subtraction (Stauffer & Grimson, 1999; Cheung & Kamath, 2005). Therefore, detecting the interesting points only in the foreground saves as much time as the omitted background area. In this chapter, the approximate median filter

Figure 4. The diagram of improved SURF for visual tracking

(McFarlane & Schofield, 1995) is used for the background modeling. The approximate median method works as follows: the background is first initialized to a static image (e.g. the first frame); then, if the difference between a pixel in the current frame and the one in the background has a value larger than the threshold, the background pixel is incremented by one, otherwise, it is decremented by one. Along with the time increasing, the background approximately converges to the median.

To build a good reference model, we need abundant feature descriptors for coping with different variations. However, the SURF point detector only locates Hessian extrema high value points in spatial and scale neighborhoods. This strict condition suits for the whole image matching, but produces few points for a human target region. In order to get more feature points in a small region of human foreground, we propose to take points with Hessian above a fixed threshold. In Figure 5, the interest points extracted by two methods are marked. Our method (in blue) produces more interest points compared to the original SURF extractor (in red). In practice, during the pose changing or partial occlusion, the target SURF interest points may be lost, leading to the divergence of the tracker. However, with the proposed abundant interest point extraction, the matching procedure becomes robust enough to deal with such difficult situations.

Relative Color Ratio and Clustering

In order to track the target, the extracted interest points from the current frame are matched to the target interest points extracted in the reference frame. The matching often yields outliers. In other words, the target interest points are matched to points not belonging to the target in the current frame. In order to alleviate the outlier matching points, the distance-based interest points clustering and color information can be exploited for good matching.

In (Strandmark & Gu, 2009), the outliers are eliminated by conserving the spatial information of the interest points through a homography transformation. The points respecting a group transformation are considered as consensus points and the remaining matched interest points are considered as outliers. The main drawback of this solution is the fact that if there is less than 4 matched points, the homography matrix cannot be computed.

In this work, we propose to first cluster the interest points according to their spatial spread. In fact, on each foreground object, the interest points are spatially close to each other. By using the distance information, these points are easily clustered into different groups corresponding to the different foregrounds. In order to enhance the distinctiveness of the clustered interest points for later selecting the target, the color information is incorporated in the descriptor vector. In each interest point region, the ratio of the blue, green and red components are computed and stored in the feature descriptors. In our experiment, in each local rectangle, blue, green, and red intensity are summed respectively in BGR channels as Sb, Sg, Sr. Then the ratio $Si/(Sb + Sg + Sr)$, is kept for color information (the index i means b, g or r).

Figure 5. Find interest points based on the foreground in the "campus" video sequence

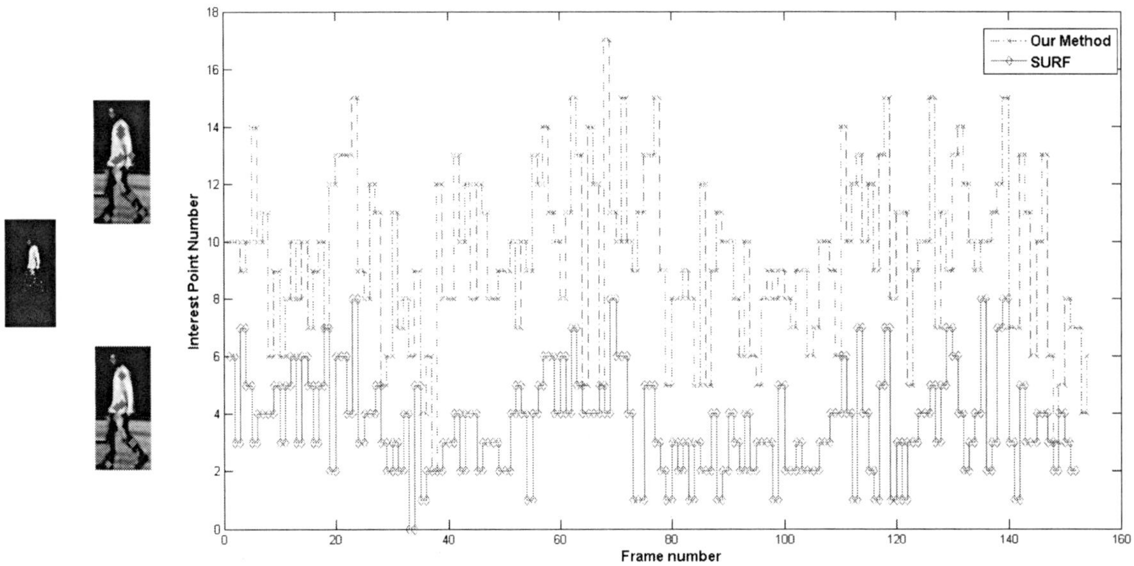

Matching and Updating

After clustering the interest points, the matching step consists of selecting the cluster representing the target. The centroïd of the selected cluster will be considered as the target position estimate. In the matching, when the minimum distance between two points is lower than the defined threshold, the two points are considered as matched pair. In Figure 6, the target of the reference model is copied to one of the following frames. All interest points are marked in red. The matched points are given additional green circles and linked together with white lines.

An important step in target tracking is the update of the reference model. The update should be performed carefully in order not to confuse the reference model with occluding objects. Once the cluster is selected as representing the target, there will be two types of points: points matching with the reference interest points and unmatched points. As the initial reference points may not reflect the varying features of the target through the time, it is reasonable to update the reference model and add new interest points. We propose to introduce a cumulative scoring strategy allowing an efficient update of the reference interest points. In the current frame, the unmatched points from the selected cluster are stored and compared with selected cluster points in the next frame. If these additional points are matching with some points in the next frame cluster, their score is incremented by 1. When the score is above a fixed threshold, which means that they have a sufficient repeatability, these points are upgraded as new reference points. In figure 6, there are five newly updated reference points marked red in the middle of the two figures. The green color means that the new reference points are matched with the interest points in the current figure. By adding new feature points, the reference model is improved. Especially, in different networked cameras viewpoints, this reference model update shows a high robustness allowing an efficient relocating of the target when a hand-off occurred between two processing cameras.

Figure 6. Matching points among the first reference figure and a captured figure

Communicating between Cameras

Instead of passing the spatial information between cameras, the distinctive local features between cameras are transmitted. Thanks to the view independent and scale resistant local features, the relay tracking camera is able to re-identify the target correctly in the next frame. One target in a scene gets around 60 interest points as reference model (including 64 dimensions haar-like features and 3 color ratio values for each interest point). Therefore, the communication load between the cameras is below 30 kbytes for one target. With the development of wireless communication, the robust feature descriptors get a broad potential to re-identify human in a camera networks. When one target disappears in a camera view, the local features (SURF features and color ratio) would be sent to its neighbor cameras. As discussed before, the SURF features are view independent. Once the target can be correctly located in another camera, the newly matching and updating step would continue to work and add novel feature points in the second camera view.

Experiments

In this section, two different networked cameras are used for the image acquisition. The type of camera A is AV1305 made by Arecont Vison and the type of camera B is Sarix12DN20 made by Pelco. Both cameras are using 640x480 resolutions. Each camera gets a processing laptop node. To test the feature performance, the cameras are positioned with non-overlapped views. The average tracking time in the laptop node (Intel dual core 2.5GHZ CPU and 4GB RAM) is around 120 milliseconds. The communication between the two cameras is done through the laptops ad-hoc wireless network.

In Figure 7, we manually choose the target to track. The reference model is automatically built based on the detected interest points and descriptors. The scene is quite challenging as many occlusions (partial and total) occur during the tracking in

Figure 7. Tracking with partial occlusions in camera A

Figure 8. Tracking with total occlusion in camera A

both networked cameras. From Figure 7, it can be noted that despite the complete occlusion caused by the tree, the target is automatically relocated just after its appearance. The red marks represent the interest points, the yellow rectangle represents the region of interest and the green circles describe the matched points during the tracking process. To compare with our proposal, RAMOSAC based visual tracking (Strandmark & Gu, 2009), which also uses the SURF features are tested on the same sequences. The bold dash lines in 3 images clearly show the deformation of the rectangle as the number of feature points decreases. In the 3rd scene, RAMOSAC begins to lose the target, while our proposal gets a correct box.

In Figure 8, the visual tracking encounters the total occlusion challenges. From left to right, the target is firstly partially occluded by a white object. By taking the use of upper body local features, the tracker correctly located the target. Latter, the target is occluded by another person with different clothes color. As the descriptors involve the color information, there are no mismatching points in the front person. After the occlusion, the algorithm succeeds to relocate the true target again.

In Figure 9, when the target disappears from the first camera A, the reference model is sent to the next camera B. In our experiment, 66 interest points and their descriptors (the total size is around 30Kb) are sent through a wireless channel link. Based on these features, the tracker identifies the target in camera B. As mentioned before, the two cameras were made by different companies and with different optical parameters. The re-identification shows that our local features are very robust to this environment and technical variations of the networked cameras. In this sequence (camera B), the target is walking towards the camera, increasing the target box scale. Based on their scale invariant nature, the used features keep the good matching points. By successively using this relaying scheme, the tracking is ensured through a network of many cameras covering complex scenes.

Figure 9. Relay Tracking with scale changing in camera B

CONCLUSION

To circumvent the complex pre-calibration in a wireless camera networks, we design a visual tracking based on local scale invariant features, namely the SURF descriptors. Due to the SURF invariance to scale and transformation in different camera views, the re-identification procedure is successfully implemented. In order to improve the robustness and fast computation in visual tracking, we propose to find abundant interest points around the extreme Hessian response regions. Color ratio information is added to the descriptors. This improves both the speed and the accuracy of the re-identification. Along with the time and the visual environment variations, an updating scheme is also proposed. The experiments show that local features distinctiveness plays a vital role in the visual tracking and detection. Future work will focus on the improvement of the speed and on the combination of different local features.

FUTURE RESEARCH DIRECTIONS

The biggest challenge for visual tracking is the variation of the target model. However, in multiple camera networks, the cameras get more information about the target in different views. This information should be shared and better exploited in order to enhance the tracker performance. To cope with the model variation in visual applications, several local features are explored. The final goal is to find fast and robust features. However, the balance between robustness and fast computation is not easy to achieve. For fast computation, fast interest point detection should be proposed. Around an interest point, what features to extract is up to the performance of the tracker. High-dimensional features lead to expensive computation. Reversely, low dimension sacrifices the distinctiveness, yielding less robust tracker. Recently, the study of the correlation between local features becomes popular. This correlation could be used to reduce the feature dimension, while keeping its robustness.

Based on robust visual tracking, the camera networks could be able to detect abnormal events, such as luggage abandon in crowded regions, forbidden region intrusion, etc. Combining low level and high level image analysis in a network of smart cameras while respecting the real-time constraints will be among the challenging research directions in this emergent multidisciplinary scientific field.

REFERENCES

Adam, A., Rivlin, E., & Shimshoni, I. (2006, June). Robust fragments based tracking using the integral histogram. In *Proceedings IEEE Conference on Computer Vision and Pattern Recognition, 1*, 798-805.

Alahi, A., Vandergheynst, P., Bierlaire, M., & Kunt, M. (2010). Cascade of descriptors to detect and track objects across any network of cameras. *Computer Vision and Image Understanding*, *114*(6), 624–640. doi:10.1016/j.cviu.2010.01.004

Bai, Y., & Qi, H. (2010). Feature-based image comparison for semantic neighbor selection in resource-constrained visual sensor networks. *EURASIP Journal on Image and Video Processing*, *2010*. doi:10.1155/2010/469563

Bay, H., Tuytelaars, T., & Gool, L. V. (2006). Surf: Speeded up robust features. *Computer Vision and Image Understanding*, *110*(3), 346–359. doi:10.1016/j.cviu.2007.09.014

Black, J., Ellis, T., & Rosin, P. (2002). Multi view image surveillance and tracking. In *IEEE Workshop on Motion and Video Computing, 0*, 169.

Calonder, M., Lepetit, V., & Fua, P. (2008). Keypoint signatures for fast learning and recognition. In *Proceedings of the 10th European Conference on Computer Vision* (pp.58–71).

Calonder, M., Lepetit, V., Fua, P., Konolige, K., Bowman, J., & Mihelich, P. (2009). Compact signatures for high-speed interest point description and matching. In *IEEE 12th International Conference on Computer Vision* (p. 357 -364).

Cheung, S.-C. S., & Kamath, C. (2005). Robust background subtraction with foreground validation for urban traffic video. *EURASIP Journal on Applied Signal Processing*, (14): 2330–2340. doi:10.1155/ASP.2005.2330

Chunhui, Z., Yan, L., Yongmei, C., & Zhaoxia, L. (2008). Research of cues selection in pedestrian detector based on Adaboost. In *IEEE Conference on Cybernetics and Intelligent Systems* (p. 546 -549).

Collins, R. T. (2003). Mean-shift blob tracking through scale space. In *Proceedings of IEEE International Conference on Computer Vision and Pattern Recognition.*

Comaniciu, D., Ramesh, V., & Meer, P. (2003). Kernel based object tracking. *IEEE Transactions on Pattern Analysis and Machine Intelligence*, *25*(5), 564–577. doi:10.1109/TPAMI.2003.1195991

Dalal, N., & Triggs, B. (2005, June). Histograms of oriented gradients for human detection. In. *Proceedings of IEEE Conference on Computer Vision and Pattern Recognition*, *1*, 886–893.

Devarajan, D., Radke, R. J., & Chung, H. (2006). Distributed metric calibration of ad hoc camera networks. *ACM Transactions on Sensor Networks*, *2*(3), 380–403. doi:10.1145/1167935.1167939

Doucet, A., Godsill, S., & Andrieu, C. (2000). On sequential Monte Carlo sampling methods for Bayesian filtering. *Statistics and Computing*, *10*(3), 197–208. doi:10.1023/A:1008935410038

Dreuw, P., Steingrube, P., Hanselmann, H., & Ney, H. (2009, septembre). Surf-face: Face recognition under viewpoint consistency constraints. In *British Machine Vision Conference*, London, UK.

Forssén, P. E., & Lowe, D. (2007, October). Shape descriptors for maximally stable extremal regions. In *IEEE international conference on computer vision*, Rio de Janeiro, Brazil.

Gabriel, P., Hayet, J. b., Piater, J., & Verly, J. (2005). Object tracking using color interest points. In *IEEE Proceedings of International Conference on Advanced Video and Signal Based Surveillance* (pp. 159-164).

Harris, C., & Stephens, M. (1988). A combined corner and edge detection. In *Proceedings of the 4th Alvey Vision Conference* (pp. 147-151).

Isard, M., & Blake, A. (1998). 1 erAugust). Condensation-Conditional density propagation for visual tracking. *International Journal of Computer Vision, 29*(1), 5–28. doi:10.1023/A:1008078328650

Laptev, I., & Lindeberg, T. (2003). Space-time interest points. In *IEEE International Conference on Computer Vision*, Nice, France.

Lee, L., Romano, R., & Stein, G. (2000). Monitoring activities from multiple video streams: Establishing a common coordinate frame. *IEEE Transactions on Pattern Analysis and Machine Intelligence, 22*(8), 758–767. doi:10.1109/34.868678

Lindeberg, T. (1998). Feature detection with automatic scale selection. *International Journal of Computer Vision, 30*(2), 79–116. doi:10.1023/A:1008045108935

Lowe, D. G. (2004). Distinctive image features from scale-invariant keypoints. *International Journal of Computer Vision, 60*(2), 91–110. doi:10.1023/B:VISI.0000029664.99615.94

Marszaek, M., & Schmid, C. (2006). Spatial weighting for bag-of-features. In. *Proceedings of IEEE International Conference on Computer Vision and Pattern Recognition, 2*, 2118–2125.

Matas, J., Chum, O., Urban, M., & Pajdla, T. (2002). Robust wide baseline stereo from maximally stable extremal regions. In. *Proceedings of British Machine Vision Conference, 1*, 384–393.

McFarlane, N. J. B., & Schofield, C. P. (1995, May). Segmentation and tracking of piglets in images. *Machine Vision and Applications, 8*(3), 187–193. doi:10.1007/BF01215814

Medeiros, H., Park, J., & Kak, A. (2008, Aug.). Distributed object tracking using a cluster based Kalman filter in wireless camera networks. *IEEE Journal of Selected Topics in Signal Processing, 2*(4), 448–463. doi:10.1109/JSTSP.2008.2001310

Mikolajczyk, K., & Schmid, C. (2005, Oct.). A performance evaluation of local descriptors. *IEEE Transactions on Pattern Analysis and Machine Intelligence, 27*(10), 1615–1630. doi:10.1109/TPAMI.2005.188

Monteiro, G., Peixoto, P., & Nunes, U. (2006). Vision based pedestrian detection using Haar-like features. In *Proceedings of Scientific Meeting on the Portuguese Robotics Festival.*

Pérez, P., Vermaak, J., & Blake, A. (2004, March). Data fusion for visual tracking with particles. *Proceedings of the IEEE, 92*(3), 495–513. doi:10.1109/JPROC.2003.823147

PETS2001. *(2001).* 2nd IEEE International Workshop on Performance Evaluation of Tracking and Surveillance.

PETS2006. *(2006).* 9th IEEE International Workshop on Performance Evaluation of Tracking and Surveillance.

Stauffer, C., & Grimson, W. (1999). Adaptive background mixture models for real-time tracking. In. *Proceedings of IEEE International Conference on Computer Vision and Pattern Recognition, 2*, 252.

Strandmark, P., & Gu, I. Y. (2009). Joint random sample consensus and multiple motion models for robust video tracking. In *Proceedings of Scandinavian Conference on Image Analysis* (pp. 450-459).

Tsai, R. Y. (1992). A versatile camera calibration technique for high-accuracy 3-D machine vision metrology using off-the-shelf TV cameras and lenses. *IEEE Journal on Robotics and Automation, 3*(4), 221–244.

Tuzel, O., Porikli, F., & Meer, P. (2006). Region covariance: A fast descriptor for detection and classification. In *Proceedings of the 9th European Conference on Computer Vision* (pp. 589-600).

Vermaak, J., Lawrence, N. D., & Pérez, P. (2003, June). Variational inference for visual tracking. In. *Proceedings of IEEE International Conference on Computer Vision and Pattern Recognition, 1*, 773–780.

Vijay, C., David, C., Andy, L., Gabriel, T., & Sam, T. …Ngai-Man, C. (2010). *Comparison of local feature descriptors for mobile visual search.* In IEEE International Conference on Image Processing(ICIP). Hong Kong, China.

Viola, P., & Jones, M. (2002). Robust real-time object detection. *International Journal of Computer Vision, 57*, 137–154. doi:10.1023/B:VISI.0000013087.49260.fb

Viola, P., Jones, M., & Snow, D. (2003, Oct.). Detecting pedestrians using patterns of motion and appearance. In. *Proceedings of IEEE Conference on Computer Vision, 2*, 734–741. doi:10.1109/ICCV.2003.1238422

Yilmaz, A., Javed, O., & Shah, M. (2006). Object tracking: A survey. *ACM Computing Surveys, 38*(4), 13. doi:10.1145/1177352.1177355

Yilmaz, A., Li, X., & Shah, M. (2004). Contour based object tracking with occlusion handling in video acquired using mobile cameras. *IEEE Transactions on Pattern Analysis and Machine Intelligence, 26*(11), 1531–1536. doi:10.1109/TPAMI.2004.96

Zhao, T., Nevatia, R., & Wu, B. (2008). Segmentation and tracking of multiple humans in crowded environments. *IEEE Transactions on Pattern Analysis and Machine Intelligence, 30*, 1198–1211. doi:10.1109/TPAMI.2007.70770

KEY TERMS AND DEFINITIONS

HOG: Histogram of oriented gradient.

Interest Point: Select point in an image, where the features are extreme in certain feature space.

Local Feature: Abstract features in local rectangular, circle or spot. Features could be color, texture, HOG, or SURF, etc.

Relay Tracking: When one camera loses the target, another nearby camera continues to track the former target.

SURF: Speed-up robust features.

Visual Human Tracking: Tracing human in a video sequences or in a real time surveillance application.

Wireless Camera: wireless sensor which is able to capture and process local images and to send the compressed data to other node by wireless communication.

Chapter 9
Object Association through Multiple Camera Collaboration for Large-Scale Surveillance System

Shung Han Cho
Stony Brook University-SUNY, USA

Kyung Hoon Kim
Stony Brook University-SUNY, USA

Yunyoung Nam
Stony Brook University-SUNY, USA

Sangjin Hong
Stony Brook University-SUNY, USA

ABSTRACT

In this chapter, we present an object association method through multiple camera collaboration for a large-scale surveillance system. The object association is achieved by locally generating homographic lines on targets in collaborating cameras. In order to maintain the object association with the insufficient separation between homographic lines due to densely populated objects, homographic points are generated in 3-D with estimated heights. The heights of targets are estimated by the linear least-squares using normal equations. The object association is confirmed by finding the pairs of the correspondences minimizing the distance between them. The proposed method is verified with real video sequences. The simulation result demonstrates that the proposed method is robust against false association because it considers all the possible pairing cases of occluded targets.

DOI: 10.4018/978-1-61350-153-5.ch009

INTRODUCTION

Recently, many researchers have much interest in a multiple camera based surveillance system (Cai, 1996; Calderara, 2005; Chang, 2000; Dockstader, 2001; Hu, 2004; Tsutsui, 2001; Utsumi, 1998). Various issues have been addressed such as optimal camera placements (Chakrabarty, 2002; Horster, 2006), calibration of multiple cameras (Lee, 2000; Senior, 2005), finding the correspondence of objects (Black, 2006; Caspi, 2006; Kang, 2003; Kelly,1995; Kyong, 2008; Li, 2002), camera handoff (Chen, 2008), tracking (Velipasalar, 2005; Tsutsui, 2001; Utsumi, 1998) etc. In a large-scale surveillance system, multiple cameras change their views flexibly for the maximized coverage and they have redundant information due to the overlapped views of multiple cameras. In order to reduce inconsistent and uncertain information, it is necessary to exchange target information for multiple camera collaboration. Moreover, the large-scale surveillance system may have the slow processing rates for multiple cameras, and it can affect the performance of objects tracking and association. The delayed processing due to the data traffic overload may create the failure of objects tracking and association. Especially, when a surveillance application needs to support wireless sensor networks, minimizing data traffic is a critical issue with multiple cameras. Therefore, efficient object association method is required for multiple camera collaboration.

Various methods are investigated for the multiple object association. They are classified into a feature based approach and a geometry based approach. Possible features used in the feature based approach are colors, histograms, heights and gaits. However, a system cannot have distinguishable features for all objects and extract them accurately. The performance of feature based approaches is severely affected by objects having similar features (Krumm, 2000; Mittal, 2003; Orwell, 1999). In geometric based approaches, virtual lines such as homographic lines or epipolar lines are often utilized. Khan (2003) proposes a method to find the correspondence of objects by using the boundaries of fields of view (FOVs). They can be automatically constructed on other cameras by using object movements. Black (2006) uses epipolar lines for the object association with known fundamental matrices. Also, Hu (2006) utilizes the principal axis of targets. The extracted principal axis of targets is transformed onto other cameras with the known homography matrix to find the correspondence of objects. However, these approaches require training to construct the boundaries of FOVs or known references to find the fundamental or homography matrix. These preprocesses restrict flexible camera movements for an autonomous surveillance system while the accurate transformation is guaranteed. Jaynes (2004) uses a network synchronization signal to calibrate multiple cameras from planar motion trajectories. The synchronization signal helps to reduce the number of trajectories that need to be compared by using the time constraint at observation points of the cameras. Fleuret (2008) also uses synchronized video streams to track people with multiple cameras by back-projecting detected objects in images into a discretized occupancy map. The method ignores the effect of synchronization issues that can frequently occurr in wireless sensor network. Kayumbi (2008) maps object trajectories with homography estimation accounting for the lens distortion. Sundaresan (2009) presents a method to track the articulated motion of humans using image sequences obtained from multiple cameras. The method focuses on the tracking of detailed human motion rather than multiple objects tracking for the surveillance system.

In order to support flexible camera movements, homographic lines are locally generated on targets in each camera and they are projected onto the other cameras (Kyong, 2008). The object association is established if projected homographic lines intersect with their corresponding targets. Because the reference ground plane is not used for transforming and projecting homographic lines, a common ground plane is not necessarily shown to all the cameras. However, the association performance is restricted by the insufficient separation of homographic lines and is degenerated for densely populated objects. In this case, the association process needs to rely on a feature based local tracking, which may lead to false or failed associations. Thus, an effective object association method is required to complement the limitation of the homographic lines based association method for a large-scale surveillance system.

In this chapter, we present a dynamic object association method through multiple camera collaboration based on the locally initiating homographic lines based approach. The limitation of the homographic lines based association method is analyzed in terms of the effectiveness in the association process. In order to alleviate the limitation of the homographic lines based association method, we introduce homographic points generated with estimated heights. The heights of objects are estimated by the linear least-squares using normal equations to minimize the effects of the different views of cameras, inaccurate camera matrices, and the unknown heights of targets. The association of targets is reestablished or confirmed by finding the pairs of corresponding targets having the minimum distance. The proposed method is verified with real video sequences and the performance is evaluated with respect to the detection performance.

BACKGROUND AND PROBLEM DESCRIPTION

The common approach for the object association is to construct the homography matrix representing the transformation relationship between two cameras. The homography matrix is constructed by known corresponding points (at least four points) (Hartley, 2003; Kayumbi, 2008) and a ground plane is commonly used because known corresponding points can be easily extracted. However, when a ground plane may not be shown to all the cameras, it is difficult to determine a reference plane for the homography matrix. Also, if the views of multiple cameras change flexibly in a large-scale surveillance system, it is not trivial to construct the homography matrix all the time.

Locally initiating homographic lines based association method is introduced to support dynamic camera movements (Kyong, 2008). Homographic lines are initiated on targets, they are projected onto the global plane and then they are transformed onto other cameras for the object association. This method assumes that association targets are the detected faces of objects when a common ground plane is not necessarily shown to all the cameras. Various techniques are available for target detection (Li, 2004; Viola, 2004). Figure 1 illustrates the locally initiating homographic lines based approach with the separation issue of homographic lines. T_i^k denotes a detected target of object O_i in camera C^k and the size of target T_i^k is denoted by r_i^k. L_i^k denotes a generated homographic line on each target and SL_i^k denotes a projected homographic line of L_i^k in other cameras. An object association is established if a projected homographic line intersects with a corresponding target distinctively. Since each target is not a point source on images, the separation of projected homographic lines should be greater than at least the twice size of each target. Thus, the object association is restricted by the required separation of projected homographic lines while the association process is dynamically initiated.

The performance of the locally generated homographic lines based association method is more degenerated by considering height uncertainty of targets, synchronization issues between multiple cameras and variance of target locations by detection performance. A projected homographic line can deviate from the centroid of a target because of them as shown in Figure 1. In order to incorporate the effect of them, the size of each target is expanded for the correct intersection with a corresponding projected

Figure 1. Illustration of an unsuccessful association of targets due to insufficient separation in a non-ideal situation, and its propagation creating the inconsistent information by a false association (the frame difference between the first frames (top) and the second frames (bottom) is 9 frames)

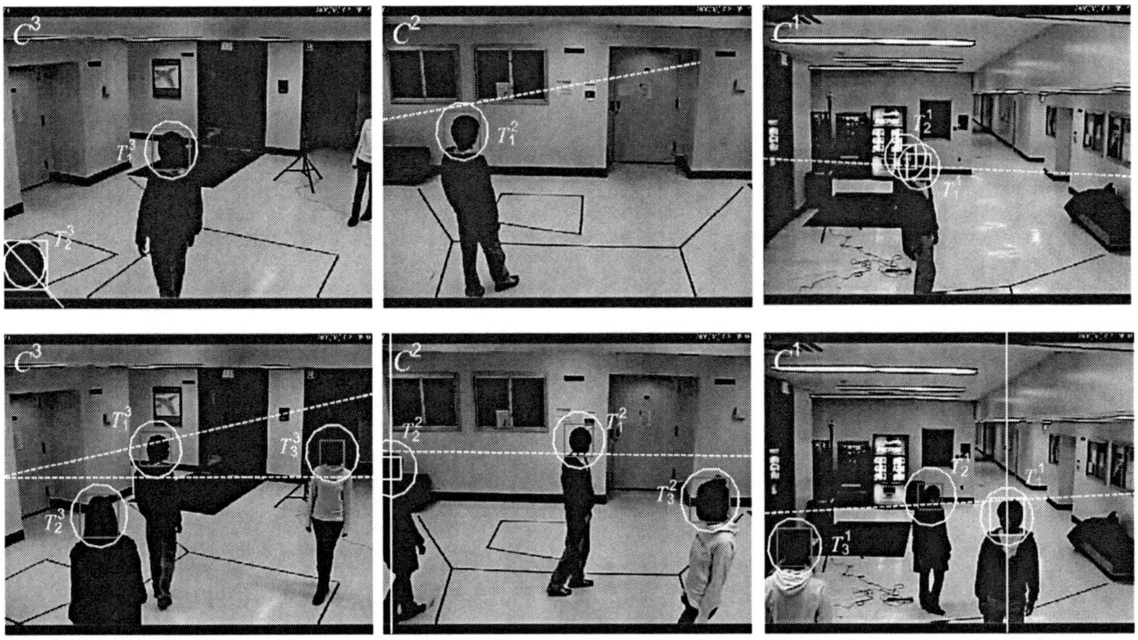

homographic line. s_{min} denotes the radius of an association circle, as the expanded size of a target. Although the expanded radius of an association circle ensures the correct intersections with corresponding homographic lines, this also increases the intersections with non-corresponding homographic lines. This may cause the association ambiguity to determine corresponding targets among cameras. Thus, projected homographic lines can become ineffective in the association process when the separation between them is smaller than the sum of radii of their association circles.

When many cameras (i.e. more than two cameras) are involved in the object association, the locally generated homographic lines based association method can be extended to support them through pair-wise collaboration. However, each pair of collaborating cameras may contradict the object association due to the insufficient separation. It may generate false associations and create the inconsistent information of uncertain association in the system. They are also propagated in time and continuously degrade the association performance. Although only one object is not associated in one instant, it affects the overall performance of object association in time. For example, Figure 1, targets are unassociated during 9 frames and they create a false association between targets T_1^1 and T_2^2 in the second frames. The system keeps the inconsistent information of the falsely associated targets until they leave the surveillance region. Moreover, when multiple cameras collaborate on object association without the sufficient separation between targets, homographic lines are unnecessarily generated on targets and projected to the other cameras without establishing object association. The association failure due to the insufficient separation wastes the computational costs of transforming homographic lines. In a distributed camera network, minimizing data to be transferred is also very important to efficiently exchange data between the camera systems without any data loss or latency. It also increases the frame rate achieved by the system and improves the performance of algorithms. In general, each target requires transformation of a local homographic line to a global homographic line or transformation of a global homographic line to a local homographic line. The number of the intersection tests is proportional to the multiplication of the number of targets and the number of transformed homographic lines. It is assumed that a pair-wise association is utilized when more than two cameras are used. Then, the computational costs for using homographic lines are represented by the number of transformations and intersection tests, C_T and C_C, respectively

$$C_T = 2I\times 2(K\text{-}1)$$

$$C_C = I^2\times 2(K\text{-}1), \quad (1)$$

where I denotes the number of detected targets in K cameras. The number of the transformation increases linearly to the number of targets and the number of intersection tests increases exponentially to the number of targets. It also exponentially increases the amount of exchanged data for target information on the network. As the performance of object association is improved, the number of targets to be associated is decreased at one time. Eventually, it decreases the amount of the exchanged data in the long term. Thus, a proper collaboration camera strategy is necessary to efficiently associate targets among different cameras as minimizing the inconsistent and uncertain information.

The proposed method uses the second strategy between two different multiple camera collaboration strategies introduced by Cho (2010) to establish the object association as reducing the computational costs and minimizing the inconsistent information. Moreover, in order to alleviate the separation issue of homographic lines, we utilize homographic points for the object association after corresponding targets are associated by locally generated homographic lines. Because the accuracy of homographic points is determined by the heights of objects, we estimate the heights of objects by the linear least-squares using normal equations. The estimated heights are determined by minimizing the effect of the different views of cameras, inaccurate camera matrices, and the unknown heights of targets. Homographic points are generated in 3-D for previously associated targets in each pair of collaborating cameras and their association is confirmed by finding the minimum distance among all the possible pairs. This enables a system to confirm the association regardless of the separation between targets.

NON-IDEAL PARAMETERS IN HOMOGRAPHIC LINES BASED ASSOCIATION

Effect of Height Uncertainty on s_{min}

When homographic lines are projected to other cameras, accurate heights of targets are initially unknown and each object has a different height. Hence homographic lines are generated through the global reference plane of an average target height. If the actual height of a target differs from the average height of a target, a projected homographic line can deviate from the centroid of a target. The deviation can affect the association performance when a projected homographic line does not intersect with the association circle with radius r_i^k of a corresponding target. Thus, adjusted radius $s_{\min,i}^k$ is defined by

$$s_{\min,i}^k = \max(r_i^k, \sigma_{h,i}^k), \quad (2)$$

where $\sigma_{h,i}^k$ denotes the possible number of pixels which deviate from the centroid of a target by the mismatched height of a target in camera C^k. $\sigma_{h,i}^k$ depends on camera configurations (i.e. locations,

Figure 2. Illustration of the effect of height uncertainty between cameras C^1 and C^2 according to the distance to an object

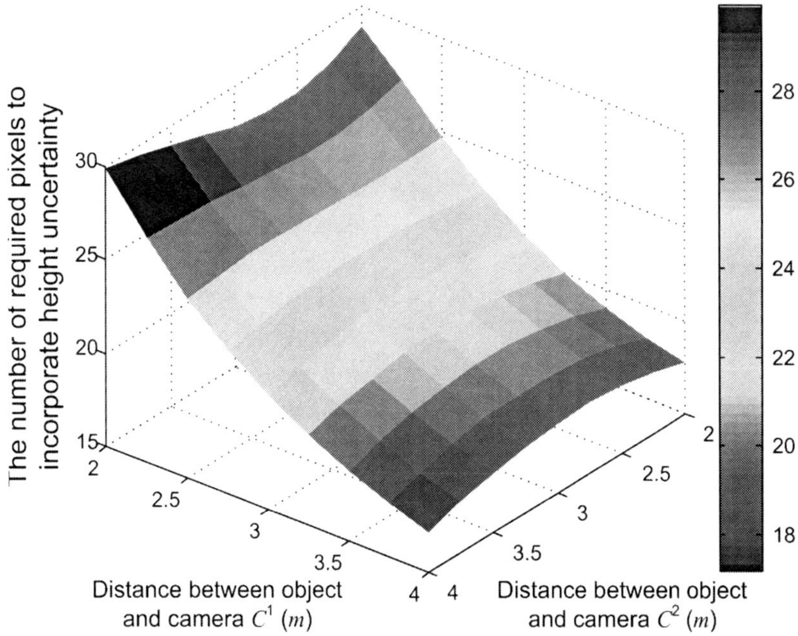

tilting angles and panning angles) generating homographic lines. Then, $\sigma^k_{h,i}$ is represented by

$$\sigma^k_{h,i} = \max_{l \neq k}(\sigma'^{k,l}_{h,i}), \quad (3)$$

where $\sigma'^{k,l}_{h,i}$ denotes the maximum number of pixels which deviate from the centroid of a target when homographic lines are generated from camera C^l to C^k with possible camera configurations.

Each value of $\sigma'^{k,l}_{h,i}$ between cameras C^k and C^l is determined by finding the maximum effect of the given amount of height uncertainty according to object locations. Figure 2 shows an example how many pixels deviate from an original point between cameras C^1 placed at ($x = 3m$, $y = 0m$, $z = 3m$) and C^2 placed at ($x = 6m$, $y = 3m$, $z = 3m$). In order to measure the number of pixels for the deviation, two homographic lines are generated from camera C^2 to camera C^1 according to object's location. One homographic line is generated with an actual height and the other is generated with an average height different from the actual height by 0.1m. The deviation is maximized when an object is close to a camera onto which homographic lines are projected.

Effect of Detection Performance on s_{min}

Another factor affecting association by homographic lines is the position of the centroid of a target obtained by detection algorithm. The centroid deviation problem affects both a camera generating homographic lines and a camera to which homographic lines are projected. When the position of a target deviates from the original position, a projected homographic line from other cameras may not cross the target within an association circle with radius r^k_i. Also, a generated homographic line on the position of a target which deviates from the original position causes the same problem. Hence, the radius of an association circle needs to be adjusted. $s_{\min}$ considering the detection performance is represented by

$$s^k_{\min,i} = \max(r^k_i, \sigma^k_{d,i}), \tag{4}$$

where $\sigma^k_{d,i}$ denotes the number of pixels to compensate for the deviation of a target by detection algorithm. The typical value of $\sigma^k_{d,i}$ is between 0 and r^k_i since the position varies within r^k_i of a target.

Effect of Frame Synchronization on $s_{\min}$

A projected homographic line may not intersect with a corresponding target within an association circle with r^k_i if objects are detected at different locations due to frame synchronization errors. A possible frame mismatch between cameras by synchronization effect is at most 1 frame. In order to incorporate the frame synchronization issue, the radius of an association circle needs to be adjusted by

$$s^k_{\min,i} = \max(r^k_i, \sigma^k_{s,i}), \tag{5}$$

where $\sigma^k_{s,i}$ denotes the possible number of pixels which deviates from the centroid of a target by the synchronization issue between cameras in camera C^k. The synchronization effect depends on the sampling period of a camera T_F and the direction toward which an object moves. Then, $\sigma^k_{s,i}$ is obtained by

$$\sigma^k_{s,i} = \max_{l \neq k}(\sigma'^{k,l}_{s,i}), \tag{6}$$

where $\sigma'^{k,l}_{s,i}$ denotes the maximum number of pixels which deviates from the centroid of a target by the synchronization issue between cameras C^k and C^l. Since the effect is maximized when the optical axes of two collaborating cameras are perpendicular to each other, only the paired case of perpendicular cameras is considered.

Figure 3 shows the synchronization effect according to sampling periods of a camera and distance between an object and a camera. Camera C^1 is placed at ($x = 0m$, $y = 3m$, $z = 3m$) and camera C^2 is placed at ($x = 6m$, $y = 3m$, $z = 3m$). An object moves in perpendicular to camera C^2 at the speed of 2 *m/s*. It is noted that the deviation of a projected homographic line is related to the relative speed of an object per each frame, not the absolute speed of an object. When the deviation of a projected homographic line is estimated by the simulation, the speed of an object is set to 2 *m/sec*. Since the frame rate varies from 0.05 *sec* to 0.2 *sec* in the simulation, it has the same effect of having the relative speed of an object, 0.1 *m/frame* to 0.4 *m/frame*. The maximum number of pixels which deviates from the original position is measured at each location and the result indicates that the deviation by synchronization is proportional to T_F and is inversely proportional to distance between an object and a camera.

Relationship between Size of Association Circle and Threshold

The three non-ideal parameters affecting the radius of an association circle are discussed in the previous section. These parameters may compensate for each other. For example, when a homographic line is generated from the position of a target which deviates from the original position by detection algorithm, a transformed homographic line with an average height of a target can be accidentally shifted to the position with an actual height of a target. A similar effect can also occur with synchronization issues. However, the system cannot predict the compensation effect by non-ideal parameters. Thus, the radius of an association circle should consider the worst effect by

$$s^k_{\min,i} = \max(r^k_i + \sigma^k_{h,i} + \sigma^k_{d,i} + \sigma^k_{s,i}). \tag{7}$$

Figure 3. Illustration of the number of pixels required to incorporate the synchronization effect according to sampling periods and distance between an object and a camera

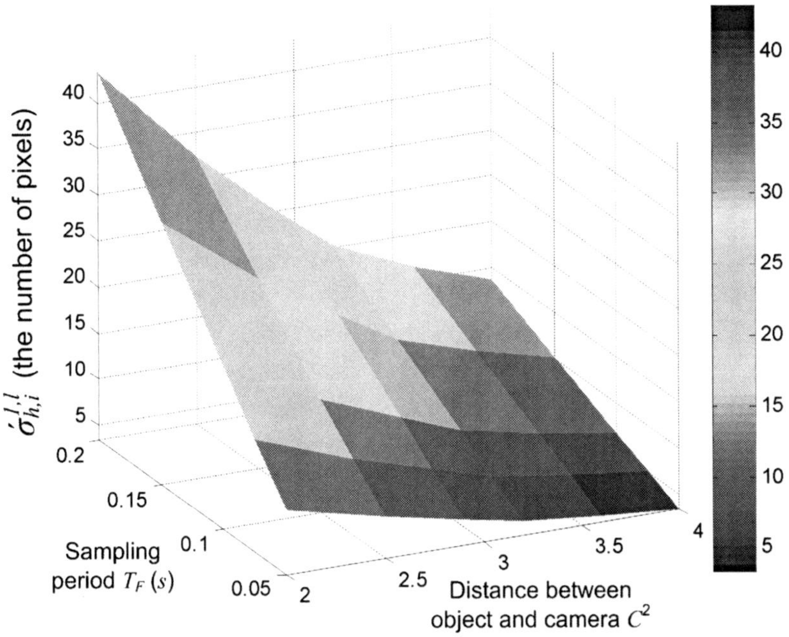

While expanded radii of association circles guarantee that targets intersect with corresponding projected homographic lines, they can degenerate association performance because homographic lines generation can be ineffective due to the insufficient separation. A threshold indicating the effectiveness of homographic lines can be represented by radii of association circles. A threshold d_{th}^{k} in camera C^k is defined as

$$d_{th}^{k} = \arg\min_{i,j}(s_{\min,i}^{k} + s_{\min,j}^{k}), \tag{8}$$

where *i,j* denote indices of neighboring targets. The smallest sum of radii of two neighboring association circles indicates effectiveness of homographic lines in an association process in camera C^k. Only when distances between neighboring targets are larger than the threshold on the image, homographic lines are generated. As the number of targets increases, the number of targets satisfying the threshold decreases. Thus, the association performance by the homographic lines based association method is highly affected by the effect of the non-ideal parameters.

MULTIPLE OBJECT ASSOCIATION

Multiple Camera Collaboration with Homographic Lines

We use the second strategy between two different multiple camera collaboration strategies proposed by Cho (2010) to establish the object association. Both the two camera collaboration strategies use homographic lines for the object association. Collaboration matrices are defined to indicate the feasibility of successful association between any two cameras. The elements in the collaboration matrices represent the required minimum separation obtained by incorporating the effect of targets height uncertainty, frame synchronization errors and detection uncertainty. The first strategy uses the collaboration matrices to select the best pair out of many cameras by using the degree of separation between homographic lines. After targets in the selected cameras are associated, the association information is propagated to unselected cameras by transforming the global information constructed from the associated targets. However, when a large number of objects are detected, the selected cameras cannot cover all the targets and the required minimum separation is hardly satisfied due to overlapped and occluded targets. The second strategy initiates the collaboration process of object association for all the pairing cases of cameras regardless of the separation. In each pair of cameras, a homographic line is generated on each target and it is projected to the camera. Then, the other collaborating camera generates homographic lines on only the intersected targets with the projected homographic lines and they are re-projected to the one camera. Since the system tests the association for each target at a time, it minimizes association ambiguity caused by homographic lines generated from all the targets. This association process is iteratively operated for all the unassociated targets in all the pairing cases of cameras.

They are compared with the basic pair-wise association in terms of association performance and computational costs. Computational costs are measured by the number of transformations and intersection tests defined in (1). Because the basic pair-wise approach initiates the association process for each pair of participating cameras, three different pairs of cameras execute the association process redundantly. In order to compare the performance between the approaches, six objects and three cameras are used for analyzing the association performance and the computational costs (Cho, 2010).

Table 1 shows the comparison of the two collaboration strategies with the basic pair-wise approach. The performance of the basic pair-wise association approach almost indicates a limit because the scheme considers all the pairing cases of multiple cameras through pair-wise collaboration. However, the effectiveness of the basic pair-wise association is affected by the redundant collaboration of the pair-wise association process among multiple cameras and it creates false associations in the system. The average false association rate by the basic pair-wise approach is about 6.2% and they are not corrected until they leave the surveillance region. It affects the average successful association rate in time. On the other hand, the second strategy reduces the average false association rate by almost a zero and improves the average successful association rate. When the second strategy is used for multiple object association, the number of successful association is greater than that of the first strategy. This is mainly because the first strategy usually selects only a pair of cameras satisfying the threshold and disregards the rest of possible association cases. On the other hand, the second strategy initiates the association process for each target and it leads to improve the association performance.

Figure 4 shows the simulation result to measure the number of transformations (i.e. C_T) and the number of intersection tests (i.e. C_C) by the association approaches according to the number of objects. The results show that the proposed methods are more efficient than the basic pair-wise approach. Also, the number of transformations and intersections tests in the first strategy is lower than that of the second strategy. This is because only two selected cameras are participating in the association process. However, in terms of the association performance, the selected cameras cannot cover occluded targets as the number of targets increases. As a result, the association performance is degraded as compared with the second strategy. Therefore, the second strategy is selected for the object association to maximize the association performance with the reasonable computational costs.

Table 1. Performance comparison of collaboration strategies

O_i	**The first strategy**			**The second strategy**			**Basic pair-wise approach**		
	Success (%)	**Failure (%)**	**False (%)**	**Success (%)**	**Failure (%)**	**False (%)**	**Success (%)**	**Failure (%)**	**False (%)**
O_1	82.3	17.7	0.0	100.0	0.0	0.0	88.3	2.9	8.8
O_2	64.9	34.3	0.8	84.8	14.4	0.8	83.3	7.9	8.8
O_3	69.0	30.2	0.8	85.2	14.8	0.0	89.1	7.8	3.1
O_4	39.3	60.7	0.0	91.6	8.4	0.0	75.0	20.2	4.8
O_5	75.8	24.2	0.0	98.5	0.0	1.5	67.0	27.2	5.8
O_6	96.1	3.9	0.0	100.0	0.0	0.0	89.6	4.7	5.7
Avg.	71.2	28.5	0.3	93.4	6.2	0.4	82.1	11.7	6.2

Figure 4. The computational costs comparison of the first strategy and the second strategy with the basic pair-wise approach

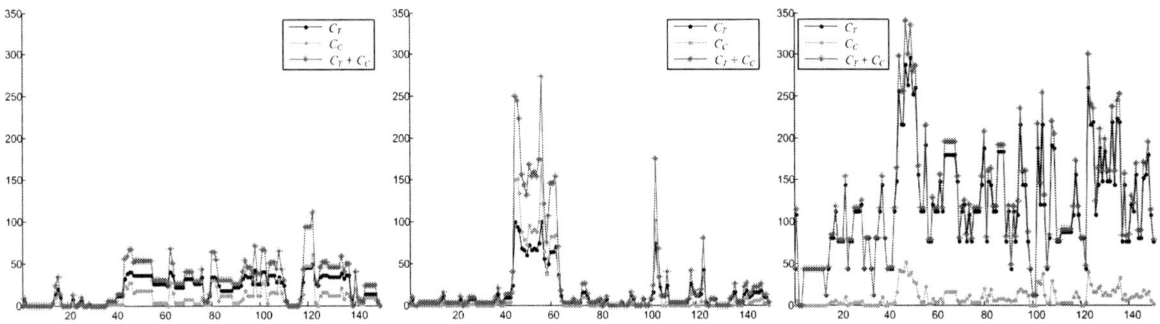

(a) The first strategy (b) The second strategy (c) Basic pair-wise approach

Point Transformation of 2D to 3D

If a target height is known, a target is transformed in 3-D with a known camera matrix. A camera perspective projection matrix is defined by

$$M = K(R\ t),$$

where K is a matrix for intrinsic parameters, R is a rotation matrix and t is a translation vector (Forsyth, 2002). Then, a world point of $P = (x,y,z,1)^T$ in the homogeneous coordinate is projected to p by

$$p = \frac{1}{z_d} \mathrm{MP},$$

where $p = (u,v,1)^T$ and z_d is a depth. If the j-th row of matrix M is denoted by M_j^T, $z_d = m_3^T \cdot P$.

When a point in 2D image is transformed to a world coordinate on a plane of $z = h$, the depth z_d should be calculated first. Suppose an inverse matrix of M is denoted by

$$M^{-1} = \begin{bmatrix} \hat{m}_{11} & \hat{m}_{12} & \hat{m}_{13} & \hat{m}_{14} \\ \hat{m}_{21} & \hat{m}_{22} & \hat{m}_{23} & \hat{m}_{24} \\ \hat{m}_{31} & \hat{m}_{32} & \hat{m}_{33} & \hat{m}_{34} \\ 0 & 0 & 0 & 1 \end{bmatrix}.$$

Then, the following relationship is established between an image point and a world point.

$$\begin{bmatrix} \hat{m}_{11} & \hat{m}_{12} & \hat{m}_{13} & \hat{m}_{14} \\ \hat{m}_{21} & \hat{m}_{22} & \hat{m}_{23} & \hat{m}_{24} \\ \hat{m}_{31} & \hat{m}_{32} & \hat{m}_{33} & \hat{m}_{34} \\ 0 & 0 & 0 & 1 \end{bmatrix} \begin{bmatrix} uz_d \\ vz_d \\ z_d \\ 1 \end{bmatrix} = \begin{bmatrix} x \\ y \\ h \\ 1 \end{bmatrix}. \tag{9}$$

The depth z_d is obtained by the third row of M^{-1} as

$$z_d = \frac{h - \hat{m}_{34}}{\hat{m}_{31}x + \hat{m}_{32}y + \hat{m}_{33}}. \tag{10}$$

Then, a coordinate (u,v) on an image is transformed onto the global plane of $z = h$ by (9). A point source is used in cameras and in global planes. If a circle is transformed in 3-D, this becomes a truncated cone because of the perspective projection property. The truncated shape is difficult to be managed by a system. The accuracy of the transformation depends on z_d and it is obtained by using a target height in (10).

Virtual Height Estimation

h is estimated in (10) to compensate for the inaccurate camera matrix and to ensure the accurate transformation. The accuracy of the point transformation is directly affected by the accuracy of a camera matrix. Intrinsic and extrinsic parameters of a camera matrix can be obtained by several calibration techniques. The calibration is a process to find parameters minimizing transformation errors between corresponding points. Thus, the accuracy of a camera matrix is also bounded by the minimization error. The only controllable parameter is the height of corresponding targets with camera matrices. (u_1^1, v_1^1) and (u_1^2, v_1^2) denote the coordinate of associated targets in cameras C^1 and C^2 respectively. Since they are projected from a same object, the following relationship is induced.

$$\begin{bmatrix} \hat{m}_{11}^1 & \hat{m}_{12}^1 & \hat{m}_{13}^1 & \hat{m}_{14}^1 \\ \hat{m}_{21}^1 & \hat{m}_{22}^1 & \hat{m}_{23}^1 & \hat{m}_{24}^1 \\ \hat{m}_{31}^1 & \hat{m}_{32}^1 & \hat{m}_{33}^1 & \hat{m}_{34}^1 \\ 0 & 0 & 0 & 1 \end{bmatrix} \begin{bmatrix} u_1^1 z_{d1}^1 \\ v_1^1 z_{d1}^1 \\ z_{d1}^1 \\ 1 \end{bmatrix} = \begin{bmatrix} \hat{m}_{11}^2 & \hat{m}_{12}^2 & \hat{m}_{13}^2 & \hat{m}_{14}^2 \\ \hat{m}_{21}^2 & \hat{m}_{22}^2 & \hat{m}_{23}^2 & \hat{m}_{24}^2 \\ \hat{m}_{31}^2 & \hat{m}_{32}^2 & \hat{m}_{33}^2 & \hat{m}_{34}^2 \\ 0 & 0 & 0 & 1 \end{bmatrix} \begin{bmatrix} u_1^2 z_{d1}^2 \\ v_1^2 z_{d1}^2 \\ z_{d1}^2 \\ 1 \end{bmatrix}, \tag{11}$$

where the superscript indicates a camera index and the values for z_{d1}^1 and z_{d1}^2 are obtained by

$$z_{d1}^1 = \frac{h_1^1 - \hat{m}_{34}^1}{\hat{m}_{31}^1 x + \hat{m}_{32}^1 y + \hat{m}_{33}^1},$$

$$z_{d1}^2 = \frac{h_1^2 - \hat{m}_{34}^2}{\hat{m}_{31}^2 x + \hat{m}_{32}^2 y + \hat{m}_{33}^2}.$$

Linear equations are obtained in terms of unknown depth by

$$Az = b \tag{12}$$

where

$$A = \begin{bmatrix} a_{11} & -a_{12} \\ a_{21} & -a_{22} \\ a_{31} & -a_{32} \end{bmatrix}, z = \begin{bmatrix} z_{d1}^1 \\ z_{d1}^2 \end{bmatrix}, b = \begin{bmatrix} \hat{m}_{14}^2 - \hat{m}_{14}^1 \\ \hat{m}_{24}^2 - \hat{m}_{24}^1 \\ \hat{m}_{34}^2 - \hat{m}_{34}^1 \end{bmatrix}, \tag{13}$$

and $a_{ij} = \hat{m}_{i1}^j u_1^j + \hat{m}_{i2}^j v_1^j + \hat{m}_{i3}^j$.

A matrix z is obtained by the linear least-squares using normal equations (Hartley, 2003).

$$z = (A^T A)^{-1} A^T b. \tag{14}$$

Since h is used to minimize the distance between transformed points from associated targets, it is not the actual height of an object in this chapter. Thus, heights obtained from (10) with z are denoted by virtual heights. Estimated *virtual heights* are used to transform tracked targets into points in 3-D.

Homographic Point Based Association

The distance between transformed points is used to confirm the association of tracked targets. The centroid of the targets is transformed into points in 3-D with previously estimated virtual heights. The 2-D Euclidean distance between transformed points $gp^m_{i,u}$ and $gp^n_{j,v}$ in 3-D is defined by $d^{m,n}_{i,u;j,v}$ where $gp^k_{i,j}$ is a transformed point in 3-D from p^k_i with h^k_j and p^i_k is the centroid of T^k_i and. The association of targets is confirmed only when $d^{m,n}_{i,u;j,v}$ is smaller than the predefined threshold d_{th}. The typical value of d_{th} is the average size of objects.

Figure 5 shows the difference between when the average height of targets is used and when each virtual height of targets is estimated by (14). As shown in Figure 5(a), when homographic points are transformed with the average height, the Euclidean distance between them is greater than their s_{min}. This can cause an association failure. Thus, the average height is not appropriate for confirming the association in densely populated objects. On the other hand, Figure 5(b) and Figure 5(c) show that corresponding points are placed at almost equivalent positions and their Euclidean distances are less than their s_{min}. The difference between Figure 5(b) and Figure 5(c) is that each height is estimated based on corresponding targets at the current sampling time in Figure 5(b) and at the previous sampling time in Figure 5(c). Dotted colored circles are transformed homographic points from the other cameras. These are generated to show the difference between two figures more clearly. In Figure 5(c), transformed points are slightly different since each height is estimated from the previously associated targets.

Association Mechanism in Occluded Situation

The homographic point based association is also applied to a case where targets are partially occluded. It is assumed that occlusions occur when the distance between targets is smaller than the sum of their s_{min}. Since occluded targets can be falsely tracked and their information can be exchanged, their association information needs to be confirmed by homographic points together. We consider all possible cases of exchanging virtual heights between occluded targets.

It is assumed that I targets are occluded in camera C^m and J targets are previously associated with I targets or occluded by them in camera C^n. Finding the correspondence is to associate a target with a correct virtual height. The different parings of targets are achieved by the permutation of virtual heights. The total number of pairing cases is $(I!)\times(J!)$ where ! is a factorial. At each pairing case, a system confirms the association of targets having the same index of a virtual height in each camera. The total number of indexes is same as I. Targets are associated only when the distance of their transformed points is smaller than the predefined threshold. A correct pairing λ is found by selecting a pairing case in which the sum of distances between transformed points is minimized.

$$\lambda = \arg\min_k \sum_{l=1}^{I} d^{m,n}_{s^k_l}, \tag{15}$$

where s^k_l denotes $\{i,l,j,l\}$ with index l of a virtual height in pairing case k. The proposed method confirms the association information of occluded targets with virtual heights. If virtual heights are exchanged by the false tracking, the distance between transformed points increases. For example, two targets in camera k can have either h_1^k or h_2^k when $h_1^1, h_2^1, h_1^2, h_2^2$ denote previously estimated

Figure 5. The comparison of generated points according to height information with respect to multiple object association (the left frame is camera $C^1$and the right frame is camera C^2, solid red circles are homographic points generated in camera $C^1$and sold blue circles are homographic points generated in camera C^2)

(a) Generated spheres on the global plane with an average height

(b) Generated spheres on the global plane with estimated heights

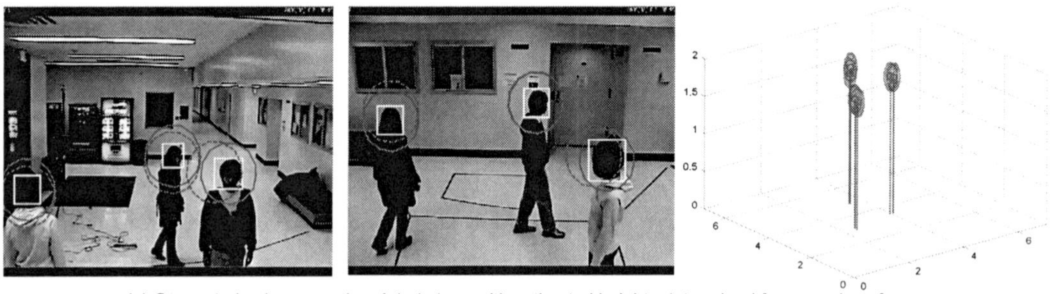

(c) Generated spheres on the global plane with estimated heights determined from previous frames

Table 2. Possible association cases with virtual heights.

	T_1^1	T_2^1	T_1^2	T_2^2
$case_1$	h_1^1	h_2^1	h_1^2	h_2^2
$case_2$	h_1^1	h_2^1	h_2^2	h_1^2
$case_3$	h_2^1	h_1^1	h_1^2	h_2^2
$case_4$	h_2^1	h_1^1	h_2^2	h_1^2

virtual heights of targets. Then, the system has four possible cases for object association as shown in Table 2.

The distance of each pairing case is represented by

$$case_1 : d^{1,2}_{1,1;1,1} + d^{1,2}_{2,2;2,2},$$
$$case_2 : d^{1,2}_{1,1;2,1} + d^{1,2}_{2,2;1,2},$$
$$case_3 : d^{1,2}_{2,1;1,1} + d^{1,2}_{1,2;2,2},$$
$$case_4 : d^{1,2}_{2,1;2,1} + d^{1,2}_{1,2;1,2}.$$

The final decision is made by

$$\arg\min_k case_k, 1 \le i \le 4.$$

Algorithm

Algorithm explains a hybrid association method using homographic points and lines. The proposed system checks if targets are previously associated or not. If targets are previously associated, the association of them is confirmed by generating homographic points with their virtual heights. After the system maintains the previously associated objects, it establishes new associations by generating homographic lines. Since homographic lines are less sensitive to target heights, homographic lines are generated with the average height of targets. If targets are mutually crossed by their transformed homographic lines, they are associated. The system estimates virtual heights for associated targets.

Algorithm: The association algorithm

```
Input: d_th,
Coordinates of targets
Previous association and virtual heights
repeat
    for k = 1 to K dofor i^k = 1 to I^k doif T^k_{i^k} is associated at n-1frame then
                              Confirm the association of targets by
using homographic points with their virtual heights
                         end
                  end
    end
       for k = 1 to K dofor i^k = 1 to I^k doif T^k_{i^k} is not associated then
Generate a homographic line on T^k_{i^k} and transform it onto other cameras
                          end
                 end
    end
    if Targets are mutually crossed over by homographic lines generated from
them then
           They are associated with each other
    end
    Estimate virtual heights for associated targets
until System stops
```

SIMULATION AND ANALYSIS

Simulation Setup

Figure 6 illustrates a simulation setup with six objects and six cameras for analyzing association performance. Camera C^1 is placed at (x = 3.65m, y = 0m, z = 2.37m) with tilting angle 86° and panning angle 0°, camera C^2 is placed at (x = 0m, y = 3.5m, z = 2.45m) with tilting angle 76° and panning angle 90°and camera C^3 is placed at (x = 1.83m, y = 7.32m, z = 2.37m) with tilting angle 76° and panning angle 156°. Videos are captured at all cameras with a resolution of 704 × 480 pixels at approximately fifteen frames per second and the proposed method is evaluated with the captured videos. The total number of frames is 150 and average height of objects is 1.7m. It is assumed that targets are detected perfectly and they can be falsely tracked. Threshold d_{th} to check the possibility of the correspondence is set to 0.5m by considering the average size of objects. Detected targets are initially associated by the homographic lines based association method. Once the correspondences of detected targets are confirmed by checking the cross of the transformed homographic lines, their association information is managed by the proposed method.

Association Performance

Figure 7 shows estimated heights of each target as a function of sampling time. The average height of targets is shown when objects are not shown on both cameras. Since virtual heights are estimated to compensate for inaccurate camera information and mismatched target positions, they vary at every sampling time.

Figure 6. The illustration of simulation setup with six objects trajectory and three cameras

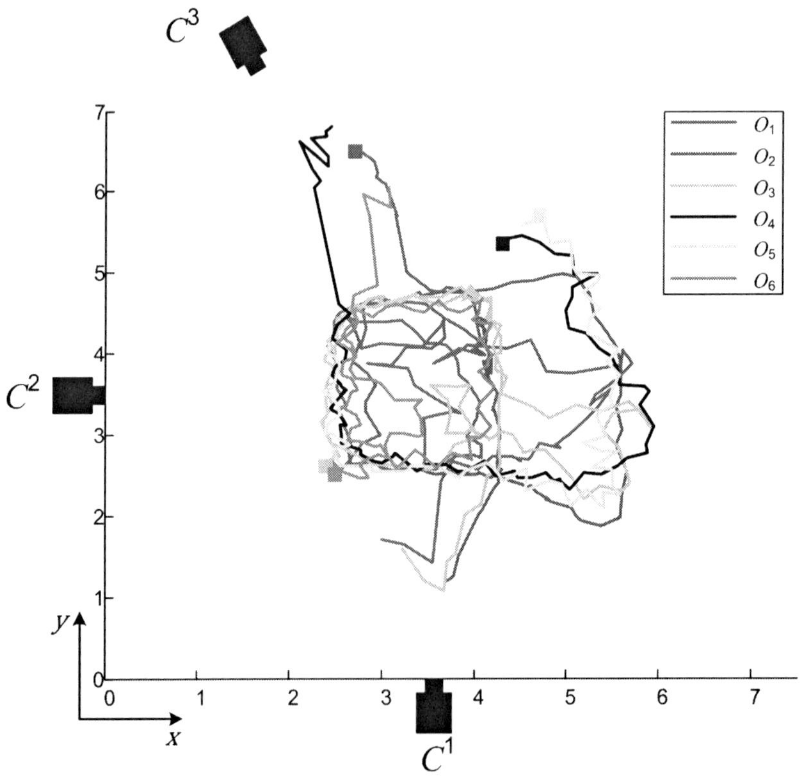

Figure 7. Estimated virtual heights of detected targets in each camera as a function of sampling time

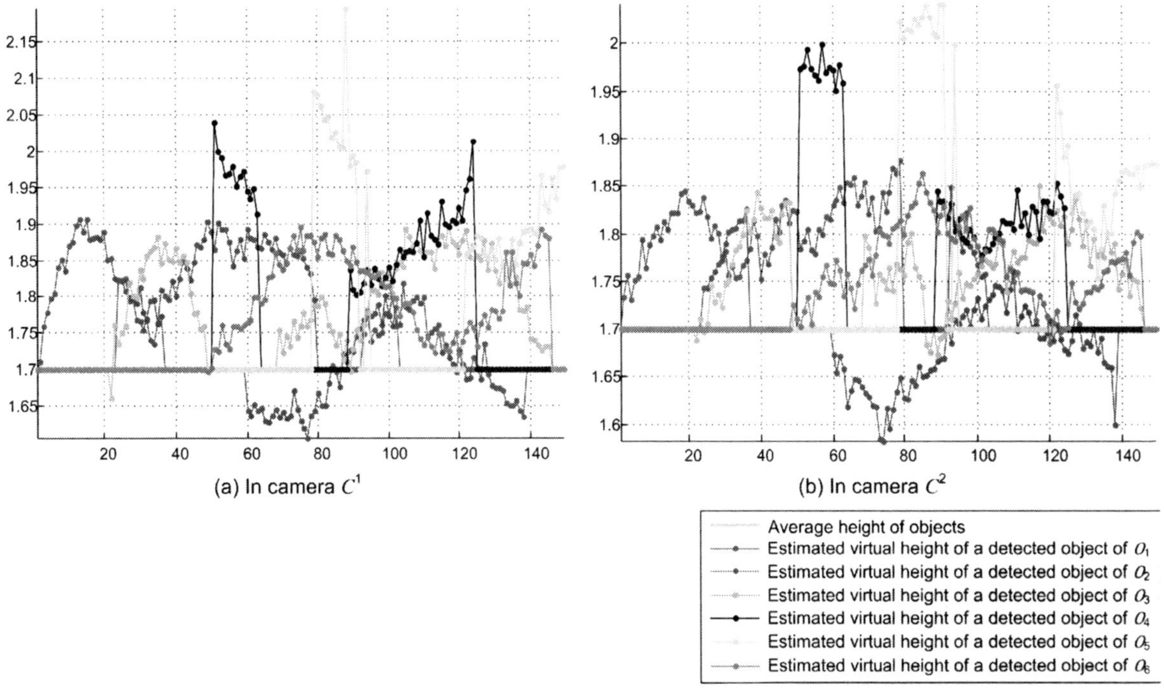

Figure 8 shows the simulation result how estimated heights are effective for object association. According to the simulation result, global points generated from corresponding targets have the smallest 2D Euclidean distance. For example, the thick red line indicates the 2D Euclidean distance generated from the corresponding targets of object O_1 with the estimated height in Figure 8(a). The other lines are the 2D Euclidean distance between points generated from the detected target of O_1 in camera C^1 and other detected targets in C^2 with the estimated virtual height of O_1. In all the subfigures, each object has the smallest 2D Euclidean distance between global points generated from corresponding objects.

Figure 9 shows the performance comparison of two cameras and three cameras. Since the homographic lines based association method is restricted by the separations of homographic lines, it leads to delay association establishment with two cameras until the enough separation is guaranteed. However, when three cameras are used for multiple object association, association is established without the delay due to the different views of different pairing cases. While the failed association rate is almost similar to the unconfirmed association rate with two cameras, the proposed method with three cameras can confirm the association of targets by multiple pairing cases. Thus, the association is more efficiently maintained even if the association of targets cannot be confirmed in some pairing cases. If N cameras are used for multiple object association, the system can use every possible pairing case or predetermined pairing cases to estimate virtual heights of targets.

Effect of Detection Performance

The proposed method is affected by detection performance because targets can deviate from original positions due to detection accuracy. If virtual heights are estimated with the shifted positions, generated points with them also deviate

Figure 8. The simulation of the 2D Euclidean distance between global points generated from local points at each sampling time (The thick colored line (red in (a), blue in (b), green in (c), black in (d), cyan in (e), and magenta in (f)) indicates the Euclidean distance between corresponding points for object O_i)

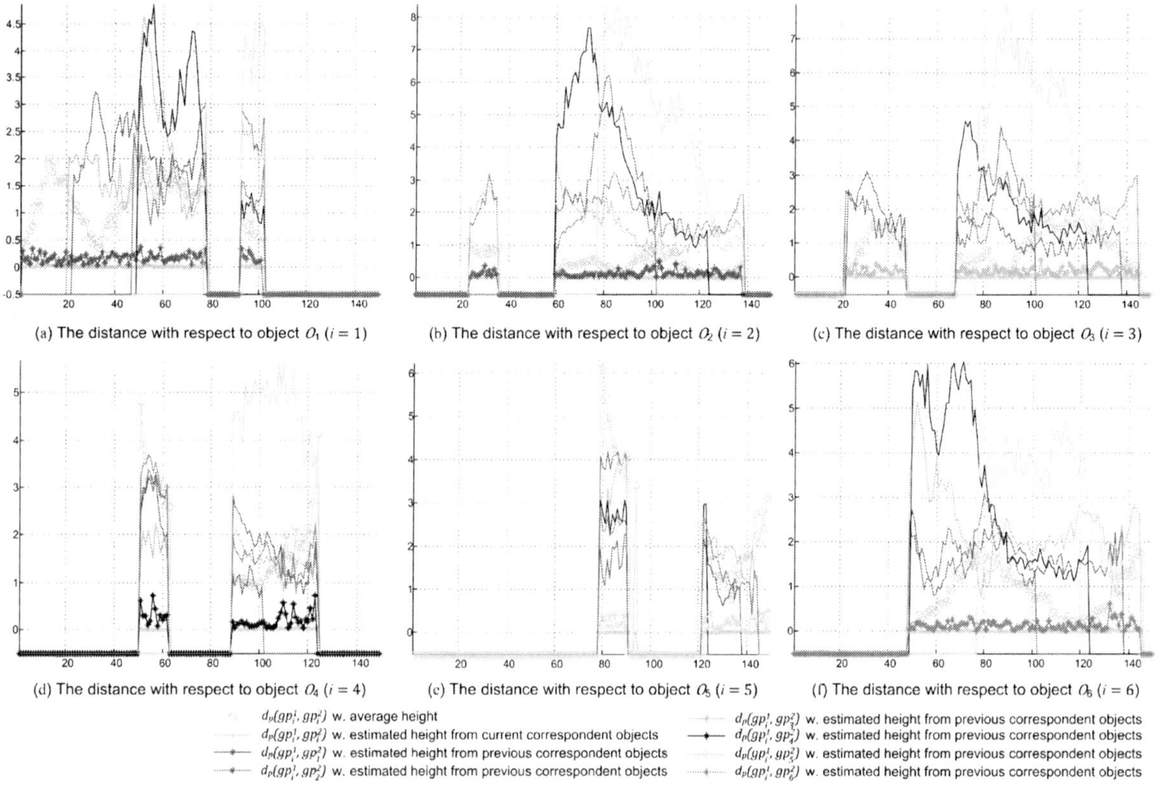

Figure 9. Performance comparison of two cameras and three cameras

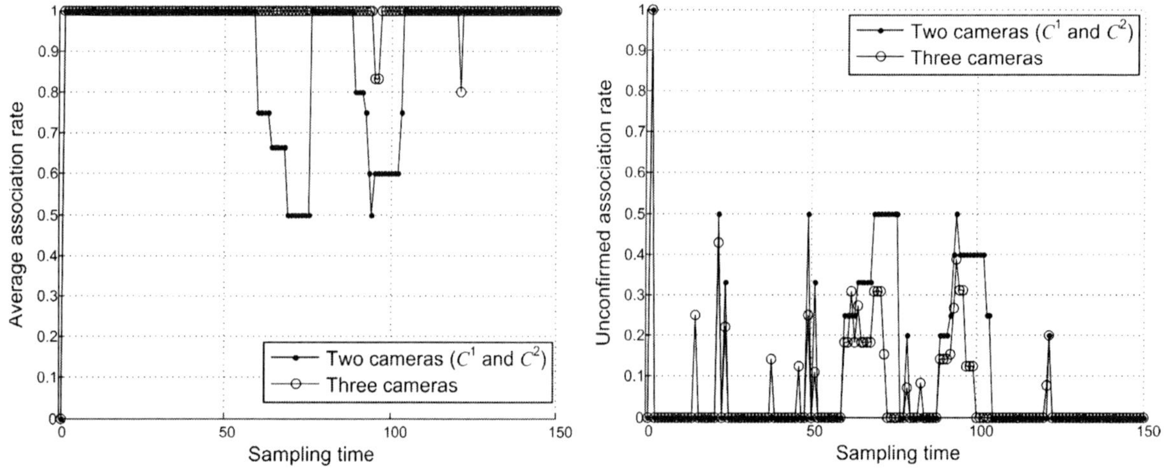

(a) Average association rate

(b) Unconfirmed association rate

from their original positions. They may not satisfy the threshold and their association may not be confirmed. Moreover, when targets are occluded, they may be falsely associated because generated points deviate from their original positions. The amount of pixels for the deviation in an x- coordinate and a y-coordinate is denoted by $d_{err,x}$ and $d_{err,y}$, respectively. It is assumed that they are independent of each other. The amount of the deviation depends on target size r_i^k of T_i^k because a target position can be varied within the target size. In the performance simulation, the amount of the deviation is generated from the uniform distribution by $d_{err,x} \sim U(0,\ r_i^k \cdot \sigma_d))$ and $d_{err,y} \sim U(0, r_i^k \cdot \sigma_d)$ where σ_d denotes an error percentage. σ_d varies from 0% to 25% in the simulation. The simulation is repeated 100 times and the results are average for the randomly generated deviation. Figure 10 shows the simulation results of the association performance according to the detection performance. As σ_d increases, false association rate and unconfirmed association rate also increase. However, they can be eventually confirmed by homographic points as long as their previous association is available. If a false association is doubted by the distance between global points, they can also be checked by generating homographic lines. The results demonstrate that the association performance is bounded by the detection performance.

CONCLUSION

We present a dynamic association method through multiple camera collaboration for a large-scale surveillance system. Targets are initially associated by the multiple camera collaboration strategy proposed by Cho (2010) to reduce the computational costs. In order to alleviate the separation issue of homographic lines, homographic points are introduced for previously associated targets and their heights in 3-D are estimated by the linear least-squares using normal equations. The correspondences of objects are confirmed by finding the pairs of corresponding targets having the minimum distance between generated points. Considering all the possible cases makes the proposed method robust against false tracking. The simulation demonstrates that the association is effectively confirmed when targets are occluded. However, the association performance is restricted by the detection performance because the position of a target can deviate from the original position. Thus, the proposed method needs to be improved to consider the effect of the detection performance for densely populated objects.

ACKNOWLEDGMENT

This research is supported by the International Collaborative R&D Program of the Ministry of

Figure 10. The effect of detection performance

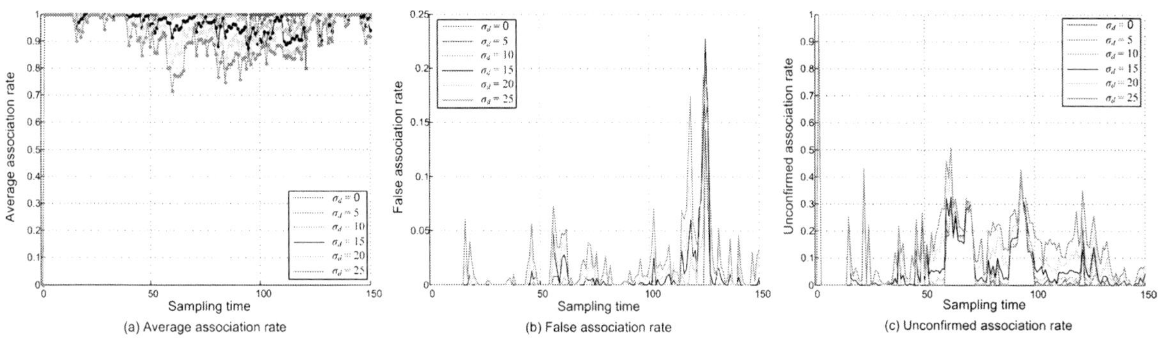

(a) Average association rate (b) False association rate (c) Unconfirmed association rate

Knowledge Economy (MKE), the Korean government, as a result of Development of Security Threat Control System with Multi-Sensor Integration and Image Analysis Project, 2010-TD-300802-002.

REFERENCES

Black, J., & Ellis, T. (2006). Multi camera image tracking. *Image and Vision Computing*, *24*(11), 1256–1267. doi:10.1016/j.imavis.2005.06.002

Cai, Q., & Aggarwal, J. (1996). Tracking human motion using multiple cameras. In. *Proceedings of the International Conference on Pattern Recognition*, *3*, 68–72. doi:10.1109/ICPR.1996.546796

Calderara, S., Prati, A., Vezzani, R., & Cucchiara, R. (2005). Consistent labeling for multi-camera object tracking, In *Proceedings of the Image Analysis and Processing (ICIAP '05)* (Vol. 3617, pp. 1206-1214).

Caspi, Y., Simakov, D., & Irani, M. (2006). Feature-based sequence-to-sequence matching. *International Journal of Computer Vision*, *68*(1), 53–64. doi:10.1007/s11263-005-4842-z

Chakrabarty, K., Iyengar, S. S., Qi, H. R., & Cho, E. C. (2002). Grid coverage for surveillance and target location in distributed sensor networks. *IEEE Transactions on Computers*, *51*(12), 1448–1453. doi:10.1109/TC.2002.1146711

Chang, T., Gong, S., & Ong, E. (2000). Tracking multiple people under occlusion using multiple cameras. In *Proceedings of the British Machine Vision Conference* (pp. 566-575).

Chen, C.-H., Yao, Y., Page, D., Abidi, B., Koschan, A., & Abidi, M. (2008). Camera handoff with adaptive resource management for multi-camera multi-target surveillance. In *Proceedings of the International Conference on Advanced Video and Signal Based Surveillance* (pp. 79-86).

Cho, S. H., Nam, Y., & Hong, S. (2010). Multiple camera collaboration strategies for dynamic object association. *KSII Transactions on Internet and Information Systems*, *4*(6), 1169–1193.

Dockstader, S. L., & Tekalp, A. M. (2001). Multiple camera tracking of interacting and occluded human motion. *Proceedings of the IEEE*, *89*(10), 1441–1455. doi:10.1109/5.959340

Fleuret, F., Berclaz, J., Lengagne, R., & Fua, P. (2008). Multicamera people tracking with a probabilistic occupancy map. *IEEE Transactions on Pattern Analysis and Machine Intelligence*, *30*(2), 267–282. doi:10.1109/TPAMI.2007.1174

Forsyth, D. A., & Ponce, J. (2002). *Computer Vision: A Modern Approach (Tech. Ref.)*. Upper Saddle River, NJ: Prentice Hall Professional.

Hartley, R., & Zisserman, A. (2000). *Multiple view geometry in computer vision*. New York, NY, USA: Cambridge University Press.

Horster, E., & Lienhart, R. (2006). In R. Lienhart (Ed.), Approximating optimal visual sensor placement, In *Proceedings of the International Conference on Multimedia and Expo* (pp. 1257-1260).

Hu, W., Hu, M., Zhou, X., Tan, T., Lou, J., & Maybank, S. (2006). Principal axis-based correspondence between multiple cameras for people tracking. *IEEE Transactions on Pattern Analysis and Machine Intelligence*, *28*(4), 663–671. doi:10.1109/TPAMI.2006.80

Hu, W., Tan, T., Wang, L., & Maybank, S. (2004). A survey on visual surveillance of object motion and behaviors. *IEEE Transactions on Systems, Man and Cybernetics. Part C, Applications and Reviews*, *34*(3), 334–352. doi:10.1109/TSMCC.2004.829274

Jaynes, C. (2004). Multi-view calibration from planar motion trajectories. *Image and Vision Computing, 22*(7), 535–550. doi:10.1016/j.imavis.2003.09.008

Kang, J., Cohen, I., & Medioni, G. (2003). Continuous tracking within and across camera streams. In. *Proceedings of the Conference on Computer Vision and Pattern Recognition, 1*, 267–272.

Kayumbi, G., & Cavallaro, A. (2008). Multiview trajectory mapping using homography with lens distortion correction. *EURASIP Journal on Image and Video Processing, 2008*, 145715. doi:10.1155/2008/145715

Kelly, P. H., Katkere, A., Kuramura, D. Y., Moezzi, S., & Chatterjee, S. (1995). An architecture for multiple perspective interactive video. In *Proceedings of the International Conference on Multimedia* (pp. 201-212).

Khan, S., & Shah, M. (2003). Consistent labeling of tracked objects in multiple cameras with overlapping fields of view. *IEEE Transactions on Pattern Analysis and Machine Intelligence, 25*(10), 1355–1360. doi:10.1109/TPAMI.2003.1233912

Krumm, J., Harris, S., Meyers, B., Brumitt, B., Hale, M., & Shafer, S. (2000). Multi-camera multi-person tracking for easyliving. In *Proceedings of the International Workshop on Visual Surveillance* (pp. 3-10).

Kyong, Y., Cho, S. H., Hong, S., & Cho, W.-D. (2008). Local initiation method for multiple object association in surveillance environment with multiple cameras. In *Proceedings of the International Conference on Advanced Video and Signal Based Surveillance* (pp. 348-355).

Lee, L., Romano, R., & Stein, G. (2000). Monitoring activities from multiple video streams: Establishing a common coordinate frame. *IEEE Transactions on Pattern Analysis and Machine Intelligence, 22*(8), 758–767. doi:10.1109/34.868678

Li, J., Chua, C. S., & Ho, Y. K. (2002). Color based multiple people tracking. In *Proceedings of the International Conference on Control, Automation, Robotics and Vision* (Vol. 1, pp. 309-314).

Li, S. Z., & Zhang, Z. Q. (2004). FloatBoost learning and statistical face detection. *IEEE Transactions on Pattern Analysis and Machine Intelligence, 26*(9), 1112–1123. doi:10.1109/TPAMI.2004.68

Mittal, A., & Davis, L. S. (2003). M2Tracker: A multi-view approach to segmenting and tracking people in a cluttered scene. *International Journal of Computer Vision, 51*(3), 189–203. doi:10.1023/A:1021849801764

Orwell, J., Remagnino, P., & Jones, G. A. (1999). Multi-camera color tracking. In *Proceedings of the International Workshop on Visual Surveillance* (pp. 14-21).

Senior, A., Hampapur, A., & Lu, M. (2005). Acquiring multi-scale images by pan-tilt-zoom control and automatic multi-camera calibration. In. *Proceedings of the IEEE Workshops on Application of Computer Vision, 1*, 433–438.

Sundaresan, A., & Chellappa, R. (2009). Multi-camera tracking of articulated human motion using shape and motion cues. *IEEE Transactions on Image Processing, 18*(9), 2114–2126. doi:10.1109/TIP.2009.2022290

Tsutsui, H., Miura, J., & Shirai, Y. (2001). Optical flow-based person tracking by multiple cameras. In *Proceedings of the International Conference on Multisensor Fusion and Integration for Intelligent Systems* (pp. 91-96).

Utsumi, A., Mori, H., Ohya, J., & Yachida, M. (1998). Multiple-human tracking using multiple cameras. In *Proceedings of the International Conference on Automatic Face and Gesture Recognition* (pp. 498-503).

Velipasalar, S., & Wolf, W. (2005). Multiple object tracking and occlusion handling by information exchange between uncalibrated cameras. In *Proceedings of the IEEE International Conference on Image Processing* (Vol. 2, pp. II - 418-21).

Viola, P., & Jones, M. J. (2004). Robust real-time face detection. *International Journal of Computer Vision*, *57*(2), 137–154. doi:10.1023/B:VISI.0000013087.49260.fb

Chapter 10
High-Level Information Fusion in Visual Sensor Networks

Juan Gómez-Romero
University Carlos III of Madrid, Spain

Jesús García
University Carlos III of Madrid, Spain

Miguel A. Patricio
University Carlos III of Madrid, Spain

José M. Molina
University Carlos III of Madrid, Spain

James Llinas
University at Buffalo, USA

ABSTRACT

Information fusion techniques combine data from multiple sensors, along with additional information and knowledge, to obtain better estimates of the observed scenario than could be achieved by the use of single sensors or information sources alone. According to the JDL fusion process model, high-level information fusion is concerned with the computation of a scene representation in terms of abstract entities such as activities and threats, as well as estimating the relationships among these entities. Recent experiences confirm that context knowledge plays a key role in the new-generation high-level fusion systems, especially in those involving complex scenarios that cause the failure of classical statistical techniques –as it happens in visual sensor networks. In this chapter, we study the architectural and functional issues of applying context information to improve high-level fusion procedures, with a particular focus on visual data applications. The use of formal knowledge representations (e.g. ontologies) is a promising advance in this direction, but there are still some unresolved questions that must be more extensively researched.

DOI: 10.4018/978-1-61350-153-5.ch010

INTRODUCTION

This chapter provides an overview of the nature of Information Fusion (IF) as a process, issues regarding the design and implementation of IF systems, and the special functions and algorithmic methods typically employed in IF processes. In addition, we focus on what is called "High Level Information Fusion" (HLIF), meaning those inferences developed by IF systems that are at a higher level of abstraction, from which the terminology derives.

We argue that there are four categories of information that can be applied to any (IF) problem: observational data, a priori knowledge models, inductively learned knowledge, and contextual information. For a broad class of applications, many IF processes and systems have been designed to work largely on the first two types of information; these are the class of systems built on a deductive-model foundation and that largely employ real-time observational data (obtained from various sources, usually sensors or instrumentation) in a scheme to more or less match the data against the models. These approaches can work well for what could be called well-behaved and well-studied problem domains but cannot be expected to work in problems where the "world-behavior" is very complex and unpredictable. Context can be defined as the set of circumstances surrounding a situation of interest that are potentially of relevance to its completion (Henricksen, 2003). In some applications, these contextual influences are not only important, but they may even be critical to understanding and interpretation, and they need to be considered.

One such type of applications are those involving video surveillance and monitoring applications –where both complex and unpredictable behavior can be expected, and where the contextual physical environment can be a prime driver or constraint to such behavior and to the system observational capability. Video applications often occur within a networked system framework, employing several cameras or other visual devices that are optimally emplaced and connected by data and control links to form a sensor network. Thus, IF techniques are required to collect, fuse, and interpret visual data. Hence, we comment briefly on the general issues in defining, designing, and implementing IF in general sensor networks, and particularly, in Visual Sensor Networks (VSNs).

In addition to describing the nature of the IF process, we also elaborate on the four informational and knowledge elements typically used to develop the IF approach, as were mentioned above. It is important we feel to understand the opportunities and constraints related to the employment of each of the components. Because of our assertion that Contextual Information is a critical informational component for video and vision applications, we further elaborate on our interpretation of the usage modes for such information.

We then focus on a detailed description of a Computer Vision application and describe issues in defining an architectural framework, strategies for dealing with contextual information, and reasoning methods for scene understanding. It is "Scene Understanding" in these applications that is the result of the HLIF process that we describe. The notion of Scene Understanding is application-dependent and must be defined in each case, but such notions usually imply a complex, high-dimensional state of the world that is to be estimated. In the most abstract definition, it can be described as a set of Entities in a set of Relationships. In the application spaces of usual interest, the Entities can be physical objects, events, and behaviors, and the relations can be of wide description. Thus, the notion of what is meant by a "Scene" is in fact a concept at a high level of abstraction, and assembling the component estimates into a Scene state estimate is a complex process. Accordingly, we propose the use of ontologies as a proper formalism to represent and reason with scene Entities and Relationships in the evolution from low-level acquired data to high-level scene descriptions. We also review some ontology-based approaches to HLIF, with a particular focus on visual HLIF.

The chapter is organized as follows. First, we overview the nature of Information Fusion and introduce the JDL process model, the canonical model to describe Information Fusion systems. In this section, we also present the challenges of Distributed Information Fusion in sensor networks. Next, we specifically focus on Visual Sensor Networks and describe the tasks that need to be faced in this kind of sensor networks from an Information Fusion perspective, as well as some previous approaches and techniques to solve them. In section 4, we discuss on the four categories of knowledge that can be applied in developing any IF approach. In section 5, we focus on the definition and the role of contextual information in HLIF; more precisely, we depict two frameworks for "a priori" and "a posteriori" exploitation of context. In section 6, we propose an architecture and a knowledge representation model for high-level context-based Information Fusion in Visual Sensor Networks. We also explain the abductive reasoning processes that are performed to achieve scene understanding from perceived and contextual information. These aspects are illustrated with some examples in a surveillance scenario. The chapter concludes with a section on the research issues related to contextual exploitation in sensor network-based IF applications. This section offers some ideas for a research agenda in this area.

INFORMATION FUSION IN SENSOR NETWORKS

Information Fusion: A Short Overview

Information fusion (IF) techniques combine data from multiple sensors, and other information and knowledge, to achieve improved accuracies and more specific inferences or estimates than could be achieved by the use of a single sensors or information sources alone[1]. It is important to understand thus that no matter the methods and sophistication of the IF process that its output is an imperfect estimate with some degree of uncertainty (but hopefully less than the estimation uncertainty of other possible techniques). The concept of multisensor IF is hardly new. Humans and animals have evolved the capability to use multiple senses to improve their ability to survive. Applications for multisensor IF are widespread. Military applications include: automated target recognition (e.g., for smart weapons), guidance for autonomous vehicles, remote sensing, battlefield surveillance, and automated threat recognition systems, such as identification-friend-foe-neutral (IFFN) systems.

The most fundamental characterization of IF involves a layered transformation between the observed energy or parameters related to an observable entity of some type (this energy provided by multiple sensors as input), and an estimate or inference (produced by fusion estimation and/or inference processes) regarding the location, characteristics, and identity of an entity, and an interpretation of the estimated entity in the context of a surrounding environment and relationships to other entities (Hall & Llinas, 1997). The definition of what constitutes an entity depends upon the specific application under consideration (e.g., an enemy aircraft for a tactical air-defense application, or the location and characteristics of a tumor in a medical diagnosis application).

Despite the benefits of improved system operation and application effectiveness by using multiple sensors and fusion processes, actual implementation of effective IF systems is far from simple. In many or most cases, the fusion process is a value-adding process that is often added to or integrated into some overarching system concept; it is often said that "we don't build fusion systems; we add fusion processes into systems". This implies that there may be existing, legacy functions and components, existing developmental

Figure 1. The JDL Functional Model of the Information Fusion Process. (Adapted with permission fromLiggins, Hall, & Llinas, 2009)

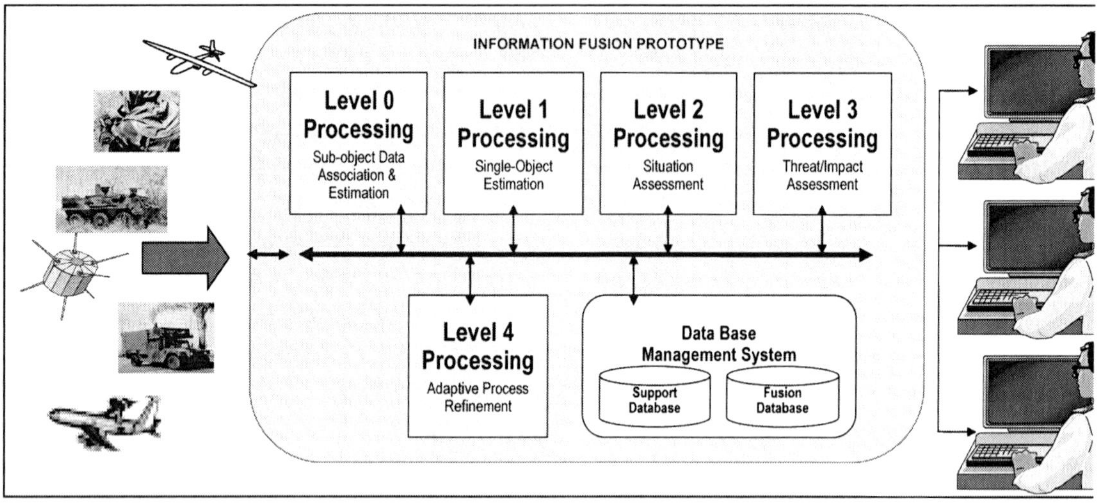

standards, design standards, etc. that the fusion process must adapt to. It also imposes a requirement to assess the added benefits of implementing a fusion process via some type of comparative analysis against some baseline system that may not incorporate fusion, or incorporates a "lesser" degree of fusion (e.g. fusion with fewer sensors), to assess the cost-benefit.

One of the descriptive, functional models of the information fusion process is the JDL[2] process model that is intended to be very general and useful across multiple application areas (Liggins, Hall, & Llinas, 2009); this process is shown in Figure 1.

While the boundaries of the IF process are fuzzy and case-dependent, generally speaking the input boundary is usually at the post-detection, extracted parameter level of signal or observational processing. The output of the IF process is (ideally) a minimally ambiguous identification and characterization (viz., location and attributes) of individual entities, as well as a higher level interpretation of those entities in the context of the application environment. As noted previously, a good way to think about the process is as an estimation process that needs to address the stochastic aspects of the various error-generating factors of any observational process or of other amplifying data, to include errors and inaccuracies in contextual information. As shown, the process conceptually involves several "Levels" of processing; these Levels were conceptualized simply to distinguish different types of complexity and abstraction that may exist in any problem, and (an important point) that any complete fusion process will involve an integrated set of functioning solution techniques. It is important to understand that Figure 1 is not an architecture but a pedagogical diagram to convey the basic concepts of IF processing. The basic Levels involve:

- **Level 0 Processing (Sub-object Refinement).** This process involves those methods to estimate the existence and features of structures of interest that may be discernible before the declaration of a named entity can be realized (e.g., segmentation in imagery)
- **Level 1 Processing (Object Refinement).** This process combines locational, parametric, and identity information to achieve

Figure 2. Example of Fusion Node

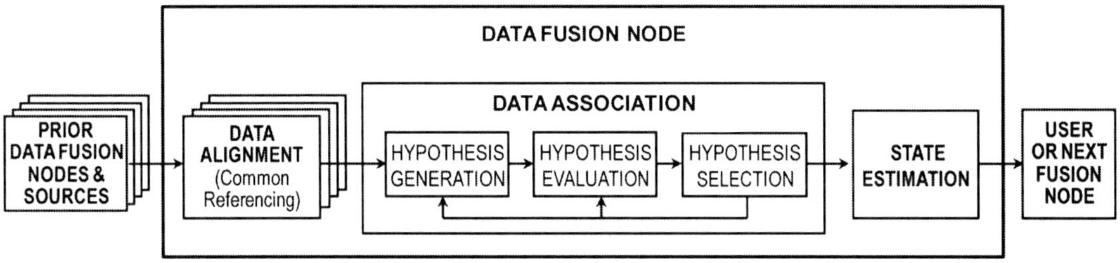

refined representations of individual objects or entities (e.g., time histories (tracks) of the kinematic properties of an object, i.e., location, velocity, etc.)

- **Level 2 Processing (Situation Refinement (similar to Scene Understanding)).** Level 2 processing develops a description of current relationships among objects and events in the context of their environment (e.g., a cyber-security related situation such as "coordinated virus attack on subnetwork X")
- **Level 3 Processing (Threat or Impact Refinement).** Level 3 processing focuses on what could be called special situations that relate to some type of threatening, critical, or otherwise special world states (situations, scenes, etc.) that are of interest for awareness or for decision and action-taking. Another way to think about Level 3 is developing a fused estimate of the Impact of that special situation to the application of interest.
- **Level 4 Processing (Process Refinement).** Level 4 processing may be considered a meta-process, i.e., a process concerned about other processes; it can be thought of as the "control law" of the overall fusion process, concerned with invoking any feasible capability needed to improve the process (one example is to control the sensing or observational process in ways that improve the fused estimate that is the output of the process).

Another way to understand the IF process is by consideration of a typical "Fusion Node". An example is shown in Figure 2.

Note that the Fusion Node (FN) nominally accepts either sensor type input from some input source or an estimate (fused or otherwise formed) from some prior FN or processing node. Other variations could be possible such as control inputs or inputs from a human operator, etc. but for this discussion we assert that the inputs are either sensor data (measurements) or fused estimates from a prior FN; the data are batched according to some batching logic (not shown here). In this characterization, the FN processing operations involve three basic functions:

1. **Data alignment (also known as Common Referencing):** Here, a variety of what could be called normalization operations are performed, ranging from the simple such as coordinate or units transformations or more complex such as uncertainty transformations; the goal of such operations is to establish common representational and semantic forms for the composite input.
2. **Data association:** Here, the multiple inputs of either estimates or measurements are examined in order to determine which (hypothetical) entity that the system believes to exist they are associated to or come from. By gauging this, the association process culminates in assigning (partitioning) the data to the on-going fusion estimation processes for each such entity. Data association

comprises three sub-functions as shown; the term "hypothesis" shown here should be understood as an association hypothesis, that is one feasible association of the measurements or estimates to the entities for which state estimates are being formed.

a. **Hypothesis Generation:** This is a step that nominates the possible causes of the measurements or the estimates coming in; an example for a measurement would be a False Alarm, i.e., establishing that in the IF designer's view, it is plausible that a measurement is not valid (from a real entity) but is an artifact. Note that it is only the nominated hypotheses to which associations can be made.
b. **Hypothesis Evaluation:** This is a step where the notion of "closeness", of say a measurement to its expected or predicted value, is judged by the computation of some association score or metric. This operation has to take into account both any measurement errors as well as any prediction errors, and so it is typical that these metrics are of a statistical nature.
c. **Hypothesis Selection:** in this final step, the information at hand can be conceptualized as a matrix of say labeled measurements and labeled association hypotheses, with the matrix populated by the "association scores" just computed in the Hypothesis Evaluation step. The problem is to determine the optimal allocation of the measurements to the hypothesized entities. In operations research, such optimization problems are called "assignment" problems since the challenge is to find an optimal assignment of the measurements. (The fact that scores have been computed for each measurement to entity pair does not necessarily lead to an obvious solution since there may be other measurements competing for the same assignment; this is a problem in combinatorial optimization, to find the best assignment that yields the best composite or measurement-set score.)

What really happens in effect at the end of Data Association is that the measurements are assigned to fusion-capable estimation algorithms that can then "digest" the multiple inputs to form an updated, fused, improved estimate of whatever is being estimated for that entity. Consider multiple frames of imagery that are being used to generate a fused estimate of an object A's shape. In the above operations, it would have been determined that certain segments of multiple images are best associated to object A, and those composite segments would be given to a shape-estimating algorithm to generate the fused shape estimate.

3. **State estimation:** This stage is just that, the step where whatever type of estimate—often about entity attributes at fusion Level 1 (e.g., kinematic properties such as location, velocity, and classification attributes such as color, etc. or a direct estimate of identity)—or about the existence of events or of certain behavioural activity of the estimates are of a Level 2 type.

Sensor Networks and Information Fusion

In addressing any topic incorporating the phrase "sensor network", one is faced with the challenge of specifying the type(s) of network being addressed. While literature search seems to reveal no standard taxonomy of sensor-network types, there are these variations mentioned:

- **Mobile Ad Hoc Networks:** A mobile ad hoc network (MANET) is a kind of wireless ad hoc network, and is a self-configuring network of mobile routers (and associated hosts) connected by wireless links—the union of which forms an arbitrary topology. The routers are free to move randomly and organize themselves arbitrarily; thus, the network's wireless topology may change rapidly and unpredictably. Alternatively, a MANET may be considered as a mobile version of an ad hoc network.

- **Wireless Sensor Network:** A wireless sensor network (WSN) is a wireless computer network consisting of spatially distributed autonomous devices using sensors to cooperatively monitor physical or environmental conditions, such as temperature, sound, vibration, pressure, motion or pollutants, at different locations.
- **Mobile Wireless Sensor Network:** A wireless sensor network in which the nodes are mobile.
- **Ad Hoc Wireless Sensor Network:** A wireless sensor network in which the nodes self-organize.
- **P2P Network:** A network without the notion of clients or servers, but only equal peer nodes that simultaneously function as both clients and servers.

In each flavor of network there are different system parameters that can affect the strategy for and viable realizations of different HLIF fusion processes. The topological dynamics of MANETs can possibly influence what is observable in the world, the bandwidth limitations of WSNs can affect what is communicable between nodes and thus the nature of what is being fused, the mobility and self-organizing factors of Mobile and Ad Hoc nets interplay with observability issues and the nature of what can be aggregated at any node, and the insertion of fusion processes impacts the fundamental homogeneity of the P2P network concept.

To at least give some sense of the nature of IF operations in sensor networks without addressing all of the net-specific factors in these network variations, we take the point of view regarding Distributed IF (DIF). In the most fundamental perspective, consider any Node in some DIF/Network architecture; this Node may only fuse two types of information—that which it "owns" (e.g., the data from local, sometimes called "organic" sensors)—and that which comes to it somehow from the network. The "somehow" is important; a Node can possibly receive information it has requested, or information sent by prior agreement from another Node (in a "Publish/Subscribe" protocol), or according to any "Information-Sharing Strategy (ISS)" that has been developed for the network. There is thus a critical interdependency between whatever the ISS protocol is and what DIF processes at various Nodes can achieve. Note that in the above network variations the feasible ISS protocols may for example be based in part on optimal energy (Node battery) power management concerns; rates and degree of intermodal communications may be energy-bound. Further, ISS´s may also involve non-technical aspects such as imputed by Nodal priorities of some type, or command authority in military applications, etc. As a result, the constraints of any given ISS may result in some degree of compromise in what can be achieved in the corresponding DIF processing operations.

DIF can also be susceptible to what is variously called the "double-counting" problem or "data incest" problem resulting from network topologies that have cycles in them, allowing redundant intermodal messages to flow into any DIF fusion Node, i.e., situations where duplicative evidential data arrives at a DIF fusion Node unbeknownst to that Node (Julier, 2009). Such issues impute requirements for appropriate meta-data tagging so that data "pedigree" can be tracked in the network (Arambel, 2008). Similar effects can be encountered because many IF algorithms are purposely built to be recursive, to deal with the dynamic world of interest. This means the algorithm outputs have a built-in memory of the effects of prior data. In the same way as just mentioned then, if IF algorithm outputs are shared across the nodes, similar double-counting effects can corrupt the correctness of the fused results at the DIF Node. Depending on the specifics of the network, and in particular what each Node knows about the Nodes from which it can receive information, there may be yet other effects on DIF process design. In large networks for example, it is unreasonable for any DIF Node to have full information about Nodes at far distance from it; in fact many network design strategies have Nodes only knowing Nodal details about their one-hop-away neighbors. This can impute a layered DIF process at any DIF Node, in which there may be at least two IF layers, in essence of different quality, to fuse data from Nodes that the DIF Node knows the details of (thus a notional high-quality fusion process) and to fuse data from Nodes about which it has less information (such as not possibly knowing certain sensor data details).

Another general problem that can happen in virtually any IF system but is more likely in DIF applications is what is sometimes called the "Out of Sequence Measurement" or OOSM problem (Shi, Wan, & Ge, 2009). This is fundamentally the problem for system/process design of how to handle the problem of retrospective corrections to fused estimates when just-arriving data or information reveals new information related to a prior time period. Often, to allow such retrospective corrections to IF-based inferencing, one must choose the largest time window over which such adjustments will be allowed, and then the system will have to store all data across such intervals to allow the current inference to be undone backward in time in accordance with the new information from that prior time point. Such problems are more likely in sensor networks due to intermodal communication delays and other factors involved in ISS management.

So it can be appreciated that what IF can be done in any sensor network can be impacted by a variety of factors that reflect the peculiarities of that network but that in addition there are some generic DIF concerns that apply for all sensor networks.

INFORMATION FUSION IN VISUAL SENSOR NETWORKS

Modern Visual Sensor Networks (VSNs) involve the deployment of a large number of cameras and the management of geographically-spread monitoring points covering a wide area. Although the existence of multiple cameras inevitably increases the complexity of the systems, it also improves the results, since multiple viewpoints can be combined. Additionally, distribution increases system robustness and fault tolerance, since the same information may be captured and replicated at different points of the network. In this state of affairs, IF is not only a convenience but a requirement; local data acquired by distributed video cameras must be combined to obtain a global understanding of the current scenario. Essentially, DIF in VSNs is concerned with the following tasks (Castanedo, García, Patricio, & Molina, 2009):

- Camera calibration, or common referencing, i.e., the process of transforming from the local coordinates of each camera to a global coordinates space. This translation may include a reconstruction step to obtain a 3D representation of the 2D image. The most used methods for camera calibration are those proposed by Tsai (1987), Heikkila (1997), and Zhang (2000).
- Object detection, i.e., to identify interesting objects in the sequence of images provided by the camera. There are various usual approaches to the detection of moving objects, e.g.: temporal differencing

–based on calculating the pixel-by-pixel difference of various consecutive frames (Lipton, Fujiyoshi, & Patil, 1998); background subtraction –based on subtracting the current snapshot pixel values with a predefined background image (Piccardi, 2004); statistical methods –a variation of basic background subtraction method based on the differences of additional statistical measures (Wang, Hu, & Tan, 2003); or optical flow –based on the computation of the flow vectors of moving objects over time (Barron, Fleet, & Beauchemin, 1994).

- Object tracking, i.e., to estimate the number of objects in a continuous scene, together with their instantaneous location, kinematic states, and other characteristics. Object tracking, which is a particular case of state estimation, has been traditionally tackled by applying statistical prediction and inference methods; for instance, widely-used tracking methods in general DIF are distributed MHT (Chong, Mori, & Chang, 1990), distributed JPDA (Chang, Chong, & Bar-Shalom, 1986), Covariance Intersection / Covariance Union (Julier & Uhlmann, General decentralized data fusion with Covariance Intersection, 2001), and distributed Kalman filter (Olfati-Saber, 2007). As we explain below, these methods may be insufficient in VSNs and require the incorporation of additional information and knowledge in the DIF process.
- Object and activity recognition, i.e., to determine the type of an object (e.g., car, human, aircraft) or the type of an activity (e.g., approaching, walking, maneuvering). Recognition can be seen as a probabilistic reasoning problem, in which case it is tackled through probabilistic models (Markov models, Bayesian networks, etc.), as in (Hongeng, Nevatia, & Bremond, 2004); or as a classification problem, in which case it is tackled through pattern recognition techniques (neural networks, self-organizing maps, etc.), as in (Hu, Xie, Tan, & Maybank, 2004).
- Process enhancement, or active fusion, i.e., to implement suitable mechanisms that use the more comprehensive interpretation of the current situation obtained after fusing data to improve the development of the previous tasks. Generally speaking, process enhancement consists in improving a fusion procedure by using feedback generated at a more abstract level; for instance, the behavior of a tracking algorithm can be changed once a general interpretation of the scene has been inferred –if we recognize that an object is moving out of the camera range through a door a object, the tracking procedure can be informed to be ready to delete this track in the near future.

These tasks can be largely classified according to the JDL model depicted in section 2.1. Camera calibration is a necessary task that must be performed before any other procedure, and hence it can be considered as a Level 0 –or even pre-Level 0– process. Depending on the nature of the object detection process, it can be seen as a Level 1 or, more likely, as a Level 0 procedure, since it aims at estimating the existence of an object, but not its properties. Object tracking and recognition are typical Level 1 procedures, given that it is concerned with the estimation of properties at object level. Some authors have discussed the convenience of taking special consideration to the fusion information and procedures between Level 1 and Level 2 processes defining a Level 1-1/2 (Das, 2008), given that the estimation of object properties frequently has a close dependence to the relations with other objects, an idea that partially overlaps with the notion of Level 2 Situation Refinement. This may be the case of object tracking, so this distinction must be noticed. Clearly, activity recognition, in its basic form, is a Level 2 procedure, because it aims at inferring the relations between objects in terms of a scene interpretation. If this process implies determining if the situation is a threat or not, it can be classified as a Level 3 task. Level 2 and Level 3 fusion are usually used to delimit the boundaries of HLIF. Last but not least, process enhancement is a Level 4 procedure, because it is concerned with providing all the possible means to improve the overall fusion procedure, from data sampling to activity recognition.

System communication and scalability are non-functional requirements of DIF in VSNs. Naturally, distributed nodes of the VSN must be properly communicated, but in several scenarios, communication simply understood as information transmission may not be enough, and coordination and/or cooperation between nodes may be demanded in the form of an ISS protocol. For example, when adjacent regions are watched by different cameras, they can take advantage of information handover when an object moves from one area to another, in which case coordinated tracking plans might be implemented (Castanedo, García, Patricio, & Molina, 2010). Scalability of the global system must be guaranteed with independence of the specific sensors and their configuration, which can be difficult when new and heterogeneous cameras are incorporated to build a large and scattered circuit. The cornerstone of effective ISS protocols in such distributed architectures is the use of common and well-defined communication vocabularies. The use of a common language facilitates component interoperability, thus allowing systems to decouple the internal implementation of sensor and their communication features. Ontologies (see section 4.2) can be successfully used to define de content language of the messages of the VSN, since their sharing and reusing features make them especially suitable to develop flexible and extensible solutions.

The accomplishment of DIF tasks in VSNs is strongly affected by the great variety of situations in which these technologies are being applied. Visual sensors are different from traditional sensors (radar, microphones, etc.), since they provide much more information while need more computational resources. As mentioned, classical statistical methods are insufficient when dealing with complex scenarios where illumination changes, occlusions from other objects or persons, and shadows and reflections appear. These methods are based on very strong statistical assumptions about the video data (e.g., Gaussian linear dynamic models of targets and noise), which unfortunately do not hold in most application domains. In addition, most of them rely on expensive computational costs that may be unaffordable in real time video applications.

The solution to these problems is to incorporate knowledge about the observed entities not directly obtained by the cameras to the image-processing procedures, i.e., to exploit additional information to achieve successful DIF –mainly, context knowledge. This *complementary* knowledge can be conveniently managed with formal representations, such as ontologies, which allow representing and reasoning with well-defined context descriptions. In the next section, we discuss on the types and the nature of the information and knowledge sources that can be incorporated to IF processes, whereas in section 5 we study in detail the role of context knowledge in HLIF. Particularly, we will show in section 6 an approach to an architecture and a framework for HLIF in VSNs based on the use of context knowledge represented with formal ontologies.

EXPLOITATION OF INFORMATION AND KNOWLEDGE IN INFORMATION FUSION

As was commented in the Introduction, there are only four types of information or knowledge that can be applied in developing any IF approach: observational data, a priori knowledge models, inductively learned knowledge, and contextual information. Each of these categories provides a distinctive type of input or support to the design of an IF process.

Observational Data

Observational data are the fundamental data from the dynamic, changing world of interest as collected from some observational capability, often an electro-mechanical sensor of some type[3]. At the basic level of fundamental physics, this is about either energy emanating from an entity (collected, observed by a passive sensor) or energy reflected from an entity that has been illuminated by an active, emitting sensor. Here we define that these data are about the observable entities in the world that are of interest.

In developing an (automated) multisensor IF system, it is clear that IF process designers need to know sensor characteristics very well, as the observational data are one of the important forms of input to the IF software process (and so, a "garbage-in, garbage-out" system). By and large, electro-mechanical sensors provide attribute and feature data about entities (such as color and shape in vision systems, and kinematic properties from radar), but do not provide *prima facia* evidence about inter-entity relationships. (In fact the philosophical literature argues that relationships are only discerned by a comparative-type processes that examine relative attributes from the entities of interest (Brower, 2001).) This limitation is one factor that makes the high-level fusion inferencing difficult due in part to the combinatorics of feasible relationships among entities and their attributes, comparing the extensive combinatorial set of all entity-entity attribute sets.

There are clearly many sensor parameters and operating characteristics that have to be understood and factored into the way any IF system would exploit such data, such as: resolution, sampling rate, detection performance, space-time pointing capability, etc. Observational conditions (a contextual factor) also have to be known. Critically important in the analysis and understanding of these inputs are the error characteristics of each sensor. This type of information is central to the formation of inter-sensor data association and the computation of the association scores is discussed in section 2.1.

A Priori Knowledge Models and Ontologies

We take the assumption that the world of interest is a dynamically-changing one and that "staring" problems involving the accumulation of observational data over a static world are not of interest. If we are trying to infer/estimate the existence of a "scene" or "situation", a HLIF product, we must understand what temporal resolution we need, since the scenes and situations are always changing. It can be argued in fact that if behaviors are components of scenes/situations, and if it is agreed that behaviors can be defined as actions over time, then placing "a" behavior as part of a scene imputes that scene development consumes time, so that the notion of a "snapshot" of a situation is inconsistent. It should be clear that a layered knowledge/model system will be required to support the layered fusion to the "Level 2" capability of figure 1, since the observational data by and large only provide discrete entity attributes; all of the relational constructs of interest need to be modeled a priori—in terms of their temporal evolution— so that the situational estimate can be formed by, in effect, hypothesis testing of these models against the observational (and other) data. Most IF systems are built on this type of deductive framework although certain environments demand more than just deductive models can provide.

Closely related to but different than models that model dynamical behaviors of interest in a world scene is the development of an ontology for the expected world of interest. Even the notions and definitions of the term ontology have become controversial in the last years in part because there is conflict regarding the degree of formalization required, which can range from an engineering-based to a philosophically-based viewpoint. We will not engage in this discussion but agree that an appropriate ontology is very often a helpful a priori component of the knowledge upon which an IF system can be developed. Ontologies aid in normalization of semantic meaning which aids in interoperability and specified ontologies can readily be made machine-readable today so that they can be exploited in the IF system if the opportunity arises. The engineering-type definition is "An explicit formal specification of how to represent the objects, concepts and other entities that are assumed to exist in some area of interest and the relationships that hold among them" (Staab & Studer, 2009), whereas the philosophical definition is "An ontology is a formal explicit specification of a shared conceptualization for a domain of interest" (Gruber, 1993), with all formal philosophical rigor to follow. From the logician's perspective, ontologies are Description Logics (DLs) knowledge bases, i.e., knowledge representations expressed in one of these well-defined subsets of first order logics (Baader, Calvanese, McGuiness, Nardi, & Patel-Schneider, 2003). But as HLIF is concerned with entities and relationships, it should be clear that ontologies have a role to play in HLIF development; much more will be said on this for the vision applications in section 6.

Learned Knowledge

In those cases where the required a priori knowledge for deductively-based HLIF process development cannot be formed—this can be due to a variety of factors—one alternative is to try and excise the knowledge through online machine learning processes operating on the observational and other data. These are procedural and algorithmic methods for discovering relationships among and behaviors of entities of interest. A wide variety of alternative machine learning strategies exists (e.g., see (Bishop, 2004)). Of course, it is likely that some a priori deductive knowledge will be applicable, and it can always be argued that the real-world problem domain may travel out of the bounds of a priori knowledge in any case, so for many applications the best approach, if workable, will be some combination of a priori and runtime-learned knowledge working in cooperation. This of course generates another problem involving dynamic knowledge management (which could be considered a "Level 4" process refinement requirement), as well as possible conflict resolution techniques, among possible other effects. As IF applications increase in sophistication and complexity, another system design question revolves about the role of humans in IF process operations. Clearly there is a cost-benefit tradeoff involved in trying to develop fully-automated algorithmic IF processes for complex problems where the insertion of human intelligence at some point in the process may be a much more judicious choice.

Contextual Information

Context and the elements of what could be called Contextual Information are both concepts that have had a lot written about them –e.g., see (Zimmermann, Lorenz, & Oppermann, 2007) and (Dey, 2001). Definitions abound but we choose one from (Henricksen, 2003), *"The context of a task is the set of circumstances surrounding it that are potentially of relevance to its completion."*, because of its task-relevance; we think here of a fusion or estimating/inferencing task whose "completion" we take to mean the development of a best-possible estimate. We also choose to add the notion of context as background, i.e., not the specific entity, event, or behavior of prime interest but that information which is influential to the formation of a best estimate of these items. So re-phrased we may say this definition as: "The context of an IF estimation task is the set of background circumstances (the circumstances perhaps equal to but minimally derivable from appropriate contextual information) surrounding it that are of potential relevance to its optimal formation". Context can aid inference development in at least two ways: (1) by providing an independent (from the core IF system data) basis for consistency-checking, i.e., checking that the asserted inference is consistent with the relevant context, and (2) by providing a basis for the amplification of an inference by adding explanatory aspects when the inference is consistent with the context. Context and contextual information are broad notions; the challenge here is to address the question of what (of all that may be available) is relevant to the clarification or improvements of an IF estimate built from the core observational and other core data.

THE ROLE OF CONTEXTUAL INFORMATION IN HIGH-LEVEL INFORMATION FUSION: UNDERSTANDING AND USING CONTEXT

As remarked in the Introduction, we elaborate here on the use of Contextual Information largely because of its criticality in vision/video applications; recent experience has shown that Contextual Information is a critical factor in modern defense applications involving counter-terrorism and counter-insurgency applications. Following the definitions stated in the previous section, contextual information in HLIF is that information that can be said to "surround" a situation of interest in the world. It is information that aids in understanding the (estimated) situation and also aids in reacting to the situation, if a reaction is required. It can be seen as a set of constraints to a reasoning process about a situation; Kandefer and Shapiro (2008) define it this way: "the structured set of variable, external constraints to some (natural or artificial) cognitive process that influences the behavior of that process in the agent(s) under consideration." Contextual Information can be relatively or fully static or can be dynamic, possibly changing along the same timeline as the situation. It is also likely that the full characterization and specification of Contextual Information may not be able to be known at system/algorithm design time, except in very closed worlds.

Thus, we envision an "a priori" framework of exploitation of Contextual Information that attempts to account for the effects on situational estimation of that Contextual Information (CI henceforth) that is known at design time; there is a question of the ease or difficulty involved in integrating CI effects into a fusion system design or into any algorithm designs. This issue is influenced in part by the nature of the CI and the manner of its native representation, e.g., as numeric or symbolic, and the nature of the corresponding algorithm. Strategies for a priori exploitation of CI may thus require the invention of new hybrid methods that incorporate whatever information an algorithm normally employs in estimation (usually observational data) with an adjunct CI exploitation process. Note too that CI may, like observational data, have errors and inconsistencies itself, and accommodation of such errors is a consideration for hybrid algorithm design. In this case then, we have a notional processing operation as shown in Figure 3.

Similarly, we envision the need for an "a posteriori" CI exploitation process, due to at least two factors:

1. That all relevant CI may not be able to be known at system/algorithm design time, and may have to be searched for and discovered at runtime, as a function of the current situation estimate.
2. That such CI may not be of a type that was integrated into the system/algorithm designs at design time and so may not be able to be easily integrated into the situation estimation process.

In this case then we envision that at least part of the job of posteriori CI exploitation would be of a type that checks the consistency of a current situational hypothesis with the newly-discovered

Figure 3. Notional Process Flow for "A Priori" CI Exploitation

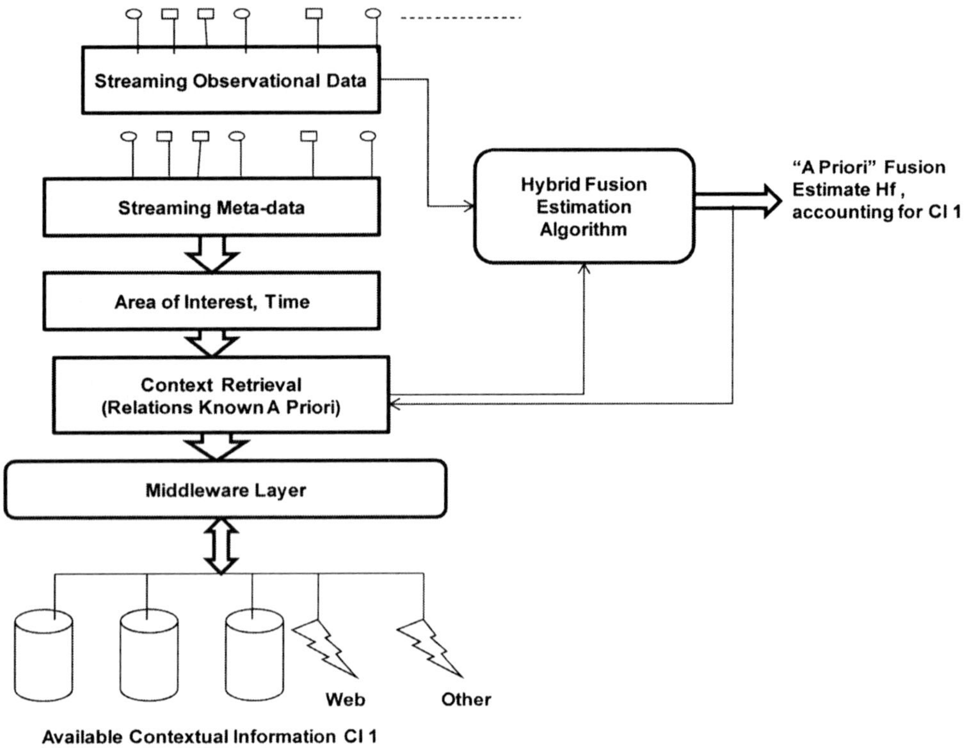

(and situationally-relevant) CI, but also – if the hypothesis is in fact consistent – adds some explanatory aspects to the declared hypothesis.

There are yet other system engineering issues. The first is the question of accessibility; CI must of course be accessible in order to use it, but accessibility may not be a straightforward matter in all cases. One question is whether the most-current CI is available; another may be that some CI is controlled or secure and may have limited availability. The other question is one of representational form. CI data can be expected to be of a type that has been created by "native" users – for example weather data, important in many fusion applications as CI, is generated by meteorologists, for meteorologists (not for fusion system designers) – thus, even if these data are available, there is likely to be a need for a "middleware" layer that incorporates some logic and algorithms to both sample these data and shape them into a form suitable for use in fusion processes of various type. In even simpler cases, this middleware may be required to reformat the data from some native form to a useable form.

In spite of some a priori mapping of how CI influences or constrains the way in which situational inferences or estimates can be developed, which may serve certain environments, the defense and security type applications, with their various dynamic and uncertain types of CI, demand a more adaptive approach. Given a nominated situational hypothesis H_f from a fusion process or "engine" (that might already have accounted for some CI, as explained above), the first question is: what CI type information is relevant to this hypothesis? As cited by Kandefer and Shapiro (2008), "The relevancy problem is defined by Ekbia and Maguitman (2001) as 'the problem of identifying and using properly [only] the information that should exert an influence on our beliefs, goals,

Figure 4. Notional Process Flow for "A Posteriori" CI Exploitation

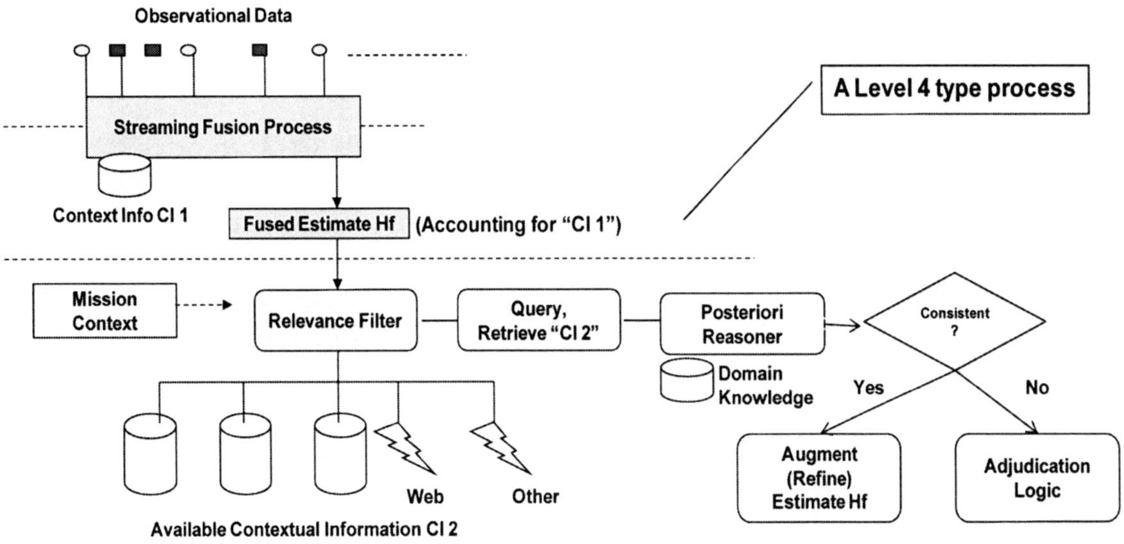

or plans'." Said otherwise, relevant CI is only that information that influences our interpretation or understanding of H_f. Presuming a "relevancy filter" can be crafted, a search function would explore the available or retrievable CI and make this CI available to a "posteriori" reasoning engine. That reasoning engine would then use a CI-guided subset of Domain Knowledge, and the retrieved CI to reason over H_f to first determine consistency of H_f with the relevant CI. If it is inconsistent, then some type of adjudication logic will need to be applied to reconcile an inconsistency between the fusion process that produced H_f and the posteriori reasoning process that judges it as inconsistent. If however H_f is judged as consistent with the additional CI, an expanded interpretation of H_f could be developed, providing a deeper situational understanding. This processing flow is depicted in figure 4 below. As noted in the figure, this overall process, which can be considered a "Process Refinement" operation, would be a so-called "Level 4" process in the context of the JDL Data Fusion Process Model.

VISUAL SENSOR NETWORK APPLICATIONS: CONTEXT-AWARE SURVEILLANCE SYSTEMS

A prototypical VSN application is surveillance with distributed camera networks. Modern surveillance systems have been named *third-generation surveillance applications*, a term that designates systems that resemble the nature of the human intelligent process of surveillance, which activates certain cognitive abilities, and that satisfy the requirements of modern surveillance, which are among others the management of a large number of cameras, the geographical spread of resources, and the need of many monitoring points (Regazzoni, Ramesh, & Foresti, 2001; Valera & Velastin, 2005). The ultimate goal of third-generation surveillance systems is to automatically achieve a high degree of understanding of the scene from multiple observations to barely require operator attention while cutting component cost. The existence of multiple cameras requires the development of IF procedures to integrate data generated at different locations and reasoning techniques to obtain a high-level and global interpretation of the scene.

CI plays a key role in the development of third-generation surveillance systems. Consistently to the definition presented in the previous sections, we can consider CI in this kind of applications as any external piece of knowledge used to complete the quantitative data about the scene computed by straightforward image-analysis algorithms can be considered context. As in (Kandefer & Shapiro, 2008), CI is therefore an "external constraint" (since it is not directly acquired by the primary system sensors) that "influences the behavior" of the fusion process (since it is used to guide and support DIF). Adapting the characterization by Bremond and Thonnat (1996), four sources of CI must be taken into account in visual DIF: (1) the scene environment: structures, static objects, illumination and other behavioral characteristics, etc.; (2) the parameters of the recording: camera, image, and location features; (3) historic information: past detected events; (4) soft information provided by humans.

This approach to the notion of CI may seem imprecise, but this is intentional. It is especially difficult to clearly distinguish which information is context and which information is perception, either from a computational or a cognitive perspective, in visual applications. Hence, we propose to create a symbolic model that blends perceived and contextual information to obtain an integrated description of the scenario. This knowledge model is encoded with formal ontologies, which are used with two purposes: representing the entities of the domain of interest, i.e., the objects and the relations (contextual and perceived) of the scene; and reasoning with them to perform DIF tasks, i.e., to derive implicit knowledge from the explicit facts to accomplish high-level scene recognition and to provide feedback to improve lower-level image processing procedures. From a practical perspective, this avoids the necessity of explicitly distinguishing between CI and other types of knowledge. In the remainder of this section, we firstly present some related research works based on the use of formal knowledge representations for CI exploitation in information fusion. Secondly, we describe a general architecture for HLIF in VSN and explain how knowledge is represented and exploited in such framework.

Related Work in Ontology-Based CI Exploitation and IF in VSNs

Symbolic models have been applied to acquire, represent, and exploit knowledge in IF, and particularly in visual IF. In the last years, the interest in ontologies has increased considerably (Nowak, 2003), and its use is becoming more and more frequent. Nevertheless, despite the fact that most of the current approaches combine contextual and perceptual information, they do not explicitly describe how context is characterized and integrated in the fusion loop.

Previous ontology-based fusion researches can be classified according to the levels defined by the JDL model. At image-data level (i.e., JDL level 0), one of the most important contributions is COMM (Core Ontology for MultiMedia), an OWL ontology to encode MPEG-7 data (Arndt, Troncy, Staab, Hardman, & Vacura, 2007). Similarly, the Media Annotations Working Group of the W3C is working in an OWL-based language for adding metadata to Web images and videos (Lee, Bürger, & Sasaki, 2009).

Other proposals are targeted at modeling video content at object level (i.e., JDL L1). For example, a framework for video event representation and annotation is described in (François, Nevatia, Hobbs, Bolles, & Smith, 2005). This framework includes two languages, namely VERL (Video Event Representation Language) and VEML (Video Event Markup Language), used to describe processes and to markup video sequences, respectively. Other authors have discussed and improved this approach to support the representation of uncertain knowledge (Westermann & Jain, 2007). The research work in (Kokar & Wang, 2002) presents a symbolic representation for the data managed by a tracking algorithm. Similarly, recently it has been a first approximation to the development of an ontology for tracking data (Snidaro, Belluz, & Foresti, 2007).

Scene interpretation issues (i.e., JDL L2 and L3) are being dealt with ontologies as well. In (Neumann & Möller, 2008), a proposal for scene interpretation based on Description Logics and supported by the reasoning features of RACER inference engine is presented. The problem of representing high-level semantics of situations with a computable formalism is also faced in (Kokar, Matheus, & Baclawski, 2009), where an ontology encoding Barwise's situation semantics is developed. The approach in (Aguilar-Ponce, Kumar, Tecpanecatl-Xihuitl, & Bayoumi, 2007) defines a multi-agent architecture for object and scene recognition in VSNs. In addition, the authors propose the use of an ontology to communicate information between task-oriented agents.

All these research works focus on contextual scene recognition, but as previously mentioned, it is also interesting to apply this knowledge to refine image-processing algorithms (which corresponds to JDL L4), and particularly trackers. A preliminary approach to this topic has been presented in (Gómez-Romero, Patricio, García, & Molina, 2009b). In this paper, the authors overview some aspects of the current state of this research and discuss its contributions from an architectural and knowledge management point of view. In the next section, we elaborate on this approach to analyze general problems and solution to CI exploitation in HLIF of VSNs.

Architecture for High-Level Context-Based Visual Information Fusion

According to the description of the flow processes for CI exploitation discussed in section 5, VSNs will favor *a posteriori* schemas, since they facilitate the incorporation of CI at runtime in relation to the current situation estimate and, which is more important, CI cannot be easily integrated into the situation estimation procedures. This is especially relevant in the case of tracking, in which our proposal focuses, because it is not clear how CI can be directly applied within the classical tracking algorithms and, when possible, it entails the development of ad hoc solutions that can be hardly extended to similar domains, which rockets application costs. The matter is that there is a considerable gap between the CI that can be exploited to improve tracking and the form in which information can be provided to change the behavior of a tracking algorithm; the former is usually defined at a high abstraction level (e.g. "a person is hidden behind a column"), whereas the latter is usually a numerical parameter of a statistical procedure (e.g. noise models or variance matrices). An "A Priori" model of this correspondence would entail to *hardwire* the CI management logic to the particular software. Consequently, we adapt the notional process flow for "A Posteriori" CI exploitation depicted in figure 4 to decouple low (classical) and high-level (context-based) DIF procedures.

The general architecture of a CI-enhanced Fusion Node in a VSN is depicted in Figure 5. The schema shows the tracking system (the GTL, General Tracking Layer) and, built upon it, the context-based extension for HLIF (the CIL, Context Information Layer). Interestingly enough, the GTL and the CIL may not be in the same node of the VSN, in such a way that the HLIF may be performed not from the data obtained by a single node, but from the fused data obtained after combining the track results of independent nodes. The communication between the GTL and the CIL is performed through the generic GTL/CIL interface, which defines method to update the CIL model according to the GTL tracking data and to consult recommendations calculated by the CIL. This interface acts as a middleware that guarantees the interoperability and the independence of both layers.

The GTL is an implementation of a classical tracking procedure –we will assume the tracker described by Patricio et al. (2008), but the HLIF

Figure 5. Architecture of a HLIF node of a VSN

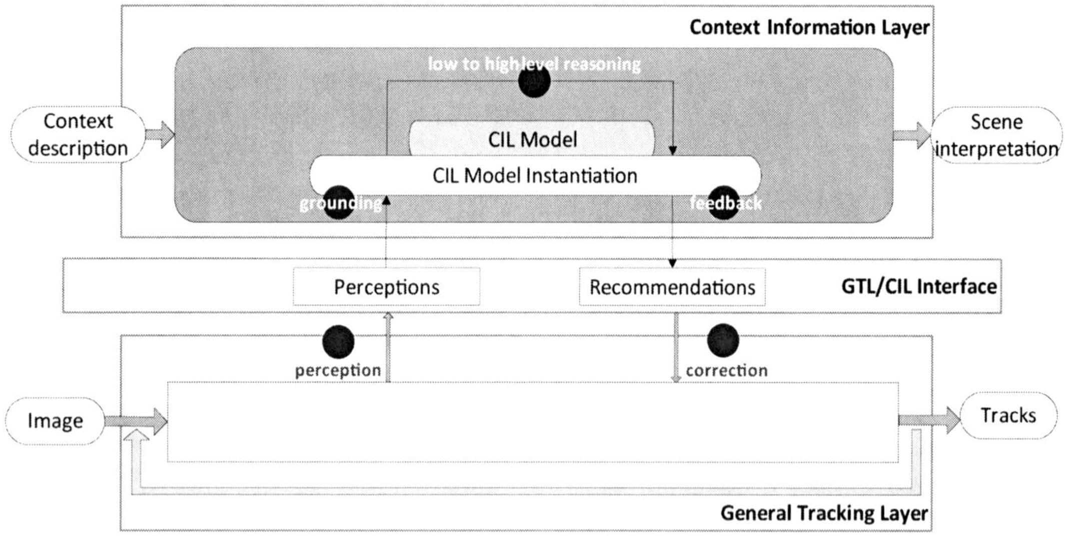

architecture is independent of the concrete algorithms. It sequentially executes various image-processing algorithms with a video sequence in order to track all the targets within the local field. The GTL is arranged in a pipelined structure of several modules, which correspond to the successive stages of the tracking process: (1) detection of moving objects; (2) blob-to-track multi-assignment; (3) track initialization/deletion; (4) trajectory analysis.

Essentially, the CIL manages the CIL model, which is the abstract representation of the scene. The CIL receives from the GTL tracking data, reasons with it, and provides as a result a high-level description of the situation. The CIL model is implemented as a set of interrelated ontologies organized according to the JDL model. The contents and the use of these ontologies are described in section 6.3. The information flow in the CIL is as follows (see Figure 6):

1. The GTL calls the GTL/CIL interface methods when tracks are created, deleted, or updated.
2. Updated track information is conveniently transformed into a symbolic representation, and the CIL model is updated with it. This triggers a reasoning procedure developed in four steps, which leads to a high-level representation of the scene from the low-level tracking data. In this process, CI obtained from several sources is applied to create and validate interpretation hypothesis. The reasoning procedure clearly reflects the task levels described in the JDL model, from L0 to L3:
    a. Abstraction of acquired tracking data, i.e., transformation of tracking data into ontology instances (*grounding*).
    b. Correspondence between tracks and objects, i.e., object identification from track properties.
    c. Recognition of activities from object descriptions, i.e., scene interpretation from object properties.
    d. Evaluation of threats, i.e., activity impact assessment.
3. Recommendations are created by reasoning with the updated scene model and CI –which can be seen as a L4 task. These recommendations are provided as a feedback to the GTL/CIL interface.

Figure 6. Process flow and knowledge structure of the CIL of a HLIF node of the VSN

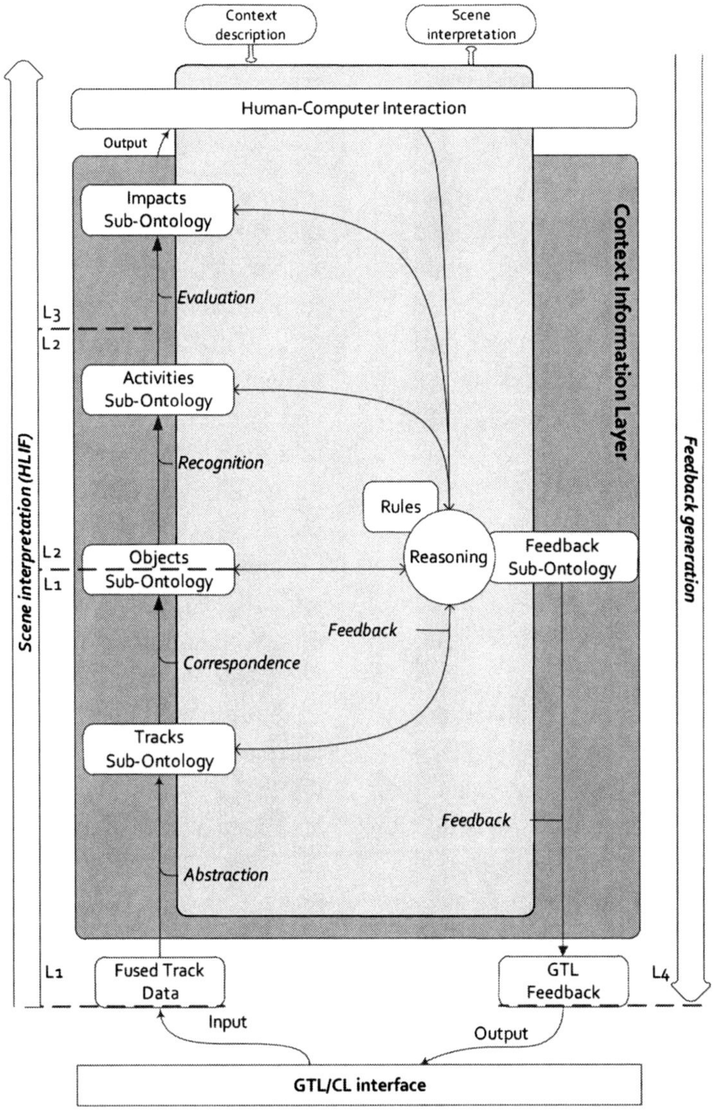

4. The GTL calls the GTL/CIL interface methods to consult recommendations from the CIL, and converts them to concrete actions to correct its behavior.

Ontological Representation of CI for Visual DIF

The knowledge of the CIL model is structured in four layers, as depicted in figure 6, from tracking data to impacts and threats. The knowledge is closely related to two modules of the architecture: camera data acquisition and feedback generation.

- **Camera data (L0).** Sequence provided by the cameras (in some processable video format).
- **Tracking data (L1).** Output of the tracking algorithm represented symbolically. Example: frames, tracks and track properties (color, position, velocity, etc.).

- **Scene objects (L1-L1/2).** Objects resulting from making a correspondence between existing tracks and possible scene objects. For example, a track can be inferred to correspond to a person (possibly by applying CI). Scene objects include static elements, which may be defined a priori, and dynamic objects, which may be defined a posteriori. Example: person, door, column, window, etc.
- **Activities (L2).** Description of relations between objects which last in time. Example: grouping, approaching, picking/leaving an object, etc.
- **Impacts and threats (L3).** Cost or threat value assigned to activities.
- **Feedback and process improvement (L4).** Abstract representation of the suggestions provided to the tracking procedure.

The levels L1-L4 corresponds to an ontology in the CIL knowledge representation model. We provide a reference version of these ontologies[4] in the standard ontology language OWL (McGuiness & van Harmelen, 2004) that must be refined in a concrete application. For instance, in a particular application it may be interesting to define a concept to represent a Mirror as a specialization of StaticObject and ReflectingObject, which are concepts of the scene objects ontology. For a more extensive description of these ontologies, see (Gómez-Romero, Patricio, García, & Molina, 2009c).

In their definition, the ontologies of the CIL model only contain descriptive knowledge; i.e., axioms defining concepts and relations. They offer a vocabulary to describe scenes that must be instantiated in a concrete application according to the entities appearing in the scene, both static and dynamic. Thus, before starting the analysis of a video sequence, it is necessary to annotate the static elements of the scenario. Annotating the scenario means to create instances of the ontologies describing scenario properties: object position and size, possible occlusions, enter and exit zones, or any other contextual knowledge that we may consider necessary (see Figure 7).

Additionally, contextual rules must be loaded into the reasoner. Ontology instances corresponding to the dynamic entities of the scene are created and deleted during the processing of the sequence as a result of the track information provided by the GTL and additional reasoning processes (see Figure 8). In the next section, we

Figure 7. Initial mark-up of the scenario[5]

Figure 8. Additional instances created during video processing

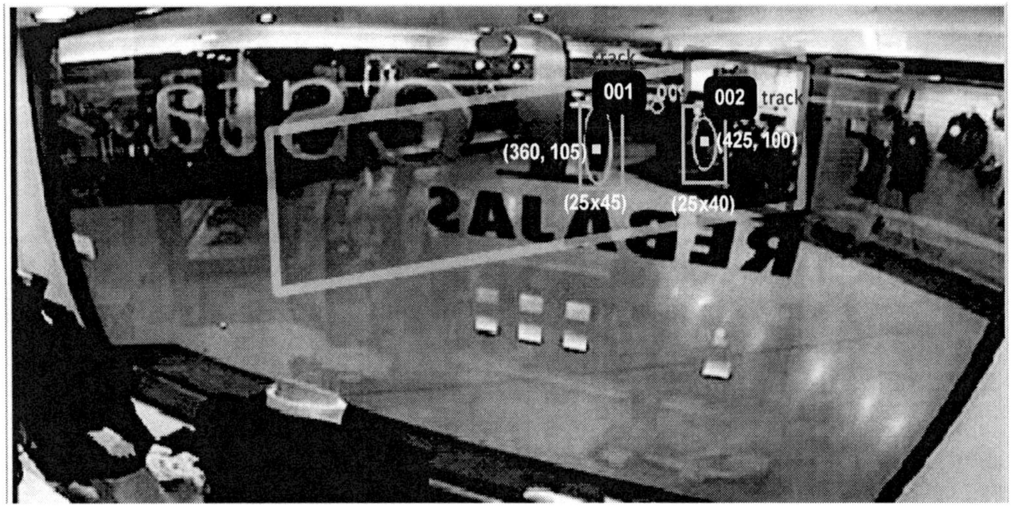

explain the role of reasoning rules in the architecture and how instances are inferred to obtain the high-level description of the scenario.

Abductive Reasoning for HLIF (Interpretation and Feedback)

Standard ontology reasoning procedures can be performed within the CIL ontologies to infer additional knowledge from the explicitly asserted facts (tracking data and CI). By using a DL inference engine, tasks such as classification or instance checking can be performed. Nevertheless, monotonicity of ontology languages forbids adding new knowledge to the models while reasoning, which is required in scene interpretation. Actually, scene interpretation is a paradigmatic case of abductive reasoning, in contrast to the DL deductive reasoning: abductive reasoning takes a set of facts as input and finds a suitable hypothesis that explains them (sometimes with an associated degree of confidence or probability). This is what is needed in this case: we want to figure out what is happening in the scene from the observed and the contextual facts. In terms of the architecture of the CIL, scene interpretation can be seen as an abductive transformation from instances of a lower level ontology to instances of a higher level ontology. Abductive reasoning is not directly supported by ontologies (Elsenbroich, Kutz, & Sattler, 2006), but it can be simulated by using customized procedures or, preferably, by defining transformation rules in a suitable query language. The RACER inference engine allows abductive reasoning (Häarslev & Möller, 2001), and therefore it is a good choice to implement the reasoning procedures within the ontologies.

In the architecture, abductive rules formally represent contextual, heuristic and common sense knowledge to accomplish HLIF and low-level tracking refinement. Accordingly, we have two types of rules: bottom-up rules and top-down rules. *Bottom-up* rules are used in scene interpretation and obtain instances of an upper-level ontology from instances of a lower-level ontology. For instance, some rules could be defined to identify objects from track measures; i.e., to obtain instances of the scene objects ontology from instances of the tracking data ontology. An example rule may be: "create a person instance when an unidentified track larger than a predefined size is detected inside a region of the image". *Top-down* rules create suggested action instances from the current interpretation of the scene, the historical data, and the predictions. These actions are used to adapt hypothesis at a lower-level to interpretations of a higher-level, which means the creation of instances of a lower-level ontology from instances of an upper-level ontology. Eventually, the top-down rules may create instances of the feedback ontology, which are retrieved and interpreted by the GTL through the GTL/CIL interface, resulting (if not discarded) in corrections to the low-level fusion procedure: tracking parameters, data structures, etc. Please notice that recommendations can be generated at different abstraction levels, either to modify instances of a lower-level ontology or to create new instances of the recommendations ontology. An example rule may recommend "to ignore a track associated to a person which is inside an area marked as a mirror". More details of the use of RACER abductive rules in the architecture can be found in (Gómez-Romero, García, Patricio, & Molina, 2009a).

FUTURE RESEARCH DIRECTIONS IN SENSOR-NETWORK-BASED CONTEXTUAL EXPLOITATION

In spite of the recent advances in contextual IF, there are several topics that require further research. Broadly speaking, the existing approaches are quite specific and it is difficult to apply the advances obtained in one application domain to another. The architecture presented in this chapter is an initial attempt to establish a common terminology and a reference organization for such systems, but it is necessary to advance in the development of frameworks that allow for the implementation of DIF applications with less effort. An IF framework should provide an adaptable infrastructure where specific procedures can be easily reused and/or integrated, especially those based on AI techniques, which are likely to play a key role in the next generation fusion applications. We envision a possible approach to a general IF multi-layer framework with: a front-end that manages hard and soft sensor inputs; an initial layer for detection, semantic labelling, and flow control, based on an intelligent repository of pluggable algorithms; a fusion layer, composed of several interrelated fusion nodes that process information at different JDL levels and incorporate CI to the process; and a presentation layer to convey the results through appropriate visualization interfaces. More discussions and realizations of general frameworks for DIF will be certainly useful to foster the creation of competitive solutions while cutting development costs in critical application areas.

There is also a promising field of study concerned with the design and the development of these architectures targeted to highly-distributed IF problems. The architecture presented in this chapter does not compel any task distribution schema, since we have assumed that IF procedures are independent, and that they can be performed by independent processes as long as proper communication mechanisms are offered. For the sake of simplicity, we have limited data alignment at tracking level, but it could be possible to combine estimations performed by single IF nodes at different levels, in such a way that the system would be able to obtain a combined view of the scenario from the detected objects or the recognized situations, instead of only the track data.

Recently, the multi-agent paradigm has been proposed to design and implement decentralized DDF schemas. The multi-agent paradigm provides a theoretical and practical framework to achieve successful communication and cooperation among the distributed components of the system. Multi-agent systems are argued to be well suited for the development of distributed systems in dynamic environments, as it occurs with VSNs. As a matter of fact, the notion of agent suits very well to the concept of intelligent camera. On the one hand, nodes in the VSNs are autonomous, in the sense that they have processing capabilities to acquire and process information in its field of view. On the other hand, the social abilities of agents provide the necessary means to implement ISS protocols and to cooperate in the overall objective of the VSN. In addition, agent-based standard communication protocols are the support to achieve interoperation with other systems. Last but not least, there are several multi-agent frameworks that hide particular communication details and provide an easy way for developing distributed systems.

In this chapter, we have proposed ontologies to represent and reason with contextual and factual knowledge for HLIF. Nevertheless, classical ontologies do not allow the representation of uncertain or vague knowledge, despite it is inherent to several fusion applications, especially when it comes to CI. We consider that it will be convenient to adapt current approaches to fuzzy and probabilistic ontologies to IF, thus facilitating the creation of knowledge bases able to deal with imprecise information. This is particularly relevant in IF systems that have textual –or, in general, subjective human-generated input data–, which are more and more frequent in cutting-edge fusion domains (e.g. modern asymmetric warfare). Modelling the vagueness and the uncertainness of this knowledge will allow its integration with traditional sensor sources and obtaining more accurate estimates and assessments.

CONCLUSION

Information Fusion aims at improving the state estimations performed by a single-node system by the combination of the observations acquired at distributed sensor nodes. Information Fusion procedures are usually classified according to the JDL model, the prevailing theory to describe fusion systems, which establishes a common vocabulary to facilitate communication and understanding among the heterogeneous specialists interested in Information Fusion, and defines a functional model to structure fusion applications. The JDL model defines five levels of processing, ranging from the less abstract signal processing procedures to the more abstract situation recognition procedures. High-level information fusion is a term that designates the achievement of a symbolic interpretation of the perceived scenario in terms of situation descriptions and assessments.

Despite the success of low-level data fusion techniques, there is still a lack of systematic approaches to high-level information fusion, mainly because there are several information sources that need to be modeled, managed, and exploited in this kind of applications. Modern applications involving high-level information fusion require the employment and exploitation of all available information due to the complexities of these problems and the frequent difficulties of achieving sufficiently-rich observational data. Contextual information can often help in the formation of fused state estimates that are improvements over those developed from observational data alone, but of course such exploitation comes at a price of complexity that has to be factored into the cost-effectiveness of any prototype design. In this chapter, we have reviewed the basic design frameworks for such contextual exploitation strategies. We have described an “a priori” and an “a posteriori” schema to the exploitation of context information; the former is targeted at applications domains where the context can be easily introduced and applied in the fusion process, whereas the latter is more appropriate when the context is not known at design time, or it cannot be directly applied to guide the fusion.

Visual sensor networks are a prototypical case of the second schema, since video cameras provide a huge amount of information that is difficult to model, to interpret, and to exploit, especially at a high abstraction level. We have studied the requirements and the issues that arise in contextual high-level fusion, and provided a firsthand architectural proposal to discuss on the specific tasks and workflow that should be considered when designing and implementing a high-level information fusion system in a visual sensor network. We have proposed the use of ontologies to represent and reason with visual knowledge in visual sensor networks, since they provide several advantages over ad hoc mechanisms: formalization, modularization, reutilization, availability of tools, etc. Information fusion is therefore seen as a model construction task accomplished by creating hypothesis through abduction on the perceived information based on the premises established by context knowledge.

However, as we have highlighted in the future research section, much remains to be done in defining and designing more robust and efficient approaches for use of contextual information, especially when visual sensor networks are involved. The fusion community needs to caucus on this issue and develop holistic research strategies that will result in unified design patterns for contextual exploitation.

ACKNOWLEDGMENT

The UC3M Team gratefully acknowledges that this research activity is supported in part by Projects CICYT TIN2008-06742-C02-02/TSI, CICYT TEC2008-06732-C02-02/TEC, CAM CONTEXTS (S2009/TIC-1485) and DPS2008-07029-C02-02.

UC3M also thanks Prof. James Llinas for his helpful comments during his stay, which has been supported by the collaboration agreement 'Chairs of Excellence' between University Carlos III and Banco Santander.

Prof. James Llinas gratefully acknowledges that this research activity is supported by a Multidisciplinary University Research Initiative (MURI) grant (Number W911NF-09-1-0392) for "Unified Research on Network-based Hard/Soft Information Fusion", issued by the US Army Research Office (ARO) under the program management of Dr. John Lavery.

REFERENCES

Aguilar-Ponce, R., Kumar, A., Tecpanecatl-Xihuitl, J. L., & Bayoumi, M. (2007). A network of sensor-based framework for automated visual surveillance. *Journal of Network and Computer Applications*, *30*(3), 1244–1271. doi:10.1016/j.jnca.2006.04.011

Arambel, P. (2008). Structured pedigree information for distributed fusion systems. *Proceedings of the SPIE* (17).

Arndt, R., Troncy, R., Staab, S., Hardman, L., & Vacura, M. (2007). COMM: Designing well-founded multimedia ontology for the Web. *Proceedings of the 6th International Semantic Web Conference (ISWC '07)* (pp. 30-43), Busan, South Korea.

Baader, F., Calvanese, D., McGuiness, D. L., Nardi, D., & Patel-Schneider, P. F. (2003). *Description logic handbook: Theory, implementation, and applications*. Cambridge, UK: Cambridge University Press.

Barron, J., Fleet, D., & Beauchemin, S. (1994). Performance of optical flow techniques. *International Journal of Computer Vision*, *12*(1), 42–77. doi:10.1007/BF01420984

Bishop, C. (2004). *Pattern recognition and machine learning*. New York, NY: Springer.

Bremond, F., & Thonnat, M. (1996). A context representation for surveillance systems. In *Proceedings of the Workshop on Conceptual Descriptions from Images at the 4th European Conference on Computer Vision (ECCV '96)*. Cambridge, UK.

Brower, J. (2001). Relations without polyadic properties: Albert the Great on the nature and ontological status of relations. *Archiv für Geschichte der Philosophie*, *83*, 225–257. doi:10.1515/agph.2001.001

Castanedo, F., García, J., Patricio, M. A., & Molina, J. M. (2009). Designing a visual sensor network using a multi-agent architecture. In *Proceedings of the 7th International Conference on Practical Applications of Agents and Multi-Agent Systems (PAAMS '09)* (pp. 430-439), Salamanca, Spain. Berlin/Heidelberg, Germany: Springer.

Castanedo, F., García, J., Patricio, M. A., & Molina, J. M. (2010). Data fusion to improve trajectory tracking in a Cooperative Surveillance Multi-Agent Architecture. *Information Fusion, 11*(3), 243–255. doi:10.1016/j.inffus.2009.09.002

Chang, K., Chong, C. Y., & Bar-Shalom, Y. (1986). Joint probabilistic data association in distributed sensor networks. *IEEE Transactions on Automatic Control, 31*(10), 889–897. doi:10.1109/TAC.1986.1104143

Chong, C. Y., Mori, S., & Chang, K. C. (1990). Distributed multitarget multisensor tracking. In Bar-Shalom, Y. (Ed.), *Multitarget-Multisensor Tracking: Advanced Applications* (*Vol. 1*, pp. 247–295). Norwood, MA: Artech House.

Das, S. (2008). *High-Level Data Fusion. London*. Boston: Artech House Publishers.

Dey, A. K. (2001). Understanding and using context. *Personal and Ubiquitous Computing Journal, 5*(1), 4–7. doi:10.1007/s007790170019

Ekbia, H. R., & Maguitman, A. G. (2001). Context and relevance: A pragmatic approach. In *Proceedings of the 3rd International and Interdisciplinary Conference on Modeling and Using Context (CONTEXT '01)* (pp. 156-169). Dundee, UK: Springer.

Elsenbroich, C., Kutz, O., & Sattler, U. (2006). A case for abductive reasoning over ontologies. In *Proceedings of the OWL Workshop: Experiences and Directions (OWLED '06)*, Athens, GA, USA.

François, A., Nevatia, R., Hobbs, J., Bolles, R., & Smith, J. (2005). VERL: An ontology framework for representing and annotating video events. *IEEE MultiMedia, 12*(4), 76–86. doi:10.1109/MMUL.2005.87

Gómez-Romero, J., García, J., Patricio, M. A., & Molina, J. M. (2009). Towards the implementation of an ontology-based reasoning system for visual information fusion. In *Proceedings of the 3rd Skövde Workshop on Information Fusion Topics (SWIFT '09)* (pp. 5-10), Skövde, Sweden.

Gómez-Romero, J., Patricio, M. A., García, J., & Molina, J. M. (2009). Context-based reasoning using ontologies to adapt visual tracking in surveillance. In *Proceedings of the 6th IEEE International Conference on Advanced Video and Signal Based Surveillance (AVSS '09)* (pp. 226-231), Genoa, Italy.

Gómez-Romero, J., Patricio, M. A., García, J., & Molina, J. M. (2009). Ontological representation of context knowledge for visual data fusion. In *Proceedings of the 12th International Conference on Information Fusion (FUSION '09)* (pp. 2136-2143), Seattle, WA, USA.

Gruber, T. (1993). A translation approach to portable ontology. *Knowledge Acquisition, 5*(2), 199–220. doi:10.1006/knac.1993.1008

Häarslev, V., & Möller, R. (2001). Description of the RACER system and its applications. In *Proceedings of the International Workshop on Description Logics (DL '01)*. Stanford University, CA, USA.

Hall, D. L., & Llinas, J. (1997). An introduction to multisensor data fusion. *Proceedings of the IEEE, 85*(1). doi:10.1109/5.554205

Heikkila, J., & Silven, O. (1997). *A four-step camera calibration procedure with implicit image correction* (pp. 1106–1112). IEEE Computer Vision and Pattern Recognition.

Henricksen, K. (2003). *A framework for context-aware pervasive computing applications* (PhD Thesis, University of Queensland).

Hongeng, S., Nevatia, R., & Bremond, F. (2004). Video-based event recognition: Activity representation and probabilistic recognition methods. *Computer Vision and Image Understanding, 96*(2), 129–162. doi:10.1016/j.cviu.2004.02.005

Hu, W., Xie, D., Tan, T., & Maybank, S. (2004). Learning activity patterns using fuzzy self-organizing neural network. *IEEE Transactions on Systems, Man, and Cybernetics, 34*(3), 1618–1626. doi:10.1109/TSMCB.2004.826829

Julier, S. (2009). Estimating and exploiting the degree of independent information in Distributed Data Fusion. In *Proceedings of the 12th International Conference on Information Fusion (FUSION '09),* Seattle, WA, USA.

Julier, S., & Uhlmann, J. (2001). General decentralized data fusion with Covariance Intersection. In Liggins, M. E., Hall, D., & Llinas, J. (Eds.), *Handbook of multisensor data fusion* (2nd ed., pp. 319–344). Boca Raton, FL, USA: CRC Press. doi:10.1201/9781420038545.ch12

Kandefer, M., & Shapiro, S. C. (2008). A categorization of contextual constraints. *Biologically inspired cognitive architectures:* Papers from the AAAI Fall Symposium (pp. 88-93). Menlo Park, CA, USA: AAAI Press.

Kokar, M., Matheus, C., & Baclawski, K. (2009). Ontology-based situation awareness. *Information Fusion, 1*, 83–98. doi:10.1016/j.inffus.2007.01.004

Kokar, M., & Wang, J. (2002). Using ontologies for recognition: An example. In *Proceedings of the 5th International Conference on Information Fusion* (2, pp. 1324-1330).

Lee, W., Bürger, T., & Sasaki, F. (2009). *Use cases and requirements for ontology and API for Media Object 1.0.* Retrieved from W3C Working Draft: http://www.w3.org/ TR/ media- annot-reqs/

Liggins, M., Hall, D., & Llinas, J. (2009). *Handbook of multisensor data fusion* (2nd ed.). Boca Raton, Florida, USA: CRC Press.

Lipton, A., Fujiyoshi, H., & Patil, R. (1998). Moving target classification and tracking from real-time video. In *Proceedings of the 4th IEEE Workshop Applications of Computer Vision (WACV '98)* (pp. 129-136). Princeton, New Jersey.

McGuiness, D. L., & van Harmelen, F. (2004). *OWL Web ontology language overview.* Retrieved from W3C Recommendation: http://www.w3.org/ TR/ owl-features/

Neumann, B., & Möller, R. (2008). On scene interpretation with Description Logics. *Image and Vision Computing, 26*, 82–101. doi:10.1016/j.imavis.2007.08.013

Nowak, C. (2003). On ontologies for high-level information fusion. In *Proceedings of the 6th International Conference on Information Fusion (FUSION 2003)* (1, pp. 657-664). Cairns, Australia.

Olfati-Saber, R. (2007). Distributed Kalman filtering for sensor networks. In *Proceedings of the 46th Conference in Decision and Control* (pp. 5492-5498). New Orleans, LA, USA.

Patricio, M. A., Castanedo, F., Berlanga, A., Pérez, Ó., García, J., & Molina, J. M. (2008). Computational intelligence in visual sensor networks: Improving video processing systems. In *Computational Intelligence in Multimedia Processing: Recent Advances* (pp. 351–377). Berlin, Heidelberg: Springer. doi:10.1007/978-3-540-76827-2_14

Piccardi, M. (2004). Background subtraction techniques: a review. In *Proceedings of the IEEE International Conference on Systems, Man & Cybernetics* (4, pp. 3099-3104). The Hague, Netherlands.

Regazzoni, C. S., Ramesh, V., & Foresti, G. L. (2001). Scanning the issue/technology. Special Issue on Video Communications, Processing, and Understanding for 3rd Generation Surveillance Systems. *Proceedings of the IEEE*, *89*(10), 1355–1367. doi:10.1109/5.959335

Shi, Y., Wan, Y., & Ge, Q. (2009). *A unified out-of-sequence measurements fusion algorithm.* 1st International Workshop on Database Technology and Applications, Hubel, China.

Snidaro, L., Belluz, M., & Foresti, G. L. (2007). Domain knowledge for surveillance applications. In *Proceedings of the 10th International Conference on Information Fusion (FUSION '07)* (pp. 1-6), Quebec, Canada.

Staab, S., & Studer, R. (2009). *Handbook on ontologies*. Berlin/Heidelberg, Germany: Springer-Verlag. doi:10.1007/978-3-540-92673-3

Tsai, R. (1987). A versatile camera calibration technique for high accuracy 3D machine vision metrology using off-the-shelf tv cameras and lenses. *IEEE Journal on Robotics and Automation*, *3*(4), 323–344. doi:10.1109/JRA.1987.1087109

Valera, M., & Velastin, S. A. (2005). Intelligent distributed surveillance systems: A review. *IEEE Proceedings - Vision. Image, and Signal Processing*, *152*(2), 192–204.

Wang, L., Hu, W., & Tan, T. (2003). Recent developments in human motion analysis. *Pattern Recognition*, *36*(3), 585–601. doi:10.1016/S0031-3203(02)00100-0

Westermann, U., & Jain, R. (2007). Toward a common event model for multimedia applications. *IEEE MultiMedia*, *14*(1), 19–29. doi:10.1109/MMUL.2007.23

Zhang, Z. (2000). A flexible new technique for camera calibration. *IEEE Transactions on Pattern Analysis and Machine Intelligence*, *22*(1), 161–195. doi:10.1109/34.888718

Zimmermann, A., Lorenz, A., & Oppermann, R. (2007). An operational definition of context. In *Proceedings of the 6th International and Interdisciplinary Conference on Modeling and Using Context (CONTEXT '07)* (pp. 558-571), Roskilde, Denmark. Berlin/Heidelberg, Germany: Springer-Verlag.

ENDNOTES

1. The literature also uses "Sensor Fusion" and "Data Fusion" but we do not enter into a discussion of the subtle and non-standardized differences between these notions and the term "Information Fusion". The fundamental functions of each of these processes are largely the same.
2. "JDL" is an abbreviation for the Joint Directors of Laboratories, a US Defense Department organization that existed in the 1980´s within which there was a Data Fusion Group that standardized many of the still-used taxonomic and functional characterizations of IF.
3. Note that in some applications, the observational capability can be of other types, to include human observers.
4. http://www.giaa.inf.uc3m.es/miembros/jgomez/ontologies/
5. The example frames have been extracted from the PETS2002 dataset. The underlying task in this dataset was to track pedestrians in indoor video sequences of a shopping mall (http://www.cvg.cs.rdg.ac.uk/PETS2002/pets2002-db.html).

Chapter 11
Quantized Variational Filtering for Bayesian Inference in Wireless Sensor Networks

Majdi Mansouri
University of Technology of Troyes, France

Hichem Snoussi
University of Technology of Troyes, France

Jing Teng
North China Electric Power University, China

Ouachani Ilham
Insitut Supérieur d'Informatique et des technologies de Communication à Hammam Sousse, Tunisia

Cédric Richard
Université de Nice Sophia-Antipolis, France

ABSTRACT

The primary focus of the chapter is to study the Bayesian inference problem in distributed WSNs with particular emphasis on the trade-off between estimation precision and energy-awareness. We propose a variational approach to approximate the particle distribution to a single Gaussian distribution, while respecting the communication constraints of WSNs. The efficiency of the variational approximation relies on the fact that the online update and the compression of the filtering distribution are simultaneously performed. In addition, the variational approach has the nice property to be model-free, ensuring robustness of signal processing. We analyze the Bayesian inference issue for several specific but representative WSN applications to elaborate the quantized variational filtering method, which can be applicable to a wider class of problems.

DOI: 10.4018/978-1-61350-153-5.ch011

Our contribution lies in the following aspects: (1) we investigate the impact of the choice of a fixed (in time) quantization level and uniform power on the quantized variational filtering (QVF) algorithm performances and propose an adaptive quantization scheme; (2) we jointly optimize the power scheduling to minimize the transmission energy consumption in WSNs; (3) we select the best candidate sensors that participate in data collection for tracking the target.

Simulation results demonstrate the significantly improved performance of our approach.

INTRODUCTION

A wireless sensor network (WSN) is a collection of a large number of sensor nodes, which provides access to information anytime, anywhere by performing high-level distributed sensing, processing and disseminating data. Each sensor node has a processing capability, multiple types of memory, various sensing units and actuators, a radio frequency (RF) transceiver as well as a power source. The configuration of sensors allows a random deployment in an ad-hoc fashion in some inaccessible terrains or during disaster relief operations. Because of the reliability, flexibility, cost-effectiveness and ease of deployment, WSNs promise to revolutionize our life in a wide range of application domains, including military, civil and ecological areas. With the unlimited applicability, WSNs have gained worldwide attention from both academia and industry in recent years. In spite of the diverse applications, WSNs face a number of unique technical challenges due to their inherent energy and bandwidth limitation, ad hoc deployment and unattended operation, etc.. Unfortunately, very little previous work on distributed systems can be applied to WSNs, since the underlying assumptions have changed dramatically. In addition, WSN protocols and algorithms must possess self-organizing and cooperating capabilities. Therefore, innovative energy-aware, scalable and robust algorithms for distributed signal processing in WSNs are highly required.

Foremost among the tasks performed by WSNs is the distributed sensing, which establishes parameterizations of physical phenomena and enables processing of information for inference and estimation (Ribeiro, Giannakis & Roumeliotis, 2006). The results of estimation or inference provide the base for decision making to manipulate the environment in return. This inference is generally based on models of what we expect to observe, where the models are designed to capture salient trends or regularities in the observed data with a view to predicting future events. Unfortunately, for the majority of real applications, the data are far too complex or the underlying processes not nearly well enough understood to design a perfectly accurate model. Therefore, the constructed model is a rough simplification of the reality, where inevitably some important aspects of the data, which cannot be exactly modeled, are considered as noise. In signal processing, the above ideas can be formalized using the concept of probability and the rules of Bayesian inference (Beal, 2003). We are concerned with data D, which are generated by a model defined by a set of unknown parameters Θ (Smidl & Quinn, 2005). The beliefs about the data are completely expressed via the parametric probabilistic observation model, $P(D| \Theta)$. Given a specific set of observed data D, the learning of uncertainty/randomness of a process is solved by constructing a distribution $P(\Theta |D)$, namely, the *a posteriori* belief about the system. The simple prescription of Bayes rule reverses the order of the conditioning:

$$P(\Theta \mid D) \propto P(D \mid \Theta)P(\Theta) \qquad (1)$$

which specifies how the prior belief, quantified by the *a priori* distribution, $P(\Theta)$, is updated according to the measured data$D$. The real-world problems always involve high-dimensional random variables with complex uncertainty ($\Theta = \{\Theta_j\}_{j=1}^{J}$). The nonparametric and sample-based estimate of uncertainty, i.e. the particle filtering (PF) method, has thus become quite popular to capture and represent the complex distribution $P(\Theta|D)$ in nonlinear and non-Gaussian environments (Arulampalam, Maskell, Gordon & Clapp, 2002). However, taking into account the stringent energy and bandwidth constraints, the propagation of a huge amount of particles has impeded the implementation of PF in WSNs.

The rest of the chapter is organized as follows. We first discuss related work and motivate the need for our proposed scheme in Section II. Section III describes the problem statement. The Bayesian estimation approach via a QVF algorithm for target tracking is described in Section VI. The sensors selection approach based on mutual information is presented in Section V. Then, the adaptive quantization scheme is presented in Section VI. Numerical simulations are shown in Section VII, and finally concluding remarks are given in Section VIII.

RELATED WORK

The problem of quantizing observations to estimate a parameter, either the target position or any other physical field (temperature, humidity,...), is different from the problem of quantizing a signal for later reconstruction (Gray, 2006). Instead of reconstructing a signal, our first objective is rather estimating the target trajectory using quantized observations. In (Goldsmith & Wicker, 2002), authors have derived the optimal number of quantization levels as well as the optimal energy allocation across bits, but they have neglected the information content relevance of measured data. The approach of (Ribeiro, Giannakis & Roumeliotis, 2006) is limited to a particular 1-bit per quantized observation. On the other hand, the work in (Teng, Snoussi & Richard, 2010) used a VF algorithm to estimate the target position based on binary sensor networks (BSN). The BSN is based on the binary proximity observation model; it consists of making a binary decision according to the strength of the perceived signal. Hence, only one bit is transmitted for further processing if a target is detected. This work has also been done considering a cluster-based scheme, where sensors are partitioned into clusters. At each sampling instant, only one cluster of sensors is activated according to the prediction made by the

VF algorithm. Resource consumption is thus restricted to the activated cluster, where intra-cluster communications are dramatically reduced. Thanks to its power efficiency, the cluster-based scheme is also considered in this paper. As only a part of information is exploited (hard binary decision), tracking in binary sensor networks suffers from poor estimation performances. There has already been a certain amount of research in the area of sensor selection for wireless sensor networks. The work (Zhao, Shin & Reich, 2002) has investigated mechanisms for selecting sensors for target tracking. Selection are made based on the conditional entropy of the posterior target position distribution. In (Li & AlRegib, 2006), (Rubin & Huang, 2006), joint sensor selection and rate allocation schemes are developed. The authors in (Quan & Sayed, 2007) propose a centralized algorithm that uses a correlated measurements by selecting the most informative sensor. Complexity results for families of utility functions are presented in (Bian, Kempe & Govindan, 2006).

The idea of using information theory in sensor management was first proposed in (Hintz, 1991) Sensor selection based on expected information gain was introduced for decentralized sensing systems in (Manyika & Durrant-Whyte, 1995). The mutual information between the current target position distribution and the predicted sensor observation was proposed to evaluate the expected information gain about the target position attributable to a candidate sensor in (Liu, Reich &

Zhao, 2003; Ertin, Fisher & Potter, 2003). On the other hand, without using information theory, Yao et. al. (Yao, Hudson, Reed, Chen & Lorenzelli, 1998) found that the tracking accuracy depends on not only the accuracy of sensors but also their locations relative to the target position during the execution of tracking algorithms.

Most of these previous work do not take into account a tradeoff between the quality of sensed data, the density node, the transmit power between one sensor and the cluster head and the power stored in nodes to select the candidate sensor that participate in data collection.

In this chapter, we propose to improve the use of the QVF by jointly estimating the target position, selecting the best sensors that participate in data association and optimizing the number of quantization bits per observation.

In fact, the QVF has been shown to be adapted to the communication constraints of sensor networks. Its efficiency relies on the fact that the online update of the filtering distribution and its compression are executed simultaneously. Firstly, we select the best sensor that provides satisfied data of the target and balances the energy level among all sensors and minimum node density in a local cluster. Then, we optimize the quantization level and estimate the target position using the QVF algorithm. The sensors selection approach is based on maximizing the mutual information function under energy constraints. This criterion defines the relevance of the sensor cooperation for target tracking. The quantization is optimized by minimizing the transmitting power under predicted Cramér-Rao bound constraint. The computation of these criteria is based on the target position predictive distribution provided by the QVF algorithm. We compared our method with the QVF algorithm under sensing range constraint and using uniform power for quantization (QVF-U), the centralized quantized particle filter (QPF) algorithm and binary variational filtering (BVF) algorithm which is based on binary proximity sensors.

MODELING AND PROBLEM STATEMENT FOR TARGET TRACKING

Quantized Observation Model

In this work, we consider a sensor network with known sensor locations. We are interested in tracking a target position $x_t = (x_{1t}, x_{2t})^T$ at each instant $t (t = 1,\ldots,N)$. Considering a 2-D space, the following notations are used:

$x_t = (x_{1,t}, x_{2,t})^T \in \Re^2$: the state vector describing the target temporal location at instant t,

- $s = (s^i_{\ 1}, s^i_{\ 2})^T \in \Re^2$: the vector describing the location of the i-th sensor, and s^i_t denotes its location at instant t,
- N: the total number of observations,
- N_s: the total number of sensor nodes,
- R_{max}: the maximum sensing range,
- R_{min}: the minimum sensing range,
- $z_t^{i,x}$: the observation between the sensor i and the target x,
- N^i_t: the quantization level used by i-th sensor to quantize the sensor's measurement during the target tracking phase,
- L^i_t: represents the number of quantization bits per observation to quantize the sensor's measurement during the target tracking phase,
- $Z_{1:t}$: all the observations collected up to instant t.

Consider the activated sensor i (the activation procedure is explained in section 4), its observation γ_t^i is assumed to be modeled by:

$$\gamma_t^i = K \, \|x_t - s^i\|^{\eta} + \epsilon_t, \qquad (2)$$

where ε_t is a Gaussian noise with zero mean and known variance σ_ε^2. The constants η and K are also assumed to be known. The sensor transmits its observation to the cluster head (CH) only if the target is detected, which is equivalent to the condition that $R_{min} \le \left\| x_t - s^i \right\| \le R_{max}$ where R_{max} (resp. R_{min}) denotes the maximum (resp. minimum) distance at which the sensor can detect the target. Based on gathered transmitted sensor information, the cluster head is in charge of processing data in order to track the target. In order to save energy, before being transmitted, the observation is quantized by partitioning the observation space into N^i_t intervals $R_j = [\tau_j, \tau_{j+1}]$, where $j \in \{1, ..., N^i_t\}$. The number $N^i_t = 2^{L^i_t}$ denotes the quantization level to be determined. It is sub-indexed by the sampling instant t since it will be optimized jointly with the online target position estimation.

The quantizer is assumed to have an uniform step $\Delta = \dfrac{\tau_{N^i_t+1} - \tau_1}{N^i_t}$, with the initial and the last thresholds set to $\tau_1 = KR^{\eta}_{min} - \sigma_{\varepsilon}$ and $\tau_{N^i_t+1} = K\dfrac{R^{\eta}_{max}}{} + \sigma_{\varepsilon}$, respectively.

The quantization rule is then given by:

$$Y^i_t = Q(\gamma^i_t) = d_j if \gamma^i_t \in [\tau_j(t), \tau_{j+1}(t)] \quad (3)$$

where, the normalized d_j is given by $d_j = \dfrac{\tau_j(t) + \frac{\Delta}{2}}{\tau_{N_{t+1}}(t) - \tau_1(t)}$, and $Q0$ is the quantization function. Figure 1 depicts a simple example for the quantized observation model.

Then, the signal received by the CH from the sensor i at the sampling instant t is written as,

$$Z^i_t = h^i_t y^i_t + n_t \quad (4)$$

Figure 1. The Quantized Proximity Observation Model is described by a simple example. With respect to the first sensor, the target is within its sensing range at instant t. Observation $y^l_t = d^l_t$ is thus transmitted to the CH. The same principle holds true for the second sensor. Concerning the third and the fourth sensors, they keep silence at instant t. The CH then assign a "zero" to the observation of them after waiting a given time slots. The situation at instant t + 1can be similarly deduced.

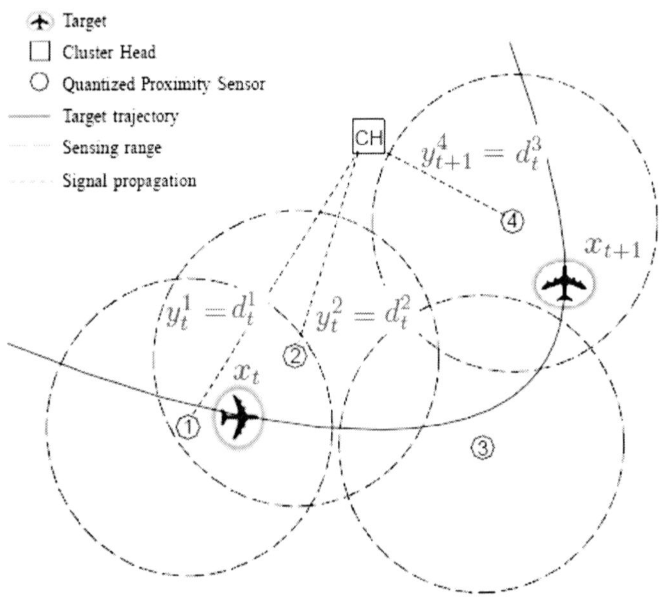

where $h_t^i = r_i^\lambda$ is the ith sensor channel attenuation coefficient at the sampling instant t, r_i is the transmission distance between the ith sensor and the CH, λ is the path-loss exponent and n_t is a random Gaussian noise with a zero mean and a known variance σ_n^2.

Figure 2 summarizes the transmission scheme occurring during the data processing.

General State Evolution Model (GSEM)

In this chapter, we employ a general state evolution model GSEM (Snoussi & Richard, 2006; Vermaak, Lawrence & Perez, 2003) (see Figure 3) instead of the kinematic parameter model (Djuri'c, Vemula & Bugallo, 2005; Yick, Mukherjee & Ghosal, 2005; Chhetri, Morrell & Papandreou-Suppappola, 2005) usually used in tracking problems. In fact, the GSEM model, introduced in (Teng, Snoussi & Richard, 2008) for visual tracking, is more practical to non-linear and non-gaussian situations where no a priori information on the target velocity is available. Considering a planar geometry, the target position x_t at instant t is assumed to follow a Gaussian model, where the expectation μ_t and the precision matrix λ_t are both random. Gaussian distribution for the expectation and whishart distribution for the precision matrix form a practical choice for the filtering implementation. The hidden state α_t is extended to an augmented state $\alpha_t = (x_t, \mu_t, \lambda_t)$, yielding an hierarchical model as follows,

$$\begin{aligned} \mu_t &\sim N(\mu_t \mid \mu_{t-1}, \bar{\lambda}) \\ \lambda_t &\sim w_{\bar{n}}(\lambda_t \mid \bar{S}) \\ x_t &\sim N(\mu_t, \lambda_t) \end{aligned} \tag{5}$$

where the fixed hyperparameters $\bar{\lambda}$, $\bar{n}$ and $\bar{S}$ are respectively the random walk precision matrix, the degrees of freedom and the precision of the Wishart distribution. Note that assuming random mean and covariance for the state x_t leads to a prior probability distribution covering a wide range of tail behaviors allowing discrete jumps in the target trajectory.

In fact, the marginal state distribution is obtained by integrating over the mean and precision matrix:

Figure 2. Illustration of the communications path-ways in a WSN: The 1st sensor makes a noisy reading γ_t^1. The quantized measurement $y_t^1 = Q(\gamma_t^1)$ with L_t^1 bits of precision is sent to the CH. The measurement z_t^1 is received by the CH, it is corrupted by an additive white Gaussian noise n_t.

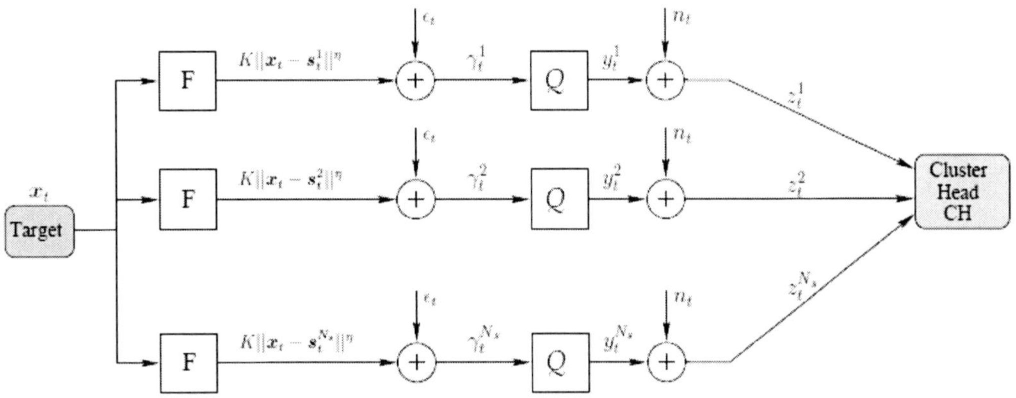

Figure 3. General state evolution model GSEM

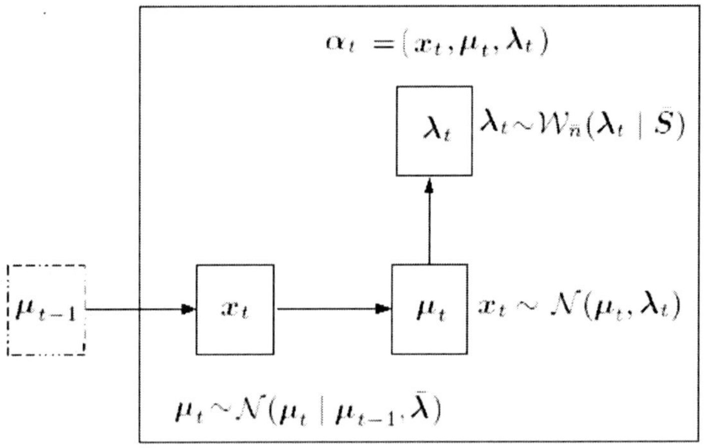

$$P(x_t \mid x_{t-1}) = \int N(x_t \mid \mu_t, \lambda_t) p(\mu_t, \lambda_t \mid x_{t-1}) d\mu_t d\lambda_t \tag{6}$$

where the integration with respect to the precision matrix leads to the known class of scale mixture distributions introduced by Barndorff-Nielsen (Barndorff-Nielsen, 1977). Low values of the degrees of freedom $\bar{n}$ reflects the heavy tails of the marginal distribution $p(x_t|x_{t-1})$.

The next section describes the Bayesian tracking solution based on a Quantized Variational Filtering algorithm.

BAYESIAN ESTIMATION VIA A QVF ALGORITHM FOR TARGET TRACKING

QVF Algorithm for Target Tracking

In this section, we assume that the quantization level is already optimized. Hence, the observation model is completely defined. The aim of this section is to describe the target position estimation procedure. According to the model (5), the augmented hidden state is now $\alpha_t = (x_t, \mu_t, \lambda_t)$. We consider the posterior distribution $p(\alpha_t \mid z_{1:t})$, where $z_{1:t} = \{z_1, z_2, , z_t\}$ denotes the collection of observations gathered until time *t*. The variational approach consists in approximating $p(\alpha_t \mid z_{1:t})$ by a separable distribution $q(\alpha_t) = \Pi_i q(\alpha_t^i)$ that minimizes the Kullback-Leibler (KL) divergence between the true filtering distribution and the approximate distribution,

$$D_{KL}(q \parallel p) = \int q(\alpha_t) \log \frac{q(\alpha_t)}{p(\alpha_t \mid Z_{1:t})} (d\alpha_t),$$

where $q(\alpha_t) = \Pi q(\alpha_t^i) = q(x_t) q(\mu_t) q(\lambda_t)$, (7)

With variational calculus, the following approximate distribution yields (Vermaak, Lawrence & Perez, 2003; Teng, Snouss & Richard, 2007a)

$$q(\alpha_t^i) \propto \exp\langle \log p(Z_{1:t}, \alpha_t) \rangle \Pi_{j \neq i} q(\alpha_t^j) \tag{8}$$

where $\langle . \rangle_{q(\alpha_t^j)}$ denotes the expectation operator relative to the distribution $q(\alpha_t^j)$.

Taking into account the separable approximate distribution $q(\alpha_{t-1})$ at time $t-1$, the filtering distribution $p(\alpha_t \mid z_{1:t})$ is sequentially approximated according to the following:

$$\hat{p}(\alpha_t \mid Z_{1:t}) = \frac{p(z_t \mid \alpha_t)\int p(\alpha_t \mid \alpha_{t-1})q(\alpha_{t-1})d\alpha_{t-1}}{p(z_t \mid Z_{1:t-1})}$$
$$\propto p(z_t \mid x_t)p(x_t \mid \mu_t, \lambda_t)p(\lambda_t)q_p(\mu_t), \tag{9}$$

with

$$q_p(\mu_t) = \int p(\mu_t \mid \mu_{t-1})q(\mu_{t-1})d\mu_{t-1}.$$

Therefore, through a simple integral with respect to μ_{t-1}, the filtering distribution $p(\alpha_t \mid z_{1:t})$ can be sequentially updated. Considering the GSEM proposed in above, the evolution of $q(\mu_{t-1})$ is Gaussian, namely $p(\mu_t \mid \mu_{t-1}) \sim N(\overline{\mu_{t-1}}, \overline{\lambda_{t-1}})$.

Defining $q(\mu_{t-1}) \sim N(\mu_{t-1}^{*}, \lambda_{t-1}^{*})$, $q_p(\mu_t)$ is also Gaussian with the following parameters,

$$q_p(\mu_t) = N(\mu_t^p, \lambda_t^p),$$
$$where \mu_t^p = \mu_{t-1}^{*} and \lambda_t^p = \lambda_{t-1}^{*\,-1} + \overline{\lambda}^{-1})^{-1}. \tag{10}$$

The temporal dependence is hence reduced to the incorporation of only one Gaussian component approximation $q_p(\mu_{t-1})$. The update and the approximation of the filtering distribution $p(\alpha_t \mid z_{1:t})$ are jointly performed, yielding a natural and adaptive compression (Chen, Hou & Sha, 2004; Yang & Sikdar, 2003). According to the Eq. (9) and Eq.(10), variational calculus leads to the following iterative solution;

$$\begin{cases} q(\mu_t) = N(\mu_t^{*}, \lambda_t^{*}) \\ q(\lambda_t) = W_2(V_t^{*}, n^{*}) \end{cases} \tag{11}$$

where the parameters are iteratively updated according to the following scheme:

$$\begin{cases} \mu_t^{*} = \lambda_t^{*-1}(\langle\lambda_t\rangle\langle x_t\rangle + \lambda_t^p \mu_t^p) \\ \lambda_t^{*} = \langle\lambda_t\rangle + \lambda_t^p \\ n^{*} = \overline{n} + 1 \\ V_t^{*} = (\langle x_t x_t^T\rangle - \langle x_t\rangle\langle\mu_t\rangle^T - \langle\mu_t\rangle\langle x_t\rangle^T + \langle\mu_t\mu_t^T\rangle + \overline{V}^{-1})^{-1} \end{cases} \tag{12}$$

Notice that <.> denotes the expectation relative to the distribution $q(.)$. The mean state and the precision matrix distributions $q(\mu_t)$ and $q(\lambda_t)$ have closed forms, such that their expectations are easily derived:

$$\begin{cases} \langle\mu_t\rangle = \mu_t^{*} \\ \langle\mu_t\mu_t^T\rangle = \lambda^{*-1} + \mu_t^{*}\mu_t^{*T} \\ \langle\lambda_t\rangle = n^{*}V_t^{*} \end{cases} \tag{13}$$

However, the target state x_t does not has closed forms. Combining the equations (8) and (9), $q(x_t)$ has the following expression:

$$\{q(x_t)\alpha N(\langle\mu_t\rangle, \langle\lambda_t\rangle)\Pi_{i=1}^{m_t} p(z_t^{i,x} \mid x_t), \tag{14}$$

Therefore, the general state evolution model (5) and the observation model (5) and (2) are naturally incorporated to update $q(x_t)$. Their distribution forms immediately suggest an Importance Sampling (IS) procedure, where samples are drawn from Gaussian distribution $N(\langle\mu_t\rangle, \langle\lambda_t\rangle)$, and are weighted according to their likelihood (taking into account the observation model):

$$\{x_t^{(k)} \sim N(\langle\mu_t\rangle, \langle\lambda_t\rangle), w_t^k \propto \prod_{i=1}^{m_t} p(z_t^{i,x} \mid x_t^{(k)}), \tag{15}$$

Hence, $q(x_t)$ can be approximated by the Monte Carlo method:

$$\{\langle x_t \rangle = \sum_{k=1}^{n} w_t^{(k)} x_t^{(k)} \tag{16}$$

As mentioned above, a standard adaptive scheme solution contains a prediction step and a update step. Besides the update of the filtering distribution $p(\alpha_t \mid z_{1:t})$, the predictive distribution $p(\alpha_t \mid z_{1:t-1})$ can also be efficiently calculated by the variational approach. In fact, by incorporating the separable approximate distribution $p(\alpha_{t-1})$ in the place of $p(\alpha_{t-1} \mid z_{1:t-1})$, the recursive adaptive scheme algorithm calculates the predictive distribution $p(\alpha_t \mid z_{1:t-1})$ in the following form:

$$\begin{aligned} \hat{p}(\alpha \mid z_{1:t-1}) &\propto \int p(\alpha_t \mid \alpha_{t-1}) q(\alpha_{t-1}) d\alpha_{t-1} \\ &\propto p(x_t \mid \mu_t, \lambda_t) p(\lambda_t) q_p(\mu_t) \end{aligned} \tag{17}$$

The exponential form solution, which minimizes the Kullback-Leibler divergence between the predictive distribution $p(\alpha_t \mid z_{1:t-1})$ and the separable approximate distribution $q_{t|t-1}(\alpha_t)$, yields Gaussian distributions for the predicted expectations, and a Wishart distribution for the target precision matrix:

$$\begin{cases} q_{t|t-1}(x_t) \propto N(\langle \mu_t \rangle_{q_{t|t-1}}, \langle \lambda_t \rangle_{q_{t|t-1}}), \\ q_{t|t-1}(\mu_t) \propto N(\mu^*_{t|t-1}, \lambda^*_{t|t-1}), \\ q_{t|t-1}(\lambda_t) \propto W_2(V^*_{t|t-1}, n^*_{t|t-1}), \end{cases} \tag{18}$$

where the parameters are updated according to the same iterative schemes (12) and (13). The target state and the activated sensors are now evaluated by the following expressions:

$$\langle x_t \rangle_{q_{t|t-1}} = \langle \mu_t \rangle_{q_{t|t-1}}, \langle x_t x_t^T \rangle_{q_{t|t-1}} = \langle \lambda_t \rangle_{q_{t|t-1}}^{-1} + \langle \mu_t \rangle_{q_{t|t-1}} \langle \mu_t \rangle_{q_{t|t-1}}^T \tag{19}$$

Compared to the particle filtering method, the computational cost and the memory requirements are dramatically reduced by the variational approximation in the prediction phase. In fact, the expectations involved in the computation of the predictive distribution have closed forms, avoiding the use of Monte Carlo integration.

Hence, the QVF provides at the sampling instant t, the predicted target position $\langle x_t \rangle_{q_{t|t-1}}$. As shown in Figure.4, based on the predicted target position $\langle x_t \rangle_{q_{t|t-1}}$, the cluster head CH_{t-1} at sampling instant *t-1* selects the next cluster head CH_t. If the predicted target position $\langle x_t \rangle_{q_{t|t-1}}$ remains in the vicinity of CH_{t-1}, which means that at least four of its slave sensors can detect the target (Chen, Hou & Sha, 2004), then $CH_t = CH_{t-1}$. Otherwise, if $\langle x_t \rangle_{q_{t|t-1}}$ is going beyond the sensing range of the current cluster, then a new CH_t is activated, based on the target position prediction $\langle x_t \rangle_{q_{t|t-1}}$ and its future tendency.

$$\begin{aligned} CH_t &= \arg\max\{\frac{\cos\theta_t^k}{d_t^k}\} \\ where\, d_t^k &= \| \langle x_t \rangle_{q_{t|t-1}} - L_{CH_t^k} \| \\ and\, \theta_t^k &= angle(\overrightarrow{\langle x_{t-1} \rangle \langle x_t \rangle_{q_{t|t-1}}}, \overrightarrow{\langle x_{t-1} \rangle L_{CH_t^k}}) \end{aligned} \tag{20}$$

where K is the number of CHs in the neighborhood of CH_{t-1} and $L_{CH_t^k}$ is the location of the k_{th} neighboring CH_t.

As illustrated in Figure 4, traditional cluster-activation rule (Yang & Sikdar, 2003) activates CH^2 to update the filtering distribution at time t, since it is the closest CH to the target prediction $x_{t/t-1}$. But according to the tendency, the target is very likely to go out of its vicinity in a short time, causing unnecessary hand-off operations. Based on the decision rule described by Eq. (17), it is CH^1 that is activated, as the target is most likely detected by it and would stay in its vicinity for a longer period. Taking into consideration the future tendency of the target, the non-myopic cluster activation rule avoids unnecessary hand-off operations compared to the traditional one. Therefore, the target tracking accuracy is ensured and the energy consumption is minimized as well. Thanks to the variational calculus, communication between CH_{t-1} and CH_t is limited to simply sending the mean and the precision of $q(\mu_{t-1})$. Therefore, the cluster-based variational filtering (VF) algorithm outperforms the classical distributed particle filtering (PF) algorithm in resource (energy, bandwidth and memory) saving, as a large number of particles and corresponding weights are maintained and propagated in the latter case. With respect to the tracking accuracy, the VF and the PF algorithms approximate the true state distribution in different ways. When calculating the integral involved in the Bayesian filtering, the PF uses a large amount of particles whereas the VF introduces hidden variables to bypass the difficulty. These random variables introduced by the VF act as links that connect the observations to the unknown parameters via Bayes law. The error propagation problem is dramatically reduced as approximation of the filtering distribution is performed during observation incorporation.

The next section is devoted to the proposed method for selecting the candidate sensors that participate in target localization.

MAXIMUM MUTUAL INFORMATION UNDER ENERGY CONSTRAINTS-BASED SENSOR SELECTION APPROACH

Since the predicted target position $x_p(t) = \langle x_t \rangle_{q_{t+1|t}}$ at time $(t+1)$ is available, it could be used to select sensors. Sensors inside the disk centered at $x_p(t)$ with radius R_{max} are pre-selected. After pre-selecting the sensors, the CH computes for

Figure 4. Prediction-based CH_t activation

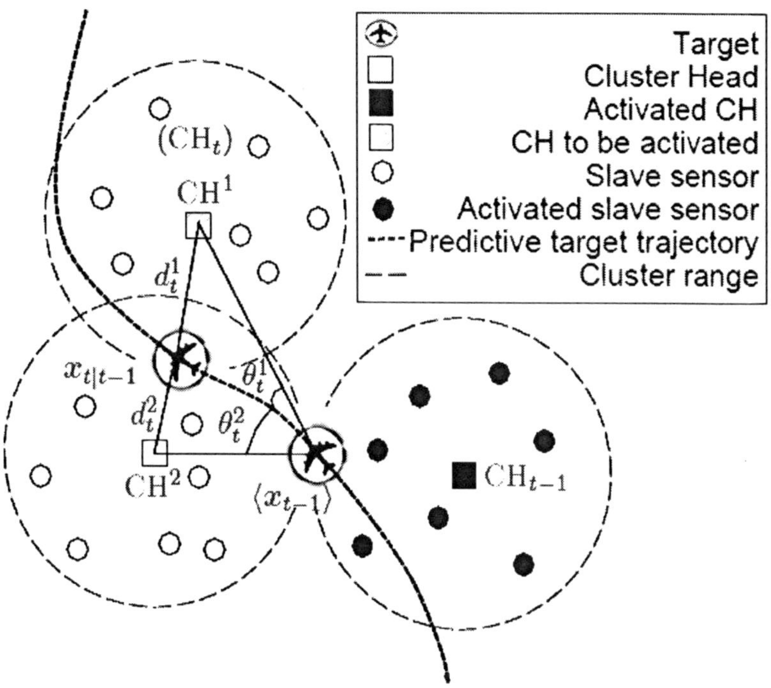

each pre-selected sensor the criterion (described in 5.1) to activate the candidate sensors. In the next subsection, we detail this criterion and the parameters that it takes into account in order to select appropriate sensors that participate in data association for tracking the target.

Maximum Mutual Information under Energy Constraints Criterion

The main idea of this criterion for sensor selection is to define the main parameters that may influence the relevance of the sensors cooperation, which are: (1) information that can be transferred from candidate sensor i ($MI(x_t,z^i_t)$) (detailed in subsection 5.1.1); (2) its energy (i) ($E(i)$) (detailed in subsection 5.1.2); (3) the transmitting energy between one sensor and the CH ($P(i)$) (detailed in subsection 5.1.3) and (4) its density ($D(i)$) (detailed in subsection 5.1.4).

The selected sensors should provide satisfied data for tracking the target while balancing the energy consumption among all sensors.

The criterion function for node i is given by:

$$\arg\max_{i=1,\dots,C} MI(x_t, z_t^2)$$

$$\text{s.t } P(i) < P_0$$

$$\text{s.t } E(i) > E_0$$

$$\text{s.t } D(i) < D_0 \qquad (21)$$

where, P_0, E_0 and D_0 (a predefined thresholds) are the maximum power, the minimum energy and the maximum density of nodes constraints. C represents the set of all pre-selected sensors.

Computation of the Mutual Information Function

The mutual information function is often used to measure the amount of information that can be obtained about a random variable by observing another one. It can be used to maximize the amount of information shared between the unknown parameter x_t and the observable variable z^i_t. The mutual information between the observation z^i_t and the source x_t is proportional to:

$$MI(x_t, z_t^i) \propto p(z_t^i \mid x_t)\log(p(z_t^i \mid x_t)) \qquad (22)$$

The likelihood function is expressed as,

$$L(s^i) = p(z_t^i \mid x_t) = \sum_{j=0}^{N_t^i-1} p(\tau_j(t) < \gamma_t^i < \tau_{j+1}(t))N(h_t^i d_j, \sigma_\epsilon^2) \qquad (23)$$

where

$$p(\tau_j(t) < \gamma_t^i < \tau_{j+1}(t)) = \int_{\tau_j(t)}^{\tau_{j+1}(t)} N(\rho_{\gamma_t^i}(s^i), \sigma_n^2) d\gamma_t \qquad (24)$$

is computed according to the quantization rule defined in Eq. (3), in which

$$\rho_{\gamma_t^i}(s^i) = K \| x_t - s^i \|^\eta, \qquad (25)$$

It is worth noting that the expression of the *MI* given in (19) depends on the target position x_t at the sampling instant t and on the activated sensor i. However, as the target position is unknown, the *MI* is replaced by its expectation according to the predictive distribution $p(x_t \mid z^i_{1:t-1})$ of the target position:

$$< MI(s^i) >= E_{p(x_t|z^i_{1:t-1})}[MI(s^i)] \qquad (26)$$

Computing the above expectation is analytically untractable. However, as the QVF algorithm yields a Gaussian predictive distribution $N(x_t; \mu_{t/t-1}, \lambda_{t/t-1})$, expectation (23) can be efficiency approximated by a Monte Carlo scheme:

$$< MI(s^i) > \simeq \frac{1}{J}\sum_{j=1}^{J} MI(\tilde{x}_t^j, s^i), \qquad (27)$$
$$\tilde{x}_t^j \sim N(x_p(t); x_{t/t-1}, \lambda_{t/t-1})$$

where x_t^j is the j*th*drawn sample at instant *t*, and *J* is the total number of drawn vectors x_t.

Computation of the Lifetime Energy

The energy *E* is an important parameter in a resource limited network such as the WSN. It is generally seen as the most important parameter. Indeed, the remaining battery level appears to be the most important thing in this parameter but it is not the only one, it is given by the following equation:

$$E(i) = A(i) \times \text{APS} \qquad (28)$$

Where *E*(*i*) is the energy parameter of the sensor *i*, *A* is the available or remained power in the battery and APS is the administrator power strategy parameter, which is a percentage defined by the administrator depending on the application. As presented in the Eq. (28), the energy *E*could be just the remaining battery level if no administrator power strategy has been defined (APS = 1). In addition, lower is the APS longer is the life time of the WSN.

Computation of the Sensor Node Transmitting Power

The total amount of required transmission power used by the ith sensor within a cluster (Cui, Goldsmith & Bahai, 2005) is proportional to:

$$P^i(t) \propto d_i^{\lambda}(N_t^i - 1)$$
$$\equiv \| s^i - L_{CH_t} \|^{\lambda} (N_t^i - 1) \qquad (29)$$

where, d_i is the transmitting distance (meters) between the CH and the ith sensor, L_{CH_t} is the location of the CH at the sampling instant *t* and λ is the path loss exponent.

Computation of the Density of Nodes

The network density varies from one deployment to another and from one node to another within the same deployment depending on the node distribution. Indeed, the more an agent has neighbors the less is the importance of its cooperation. For simplicity's sake, the density for sensor *i* is computed by the Eq. (30),

$$D(i) = \frac{N_{th}^i}{N_{re}^i} \qquad (30)$$

where N_{th} is the theoretical number of nodes and it is given from the ideal distribution of the nodes or the grid distribution. N_{th} corresponds to the number of nodes within the R_{max} range. N_{re} is the number of the one hop neighbors, which means the nodes within the radio range of the node in question.

The next section presents the adaptive quantization by minimizing the transmitting power under predicted Cramér-Rao constraint.

ADAPTIVE QUANTIZATION OPTIMIZATION

At sampling time *t-1*, the selected cluster head CH_{t-1} executes the QVF algorithm and provides the Gaussian predictive distribution $N(x_{t/t-1}, \lambda_{t/t-1})$. The predicted position allows the selection of the cluster to be activated as described in section 4. Furthermore, this target position is used by the CH_{t-1} to give the optimal quantization level $\hat{N}_t^i$ minimizing the transmitting power under predicted Cramér-Rao constraint. This optimal quantization level is then transmitted to the CH_t

before being broadcasted to the activated sensors so that they use it to quantize their observations. These quantized observations are then used by the CH_t to execute the QVF algorithm at the sampling instant t. In the following, we describe the technique in details.

Predicted Cramér-Rao Bound (PCRB)

The Cramér-Rao bound is often used to evaluate the efficiency of a given estimator. In its simplest form, the bound states that the covariance of any estimator is at least higher as the inverse of the Fisher Information (FI) matrix. The FI matrix is a quantity measuring the amount of information that the observable variable z_t carries about the unknown parameter x_t. The FI matrix elements at the sampling instant t are given by:

$$[FI(x_t)]_{l,k} = E_{z_t|x_t}\left[\frac{\partial \log(p(z_t \mid x_t))}{\partial x_{l,t}} \frac{\partial \log(p(z_t \mid x_t))}{\partial x_{k,t}}\right], \quad l,k \in \{1,2\} \times \{1,2\} \tag{31}$$

where z_t denotes the observation at the sampling instant t, $x_t = [x_1,x_2]^T$ is the unknown 2×1 vector to be estimated, and $E_{z_t|x_t}[.]$ denotes the expectation with respect the likelihood function $p(z_t|x_t)$.

As shown in (Mansouri, Ouachani, Snoussi & Richard, 2009), the derivative of the log-likelihood function can be expressed as:

$$\begin{aligned}\frac{\partial \log(p(z_t^i \mid x_t))}{\partial x_{l,t}} &= \frac{\eta K}{\sqrt{2\sigma_n^2}}(x_{l,t} - s_{l,i}) \| x_{l,t} - s_{l,i} \|^{\eta-2} \times \\ &\sum_{k=1}^{N_t^i}\left[\exp\left(-\frac{1}{2}\frac{(\tau_k - \rho_{\lambda_t^i}(x_t))^2}{\sigma_n^2}\right) - \exp\left(-\frac{1}{2}\frac{(\tau_{k+1} - \rho_{\lambda_t^i}(x_t))^2}{\sigma_n^2}\right)\right] \\ &\times \exp\left(-\frac{1}{2}\frac{(z_t(k) - h_t(k)d_k)^2}{\sigma_\epsilon^2}\right) \Big/ \sum_{k=1}^{N_t^i}\left[erfc\left(\frac{\tau_k - \rho_{\lambda_t^i}(x_t)}{\sqrt{2\sigma_n^2}}\right) - \right. \\ &\left. erfc\left(\frac{\tau_{k+1} - \rho_{\lambda_t^i}(x_t)}{\sqrt{2\sigma_n^2}}\right)\right] \times \exp\left(-\frac{1}{2}\frac{z_t(k) - h_t(k)d_k)^2}{\sigma_\epsilon^2}\right)\end{aligned} \tag{32}$$

Substituting expression (29) in (28), the FI matrix is easily computed by integrating over the likelihood function $p(z_t|x_t)$ at the sampling instant t.

We can notice, that the expression of the FI given in (28) depends on the target position x_t, on the quantization level N_t. However, as the target position is unknown, the FI is replaced by its expectation according to the predictive distribution $p(x_t|x_{1:t-1})$ of the target position:

$$< FI(x_t) >= E_{(x_t|z_{1:t-1})}[FI(x_t)] \tag{33}$$

Computing the above expectation is analytically untractable. However, as the QVF algorithm yields a Gaussian predictive distribution. Expectation (30) can be efficiently approximated by a Monte Carlo scheme:

$$\begin{cases} < FI(x_t) >\simeq \frac{1}{L}\sum_{l=1}^{L} FI(\tilde{x}_t^l) \\ where, \\ \tilde{x}_t^l \sim N(x_p(t); x_{t/t-1}, \lambda_{t/t-1}) \end{cases} \tag{34}$$

Optimizing the Quantization

According the Eq.(26), the optimal resource allocation problem is expressed as an optimization of the quantization level N_t^i over the field Z_+ subject to an average estimation error constraint M_0,

$$\min_{N_t^i \in z+} \sum_{i=1,\dots,N_a} P^i(t)^2$$

$$s.t CRB(x_t, N_t^i) < M_0 \tag{35}$$

where N_a is the number of activated sensors at instant t. The Lagrangian formulation of this constrained optimization yields the following expression to be minimized:

$$L = \sum_i r_i^{2\lambda}(N_t^i - 1)^2 + (\mu CRB(x_t, N_t^i) - M_0) \quad (36)$$

where μ is the Lagrange multiplier constant.

Then, the optimal value of $\hat{N}_t^i$ is a solution of the following equation:

$$f(\hat{N}_t^i) = \frac{\partial L}{\partial \hat{N}_t^i} = (\hat{N}_t^i - 1) - g(\hat{N}_t^i) = 0 \quad (37)$$

where $g(\hat{N}_t^i)$ denotes $-\frac{\mu}{2r_i^{2\lambda}} \frac{\partial CRB(x_t, \hat{N}_t^i)}{\partial \hat{N}_t^i}$.

Applying the Newton-Raphson procedure yields the following sequence converging to the solution of (37):

$$\hat{N}_t^i(n+1) = \hat{N}_t^i(n) - \frac{f(\hat{N}_t^i(n))}{f'(\hat{N}_t^i(n))} = \hat{N}_t^i(n) - \frac{(\hat{N}_t^i(n) - 1) - g(\hat{N}_t^i(n))}{1 - g'(\hat{N}_t^i(n))} \quad (38)$$

The overall proposed algorithm is summarized in Algorithm 1 (Figure 5).

SIMULATION RESULTS ANALYSIS

The performance of the tracking algorithm can be essentially assessed by the tracking accuracy (detailed in subsection 7.1), by the Root Mean Square Error (RMSE) (detailed in subsection 7.2), and by computing the energy expenditure during the whole tracking process (detailed in subsection 7.3).

Figure 5.

Algorithm 1: Pseudo-code of the proposed algorithm

— Initialization:

1) Select the candidate sensors which exist within R_{max} range.
2) Quantize sensors' measurements using an uniform power.
3) Execute the VF algorithm.

— Iterations:

1) Select sensors according to Sub. V
2) Compute the predicted target distribution.
3) Compute the mutual information function based on the predicted target position using the Eq.(22).
4) Select the optimal candidate sensors according to Section V.
5) Compute the Cramér-Rao bound based on the predicted target position using the Eq.(31).
6) Compute the optimal level quantization by minimizing the transmitting power under Cramér-Rao bound constraint using the Eq.(38).
7) Quantize sensors' measurements using the optimal level quantization.
8) Execute the QVF algorithm.

We refer the approach that uses the QVF algorithm with activating sensors in the circle of radius R_{max} for target tracking and using uniform power for quantization; to as QVF-U.

In the following, we compare the tracking accuracy of the proposed adaptive quantized variational filter (AQVF) algorithm, with QVF-U algorithm, the binary variational filtering (BVF) algoritthm (Quan & Sayed, 2007) and the centralized quantized particle filter (QPF) algoritthm. All the simulations shown in this chapter are implemented with Matlab version 7.1, using an Intel Pentium CPU 3.4 GHz, 1.0 G of RAM PC. The system parameters considered in the following simulations are: $\eta = 2$ for free space environment, the constant characterizing the sensor range is fixed for simplicity to $K = 1$, the cluster head noise power is $\sigma_n^2 = 0.05$, the total number of sensors is $N_s = 100$, the total sampling instants is $N = 100$, the sensor noise power is $\sigma_\varepsilon^2 = 0.01$, the maximum sensing range R_{max} (resp. the minimum sensing range R_{min}) is fixed to 10 m (resp. 0 m). 200 particles were used in AQVF, QVF-U, BVF and QPF algorithms.

Analysis of Tracking Accuracy

To show the efficiency of the proposed algorithm, we compare it with four previously proposed methods. The quantized proximity observation model, formulated in Eq. (3), was adopted for all the algorithms, except for the BVF algorithm, which is based on the binary proximity sensors.

One can notice from Figure 6.a) that, even with abrupt changes in the target trajectory, the desired quality is achieved by the AQVF algorithm and outperforms the QVF-U algorithm. Figure 6.b) compares their tracking accuracies in terms of Root Mean Square Error (RMSE):

$$RMSE = \sqrt{E((x - \hat{x})^2)}. \qquad (39)$$

where x_t (resp. $\hat{x}_t$) is the true trajectory (resp. the estimated trajectory).

AQVF and BVF performances are compared in Figure 7. The results confirm the impact of neglecting the transmission energy consumption in a WSN and the information relevance of sensor measurements. Unlike the QPF algorithm, the temporal dependence in the QVF is reduced to one single Gaussian statistic instead of a huge number of particles. On the other hand, the precision of the PF algorithm depends on the choice of the importance sampling distribution. The QVF yields an optimal choice of the sampling distribution over the target position x_t by minimizing the KL divergence. In fact, variational calculus leads to a simple Gaussian sampling distribution whose parameters (estimated iteratively) depends on the observed data. As can be expected, with the amount of particles increasing, the QPF algorithm demonstrates much more accurate tracking at the cost of a higher computation complexity. In particular, the computation time grows proportionally to the increment of the number of particles. The tracking accuracies of the QPF is compared to that of AQVF in Figure 8. The smaller RMSE of the AQVF in comparison with the two approximation methods confirms once again the effectiveness of the AQVF algorithm in terms of tracking accuracy.

Root Mean Square Error (RMSE) Analysis

The Root Mean Square Error (RMSE) of the above algorithms may depend on several factors such as the transmitting power between the candidate sensors and the CH, the nodes density, the sensing range, the pass losses as well as the sensor noise variances. The purpose of this subsection is to study the impact of these factors when comparing the performance of the above mentioned tracking algorithms. Figure 9.a) shows the variation of the RMSE with respect to the nodes density varying in {50,…,200}. As can be expected, the RMSE

Figure 7. (a) Tracking accuracy between AQVF and BVF methods; (b) Mean Square Error comparison between AQVF and BVF methods

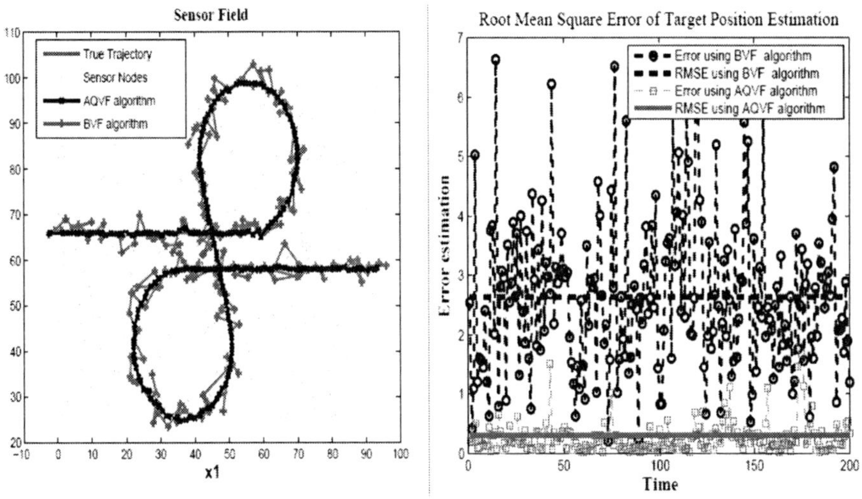

(a) (b)

decreases for all the algorithms when the nodes density increases. One can also note that the proposed AQVF technique outperforms all the other techniques when varying the nodes density. It is also worth noting that its RMSE decreases more sharply than the other filtering techniques. Figure 9.b) plots the RMSE versus the nodes noise variances varying in {0,…,0.25} and Figure 10.a) plots the RMSE with respect to the nodes transmitting power (varying in {50,…,200}). From the Figure 10.b), we can show that when sensing range varying in{5,…,13}, the error estimation decreases. These results confirm that the proposed method

Figure 8. (a) Tracking accuracy between AQVF and QPF methods; (b) Mean Square Error comparison between AQVF and QPF methods

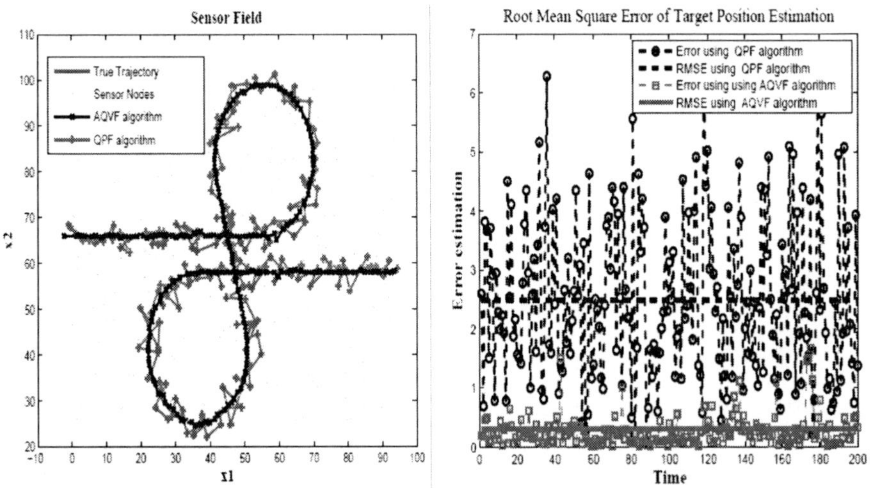

(a) (b)

Figure 6. (a) Tracking accuracy between QVF-U and AQVF methods; (b) Mean Square Error comparison between QVF-U and AQVF methods

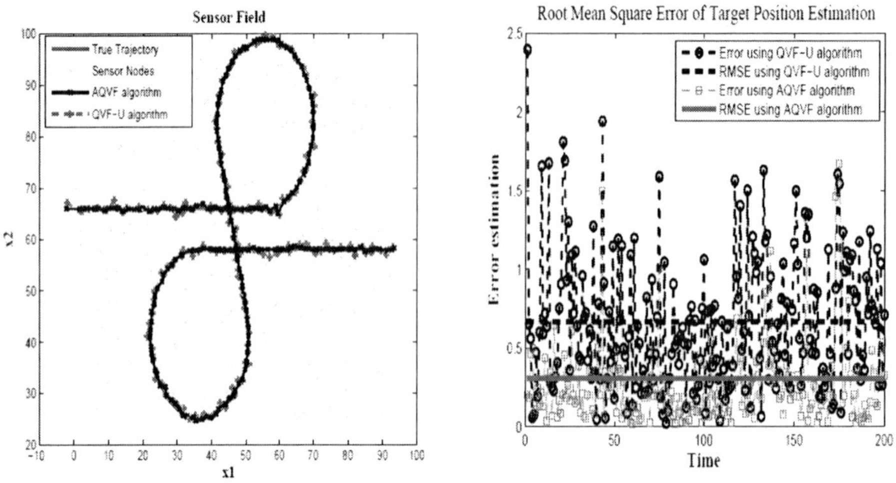

(a) (b)

outperforms the classical algorithms when varying the simulation conditions.

Energy Analysis

The energy consumption is computed following the model proposed in (Heinzelman, Chandrakasan & Balakrishnan, 2000); in which we assume that the communication between the active sensors is via single hop and that the energy consumed in scheduling and computing can be neglected relative to the energy consumed during communications. The communication energy is mainly consumed during three phases: emission phase, radio communication phase and the reception phase. The transmitting power consumed at sensor (i), while transmitting data to the cluster head is given by:

$$E_T = \varepsilon_e + L^i_{\varepsilon_a r_i^\lambda} \tag{40}$$

where ε_a is the energy dissipated in Joules per bit per m^2, ε_e is the energy consumed by the circuit per bit.

The power consumed at sensor (i) when receiving data from the cluster head is given by:

$$E_R = L^i_{\varepsilon_r} \tag{41}$$

Similarly, the power consumed in sensing is defined by:

$$E_s = L^i_{\varepsilon_s} \tag{42}$$

where ε_s is the energy expending parameter for sensing one bit of data.

Considering the energy model parameters, we choose $\varepsilon_a = 100pJ / bit / m^2$, $\varepsilon_e = 50nJ / bit$, $\varepsilon_r = 135nJ / bit$ and $\varepsilon_s = 50nJ / bit$ (Teng, Snoussi & Richard, 2007b). Let N_p denote the number of particles, $N_w = N_p$ the number of corresponding weights, L_{opt} the optimal bits number obtained by AQVF, and L_f the fixed bits number. The nodes communication of these algorithms were compared when a hand-off operation occurred. One can notice in Table 1 that the first components $L_f N_a$ (N_a is the number of activated sensors) are equal for the QVF-U and QPF algo-

Figure 9. (a) RMSE vs. Nodes density varying in {50,..., 200}; (b) RMSE vs Nodes noise variances varying in {0,..., 0.25}

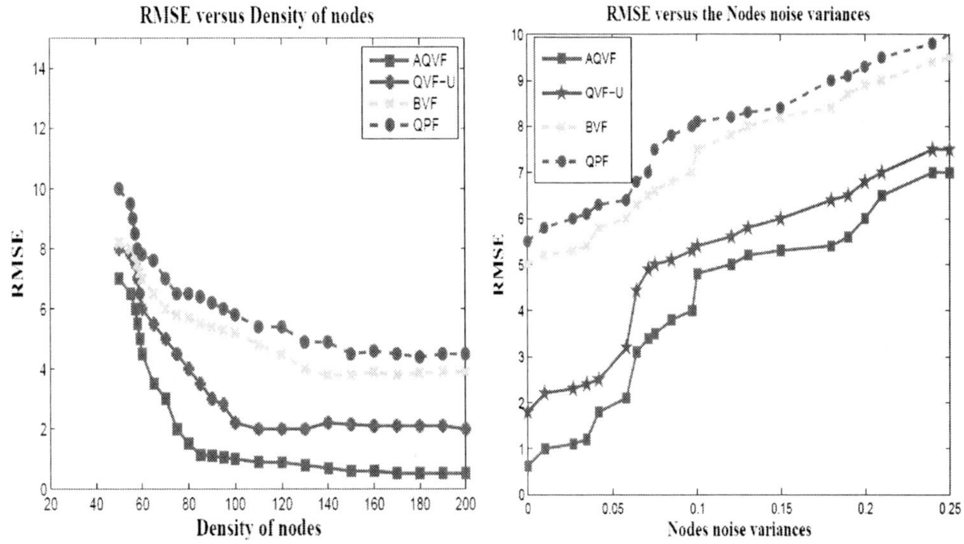

(a) (b)

rithms, since each activated sensor that detects the target transmits L_f bits of information to the CH. While for the AQVF algorithm, each activated sensor which detects the target can transmit the optimal bits number L_{opt}. However, for the BVF, each activated sensor transmits one bit of information to the CH. By approximating the filtering distribution with a single Gaussian sta-

Figure 10. (a) RMSE vs. Transmitting power varying in {50,..., 200}. (b) RMSE vs. Sensing range varying in {5, ...,3}

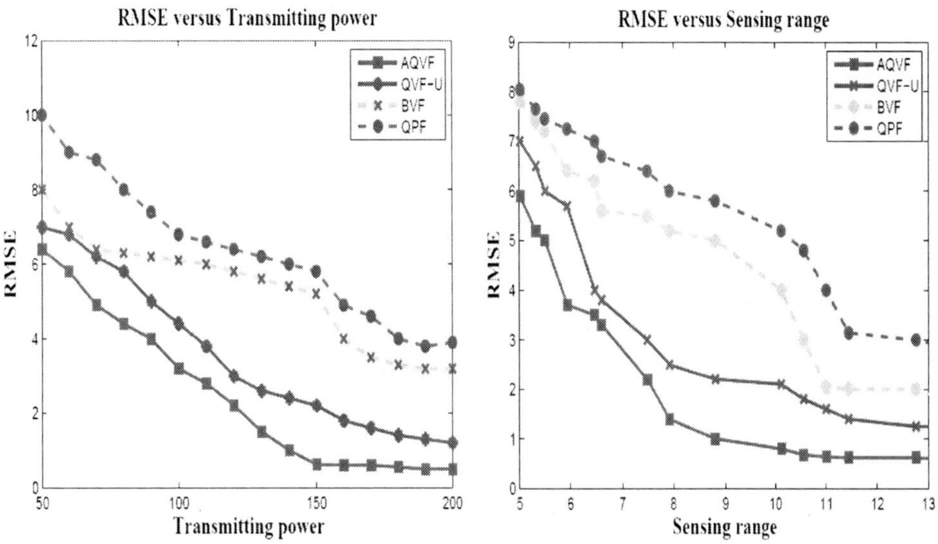

(a) (b)

tistic, the VF and GPF just have to transmit the expectation and the precision matrix. Nodes communication of the QPF mainly lies in communicating the particles and their corresponding weights, which is much greater than that of the VF and GPF. The nodes communication is shown in Table 1. From Figure 11, Figure 12 and Figure 13, we can see that our protocol successfully balances the trade-off between the energy consumption and the tracking accuracy even with several abrupt changes in the trajectory for all algorithms.

These results confirm that the proposed method outperforms the classical algorithms in terms of energy expenditure during the whole tracking process.

Table 1. Nodes Communication Comparison

Method	Nodes Communication (bits)
AQVF	$L_{opt}N_a+32+64$
QVF-U	$L_fN_a+32+64$
BVF	$N_a+32+64$
QPF	$L_fN_a+32N_p+64N_w$

We have also applied our proposed method for tracking multiple target. In the following, we compare the tracking accuracy of our adaptive scheme for multiple target tracking, the binary variational filtering (BVF) (Djuri'c, Vemula & Bugallo, 2005), and the centralized quantized particle filter (QPF) (Quan & Sayed, 2007). The system parameters considered in the following simulations are: the total sampling instants for Target 1 is $w_1 = 50$, for Target 2 is $w_2 = 100$ and for Target 3 is $w_3 = 100$. One can notice from Figure 14) and Figure 15) that the best tracking quality is ensured by the adaptive AQVF algorithm compared to the two BVF and QPF algorithms for three targets (Target1, Target 2 and Target 3). As we can observe, the estimated trajectories using the proposed method represent well the real trajectories, even with abrupt changes in the targets trajectories. Indeed, the performance of the tracking is assessed according to the Root Mean Square Error (RMSE), which is presented in Table 2. Smaller is the RMSE higher is the accuracy of the estimated trajectory. Thus, the obtained results (Table 2) prove that our tracking method has an average precision of 0.39 meter

Figure 11. Energy consumption comparison between AQVF and QVF-U algorithms for a) L = 3. b) L = 4

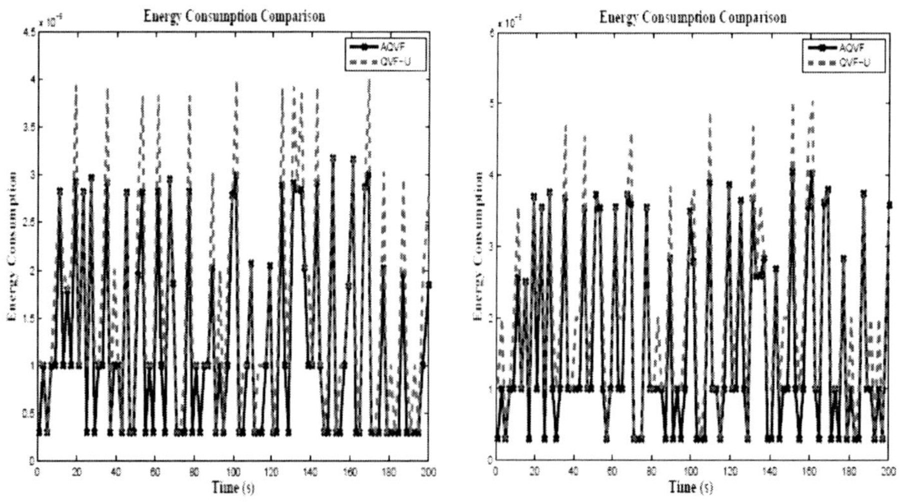

(a) (b)

Figure 12. Energy consumption comparison between AQVF and BVF algorithms for a) L = 3. b) L = 4

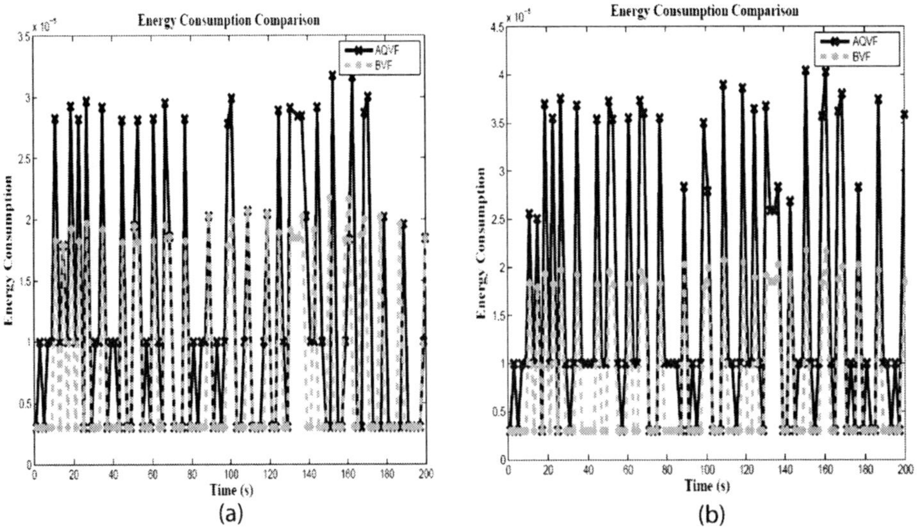

approximately, while the maximum error is never more than 0.44 meter.

CONCLUSION

The main objective of the chapter is to implement Bayesian inference techniques in distributed wireless sensor networks (WSNs) by the variational filtering method. The trade-off between estimation precision and energy-awareness is the primary focus of the chapter. In order to consider the inherent cost in distributed signal processing, we have examined two important issues: the minimal size of message, in bits, of a representation of the observed data, and the approximation of the sample-based estimates yielded by the inference algorithm. The cluster head determination and the best sensors

Figure 13. Energy consumption comparison between AQVF and QPF algorithms for a) L = 3. b) L = 4

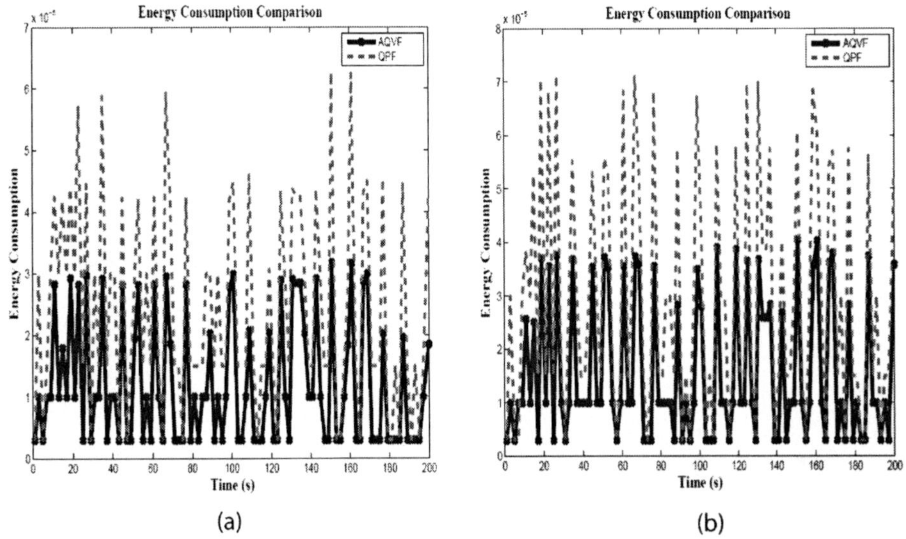

Figure 14. Tracking accuracy between AQVF and BVF methods

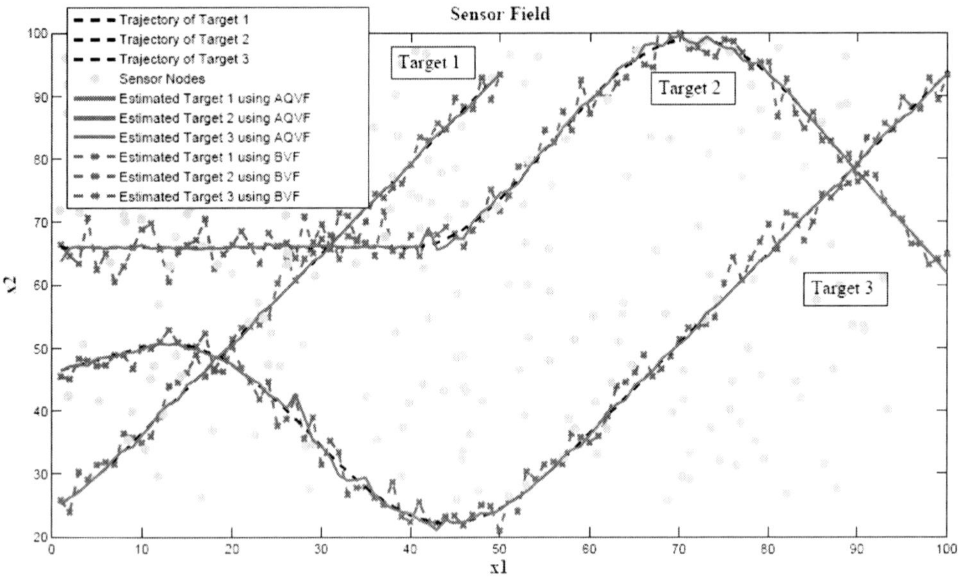

selection were studied as well. The contributions of the chapter consisted in the following points: By adopting the QVF algorithm, the update and the approximation of filtering distribution were jointly performed, allowing a lossless compression through the sensor network. In terms of resource consumption, the temporal dependence has been reduced to only one single Gaussian statistic by the QVF, favoring distributed signal processing in WSNs. A general state evolution model has been proposed to describe the target state, which introduces hidden variables to help parameterizing

Figure 15. Tracking accuracy between AQVF and QPF methods

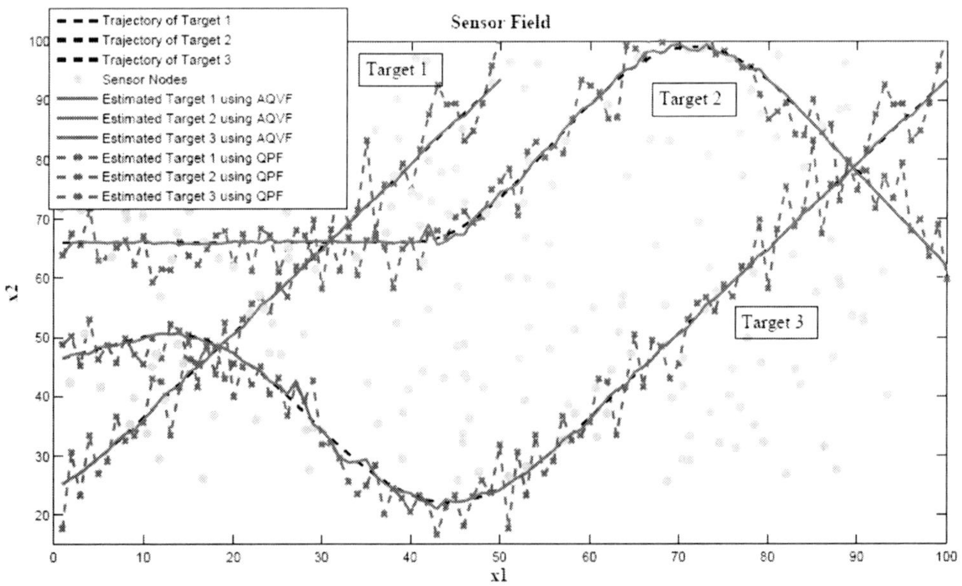

Table 2. RMSE Comparison

Methods	RMSE: Target 1	RMSE: Target 2	RMSE: Target 3
AQVF	0.38m	0.35m	0.44m
BVF	2.4m	2.12m	2.52m
QPF	4.15m	4.11m	4.25m

the filtering distribution. Unlike the traditional kinematic models, no assumption was made on the target velocity or acceleration. Instead, assuming random mean and covariance for the target state have led to a prior probability distribution covering a wide range of tail behaviors, allowing discrete jumps in the target trajectory. As for economical reasons, in the hardware layer, the deployment of quantized sensors saves greatly energy; in the software layer, the adaptive QVF algorithm decreases the information exchanged between CHs. Based on the prediction made by the QVF algorithm, and taking into account the longevity and robustness of WSNs, a robust sensors selection method has been proposed. It consisted of activating the candidate sensors reactively to maximize observation data relevance and to evenly distribute energy consumption.

The proposed method provides not only the estimate of the target position using the QVF algorithm but also selects the best candidate sensors that participate in data collection for tracking the target and optimize the number of quantization bits per observation. The sensors selection approach is based on maximizing the mutual information function under energy constraints. This criterion defines the main parameters that may influence the relevance of the sensor cooperation. The quantization is optimized by minimizing the transmitting power under predicted Cramér-Rao bound constraint. The computation of these criteria is based on the target position predictive distribution provided by the QVF algorithm. We compared our method with the QVF algorithm under sensing range constraint and using uniform power for quantization, the centralized quantized particle filter QPF algorithm and BVF algorithm which is based on binary proximity sensors. The proposed method is also applied for simultaneous tracking of several moving targets. Simulation results demonstrate the significantly improved performance of our approach.

REFERENCES

Arulampalam, M. S., Maskell, S., Gordon, N., & Clapp, T. (2003). A tutorial on particle filters for online nonlinear/non-Gaussian Bayesian tracking, *Signal Processing. IEEE Transactions on, 50*(2), 174–188.

Barndorff-Nielsen, O. (1977). Exponentially decreasing distributions for the logarithm of particle size. In *Proceedings of the Roy. Soc.* (pp. 401–419).

Beal, M. (2003). *Variational algorithms for approximate Bayesian inference*. Citeseer.

Bian, F., Kempe, D., & Govindan, R. (2006). Utility-based sensor selection. *Information Processing in Sensor Networks (IPSN 2006). The 5th International Conference on* (pp. 11–18).

Chen, W., Hou, J., & Sha, L. (2004). Dynamic clustering for acoustic target tracking in wireless sensor networks. In *Proceedings of the IEEE Transactions on Mobile Computing* (pp. 258–271).

Chhetri, A., Morrell, D., & Papandreou-Suppappola, A. (2005). Energy efficient target tracking in a sensor network using nonmyopic sensor scheduling. *8th International Conference on Information Fusion, 2005.*

Cui, S., Goldsmith, A., & Bahai, A. (2005). Energy-constrained modulation optimization. *Wireless Communications. IEEE Transactions on, 4*(5), 2349–2360.

Djuri'c, P., Vemula, M., & Bugallo, M. (2005). Tracking with particle filtering in tertiary wireless sensor networks. In *Proceedings of the 2005 IEEE International Conference on Acoustics, Speech, and Signal Processing* (pp. IV-757-IV-760).

Ertin, E., Fisher, J., & Potter, L. (2003). Maximum mutual information principle for dynamic sensor query problems. In *Proceedings of the 2nd International Conference on Information Processing in Sensor Networks* (pp. 405-416). Berlin/Heidelberg, Germany: Springer-Verlag.

Goldsmith, A., & Wicker, S. (2002). Design challenges for energy-constrained ad hoc wireless networks. *Wireless Communications, IEEE*, *9*(4), 8–27. doi:10.1109/MWC.2002.1028874

Gray, R. M. (2006). Quantization in task-driven sensing and distributed processing. *Proceedings of the 2006 IEEE International Conference on Acoustics, Speech and Signal Processing* (pp. V1049-V1052).

Heinzelman, W., Chandrakasan, A., & Balakrishnan, H. (2000). Energy-efficient communication protocol for wireless microsensor networks. *Proceedings of the 33rd Annual Hawaii International Conference on System Sciences* (pp. 8020).

Hintz, K. (1991). A measure of the information gain attributable to cueing. *Systems, Man and Cybernetics. IEEE Transactions on*, *21*(2), 434–442.

Li, J., & AlRegib, G. (2007). Rate-constrained distributed estimation in wireless sensor networks. *Signal Processing. IEEE Transactions on*, *55*(5), 1634–1643. doi:10.1109/TSP.2006.890823

Liu, J., Reich, J., & Zhao, F. (2003). Collaborative in-network processing for target tracking. *EURASIP Journal on Applied Signal Processing*, *2003*, 378–391. doi:10.1155/S111086570321204X

Mansouri, M., Ouachani, I., Snoussi, H., & Richard, C. (2009). Cramer-Rao bound-based adaptive quantization for target tracking in wireless sensor networks. *IEEE/SP 15th Workshop on Statistical Signal Processing (SSP '09)* (pp. 693–696).

Manyika, J., & Durrant-Whyte, H. (1995). *Data fusion and sensor management: A decentralized information-theoretic approach*. Upper Saddle River, NJ, USA: Prentice Hall PTR.

Quan, Z., & Sayed, A. (2007). Innovations-based sampling over spatially-correlated sensors. In *Proceedings of the 2007 IEEE International Conference on Acoustics, Speech and Signal Processing* (pp. III-509-III-512).

Ribeiro, A., Giannakis, B., & Roumeliotis, S. I. (2006). *SOI-KF: Distributed Kalman filtering with low-cost communications using the sign of innovation, 54*(12), 4782–4795.

Rubin, I., & Huang, X. (2006). Capacity aware optimal activation of sensor nodes under reproduction distortion measures. *Military Communications Conference (MILCOM '06), IEEE* (pp. 1–8).

Smidl, V., & Quinn, A. (2005). *The variational Bayes method in signal processing*. Berlin/Heidelberg, Germany: Springer-Verlag.

Snoussi, H., & Richard, C. (2006). Ensemble learning online filtering in wireless sensor networks. In *Proceedings of the IEEE ICCS International Conference on Communications Systems* (pp. 1-5).

Teng, J., Snoussi, H., & Richard, C. (2007a). Binary variational filtering for target tracking in sensor networks. *IEEE/SP 14th Workshop on Statistical Signal Processing (SSP '07)* (pp. 685–689).

Teng, J., Snoussi, H., & Richard, C. (2007b). Prediction-based proactive cluster target tracking protocol for binary sensor networks. *IEEE International Symposium on Signal Processing and Information Technology* (pp. 234-239).

Teng, J., Snoussi, H., & Richard, C. (2008). Prediction-based proactive and reactive cluster managements for target tracking in binary sensor networks.

Teng, J., Snoussi, H., & Richard, C. (2010). Decentralized variational filtering for target tracking in binary sensor networks. *IEEE Transactions on Mobile Computing*.

Vermaak, J., Lawrence, N., & Perez, P. (2003). Variational inference for visual tracking. In *Proceedings* of the *2003 IEEE Computer Society Conference on Computer Vision and Pattern Recognition* (pp. I-773-I-780).

Yang, H., & Sikdar, B. (2003). A protocol for tracking mobile targets using sensor networks. In *Proceedings of the 1st IEEE International Workshop on Sensor Network Protocols and Applications* (pp. 78-87).

Yao, K., Hudson, R., Reed, C., Chen, D., & Lorenzelli, F. (1998). Blind beamforming on a randomly distributed sensor array system. *IEEE Journal on Selected Areas in Communications*, *16*(8), 1555–1567. doi:10.1109/49.730461

Yick, J., Mukherjee, B., & Ghosal, D. (2005). Analysis of a prediction-based mobility adaptive tracking algorithm. In *Proceedings of the 2nd International Conference on Broadband Networks (BroadNets '05)* (pp. 753-760).

Zhao, F., Shin, J., & Reich, J. (2002). Information-driven dynamic sensor collaboration for tracking applications. *IEEE Signal Processing Magazine*, *19*(2), 61–72. doi:10.1109/79.985685

APPENDIX A. VARIATIONAL CALCULUS

Assuming that the approximate distribution for the mean μ_{t-1} follows a gaussian model $\left(q(\mu_{t-1}) \sim N(\mu^*_{t-1}, \lambda^*_{t-1})\right)$ and taking into account the Gaussian transition of the mean $(p(\mu_t|\mu_{t-1}) \sim N(\mu_{t-1}, \bar{\lambda}))$, the predictive distribution of μ_t t is given by:

$$
\begin{aligned}
q_p(\mu_t) &= \int p(\mu_t \mid \mu_{t-1}) q(\mu_{t-1}) d\mu_{t-1} \\
&\sim N(\mu^*_{t-1}, (\lambda^{*\,-1}_{t-1} + \bar{\lambda}^{-1})^{-1}).
\end{aligned}
\tag{42}
$$

Let denote μ_t^pt and λ_t^p respectively the mean and the precision of the Gaussian distribution $q_p(\mu_t) \sim N(\mu_t^p, \lambda_t^p)$. According to the equation (8), the approximate distribution $q(\mu_t)$ is expressed as:

$$
\begin{aligned}
q(\mu_t) &\propto \exp\left\langle \log p(z_{1:t}, \alpha_t) \right\rangle_{q(x_t)q(\lambda_t)} \\
&\propto \exp\left\langle \log p(\alpha_t \mid z_t) \right\rangle_{q(x_t)q(\lambda_t)} \\
&\propto \exp\left\langle \begin{matrix} \log p(z_t \mid x_t) + \log p(x_t \mid \mu_{t,}\lambda_t) \\ + \log p(\lambda_t) + \log q_p(\mu_t) \end{matrix} \right\rangle_{q(x_t)q(\lambda_t)}
\end{aligned}
\tag{43}
$$

Therefore,

$$
\begin{aligned}
q(\mu_t) &\propto q_p(\mu_t) \exp\left\langle \log p(x_t \mid \mu_t, \lambda_t) \right\rangle_{q(\infty_t)q(\lambda_t)} \\
&\propto q_p(\mu_t) \exp\left\langle -\frac{1}{2}(x_t - \mu_t)^T \lambda_t (x_t - \mu_t) \right\rangle_{q(\infty_t)q(\lambda_t)} \\
&\propto q_p(\mu_t) \exp -\frac{1}{2}\{tr[\langle \lambda_t \rangle_{q(\lambda_t)} \langle (x_t - \mu_t)^T (x_t - \mu_t) \rangle_{q(\infty_t)}]\} \\
&\propto \exp -\frac{1}{2}[(\mu_t - \mu_t^p)^T \lambda_t (\mu_t - \mu_t^p) - 2\mu_t^T \langle \lambda_t \rangle \langle x_t \rangle + \mu_t^T \langle \lambda_t \rangle \mu_t],
\end{aligned}
\tag{44}
$$

yielding a Gaussian distribution $q(\mu_t) \sim N(\mu_t^*, \lambda_t^*)$. The first and the second derivatives of the logarithm of $q(\mu_t)$ are expressed as:

$$
\begin{aligned}
\frac{\partial \log(q(\mu_t))}{\partial \mu_t} &= -\frac{1}{2}[2\lambda_t^p(\mu_t - \mu_t^p) - 2\langle \lambda_t \rangle \langle x_t \rangle + 2\langle \lambda_t \rangle \mu_t], \\
\frac{\partial \log(q(\mu_t))}{\partial \mu_t \partial \mu_t^T} &= -\lambda_t^p - \langle \lambda_t \rangle,
\end{aligned}
\tag{45}
$$

the precision λ_t^* t and the mean μ_t^* of $q(\mu_t)$ are obtained as follows:

$$\lambda_t^* = \langle \lambda_t \rangle + \lambda_t^p, and \mu_t^* = \lambda_t^{*-1}(\langle \lambda_t \rangle \langle x_t \rangle + \lambda_t^p \mu_t^p. \tag{46}$$

The approximate separable distribution corresponding to λ_t can be computed following the same reasoning as above:

$$\begin{aligned} q(\lambda_t) &\propto \exp \langle \log p(\alpha_t \mid z_t) \rangle_{q(\infty_t)q(\mu_t)} \\ &\propto \exp \langle \log p(z_t \mid x_t) + \log p(x_t \mid \mu_t, \lambda_t) + \log p(\lambda_t) + \log q_p(\mu_t) \rangle_{q(\infty_t)q(\mu_t)} \\ &\propto p(\lambda_t) \exp \langle \log p(x_t \mid \mu_t, \lambda_t) \rangle_{q(\infty_t)q(\mu_t)} \\ &\propto W_2(\bar{V}, \bar{n}) \mid \lambda_t \mid^{\frac{1}{2}} \exp - \frac{1}{2} \{ tr[\lambda_t \langle (x_t - \mu_t)^T (x_t - \mu_t) \rangle_{q(\infty_t)q(\mu_t)} \} \\ &\propto \mid \lambda_t \mid^{\frac{n+1-(2+1)}{2}} \exp - \frac{1}{2} \{ tr[\lambda_t (\langle x_t x_t^T \rangle - \langle x_t \rangle \langle \mu_t \rangle^T - \langle \mu_t \rangle \langle x_t \rangle^T + \langle \mu_t \mu_t^T \rangle + \bar{V}^{-1})] \}, \end{aligned} \tag{47}$$

which yields a Wishart distribution $W_2(V^*, n^*)$ for the precision matrix λ_t with the following parameters:

$$\begin{cases} n^* = \bar{n} + 1, \\ V^* = (\langle x_t x_t^T \rangle - \langle x_t \rangle \langle \mu_t \rangle^T - \langle \mu_t \rangle \langle x_t \rangle^T + \langle \mu_t \mu_t^T \rangle + \bar{V}^{-1})^{-1}. \end{cases} \tag{48}$$

Finally, the approximate distribution $q(x_t)$ has the following expression:

$$\begin{aligned} q(x_t) &\propto \exp \langle \log p(\alpha_t \mid z_t) \rangle_{q(\mu_t)q(\lambda_t)} \\ &\propto \exp \langle \log p(z_t \mid x_t) + \log p(x_t \mid \mu_t, \lambda_t) + \log p(\lambda_t) + \log q_p(\mu_t) \rangle_{q(\mu_t)q(\lambda_t)} \\ &\propto p(z_t \mid x_t) \exp \langle \log p(x_t \mid \mu_t, \lambda_t) \rangle_{q(\mu_t)q(\lambda_t)} \\ &\propto p(z_t \mid x_t) \exp - \frac{1}{2} \{ tr[\langle \lambda_t \rangle_{q(\lambda_t)} \langle (x_t - \mu_t)^T (x_t - \mu_t) \rangle_{q(\mu_t)}]\} \\ &\propto p(z_t \mid x_t) N(\langle \mu_t \rangle, \langle \lambda_t \rangle), \end{aligned} \tag{49}$$

which does not have a closed form. Therefore, contrary to the cases of the mean μ_t and the precision λ_t, in order to compute the expectations relative to the distribution $q(x_t)$, one has to resort to the importance sampling method where samples are generated according to the Gaussian $N(\langle \mu_t \rangle, \langle \lambda_t \rangle)$ and then weighted according to the likelihood $p(z_t|x_t)$.

Section 4
Hardware Technology and Applications in VSN

Chapter 12
Using Line Cameras for Monitoring and Surveillance Sensor Networks

Jiang Yu Zheng
Indiana University-Purdue University, USA

ABSTRACT

The innovative combination of wireless sensor network (WSN) technology with visual monitoring and surveillance technology in computer vision has been emerging as an important new paradigm. This emerging technology will play a crucial role in visual monitoring and surveillance for automatic object detection and tracking in applications such as real-time traffic monitoring and control, vehicle parking control, intrusion detection, security surveillance, military battlefield monitoring, and so on. Compared to traditional WSNs with scalar sensing data, the development of WVSNs presents much greater challenges in terms of node's computation power, storage, wireless bandwidth capacity and energy conservation due to the processing and transmission of the huge amount of two-dimensional (2D) image data. We introduce the use of linear CCD sensors for wireless sensor network here. It reads temporal data from a CCD array continuously and stores them to form a 2D image profile. Compared to most of the sensors in the current sensor networks that output temporal signals, it delivers more information such as color, shape, and event of a flowing scene. On the other hand, it abstracts passing objects in the profile without heavy computation and transmits much less data than a video from normal cameras. This paper focus on several unsolved issues of line sensors in capturing targets in the 3D space such as sensor setting, shape analysis, robust object extraction, and real time background adapting to ensure long-term sensing and visual data collection via networks. All the developed algorithms are executed in constant complexity for reducing the sensor and network burden. A sustainable visual sensor network can thus be established in a large area to monitor passing objects and people for surveillance, traffic assessment, invasion alarming, etc.

DOI: 10.4018/978-1-61350-153-5.ch012

INTRODUCTION

Many works of sensor network have focused on low-power sensors and reliable data transmission. Most sensor receptors obtain a temporal signal and send data to a centralized server for analysis and visualization via wireless sensor networks (WSN). For visual information, however, the temporal two-dimensional data from a camera to transmit are tremendous. Different from broadcasting and multicasting of a video via the Internet that can eventually use the peer-to-peer strategy to reduce overall data transmission on the network, transmitting different videos to a server simultaneously in a unicast style is more constrained by the network bandwidth and the server capacity.

To diminish the data sent in a sensor network, one alternative way is to reconsider linear CCD sensors, which were used in the early stage of digital imaging (Zheng & Tsuji, 1989, 1990; Taniguchi & Seki, 1994) and are still widely used in scanners and terrain data acquisition (Gupta & Hartley, 1997). In sports areas, static line cameras are used for the time puncture of racing cars, boats, and athletes at destinations, capturing the time of arrival from the line sensor image. Similar use is in traffic monitoring (Zhu, Xu, Yang, Shi & Lin, 2000) with a static line sensor mounted at some high position above the road. A vehicle-borne line sensors have also been used in scanning environments for panoramic views and route panoramas (Zheng, 2003; Chitnes, Liang, Zheng, Pagano & Lipari, 2009; Zheng & Shi, 2006) where the line sensor is rotated or translated so that the accumulated sets of line data form long images representing surrounding scenes or side scenes along a street. A handheld line sensor further sweeps objects or environments to obtain texture map for graphics models (Rademacher & Bishop, 1998). However, they have not been explored in the frame of sensor network applied to broader areas. In many applications such as counting passing people in an entertainment park, station, store or exhibition site (MacDorman, Nabuta, Koizumi & Ishiguro, 2007), alarming invasion across a border or a critical facility and monitoring a traffic flow, the detailed moving directions, speeds and trajectories of passing objects are already obvious or less interested. Rather than processing normal video, we use line sensors for counting passing targets for saving computation cost and extend the area of monitoring via networks, which is new in sensor network distributed in a large 3D space. To achieve more general sensing tasks and a variety of dynamic environments in a sensor network, we need to explore the sensor setting conditions such as the viewing angle, distance, alignment, and sampling rate, and discover the properties of the sensed images. How to locate sensors and how to analyze sensor data to obtain meaningful information are essential topics to be investigated.

Degrading from 2D images, the 1D temporal data are captured continuously by a CCD array to form a 2D image temporally. The sensor delivers visual information to a server in real time and requires much less bandwidth than a closed-circuit television (CCTV) system. If penetrating objects are pursuing a simple movement, the resulting 2D profile contains information on shape, time, and identities (Zheng & Tsuji, 1989, 1990; Nakanixhi & Ishii, 1992; Taniguchi & Seki, 1995; Yoshioka, Nakaue & Uemura, 1999), but requires less redundant processing than normal video (Gupte & Papanikolopoulos, 2002; Li, Huang, Gu & Tian, 2003). It is also more intuitive to examine the history of passing entities in the line sensor produced image than a sequential signal and a video sequence, which serves the purpose of indexing of surveillance data.

This chapter focuses on algorithms and sensing modes of a line sensor network in the aspect of computer vision for counting object flows through spatial channels or guard lines. We discuss how to use the line camera in the sensor network and what effects or information can obtained from a set of line sensors. We propose efficient algorithms in constant complexity for data processing over time such that the sensors can send long archives of passing targets via networks. Several critical problems solved in this work are the sensor alignment for capturing object shapes, robust moving object extraction and background updating, and data transmission over the wired/wireless network. In details, the new achievements in this paper are as follows:

- Giving geometry analysis of the 2D shapes projected from transitive 3D motion by using a 1D CCD array, in order to guide proper sensor alignments and flexible view rectification;
- Setting a line sensor network to extract passing objects at critical locations and transmit summary data to a central server in real time;
- Preparing sensor-based functions to deal with changeable environments such as varying illumination and weather conditions, and temporal static scenes. A constant order median filtering algorithm is developed for real-time background adapting. No complicated computation will be left over to the central server via the network.

Although this passive sensor is incapability of capturing complex movements of targets, and its image quality from a camera with a low sampling rate cannot compete with that from a normal 2D camera, it is still a choice for the traffic monitoring on roads and construction sites, invasion detection at critical facilities and borders, people counting in stores and event sites, etc. due to its limited transmit data and low cost in computation. The inherent property of line sensors in profiling passing flows significantly reduces the computation in 2D images such that sensor networks can be deployed widely for the purposes of security, disaster preparation, business, and environment assessment.

Section 2 introduces the line sensor generated image, the contained information, and the extraction of passing objects from the sensor data. Section 3 generalizes the line sensors and analyzes the shape deformation in the line sensor images, based on which dynamic objects can be visualized. Section 4 deals with long-term sensing robust to background changes, in which a median filtering algorithm with constant complexity is developed. Section 5 gives experiments of the real sensing and data transmission over the network.

PROFILING DYNAMIC FLOW

Temporal and Spatial Events

A static video camera is set to monitor a gate, a road, or a river through which dynamic objects may pass (Figure 1). Aligning a sampling line, $\boldsymbol{l}_s$, in the camera frame determines a plane of sight P in the 3D space with the camera focal point. The line is sampled at a fixed rate (normally video rate or higher) and the captured 1D pixel arrays are copied continuously to another image memory $I(t, y)$, where $y \in [0, h]$ is the coordinate on $\boldsymbol{l}_s$ and t is the time. This forms a dynamic projection image (Zheng & Tsuji, 1989, 1998; Chitnes, Liang, Zheng, Pagano & Lipari, 2009), i.e., a 2D line sensor image. As an object passes through the plane of sight, its different parts are exposed to the sampling line in order so that it leaves a shape in the resulting image; P must be set to cut the moving flow for recording shape segments consecutively. Objects should have approximately a constant moving direction confined by a road, stairway, gate, or riverbank. For hardware devices, a real line sensor is also available for selection if not using a video camera connected to a computer. In such a case, the plane of sight must be aligned manually rather than picking a sampling line digitally on the frame of video input.

A static background is projected onto the sampling line constantly. The same color distribution lasts along the time axis. The parallel patterns are thus visible in the generated $I(t,y)$. Figure 2

Figure 1. Principle in generating a line sensor image

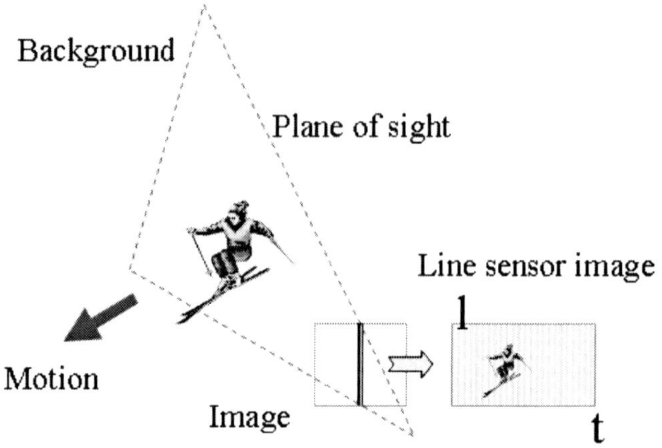

Figure 2. A sampling line in the field of view and its generated line sensor image. (left) An image of the view with the vertical sampling line depicted. (right) Output profile recording passing people both in front and in the hallway behind the window.

gives an example to record people passing through a yard. Different from an active ray ejected to a receiver (e.g., laser or radar) and a passive infrared sensor that outputs 1D signal pulses, a line sensor generates a 2D profile of passing subjects from the 1D pixel array, which can be stored and accessed as an index of passing targets. Some defect on passing people is yielded from undersampling, which will be described in section 3.2. On the other hand, by reducing one dimension in the data, the transmission is remarkably reduced as compared to video transmission over the cable network, which inherently makes the real time transmission possible on wireless sensor networks.

Dynamic Object Detection

The passing objects are distinct from the background not only in color, but also in shape or structure since the background is composed of horizontal stripes along the time axis. This provides a chance to detect visual events passing the sampling line. Transmitting extracted dynamic shapes only further reduce the data size because it neglects the static sections of background in the entire archive. The object detection now becomes a problem to extract patterns from horizontal stripes.

To achieve this goal, the object detection and transmission of line sensors must satisfy the real time obligation. In extracting dynamic objects, we take both the temporal differentiation and background colors into consideration. At each height y, we differentiate the color value along the time axis by

$$I_t(t, y) = \partial I(t,y)/\partial t \quad (1)$$

which is an differential image of the line sensor image $I(t,y)$, producing the boundaries of passing objects. This is done by normal filtering operator, e.g., Sobel operator in horizontal direction. On the other hand, subtracting the lines in $I(t,y)$ with the single line background color distribution $b(y)$ on the sampling line as

$$I_b(t, y) = \| I(t, y) - b(y) \| \quad (2)$$

we obtain another image $I_b(t,y)$ where high values in the image contains the pixels belong to a passing object. The collection of those points in $I(t, y)$ shows the dynamic object occupied regions. Using only a single source above may not enclose a correct shape for a dynamic object, since the object and background colors may be identical partially (yielding small values in $I_b(t,y)$), or the extractable edges from the differentiation in the object area are not dense enough for filling a closed region.

For a period of background pattern in image $I(t, y)$, both $I_t(t, y)$ and $I_b(t, y)$ values are close to zero there. If an object passes, however, $I_t(t, y)$ is non-zero at boundaries and $I_b(t, y)$ is non-zero in the object occupied region except some holes. We fill such regions to enclose dynamic objects by checking if

$$\| I_t(t, y) \| > \delta_1 \text{ or } \| I_b(t, y) \| > \delta_2 \quad (3)$$

where δ_1 and δ_2 are thresholds. A detected object thus contains information on arriving time and shape for identification. At the boundary of a dynamic object, the two thresholds are basically consistent, i.e., they are the difference from the background intensity. With an adaptive background described in Sec. 4, the thresholds can be fixed for a wide range of illumination.

For detecting the one dimensional array of background pattern that expends to horizontal stripes in $I(t, y)$, we calculate a horizontal section T_s in $I(t, y)$ without any dynamic object, where the accumulations of $I_t(t, y)$ along the sampling line are close to zero, i.e., for every $t \in T_s$,

$$T_s = \{ t \mid \Sigma_y \| I_t(t, y) \| < \delta_3 \} \quad (4)$$

where δ_3 is a threshold. Averaging $I(y, t)$ horizontally in the scope of T_s, the background color on the sampling line $\boldsymbol{l}_s$ is obtained as

$$b(y) = \Sigma_t I(t, y) / \|T_s\|, \; t \in T_s \text{ and } y \in [0, h] \quad (5)$$

where $\|T_s\|$ is the length of T_s, and h is the height of image $I(t,y)$. The shape of a passing object is further related to its speed. The faster the speed, the narrower the yielded shape is. The dynamic regions with small vertical sizes are ignored as noises in the object extraction. Such noises include waving leaves on trees and ripples in rivers. On the other hand, we do not separate a shadow or glass reflection from an object because they move with the object. Different from the 2D video frame differentiation, the line sensor only involves several consecutive pixel lines in target detection, which also reduces the computation cost to one dimension.

After the dynamic regions are extracted sequentially, their occupied sections with time stamps are sent to a server through the wired/wireless connection. Figure 3 in the end of the paper shows an example to detect passing vehicles on a road. The surveillance camera is hidden at a distant location (Figure 3(a)) and a vertical sampling line generates a temporal profile of the road (Figure 3(b)). The sensor detects vehicle sections and sends to the server, where the accumulated number of vehicles in a long period is shown in Figure 3(c). We can also notice the trees in the front field are waving but they are excluded from dynamic objects due to their horizontally elongated shapes in $I(t, y)$. The field of view is large, and the sensor still obtains moving objects.

A single sampling line does not acquire the direction of penetrating. We can infer the direction from some oriented features associated with the objects. As an example, Figure 4 shows a vehicle moving across the sampling line in different directions. Because the lighting direction is fixed from the right in that period, the shadows should be

Figure 3. Detecting dynamic objects in real time. (a) Field of view with a sampling line intersecting a distant road. (b) A section of line sensor image with vehicles and their times of arrival. (c) Detected vehicles and the transmitted sections.

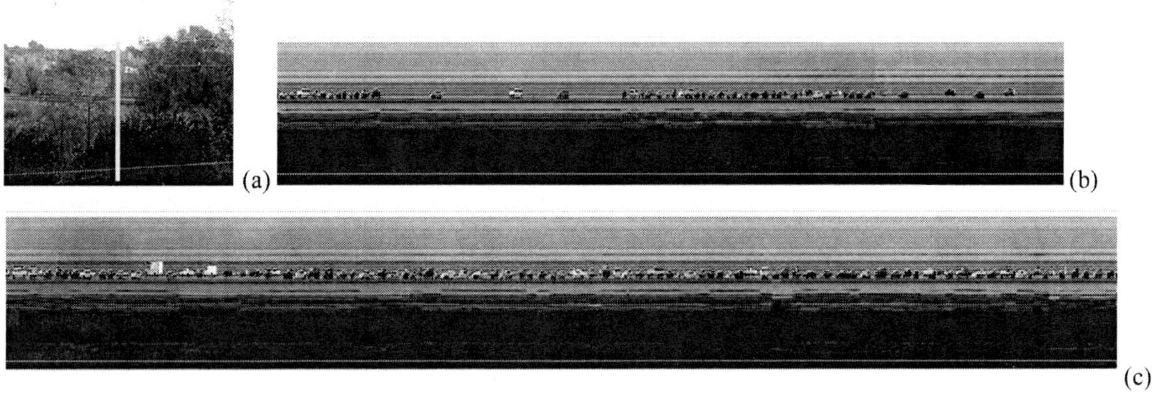

invariant at the left side of the vehicle regardless of the vehicle moving direction. The exception of the line sensor images (Figure 4(b)(d)) means the vehicle is moving in the opposite direction.

SAMPLING LINE SELECTION

Sensor Alignment and Resulting Shape

The alignment of a line sensor so far is to set the plane of sight orthogonal to a dynamic flow. This assumption may not be satisfied in the real situation because of the camera location, spots to monitor and possible occlusion. In order to take the advantages of the line sensors for various applications, this section extends the sensor setting in a general 3D space; the plane of sight may not be orthogonal to the object moving direction (Bolles, Baker & Marimont, 1987; Gupta & Hartley 1997; Yu & McMillan, 2004). This may affect object shapes in the generated line sensor image that are different from normal perspective images.

To align the sampling lines properly, we examine three orthogonal directions that determine the object shapes in $I(t, y)$. Assume that the object moving direction is $\boldsymbol{V}$, and the normal of plane of sight P is $\boldsymbol{n}$ in the camera-centered coordinate system located at focal point O. The plane of sight must satisfy $\boldsymbol{V} \cdot \boldsymbol{n} \neq 0$ for the object flow to be scanned, because the flow $\boldsymbol{V}$ must penetrate P for a relative motion even if it is not parallel to $\boldsymbol{n}$. To observe passing objects, there is another principle direction, $\boldsymbol{L}$, which can be vertical for walking people or cars if we want to catch their side views, as shown in Figure 5(a). The principle direction $\boldsymbol{L}$ can also be horizontal as in Figure 5(b), right interesting a road or river in an overhead view from a bridge, building, pole, etc. Vectors $\boldsymbol{V}$ and $\boldsymbol{L}$ determine a plane, named *front face* here, displaying the basic shape of passing objects. The direction $\boldsymbol{D}$ orthogonal to $\boldsymbol{V}$ and $\boldsymbol{L}$, i.e., $\boldsymbol{D} = \boldsymbol{V} \times \boldsymbol{L}$, is named *depth direction* in this paper. As depicted in the left column of Figure 5, the *moving*, *principle*, and *depth* directions are projected to the camera frame in perspective projection. As described in computer vision, 3D lines parallel to one of the directions of $\boldsymbol{V}$, $\boldsymbol{L}$ and $\boldsymbol{D}$, will have their image projections converge in the image plane to the corresponding vanishing point if their projections are extended. For the three sets of lines parallel to $\boldsymbol{V}$, $\boldsymbol{L}$ and $\boldsymbol{D}$, we denote their vanishing points by Q_V, Q_L, and Q_D respectively in the image plane,

Figure 4. Examples of line sensor images taking a vehicle moving in different directions from two aspects. The vehicle directions can be determined from the shadows and lit surfaces on the car under the condition that the lighting is from the right.

Viewing direction	Generated line sensor image	Object moving direction in the field of view
Side views	 (a) → t	From right to left ←
	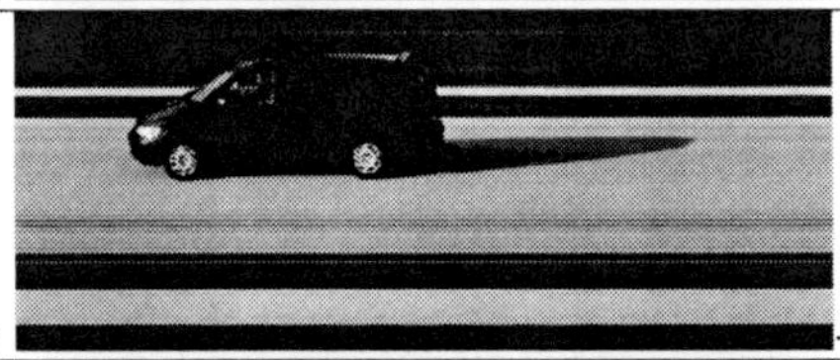 (b) → t	From left to right →
Oblique side views	 (c) → t	From right to left ←
	 (d) → t	From left to right →

which might be out of the image frame and even locate at infinity.

Our general alignment of the sampling line is to set l_s (or its extension) through vanishing point Q_L in the image plane. Therefore, the sampling line will scan line set L in order during the object movement. As shown in the left column of Figure 5, two *principle* directions on a vehicle object can be selected in orthogonal to the flow V. Either aligning the plane of sight vertically (Figure 5a) or parallel to the horizontal direction as shown in Figure 5b, the sampling line passes Q_L, or equivalently overlap in the image with one of the projected lines in L set. The projection from moving scenes to the line sensor image is thus a *temporal-perspective projection*. Under this assumption, the shape characteristics in the line sensor image are summarized as follows (depicted in the right column of Figure 5):

1. The length of an object along the time axis is inversely proportional to object speed $|V|$, while the vertical scale of the object depends on its distance from the sensor. If $|V| = 0$, the projected pixels last forever to form a background.
2. The lines parallel to V on moving objects are imaged as horizontal lines v in $I(t,y)$.

Figure 5. Shape deformations on objects in the line sensor images. Three orthogonal line sets characterize the projection and the shapes in the font surface are preserved. (a) Principle direction set vertically for side surfaces of objects. (b) Principle direction is set horizontally for viewing objects from the top.

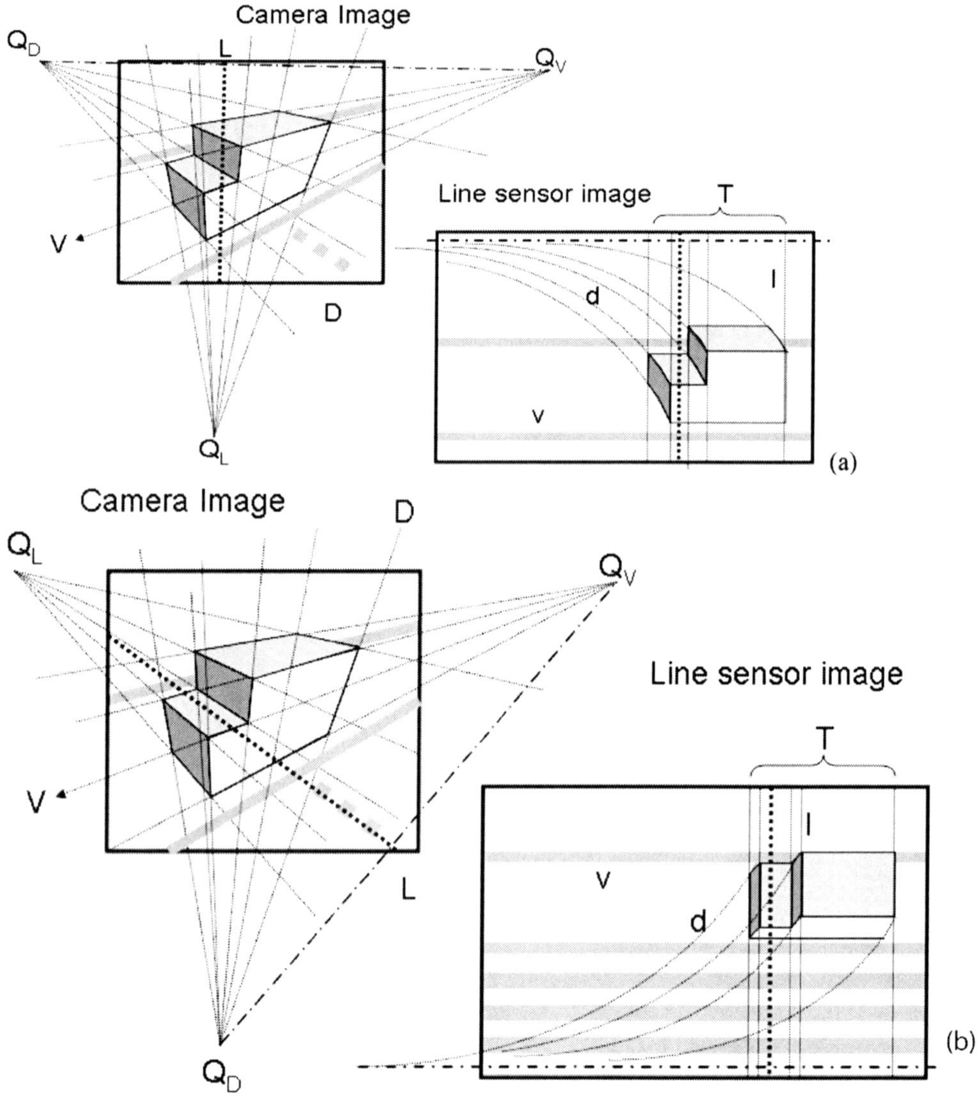

3. Line set parallel to ***L*** is projected to vertical lines *l* (perpendicular to the *t* axis) in *I(t,y)*, because each line passes plane *P* instantly. The *front face* then preserves the shape of object for object recognition and classification, regardless of additional surfaces appearing in the depth direction.
4. Line set parallel to ***D*** direction on the objects is projected to a set of hyperbolas *d* in the line sensor image. Further, these hyperbolas approach to a horizontal asymptotic line, $y = y_q$, if they are extended. If we connect vanishing points Q_V and Q_D with a line, the line intersects sampling line l_s at a point, *q* (not necessarily within the image frame confined by [0, h]). The position, y_q, of *q* on sampling line l_s determines the asymptotic line.

Properties (1-3) can be derived easily, and the proof of (4) is omitted here for the simplicity. According to properties (2) and (3), the shapes on the *front faces* are briefly preserved in resulting I(t,y), except the aspect ratio changed by property (1). This can also be verified in Figure 4, where the side surfaces of the car are all the same, in spite of the extension of front and back surfaces of the car visible in the depth direction. This provides opportunities for partial matching of multiple line sensor images from different directions, and a line sensor image with synthesized images generated from 3D object models, for object recognition.

In monitoring a wide road or a river from a high position for counting people, moving vehicles or floating boats, the principle direction can be set horizontal to avoid mutual occlusion of objects. To reduce the shape distortion along the hyperbola direction, we prefer short lengths of line $\boldsymbol{D}$ captured on objects. After identifying a flow, we examine object sizes and aspect in the image to determine the principle direction $\boldsymbol{L}$.

The flexibility in selecting the sampling line allows us to set a camera even in an indoor location far away from the monitoring location. This is the superior of a visual sensor to an infrared or laser sensor. The line sensor can also monitor a narrow area where objects even do not show their entire shape.

Sampling Rate of Line Sensors

After setting the sampling line, the change of sampling rate alters the object length along the time axis in the line sensor image. A low sampling rate may not obtain sufficient object resolution horizontally and thus is difficult for visualization and object recognition. Inversely, a high sampling rate increases the sensor computation load and data to transmit for dynamic objects. In addition, the sampling rate cannot be increased arbitrary since the CCD elements in the camera need a time cycle for accumulating irradiance and digitization. Depending on the device such as a video camera or a line sensor, the sampling rate ranges from 60 to several thousand lines per second. Assume that the maximum sampling rate available is r lines per second, and the image velocity of a passing object is v, measured by pixel/(video frame). The object length T in the captured line sensor image is then

$$T = \frac{r}{60v} length \qquad (6)$$

where *length* is the object length projected in the image frame, and the video has 60 frames per second if we capture interlaced images.

If r is small, which results in $T<length$, the object shape is *under-sampled* to the line sensor image as depicted in Figure 6a. This happens when an object passes P with a fast speed ($\|\boldsymbol{V}\|$ is large) or the object is close to the camera since the image velocity is inversely proportional to the depth in perspective projection. According to the *Nyquist* theorem, such a shortened object is possible to be recognized only if its spatial frequency along the moving direction is low, i.e., the object surface is monotonic. The under-sampled case can be found in Figure 2 where the sampling rate of the video camera is 30 frames per second.

On the other hand, if the sampling rate r is much higher than the video rate or $\boldsymbol{V}$ is slow as shown in Figure 6b, the obtained object length increases horizontally. However, this does not increase the object resolution in the line sensor image. The reason is that, at two consecutive sampling time instances, the cone through a pixel in the camera frame captures the irradiance from almost the same object point (with only a slight shift). This *overlapped-sampling* causes a temporal blur along the t axis in the line sensor image, which is called *stationary blur* in (Zheng, 2003). The extreme case is a static point in the background that is sampled repeatedly and its

Figure 6. Different coverage of consecutive PSF on an object due to different sampling rates and object moving speeds. The figures are depicted in the object-centered coordinate system where the sensor and PSF has a relative motion of –V. (a) Under-sampling, (b) Overlapped-sampling, (c) Just sampling.

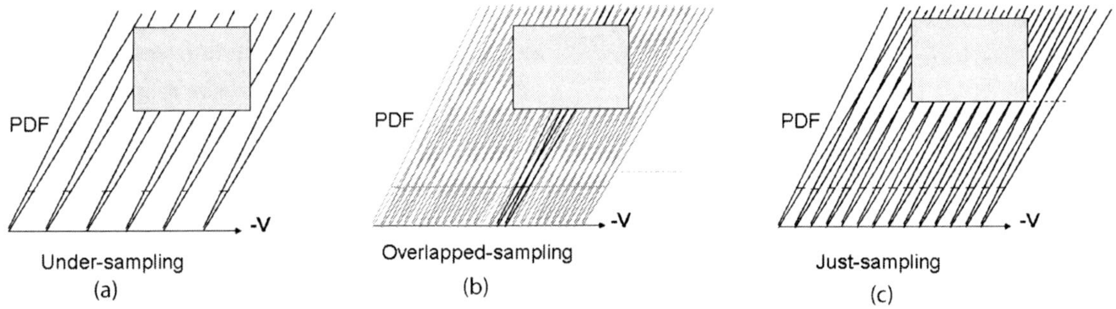

intensity/color lasts along the t axis. Although an additional spatial-temporal filtering can increase the high frequency component on the non-background objects, it will not be an ideal approach for this framework, in which we minimize the processing at the sensor level. The gain from a high-sampling rate can be fully used for increasing the object size by zooming to the scene (and raises the image velocity).

As a result, there exists an ideal sampling rate that keeps the same object resolution in the line sensor image as in a normal perspective image, i.e., *T=length*. It will not include redundant data as in a temporal-blurred image either. Such a sampling rate, r_j, namely *just-sampling rate,* is figured out at $r_j = 60v$. If a line sensor is set to monitor a class of objects with roughly similar speeds and sizes, the sampling rate is selected accordingly for acquiring a proper object resolution. The objects can be separated as moving vehicles, walking people, and escalators by measuring the length of several sample objects after their images are taken. For applications that require counting objects or alarming without concerning their shapes, a sampling rate $r<r_j$ is sufficient and this also reduces the data of transmission to the control center.

LONG-LASTING ROBUST SENSING

Adapting to Changeable Illuminations

Line sensors are required to work for days and months. The sensing thus encounters different illuminations (from shining to cloudy, from day to night), object casting shadows, waving backgrounds (water, tree, rain, snow), and temporally static objects. Compared with a video sequence, line sensors are less influenced from complicated backgrounds because of the limited 1D view. However, the sensor still needs to deal with low contrast, salt noise, and ripple texture in the background, as well as unpredictable object lengths in the generated line sensor image automatically. The background in day and night must be adapted over time (Stauffer & Grimson, 1999; Li, Huang, Gu & Tian, 2003; Bhandarker & luo, 2005).

On the temporal scale, vehicles and people pass a location in a shorter time than changes of weather, as well as the shifts of a shadow from a building and a glass reflection from a static window. As exceptions, the glass reflection from a moving car can be treated as a dynamic pattern with the car. Shadows cast by clouds are usually blurred at their edges, either due to the scattering through transparent clouds or the diffraction from opaque clouds. Even if clouds move fast, their shadow boundaries are not as sharp as that from a building. This reduces the temporal differential of intensities from sunny to cloudy captured in the line sensor image. In addition, the camera auto-exposure system maintains the global illumination, which also diminishes the abrupt intensity change on the scenes. The instant temporal differentiation applied over several pixels (<150ms) for detecting people and vehicles is hence insensitive to weather changes. The key issue now is the update of background over time under changeable illumination so that the sensors can work robustly in thresholding dynamic objects from the backgrounds.

If dynamic objects passing a pixel on l_s are not overcrowded along the time axis, the majority of the intensities collected over a long period will be dominant by the background pixel. We can thus update each pixel in the background distribution $b(y)$ by temporal median filtering along the time axis (Huang, Yang & Tang, 1979; Zhao & Taubin, 2006; Choi, Lee & Ko, 1999). A static period T_s in the line sensor image can be obtained by (4) and updated by a median filter with a window width τ, i.e.,

$$b(t,y) = \text{median}(I(t_s,y)),\ t_s \in T_s[t,\ t-\tau] \qquad (7)$$

If objects passing the sampling line are not densely over time, we can even apply the median filter to the line sensor image directly without separating the background first. Dynamic objects are considered as outliers or noises to be removed in (7). The larger the window, the more stable the filtered result is. However, the sorting cost increases tremendously for a larger τ window. Many works have improved the computation order by dividing regions and using a quick selection algorithm, the computation complexity is O(τlogτ) for each time instance (Zhao & Taubin, 2006), which is still not affordable in real time processing on a sensor. We develop an efficient algorithm on temporal data in the following to achieve real time background updating over days and months, and avoiding propagating the computation load over the network to a server.

Real-Time Background Updating

Our median filtering has a constant computation order as O(1) irrelevant to the window size, which is an order equivalent to the order in (Huang, Yang & Tang, 1979) for 2D median filtering. Although the median filtering has to sort τ pixels at each moment, the windows at consecutive instances have overlapped duration as depicted in Figure 7a. Instead of processing these redundant data, our algorithm adjusts the median value only based on the newly included and excluded pixels. To take this advantage of the common data elements, we maintain an intensity histogram *hist*(*level*) for each height *y*, where intensity $level \in [0,255]$. Assume that the median pixel, i.e., the $(\tau/2)$th smallest pixel in the window, has intensity $k \in [0, 255]$, as is virtually sorted in Figure 7b. We consider *k* as the stable background intensity. In *hist*(*level*) shown in Figure 7c, the median pixel is larger than τ/2-1 pixels shown in gray. By adjusting the position of the median pixel in *hist* in accordance with the shifting window, we will find the background intensity changing over time. This adapting of median pixel saves the cost of sorting the entire sequence.

Denote the newly included pixel by $ad = I(t, y)$ and dropped pixel by $dp = I(t-\tau-1, y)$. Denote *S* as the number of pixels with intensities less than *k*, i.e.,

$$S = \sum_{level=0}^{level=k-1} hist\left(level\right) \text{ and } S < \tau/2 \qquad (8)$$

When the window moves a pixel forward, *ad* is added to and *dp* is dropped from *hist*. As shown in Figure 7d, if *ad* and *dp* are on the same side of position *k* in *hist* either left or right, the histogram change does not affect *S*. Only when

Figure 7. Median filtering of a line sensor image in a constant computation order. (a) Redundant data coverage of consecutive windows. (b) The median pixel (at τ/2) in the sorted sequence represent the majority of intensity values, k. (c) Intensity histogram in the window where the median pixel is at position k. (d) The histogram changes when the window moves one pixel forward, but the position of median pixel does not change. (e) The histogram change may influence the position of the median pixel only when ad and dp are on different sides of k.

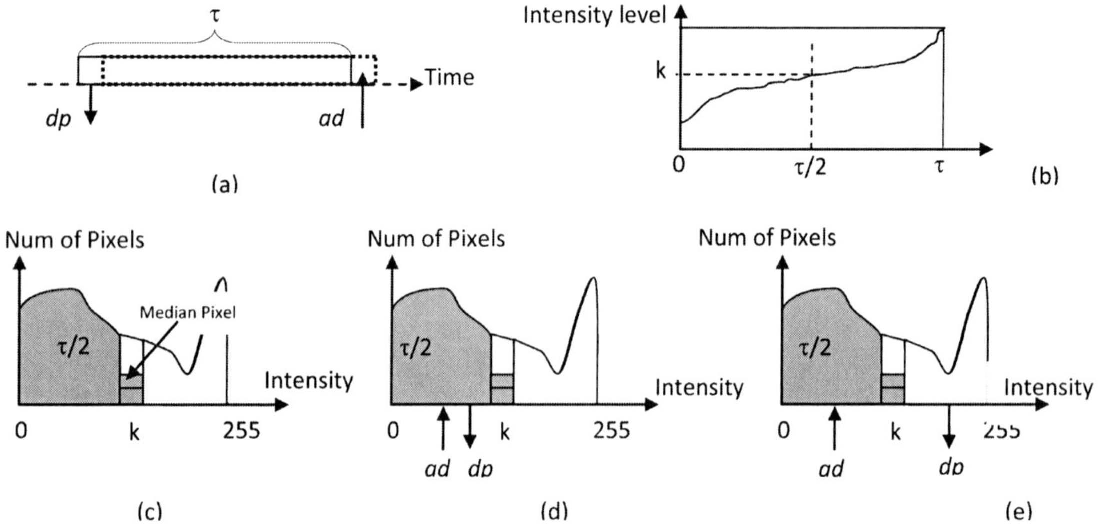

ad and *dp* are on different sides of *k*, *S* changes (Figure 7e), which may push the median pixel away from *k*. It may also keep the median pixel at *k*, if many pixels have intensity *k*. If the median pixel changes intensity, it shifts to the left if $ad<k$ and $dp\geq k$; and shifts to the right if $dp\leq k$ and $ad>k$. Based on analysis, we give pseudo code of median filtering algorithm in Figure 8. Initially, *hist*, and *S* are collected for τ lines after the sensor is turned on. It then updates the median pixel over time in response to the changes of histogram and *S*. For each instance *t*, updating *k* with a limited shift normally iterates once or at most several steps, because the background intensity will not change dramatically in real situation. Therefore, the computation order is O(1) for each instance, which is the best complexity obtained so far.

This algorithm is carried out for all heights of $I(t,y)$ at RGB channels separately to yield the background color at every time instance. The window width τ can be set very large (e.g., τ=600 pixels) to produce stable background. Figure 9 shows an example to extract dynamic vehicles under changeable illuminations. The background segments with the dynamic objects are extracted and composed. We can observe the background/illumination changes between periods.

Dealing with Other Background Events

This section deals with incidental stopping of objects at the sampling position. A vehicle may stay for a signal or in a traffic jam, and a person may stop for chat. If such a time is short, the target should still be considered as a dynamic object. Otherwise, it is merged into the background in order to separate other passing objects in front of it.

Figure 8. Algorithm for Median filtering in constant computational complexity

```
For i∈[0, 255], hist(i) = 0;
For( t ≤ τ) hist( I(t) ) ++;               // Initial histogram building
If(t == τ ){                               // Initial calculation of S
    k = 0; S = 0;
    while( S+hist[k] < τ/2 and k < 256 ) {
            S += hist[k];
            k++;
    }
}
If(t > τ) {                                // Background adapting over time
    hist[ad] ++;
    hist[dp] --;                           // histogram maintenance
    if( dp < k ) S --;
    if( ad < k ) S ++;                     // maintenance of S
    if( ad < k and k ≤ dp )                // possible shift of median pixel
        while(S≥τ/2 and k>0) {
              k --;
              S - = hist[k];
        }
    else if( dp ≤ k and k < ad )
        while( S+hist[k] < τ/2 and k < 256) {
              S += hist[k];
              k++;
        }
}
```

Figure 9. Dynamic object extraction under illuminations changing from sunny to cloudy in outdoor scenes

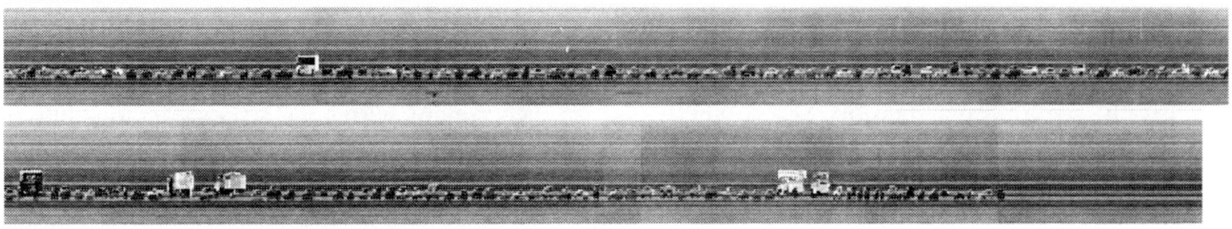

With the adapted background in previous section, a stopping object will be detected as a dynamic one if its staying duration is shorter than the half of the filtering window. Otherwise, it is identified as a background after $t_0+\tau/2$ where t_0 is the object arriving time. If the object moves out again, its remaining part not scanned by the sampling line is also extracted as the dynamic region. The detectable duration is also $\tau/2$ approximately. This effect informs that an object has detached from the background and has moved away.

To deal with background disturbance due to snow, ripples, waving leaves on trees, the extraction of dynamic objects is adapted as follows. The temporal differentiating filter is convolved with a large Gaussian filter $G_\sigma(t,y)$ and it extracts large features along the time axis. Although a 5×5 or 7×7 size filter increases the computation time, it is still in a complexity possible to handle by the sensor. Another operation on such a line sensor image is to take temporal accumulation of differential values over time. A small accumulation of $\Sigma_t \| I_t(t,y) \|$ indicates a background area, and a large value of $\Sigma_t \|I_t(t,y)\|$ identifies the region of object flow area.

SCALABLE LINE SENSOR NETWORK

Experiments

Our experiments were carried out with several video cameras connected to computers with image grabbers. If a line is not aligned with the image grid, the positions of the sampling pixels are calculated by varying integer y. For a non-integer position x in the image frame, its value is a weighted sum of two neighboring pixels according to the distances from it (Zheng, Zhou & Mili, 2006). The positions and weights are tabled for fast access. If only a single line is picked up in the camera frame for sampling, we always rotate the camera physically to align the line with the frame grid to speed up the sampling.

In the data processing, we collect temporal profile with a large circular memory $M[t,y]$, where $t \in [0,1000]$. The grabbed data is wrapped back to the beginning of M if the lines fill up M, i.e., $M[t,y] = I(t \bmod 1000, y)$. The horizontal filtering and pixel subtraction from (1) through (4) are all with the complexity of $O(1)$ for a fixed length h of the sampling line; the processing has no accumulative latency.

Only extracted segments of dynamic objects are transmitted from the computers to a central server via wired and wireless network. Degrading one dimension of the visual field can thus reveal targets in a complicated background without heavy processing, which is particularly suitable for sensor network. A low-end computer (for example, 400MHz CPU) can undertake the sensor data processing. This flexible sensor network can cover various spots in a large area. A sensor node can be added in a room for capturing an object flow at a distant street through the window, if the camera zooms up scenes.

A real line sensor available on the market has a much higher sampling rate up to 1000Hz, which widens the temporal dimension of line sensor image and increases the granularity of registration for the time of arrival. We use video cameras to serve as line sensors and read data on a sampling line at the rate of 60Hz, which is adequate to capture walking people at distance.

In a dim night, line sensors only capture headlights of vehicles and leave bodies undetected. The threshold for recognizing an object is lowered down. Similarly, Figure 10 senses a low intensity distribution in a snow day. As an example, Figure 11 further shows a rippled background in the line sensor image taken at a riverbank. Also, the line sensor can be applied to infrared cameras for round-the-clock monitoring at borders. The obtained image would have relatively dark intensities on backgrounds and bright intensities on people, vehicles (particularly on wheels), and other heat sources. The thresholds for the dynamic object detection should be set to deal with the data from the infrared CCD arrays. Figure 12 shows an experiment with four sensors facing indoor and outdoor environments in a campus to count people. The computers send detected targets to the server by socket connections (TCP/IP based) via WiFi network and standard Ethernet. The image sections ($y \in [0,h]$) containing passing objects are displayed on the server and the number is counted.

We have also tested the transmission of dynamic object occupied regions only, which are a fraction of the entire vertical lines in many figures displayed in this paper. A simple run-length coding is applied to separate the static and dynamic pixels in each vertical line. The upper bound of the transmission rate of dynamic objects is

Figure 10. Detecting dynamic vehicles and persons in a dim and snow day. (a) Field of view by a zooming camera to three streets more than 200m away. Dynamic objects pass a sampling line at different distances. (c) Screen display of progressive line sensor image where three groups of car shapes are visible along three horizontal belts, and background updating and vehicle extraction in real time.

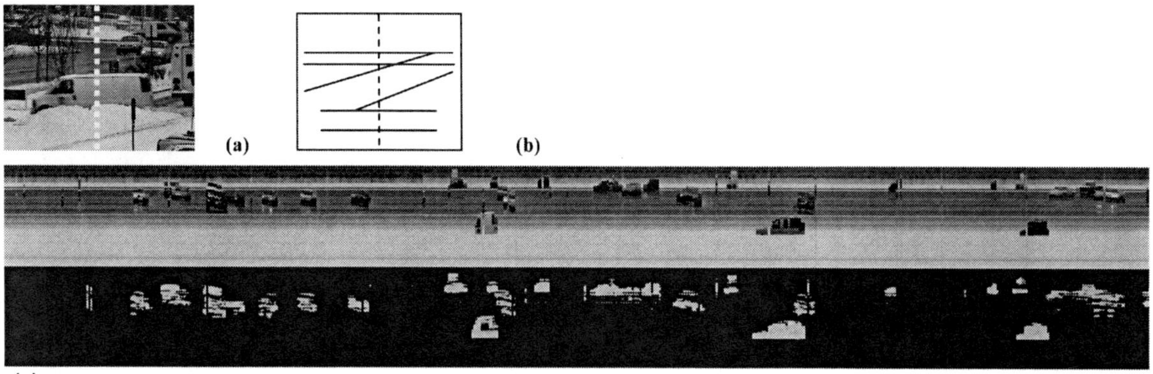

(c)

Figure 11. Line sensor monitoring a river to find flooded objects. top: Riverbank and ripples in strong wind, bottom: mild ripples at another location of the river.

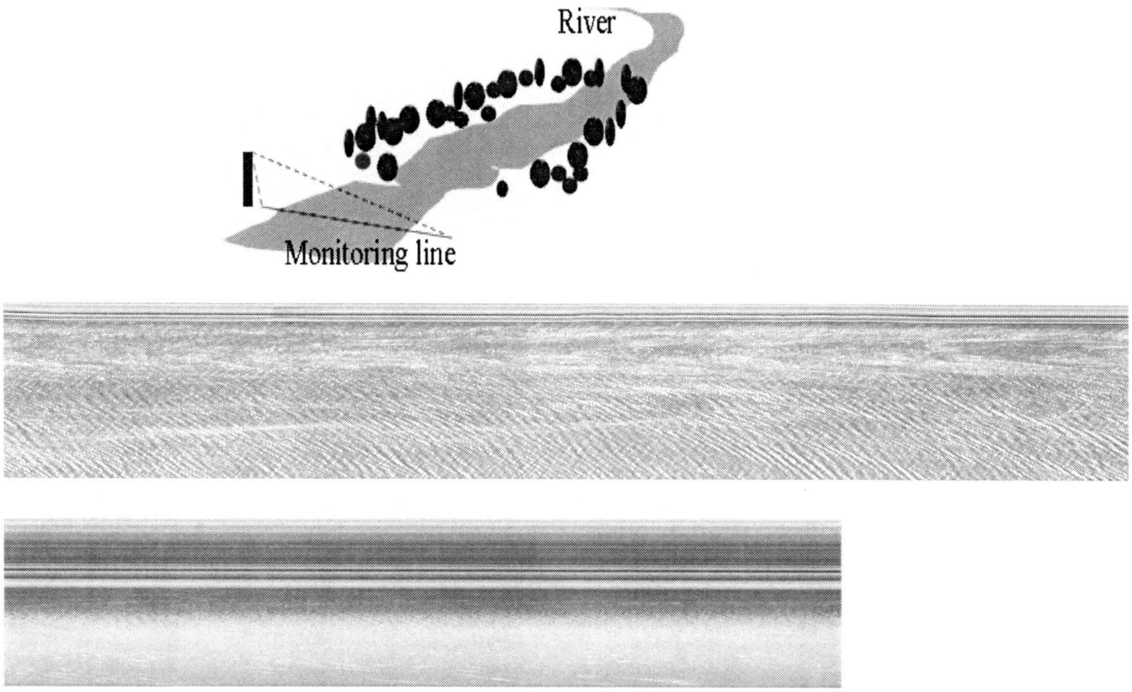

Figure 12. Line sensor network monitoring people through several locations. (Top) Views of four locations from left to right: (i) outdoor-shadow, (ii) indoor-window, (iii) outdoor-sunny, (iv) indoor under constant illumination. (Lowers) Sections containing dynamic targets are extracted and sent to a server, (i) A temporary stopping person is visible in transmitted profile, (iii) The principle direction for the sampling line is set horizontal over a road. The persons walking direction can be inferred according to the order of head-and-foot along the time axis in the line sensor image. (iv) Back-and-forth movements across the sampling line are visible.

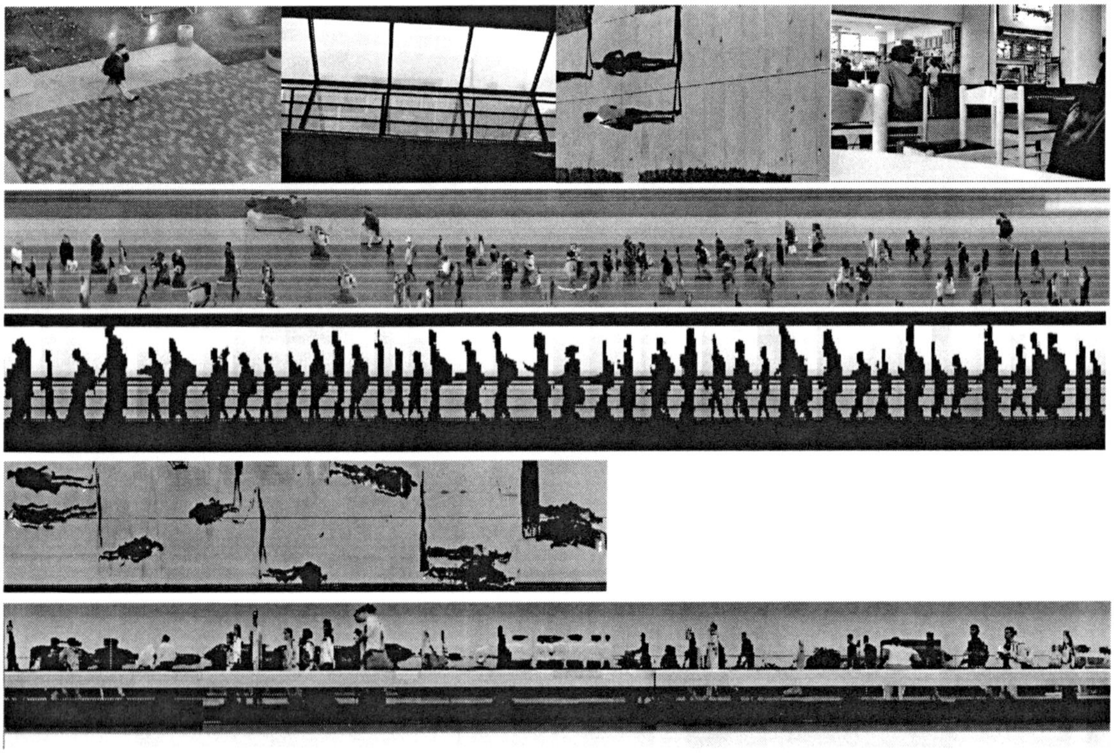

approximately 340KB for each sensor, and our experiment shows that the average data flow after dynamic region extraction is less than 100KB.

Frequency based data compression such as JPEG compression is also considerable for transmitting dynamic object occupied region in the line sensor image. However, this adds more computation to the sensor and may slow down the sampling and processing. Because the line sensor has already reduced the data size significantly from a video, the further drop of frame rate is unfavorable to quality data visualization.

After the sampling rate of sensors is fixed, the duration of passing objects determines the network traffic load. For a border surveillance where occasional penetrating happens, a sensor is silent in most of the time and the number of sensors in a WSN thus can be increased to cover a large area. On the other hand, for theme parks or stores, sensors monitoring crowds at a gate or path may constantly send out data. Nevertheless, the sensors at rest of the locations may be idle. As a result, the data transmitted on the network are proportional to the total number of targets through guard lines and gates at an instance, which is further a percentage of the total population in the area.

Based on the above estimation, a network with sensors for alarming invasion can be scaled

Figure 13. Sensor network with distributed line sensors at a large event site. Lines cross roads or channels for monitoring vehicles (yellow line) and people (black line) by line sensors are indicated.

largely in area and number of sensors. The traffic prediction of the wireless sensor network in a crowded site (museum, store, theme park, etc) only needs to consider the total number of customers. Figure 13 shows the distribution of sensors in an area for passing traffic and people, and Figure 14 is the diagram to set sensors indoor and outdoor for data collection.

Discussion

To extend the line sensor network to a large area, we examine the possible data size for transmission. The first requirement in setting a line sensor is to cover a guard line through which a planned or incidental object movement will be detected. It can at least trigger an alarm and count the passing objects. The second objective is to show or record brief shapes of targets, which further depends on object passing speeds, sampling and image processing rate p (frame/sec) of the sensor. The ideal selection of sampling and processing rate is r in order to preserves the rough aspect ratio of objects and avoids the *overlapped sampling*. If the reachable rate of the system is lower than the *just-sampling* rate, the possible tradeoff is either sacrificing the shape but keeping the object counting function, or zooming out the camera in order to reduce the object size and resolution. The zooming out approach also transmits less data of

Figure 14. Model of line sensor network in critical facilities. Circles indicate sensors at certain heights and arrows indicate the monitoring scopes of the sensors. Dotted line indicates the transmission between sensors and wireless access points.

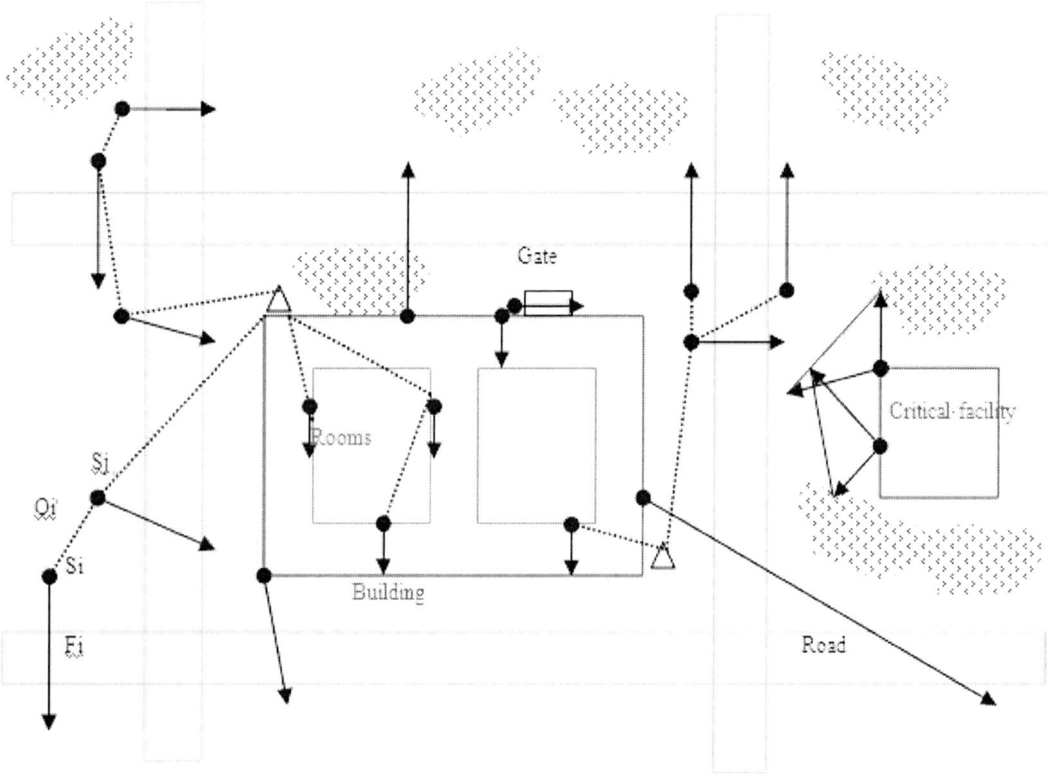

passing objects because the object heights are also reduced beside the narrowed widths.

Table 1 summarizes attributes of the line sensors as compared to signal and image sensors. Besides the advantages mentioned so far, the disadvantage in using line sensors are as follows. First, the line sensor does not provide the trajectory of a moving target as what can be tracked in a video. The line sensor achieves its efficiency by ignoring the redundant data. Hence, it scarifies the repeated shapes for tracking in the time domain. This drawback can be partially improved if two separated sampling lines can be set to check the flow. By matching targets in the generated line sensor images and detecting the times of arrivals, target traces can be figured out qualitatively. Second, a line sensor image may contain a special deformation on the object with a large depth, in case that the sampling plane is not orthogonal to the motion vectors. However, the object shapes are invariant in the front face, as long as the sampling line is along our defined principle direction. Nevertheless, a perspective image can be triggered to record the correct object shapes when a dynamic event is detected by the sampling line. The discrete images can be sent to the server afterwards in the interval when no passing objects are transmitted. Third, if the sampling rate of the line sensor is low using a normal camera, the line sensor image is under-sampled; it may yield imperfect shape of the targets. The sampling rate for real physical line sensor is sufficiently high to capture complete shapes of passing objects.

Table 1. Media of different dimensions and their delivered information

	Signal sensor	Line sensor	Image sensor
Instant output	One element, scalar	An array of *n* elements	A 2D image with n^2 elements
Total amount of data	Signal profile	Composite 2D image	Video, image sequence, data volume
Media information	Signal level change, time, strength, frequency	In addition to left, contains imperfect shape, speed, object identity information	In addition to left, contains complete shape, trajectory, motion information
Transmission and computation load	$t \times C_1$, less bandwidth demanded	$n \times t \times C_2$, extracting compact information, loadable on current network infrastructure	$n^2 \times t \times C_3$, heavy transmission load, requiring high compression rate and new generation of network
Information Quality	Robust but not intuitive	Partially accessible, brief profile, shape suffers from fast moving objects	Precise, completely visible, all types of motion, except motion blur on a fast moving target

CONCLUSION

This chapter explored intrinsic properties of line sensors for establishing a sensor network and introduced efforts to achieve robust and less deformed sensing under various environments and setting. We have achieved target extraction and background updating in algorithms with constant complexity, and transmitted dynamic data from multiple sensors to a server via wired/wireless Internet in real time. Although line sensors have weakness on image quality and acquiring limited types of object motions, i.e., mainly translation, the reduction of one dimension of views abstracts passing objects in a profile and this greatly reduces the data to compute and transmit in constructing a sustainable visual sensor network. The dynamically generated profile contains the basic information on passing targets in time and shape. The sensor has a potential to be built as low cost sensor nodes that can be widely distributed in an area. Such line sensor based network can achieve many tasks currently implemented by video cable surveillance systems such as alarming, target counting, flow recording, speed measuring, and dynamic event retrieving, while requires less infrastructure and bandwidth. The new sensor network architecture leads to substantial reduction of the current, prohibitive cost of wireless visual sensor network deployment. This cost reduction is essential for the deployment of large-scale monitoring and surveillance network for civic, national security, and military purposes.

REFERENCES

Bhandarkar, S. M., & Luo, X. (2005). Fast and robust background updating for real time traffic surveillance and monitoring, In. *Proceedings of the IEEE Conference on Computer Vision and Pattern Recognition, 2005*, 22–26.

Bolles, R. C., Baker, H., & Marimont, D. H. (1987). Epipolar-plane image analysis: An approach to determining structure from motion. *International Journal of Computer Vision, 1*, 7–55. doi:10.1007/BF00128525

Campbell, J., Gibbons, P., & Nath, S. (2005). Iris-Net: An internet-scale architecture for multimedia sensors, In *Proceedings of the 13th Aannual ACM International Conference on Multimedia* (pp. 81-88). New York, NY: ACM.

Chitnes, M., Liang, Y., Zheng, J. Y., Pagano, P., & Lipari, G. (2009). International Workshop on Modeling Analysis and Simulation of Wireless and Mobile System, In *Proceedings of the 6th ACM Symposium on Performance Evaluation of Wireless Ad Hoc, Sensor, and Ubiquitous Networks* (pp. 71-78).

Choi, K., Lee, J., & Ko, S. (1999). New autofocusing technique using the frequency selective weighted median filter for video cameras. *IEEE Transactions on Consumer Electronics*, *45*(3), 820–827. doi:10.1109/30.793616

Dalaff, C., Reulke, R., Kroen, A., Kahl, T., Ruhe, M., & Schischmanow, A. (2003). *A traffic object detection system for road traffic measurement and management* (pp. 78–83). Tuchscheerer, W.: Image and Vision Computing NZ.

Feng, W., Code, B., Kaiser, E., Shea, M., Feng, W., & Bavoil, L. (2003). Panoptes: Scalable low-power video sensor networking technologies, In *Proceedings of the 11th ACM International Conference on Multimedia* (pp. 562-571). New York, NY: ACM.

Gupta, R., & Hartley, R. (1997). Linear pushbroom cameras. *IEEE Transactions on Pattern Analysis and Machine Intelligence*, *19*(9), 963–975. doi:10.1109/34.615446

Gupte, S., Masoud, O., Martin, R. F. K., & Papanikolopoulos, N. P. (2002). Detection and classification of vehicles. *IEEE Transactions on Intelligent Transportation Systems*, *3*(1), 37–47. doi:10.1109/6979.994794

Huang, T. S., Yang, G. Y., & Tang, Y. (1979). A fast 2-dimensional median filtering algorithm, *Acoustics, Speech, and Signal Processing. IEEE Transactions on*, *27*, 13–18.

Koschan, A., Page, D., Ng, J. C., Abidi, M., Gorsich, G., & Gerhart, G. (2004). SAFFR under vehicle inspection through video mosaic building. *International Journal of Industrial Robot*, *31*(5), 435–442. doi:10.1108/01439910410551854

Li, L., Huang, W., Gu, I., & Tian, Q. (2003). Foreground object detection from videos containing complex background, In *Proceedings of the 11th ACM International Conference on Multimedia* (pp. 2-10). New York, NY: ACM.

MacDorman, K. F., Nobuta, H., Koizumi, S., & Ishiguro, H. (2007). Memory-based attention control in a distributed vision system that recognizes group activity at a subway station. *Multimedia, IEEE*, *14*(2), 38–49. doi:10.1109/MMUL.2007.39

Nakanixhi, T., & Ishii, K. (1992). Automatic vehicle image extraction based on spatio-temporal image analysis, In *Proceedings of the 11th IAPR International Conference on Pattern Recognition* (pp. 500-504).

Rademacher, P., & Bishop, G. (1998). Multiple-center-of-projection images, In *Proceedings of the 25th Annual Conference on Computer Graphics and Interactive Techniques* (pp. 199-206). New York, NY: ACM.

Song, D., & Goldberg, K. (2003). The co-opticon: Shared access to a robotic streaming video camera, In *Proceedings of the 11th ACM International Conference on Multimedia* (pp. 104-105). New York, NY: ACM.

Stauffer, C., & Grimson, W. E. L. (1999). Adaptive background mixture models for real-time tracking, In. *Proceedings of the IEEE Computer Society Conference on Computer Vision and Pattern Recognition, 1999*, 246–252.

Taniguchi H., Seki A., et al. (1994). A method of motion analysis using spatio-temporal image, *IEICE (Institute of Electronics, Information, and Communication Engineering), Transactions of, J77-D-II*(10), 2019-2026.

Yoshioka, T., Nakaue, H., & Uemura, H. (1999). Development of detection algorithm for vehicles using multi-line CCD sensor, In. *Proceedings of the International Conference on Image Processing, 1999*, 21–24.

Yu, J., & McMillan, L. (2004). General Linear Cameras. In. *Proceedings of the European Conference on Computer Vision*, *2*, 14–27.

Zhao, Y., & Taubin, G. (2006). Real-time median filtering for embedded smart cameras, In *Proceedings of the IEEE International Conference on Computer Vision Systems (ICVS '06)* (pp. 55-55).

Zheng, J. Y. (2003). Digital Route Panorama. *IEEE MultiMedia*, *10*(3), 57–68. doi:10.1109/MMUL.2003.1218257

Zheng, J. Y., & Shi, M. (2006). Removing temporal stationary blur in route panorama, In *Proceedings of the 18th International Conference on Pattern Recognition* (Vol. 3, pp. 709-713).

Zheng, J. Y., & Tsuji, S. (1989). Spatial representation and analysis of temporal visual events. *IEEE International Conference on Image Processing '89* (pp. 775-779).

Zheng, J. Y., & Tsuji, S. (1990). From anothoscope perception to dynamic vision. *IEEE International Conference on Robotics and Automation 1990* (pp. 1154-1160).

Zheng, J. Y., & Tsuji, S. (1998). Generating dynamic projection images for scene representation and understanding. *Computer Vision and Image Understanding*, *72*(3), 237–256. doi:10.1006/cviu.1998.0678

Zheng, J. Y., Zhou, Y., & Mili, P. (2006). Scanning scene tunnel for city traversing. *IEEE Transactions on Visualization and Computer Graphics*, *12*(2), 155–167. doi:10.1109/TVCG.2006.37

Zhu, Z., Xu, G., Yang, B., Shi, D., & Lin, X. (2000). VISATRAM: A real-time vision system for automatic traffic monitoring. *Image and Vision Computing*, *18*(10), 781–794. doi:10.1016/S0262-8856(99)00046-3

Chapter 13
Distributed Visual Surveillance with Resource Constrained Embedded Systems

Mangesh Chitnis
Scuola Superiore Sant'Anna, Italy

Claudio Salvadori
Scuola Superiore Sant'Anna, Italy

Matteo Petracca
Scuola Superiore Sant'Anna, Italy

Paolo Pagano
Consorzio Nazionale Interuniversitario per le Telecomunicazioni, Italy

Giuseppe Lipari
Scuola Superiore Sant'Anna, Italy

Luca Santinelli
Scuola Superiore Sant'Anna, Italy

ABSTRACT

Wireless Sensor Network (WSN) is a promising technology for distributed sensing and computation. Recently, researchers and engineers started to investigate the use of WSNs to support resource demanding real-time applications, including process control, industrial automation, video surveillance, and multimedia streaming.

Surveillance applications – like traffic monitoring, vehicle parking control, and intrusion detection – involve monitoring of the environment in order to detect and interpret the activities of relevant objects and their behaviour. These applications demand real-time images from the scene in order to collect relevant information. Multimedia Wireless Sensor Network (MWSN) provides a low-cost and flexible solution for distributed video surveillance based on low-power autonomous camera sensor nodes.

MWSN faces the challenge of reliably distributing complex imaging data over low bandwidth and resource constrained computational nodes. To deal with this problem, many researchers have proposed techniques like image compression, in-network processing, distributed statistical correlation and image prediction. However, these techniques still have to store and process huge amount of visual data on energy and resource constrained sensor nodes.

DOI: 10.4018/978-1-61350-153-5.ch013

In this chapter, we present an innovative technique of line sensor based image capturing and processing in order to detect moving objects such as vehicles. Line Sensor techniques, when used in MWSN, may achieve faster processing results with much less storage and bandwidth requirements while conserving node energy. Line Sensor based processing algorithms provide novel ways for object counting, classification and speed measurement. This solution presents itself as an ideal low-cost candidate for Intelligent Transport Systems (ITS) to monitor and control urban traffic.

INTRODUCTION

WSN is a promising technology for distributed sensing and computation. Although constrained by limited computational capabilities, sensor networks are cost-effective and have good scaling virtues. The use of a low-power radio communication protocol such as the IEEE 802.15.4 allows to eliminate the cost of cabling the sensors, and allows more flexibility in the deployment. So far, WSNs have been mainly used in applications with low-frequency sampling and little computational complexity, such as environmental monitoring in agriculture, monitoring and control of the temperature and light in home automation, etc.

Recently, researchers and engineers started to investigate the use of WSNs to support more demanding applications, including process control, industrial automation, video surveillance, and multimedia streaming. Some early attempts have been made to use WSN technology for low-quality video streaming (Kulkarni, Ganesan, Shenoy & Lu, 2005). The success of such attempts have created a new challenging area of research, Multimedia Wireless Sensor Networks (MWSN) (Akyildiz, Melodia & Chowdhury, 2007).

MWSN technology is a good candidate for use in pervasive contexts like info-mobility. The idea is to use a set of inexpensive sensors nodes equipped with low-cost cameras in city streets to monitor traffic flows, number of cars in parking lots, etc.. The information are collected by the sensors and sent through wireless communication to a *concentrator node* that aggregates the incoming data (sensor fusion). The concentrator, which is in charge of collecting data related to a specific city area, is then connected to a wide area network, and sends aggregated data to higher levels of the information system hierarchy, which monitor and control urban mobility.

The architecture of such a system is shown in Figure 1. The backbone network can be wired (e.g. ADSL or Ethernet) or wireless (e.g. WiFi, WiMax, GPRS or UMTS), depending on the specific needs. In this proposal, we are interested in the lower level of the hierarchy that includes the sensor nodes and the concentrator.

Examples of applications are:

- **Counting the number of cars passing on city streets.** This information can help in estimating the traffic flow entering in (or exiting from) a city area, and take appropriate actions to prevent congestion. While there are other methods to count the number of cars passing in a street (by magnetic sensors, photoelectric cells, etc.), we envision that the one based on WSN cameras will be more flexible, easier to install and more cost-effective.
- **Identifying the occupancy level of a parking lot in a open area.** This information can help to provide appropriate advices to drivers and guide them to the free spots. Also in this case there are many working examples of parking lots equipped with sensor systems. However, these systems are more difficult to be installed in open environments, due to high cabling cost. The use of WSN will help to lower the cost of installation and maintenance.

Figure 1. General System Architecture

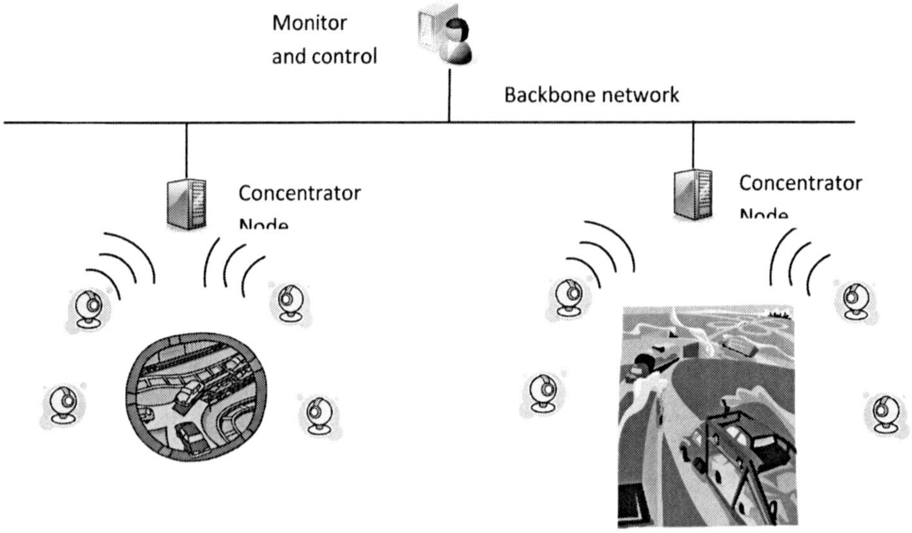

- **Counting number of people entering a building or in a room, or detect intrusion into limited access areas.** Such information can be exploited for marketing purposes (e.g. how many people pass by a corridor) or scheduling personnel (i.e. how many employees are needed at a particular time of the day). A typical solution is an embedded unit with a single counter and camera integrated. The use of WSN could increase the system robustness, especially in challenging scenarios where the error rate goes up to 30%.
- **Counting people outdoor is a state-of-the-art challenge and no product is doing that reliably these days.** The information is useful for public authorities (e.g. town-hall, police) that are interested in to distributing police agents and schedule their personnel. The use of WSN in outdoor counting will help lower main technical problems associated to this scenario, like harsh illumination conditions, weather, occlusions due to crowds, etc.
- **Monitoring industrial processes.** Video cameras are being used in industrial processes to identify defects in products, or to monitor a hazardous area. Using less expensive wireless cameras would simplify this tasks and allow more cameras to be installed with less effort and inferior cost with respect to current monitoring systems. Also, a real-time architecture for WSN would allow to *control* the manufacturing process, and de-localize control algorithms, to make them closer to the sensors, thus reducing the overall cost.

Such applications can be denoted as Visual Monitoring (VM) applications, since they are characterized by the use of vision to monitor some environmental data of interest.

STATE OF THE ART

The success of WSN in traditional applications handling scalar data has motivated many researchers to look in the direction of exploiting WSN for data and computation intensive applications such as multimedia. In last two to three years we have seen few research labs exploring the use of multimedia in WSN with help of some indigenous ways for using hardware platform and software architecture. Most of their work is related to building image sensor motes, distributed imaging wireless systems, image data compression and middleware support for distributed imaging and database applications.

Reference (Kulkarni, Ganesan, Shenoy & Lu, 2005) presents a multi-tier heterogeneous camera-based surveillance network called SensEye. It makes use of low fidelity camera sensors at the second tier to view an object and comparatively higher fidelity cameras at tier three to perform object tracking. IrisNet (Internet-scale Resource-Intensive Sensor Network Services) (Campbell, Gibbons, Nath, Pillai, Seshan & Sukthankar, 2005) is a heterogeneous wireless multimedia sensor network platform, which allows the user to perform queries over the video sensors distributed worldwide. Another multi-tier wireless multimedia sensor test bed (Wireless Multimedia Sensor Testbed, 2010) is deployed at Broadband and Wireless Networking (BWN) laboratory in Georgia Tech which makes use of scalar sensors, CMOS based camera nodes, medium-end video sensors based on Logitech webcams interfaced with Stargate platforms and pan tilt cameras installed on a robotic platform.

Sensor Andrew (Sensor Andrew, 2010) is a campus wide infrastructure deployment of heterogeneous WSN motes for sensing and control at Carnegie Mellon University. They make use of CMUcam3 (CMUcam3 datasheet version 1.02, 2007) vision sensors for computer vision applications like people tracking. Many similar test beds have been deployed at University of California, Irvine (SAFIRE, 2010; Situational Awareness from Multimodal Input, 2010) for rescue applications for giving more situational awareness using vision sensors.

These multimedia wireless sensor network test beds make use of camera-based modules either having a separate processing unit for imaging algorithms, or that are directly connected to the wireless mote where the processor is shared between networking and imaging tasks. These camera nodes make use of CMOS based cameras such as Cyclops (Rahimi, Baer, Iroezi, Garcia, Warrior, Estrin & Srivastava, 2005) having an 8-bit Atmel ATMega 128L CPU and CMUcam3 at a lower tier and Logitech webcams for high end image processing at the next higher tier. Panoptes (Feng, Kaiser, Feng & Baillif, 2005) is an example of camera sensor node where a relatively fast Stargate (Intel XScale PXA255) processor is connected to a USB camera and an 802.11b wireless card.

Stanford MeshEye (Hengstler, Prashanth, Fong & Aghajan, 2007) mote is a smart camera architecture developed for distributed intelligent surveillance. It is used to determine the position, range and size of moving objects. The platform is equipped with a low resolution VGA camera module and a CC2420 transceiver for wireless communication. A new wireless camera vision system is developed using two independent VGA color sensors on the same mote (Kleihorst, Schueler, Danilin & Heijligers, 2006) in which the two cameras view the same scene from different viewpoints and construct a high performance surveillance system using an onboard video analysis processor, 8051 microcontroller and IEEE802.15.4 enabled transceiver. Reference (Itoh, Kawahito & Terakawa, 2006) describes a tiny CMOS-based single chip sensor of size less than 5mm on a side for biomedical applications which consists of a 320 * 240 pixel array and a radio module for communication. On the other hand, motes like Citric (Chen et al., 2008) and WiCa (Kleihorst, Abbo, Schueler & Danilin, 2007) make use of an auxiliary processor attached to the camera for image processing applications.

The wireless image/video sensor network platforms mentioned above are all based on traditional two-dimensional image processing for object detection and tracking. In (Pagano, Mastasi & Liang, 2008), Pagano et al described the communication bottleneck problem in a two-dimensional image processing node. Zheng and Sinha (Zheng & Sinha, 2007) explored the use of line camera sensors in wired communication using high end video cameras. Chitnis et al. (Chitnis, Liang, Zheng, Pagano & Lipari, 2009) showed a working model of line sensor based imaging using a wireless sensor mote attached with a CMOS camera and CC2420 radio for networking.

Issues in VISUAL Wireless Sensor Networks

A WSN is a distributed network of autonomous processing units (nodes) responsible for sensing and monitoring the environment. The primary goals of a WSN are scalability, energy conservation with efficient and relevant data gathering and fusion. Towards these goals it is very important to deploy cost effective nodes with limited hardware resources and communication bandwidth.

Traditionally, WSNs are used to monitor and gather information on temperature, pressure, lighting in the surrounding environment. Such applications do not demand heavy computation and network bandwidth. Hence it is sufficient to use the existing nodes with the 250kbps bandwidth (IEEE 802.15.4 standard) for such applications.

WSNs present an engineering challenge when used in applications which demand the analysis and processing of huge amount of data in resource (processing, memory and bandwidth) limited platforms. An example of such applications is surveillance and monitoring of the environment using camera nodes and processing the acquired images.

Due to the inherent huge size of raw images, it is impossible to transmit these raw images from the place of occurrence of a certain surveillance event to the sink node in a WSN, because of the limited available bandwidth. Another limitation is that data transmission is considered to be the most significant contributing factor towards the energy consumption in a node. On the other hand, in order to transmit a processed image, the node requires significant amount of memory to store consecutive images and apply some processing algorithms to retrieve and transmit only the useful information. We now report some practical examples of such problems.

Problem 1 (Storage): A typical low-cost sensor node has a RAM size of around 30KBytes which is shared among user applications, operating system, network and device drivers.

A simple gray scale raw image of resolution 160 * 120 requires:

$160*120 = 19200$ bytes ≈ 19 Kbytes.

With this image size and memory constraints it is impossible to perform a simple image processing tasks (e.g.: object or motion detection) which requires storing at least two raw images.

Problem 2 (Communication Bottlenecks): If the node tries to avoid local processing by transmitting the entire raw image, it will suffer from low frame rate due to hardware bottlenecks. In Figure 2, we show the typical setup of a sensor network in which a node communicates to a base station via IEEE 802.15.4, highlighting all communication among the different subsystems: a data bus to connect the camera to the node processor; a serial line between the radio and the processor (again, typically 10 Mhz); the network bandwidth, limited to 250kbs; the serial line between the receiving node and the base station.

In this setup the bottleneck is given by the wireless MAC protocol, which limits the rate at which the data can be transmitted. This in turn limits the final frame rate of the image sensor. As a consequence, it is likely to miss a fast moving object whose speed is more than the effective frame rate for a wireless image sensor node.

One of the ways to deal with this problem is to use image compression algorithms like jpeg and

Figure 2. A typical flow of image data from the image sensor to the base station

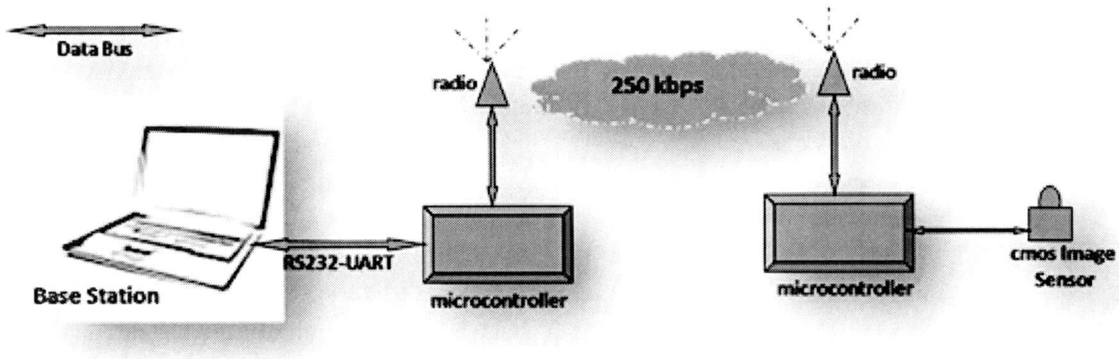

mpeg. The drawback in using this approach is the heavy computation load due the transformation process involved in compression. The compressed image has to be decompressed every time there is a need to retrieve some information over the raw image.

Depending on the accepted quality of the original image, jpeg compression is able to achieve a compression ratio varying from (2 – 100): 1. This translates to having a minimum compressed image size of 1.9 kbytes for a single image of resolution 160 * 120 pixels. Even this size places a heavy communication load over standard WSN applications, because one packet over 802.15.4 is 127 bytes, of which a maximum of 108 bytes are for payload. This means that no less than 19 packets are required to transfer one compressed image, for a total of approximately 2.3 kbytes.

Problem 3 (Computational Bottlenecks): The sensor nodes have limited clock speed (in some cases of the order of tens of MHz), limited Instruction Set Architecture (ISA), usually not including a Floating Point unit block. Therefore, the amount of complex calculations required for image processing that can be performed on a frame is severely limited.

These problems prevented us from using traditional methods of image capturing and processing in a WSN. Therefore, it is necessary to completely change perspective on the sampling and processing methods.

Line Sensors

The previous section described the unfeasibility of using classical image processing algorithms, in which images are captured and communicated among sensor nodes for low-cost sensor nodes with severely limited memory. The unfeasibility arises due to the size of image data.

Line Sensor based image capturing and communication provides an efficient computation and communication technique for WSN. The technique consists in extracting a single array from the two dimensional matrix of pixels. To reduce sampling time, the camera designed (or an existing camera can be adapted) to retrieve an image containing one single row. This line image could be aligned along the width or height of the captured image frame giving an image of resolution (width * 1, horizontal image) or (height * 1, vertical image). This paves the way to achieve a compression ratio of the order of (width | height): 1.

An analogy to understand this concept is the technique of scanning a document with a standard document scanner. Scanner devices belong to the family of line sensors, since a scanner periodically captures a single line of the target document at a time. The scanner then moves to the next line along the height of the document. If the scanner scans the document along its height then the complete document is ready at time instance δt * *height*, where δt is the time interval between successive scans, and *height* is expressed in pixels

The line sensor technique for image capturing works in a similar way. However, we are not interested in capturing a static scene, but rather in detecting moving objects. For this reason, instead of continuously moving the camera to capture the entire reference scene, we fix the camera position always capturing the same image of width (or height) equal to 1 pixel. In this case, the dynamic part of the scene refers to moving objects while the scene background remains constant. A direct effect of this phenomenon is that it is only possible to capture moving object, whereas everything static is filtered as being background. Thus, line sensor technique inherently provides support for moving target detection with minimal image processing, thus saving on computation cost.

PRINCIPLES OF LINE SENSOR

The line sensor extracts a line from the two dimensional image retrieved by the camera module. This is equivalent to focusing a sampling line (L) in the field of view (P). A continuous projection of this sampling line will result in capturing a single line over the focused plane. Any static object in (L) will appear as part of the background, and hence cannot be detected. The continuous projection of the sampling line will capture consecutive portions of an object at different time instants when the object moves across the plane of sight as shown in Figure 3.

An effective scheme to capture all the details of the moving object depends on:

1. The projection of sampling line and the moving direction of the object. The camera focus and the sampling line will form a plane of sight within the focal distance of the camera. In order to capture the shape of an object moving in direction V, the plane of sight and the object movement must satisfy the following condition: $V \cdot n \neq 0$; where n is a normal to the plane of sight in the 3D space.

Figure 3. Constructing a temporal view with line sensors

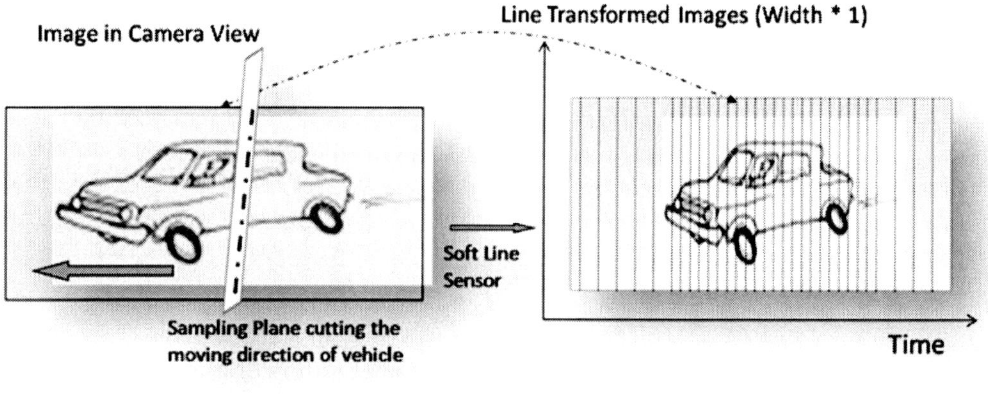

This means that the object should not move in a direction parallel to the plane of sight or the direction of line. This restriction shows that there is one specific direction of a moving object which cannot be captured with a given camera alignment. We can resolve this problem using a WSN based multi-vision system to align the camera along the three basis orthogonal vectors in a 3D space.

2. A high sampling rate at which the sampling line is projected.

The rate at which the sampling line is projected is very important to capturing the moving object. If the sampling rate is too low, it might miss an object altogether. On the other hand, a too fast sampling rate would replicate the same object portion. A proper sampling rate depends on the length and the velocity of the moving object assuming that the previous line sensor alignment condition is satisfied.

This sampling rate places an additional timing constraint on the application. The image processing and communication task needs to be finished before the next sampling period, otherwise either the sampling line has to be skipped or the processing tasks have to be terminated.

SOFTWARE BASED LINE SENSOR

A hardware-based line sensor capable of capturing one-dimensional image for high-end applications is typically very expensive. In addition, such sensors are not robust enough to be deployed in an ad-hoc manner under harsh environment conditions.

A software-based line sensor can be realized by using a popular and inexpensive two-dimension image sensor such as HV7131GP (CMOS Image Sensor with Image Processing HV7131GP, 2010).

Figure 4 displays the hardware setup for the software-based line sensor. The CMOS camera module of the HV7131GP is able to capture the image at the maximum resolution of 640*480 pixels. This image is however a two dimensional image stored within the internal memory of the camera. This camera module has a feature which allows the microcontroller to select the desired width and height of the image. The camera module only transfers the selected part of the image from its internal buffer.

Figure 4. Microcontroller camera interface

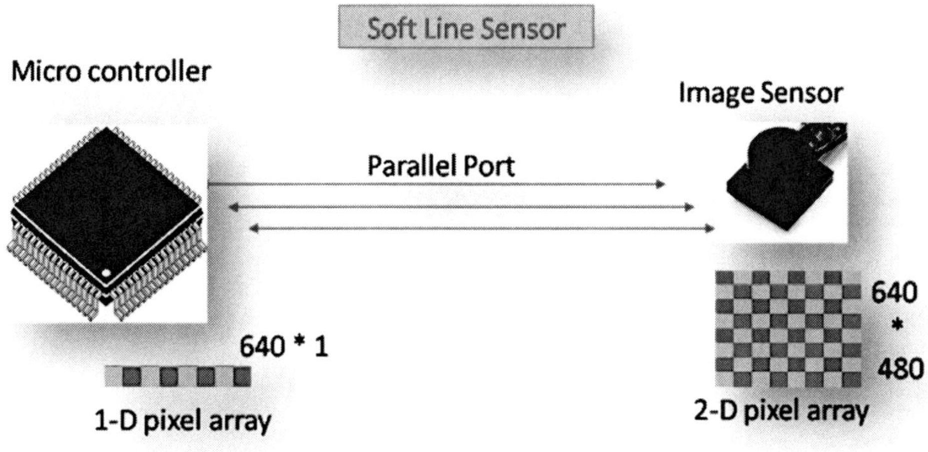

In order to obtain line images, the microcontroller selects the desired height of the image to be equal to one pixel only. This approach not only effectively realizes a software enabled line sensor for Wireless Line Sensor Networks, but also provides us with the flexibility to obtain either one or two dimensional images on demand and online in the same framework depending on the application requirement.

ONE DIMENSIONAL IMAGE PROCESSING ALGORITHMS

A suite of low-complexity algorithms for target analysis operate on line images obtained with software-based line sensors. These algorithms can detect moving objects whose speed is within the limits set by the sampling rate of line images. Collected one dimensional images can be processed locally by the sensor node, or transmitted to the base station for temporal reconstruction of the image (as shown in Figure 3).

Figure 5 depicts an example of a possible line sensor deployment in a WSN. A CMOS camera sensor communicates with the sensor node using a command/data bus. The sensor node has a Central Processing Unit (CPU) which is in charge of handling communication and imaging tasks. The result of the image processing is then forwarded via a wireless network over Radio Frequency (RF). The base station consists of a coordinator node receiving the object count from the end device. The coordinator forwards these results to a high-end machine capable of storing the results in a database for data retrieval and analysis.

To address the resource constraints in sensor nodes, we make the assumption that an image sensor node is only able to store a single array of one-dimensional line, whereas the base station (coordinator) is a more powerful device that can store multiple lines. The image buffer at the base station can be organized as a dynamic circular FIFO two-dimensional buffer, with the size of (one dimensional image width) * (window buffer height). Every new line is added at the top of the circular buffer, thus overwriting the oldest line in the window. This window can then be continuously displayed giving the impression of a scanned image.

This setup can be deployed for monitoring traffic and assisting flow control. The distributed nodes are spread across a street to capture the direction and number of passing vehicles. The information is then passed to the base station

Figure 5. Object counter with wireless sink

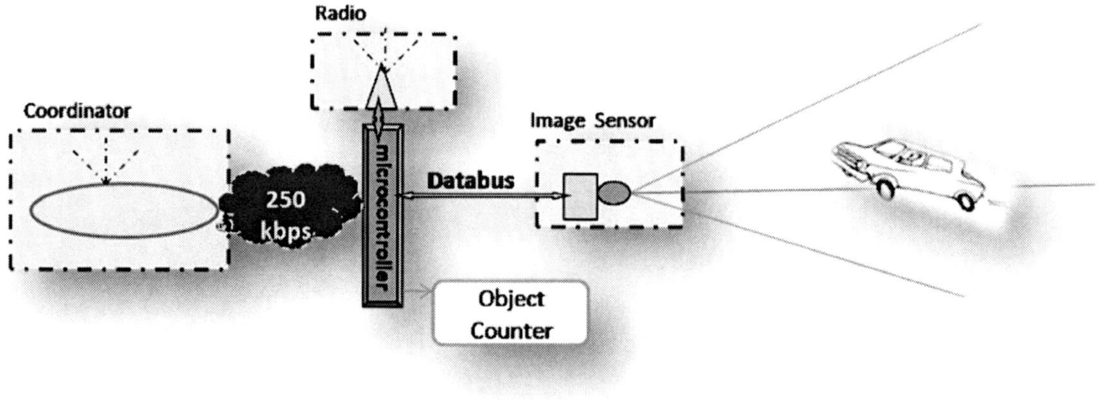

which is in charge of monitoring traffic performing real-time measurements of vehicle flow.

In the following sections, we describe the algorithms that we developed for traffic monitoring.

Background Image

The background image detection algorithm calculates the camera sight background on every input line image. In Figure 6, we show the temporal reconstruction of the background image of a sensor node.

This one-dimensional background image is maintained in a separate buffer of (width | height) pixel bytes. In order to identify a line as a background every pixel is compared with the buffered line. Every pixel in the input line is matched with the corresponding pixel in the buffered line. If each individual pixel difference is within the specified threshold then the input line is considered to be a background image.

Three different approaches could be used for maintaining the background line:

a. The new background buffer value is the average of the two lines. The moving average value helps to maintain a continuously updating background with an *O(n)* complexity. The advantage of this solution is that the algorithm needs to maintain a buffer of (width | height) pixel bytes in order to calculate the average value.
b. The new background buffer value is the median of the pixel values within a sliding window. In order to calculate the median the algorithm has to maintain a buffer size proportional to the width of the sliding window.
c. The background line is calculated once every background refreshing interval. This approach avoids the need to maintain and calculate background pixels for every captured line. However, it suffers from fact that the background could change faster than the refreshing interval and hence raise false alarms.

Foreground Image

The foreground image is computed by subtracting the average background image from every input image. The line is first compared against the background image to identify it as a foreground or background line. Once it has been identified as belonging to the foreground image, the difference image between input and background is compared to a threshold. The pixel having intensity greater than a threshold is projected as a foreground object pixel with intensity set to 255 (the background is assigned a null intensity). This threshold allows generating a binary image with noise removal (Exhibit 1).

The binary foreground image simplifies further processing of moving objects by eliminating

Figure 6. Temporal reconstruction of the background image

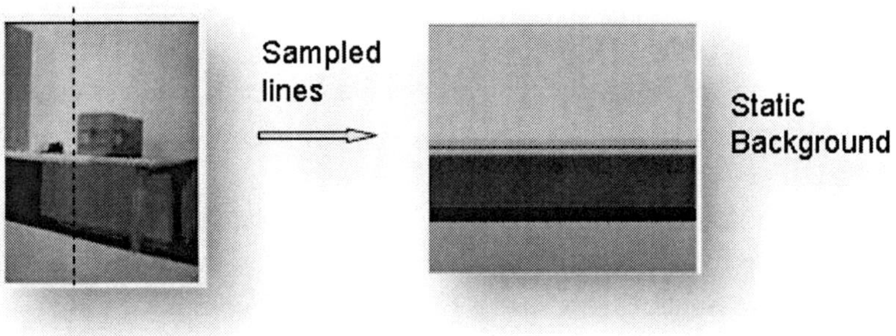

Exhibit 1. Foreground image formation

```
Collect the input line input (i) recognized as
"NOT" background in Algorithm 1
If ∀i,:   0<i< WIDTH it follows:
          |input(i)- bGavG(i)| > threshold2
then:
          input(i) = 255;
else:
          input(i) = 0.
```

unwanted data. It is also easier to compress the image data since the pixels have now a binary value (0 or 255). Also, depending on the mode of operation the data transmission can be further reduced by transmitting only the foreground pixels.

Moving Object Boundary Extraction at the Base Station

The object boundary detection is a crucial step to detect and recognize the type of moving objects. As anticipated, due to its memory limitations, a sensor node can store only one image line at a time. However, the algorithm for boundary detection acts on multiple lines: in fact, the separation of the object boundary from the object body is difficult without global information spanning multiple lines. For this reason, the boundary extraction has not been implemented on the end device, but only on the base station.

Since the base station is able to maintain a moving window of image lines with a height of 480 lines, it is possible to process multiple lines. However, classical boundary extraction algorithms that act on a 2D image cannot be easily used in our scenario, because the image data may change for the duration of image processing, whereas in our implementation the algorithm needs to be completed before a new line enters the buffer. This imposes a real-time deadline for the algorithm. Because of this restriction, we limit our processing to small portion of the moving window so as to reduce the computation time and finish before its deadline.

The algorithm works as follows. After the foreground has been separated from the background in the previous step, all the object pixels are assigned an intensity value of 255. The boundary extraction algorithm uses a connected labelling to identify 8 neighbour pixels. A pixel having 8 connected neighbours whose value is greater than 0 is identified as an interior object pixel, whose intensity is changed to a lower value. This process requires only three image lines at a time from the moving buffer. Thus, the output of this process is a moving image window having background pixels with intensity 0, boundary pixels with intensity 255 and interior pixels with intensity greater than 0 and less than 255. Again, the resulting image can be compressed by using only two bits per pixel.

Object Counting and Reporting

The object detection is based on an algorithm which processes each one dimensional image. In order to perform object detection the microcontroller needs to store and continuously update the background line, and detect the foreground, as described above. The FSM starts in the *Line Scan* state.

Let *L(i)* be the *i*-th pixel of the scanned line and *B(i)* the *i*-th pixel of the background. The difference is computed simply as:

$D(i) = L(i) - B(i)$ for all $i = 0 \ldots p$

After the foreground extraction, the object detection algorithm can be carried out on the end device by using a Finite State Machine (FSM) as shown in Figure 7. In this case, we just need to store one single line, which is scanned for a region having a set of connected pixels which would count as valid object region. If $D(i) \neq 0$ for k consecutive pixels (where k is a tunable parameter), the line is classified as part of the new object. Once a part of the sampled object is detected, the FSM moves in to *Object Start* state. In this state the FSM scans for a set of connected sampled lines each having a region of connected pixel in order to move to the final *Object detect* state.

Upon detecting a new object, the end device sends a message containing the object features to the coordinator. The object features consist of parameters such as object size in number of lines and pixels, object direction, orientation, etc. The coordinator retrieves this message and stores it temporarily for further processing such as speed measurement. The required number of connected pixels and connected lines are free parameters to be tuned for optimal settings.

Experiment (Traffic Flow Detection)

We implemented and tested a Wireless Line Sensor Network to estimate vehicle flows in a 2-lanes road using a set-up pictorially shown in Figure 8.

The line sensors are installed on lighting poles at the two sides of the road. Another mobile unit is installed in the communication range of the sensors and is in charge of aggregating the records and forwarding the result to the Internet.

The system detects the vehicle flow in the two directions following the rationale proposed in the object counting algorithm. The system is very reliable, up to about 95% with respect to human observers as shown in Figure 9. The experiment was run for a duration of 20 minutes. The deviation in vehicle detection towards the end of the experiment is due to the changing light condition which changes the dynamic background of the scene from the static average background. A topic of further research is to use a better background modelling algorithm to circumvent the problem of missed/false vehicle detection due to dynamic changes in the background line.

Speed Measurement

Another interesting application of Wireless Line Sensor Networks consists in using a pair of line

Figure 7. Finite State Machine for object detection

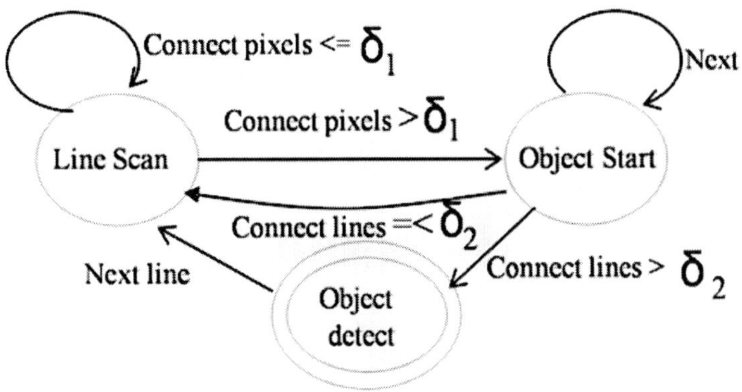

Figure 8. Networked flow detection using line sensor nodes

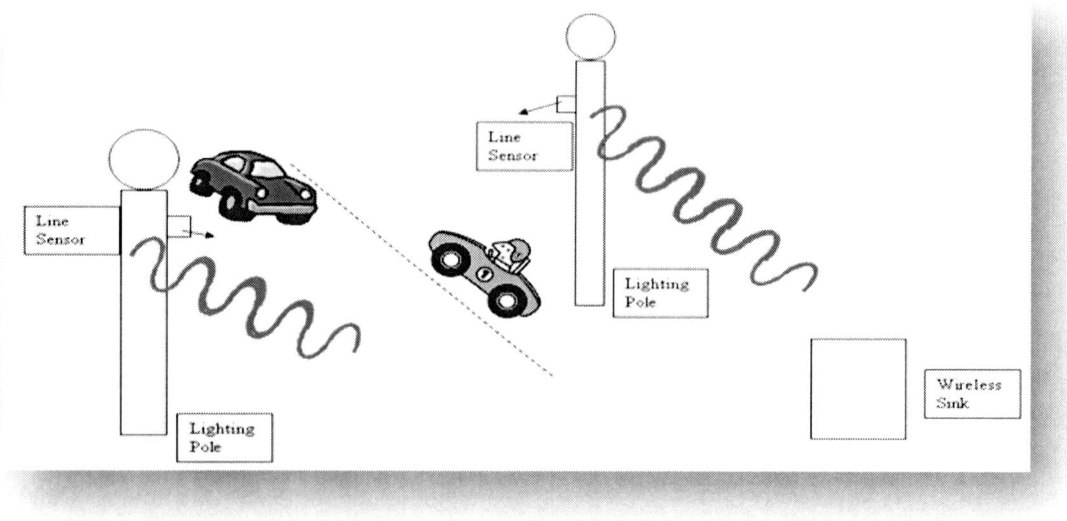

sensors nodes to measure the object speed[1]. In Figure 10 we show the general system architecture for this application.

The idea is very simple: let us set up the two cameras at a known distance along a direction parallel to the speed of incoming objects. The end device (ED) operates the object detect function described above. When a vehicle has been detected, each ED collects the foreground lines and sends a message to the coordinator, which consists

Figure 9. Vehicle Flow Detection

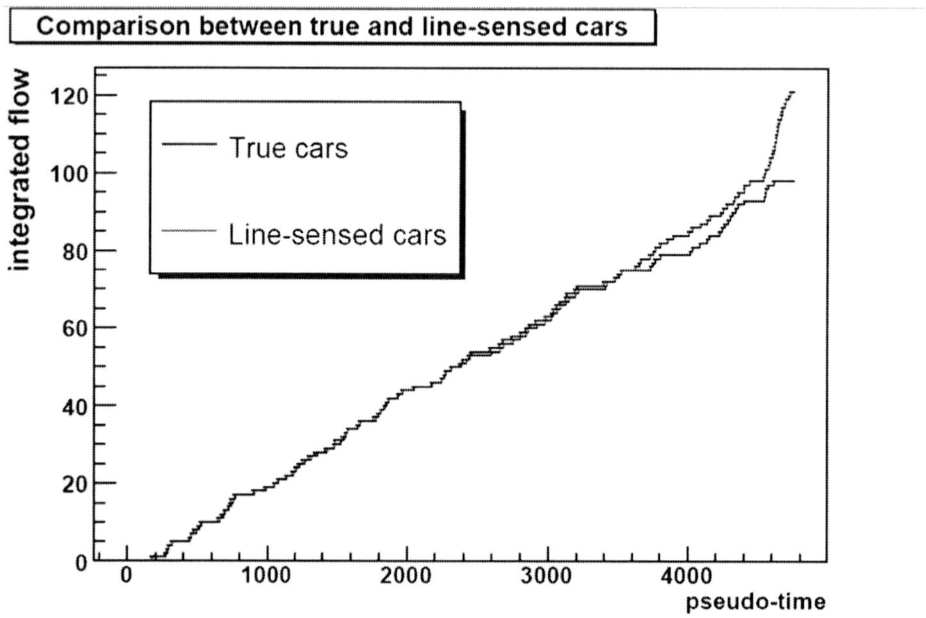

Figure 10. Speed measurement with line sensors

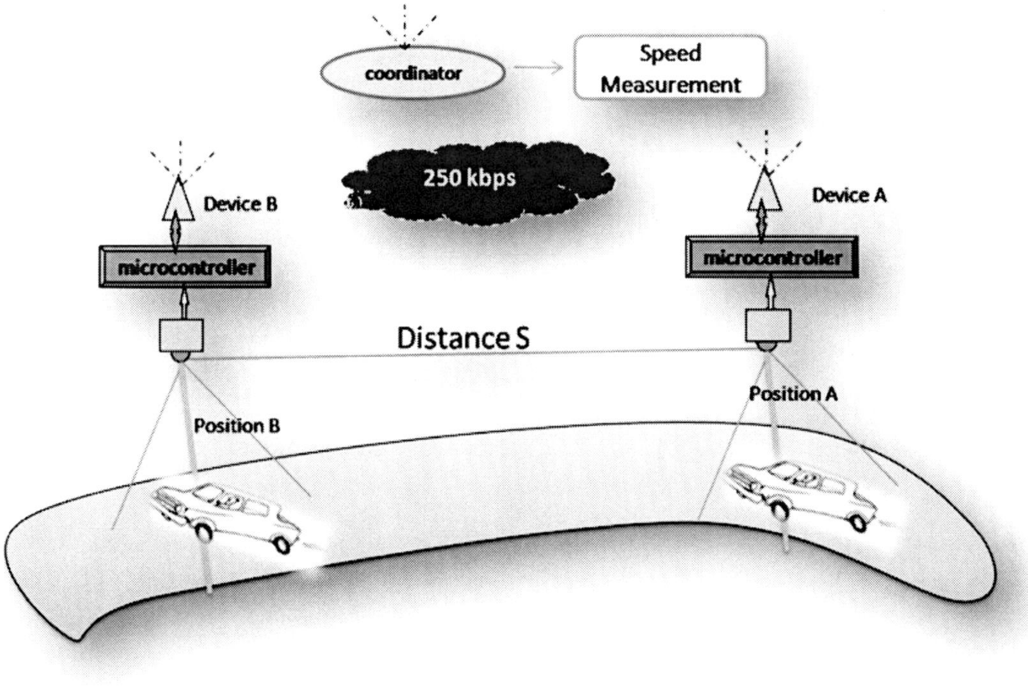

of a feature vector related to that vehicle. The coordinator collects the messages, pairs them, and computes the speed. If we neglect the uncertainties and faults of the medium access (which can be minimized making use of appropriate real-time networking approaches (Koubaa, Alves & Tovar, 2006; Huang, Pang & Hung, 2008)), the coordinator can calculate the speed of a moving object from the time interval between the reception of the two messages incoming from the two EDs.

A very naive implementation of the algorithm in the coordinator only collects the messages and computes their time separation, without any vehicle recognition. However, in realistic settings this could lead to a constant error: in fact, if one message is lost, the coordinator could start pairing messages relative to different vehicles. For this reason, the coordinator can implement a feature comparison, as shown in the flow chart with the dashed box of Figure 11. In this case, messages are paired based on the similarity between their features.

If the line sensor firmware implements the FSM proposed in Figure 7, it follows that the time delay between the *Line Scan* state and the *Object Detected* state (the time needed to analyze δ_2 lines) depends on external parameters only (e.g. the CPU speed, the camera frame rate, etc.) and in principle has no systematic uncertainty associated.

In the example depicted above, adopting a 250 kbps wireless link (like the one provided by the IEEE 802.15.4 standard for the Physical Layer), the time needed to encode and transmit an empty frame (with no payload introduced at the MAC layer) is given by:

20 bytes / 250 kbps $\approx$ 600 us

Suppose the car is moving at 100 Km/h $\approx$ 30 m/s. If the two poles are separated by 1 m, the delay in time between two signals is given by the transit time of the car, i.e. about 30 ms, while the protocol overhead is about 1.2 ms. The only

Figure 11. Speed Measurement Flow Chart

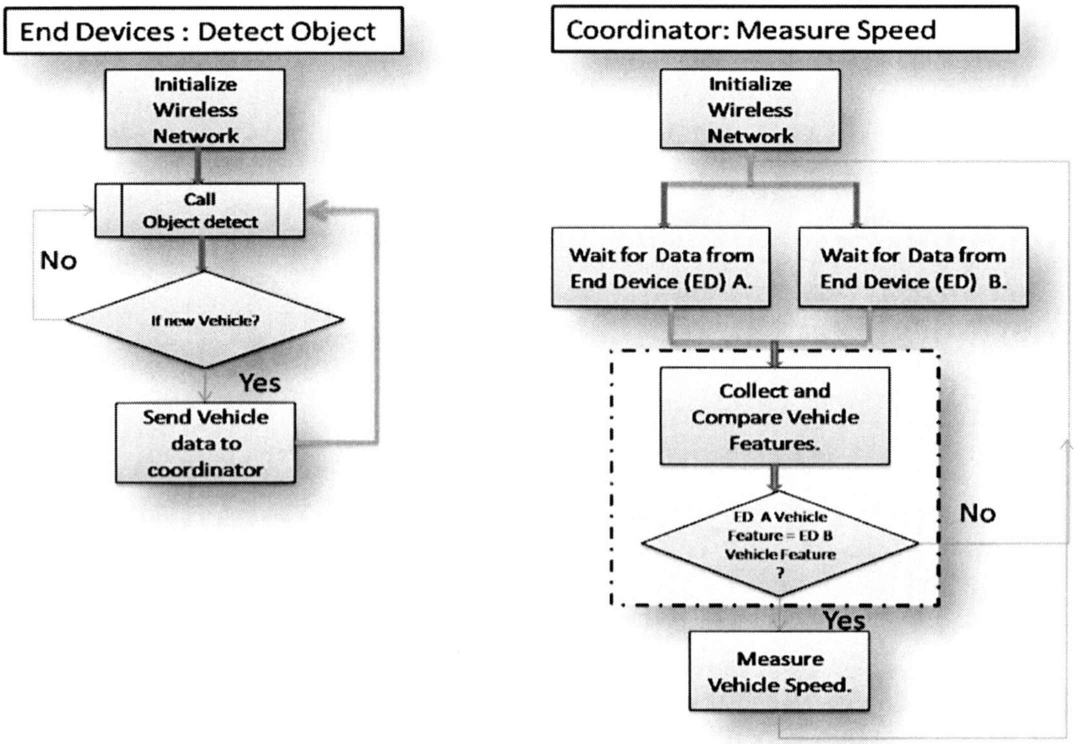

physical constraint relies in the single event detection done by each line sensor node, which must embed proper cameras to retrieve at least δ_2 lines whenever a car passes at the maximum speed.

EXPERIMENT (Speed Measurement)

In order to analyse our line sensor based model we set up an experimental test bed as shown in Figure 12. The aim of the experiment is to obtain vehicle speed with line sensor based wireless sensor network cluster and compare the results with the vehicle speed measured by a Global Positioning System (GPS) device.

The experimental wireless sensor network setup consists of two end devices as shown in Figure 12.b, each employing a CMOS camera (HV7131GP (CMOS Image Sensor with Image Processing HV7131GP, 2010)), IEEE 802.15.4 complaint transceiver (Microchip MRF24J40 (Microchip MRF24J40, 2010)) and a 32-bit PIC32 microcontroller (Microchip PIC32MX360F512L (Microchip PIC32MX360F512L, 2010)). The line sensor application runs over ERIKA (E.R.I.K.A, 2010), an operating system for real time embedded systems. The communication between the end device and coordinator is done using the uWireless implementation of the IEEE 802.15.4 specification. The coordinator as shown in Figure 12.a is a FLEX (The Flex board, 2010) demo board connected to the base station with a RS-232 serial channel. The base station runs a data acquisition application to collect data from the coordinator about vehicle detection and calculate the average vehicle speed. The line sensor device continuously monitors the lane at a fixed sampling

Figure 12. Experimental Setup. List of figures (Clockwise) a) Coordinator and base station measuring speed. b) Line Sensor device detecting vehicle c) Two line sensor devices with coordinator d) Lane used to measure vehicle speed.

line rate to detect any moving vehicle. Upon detecting a vehicle the device sends a packet to the coordinator containing information on the physical characteristics of the device. The second end line sensor device transfers the same information to the coordinator after a time interval depending on the speed of the detected vehicle.

The coordinator in turn transfers this information to the data acquisition tool running in the base station. The base station calculates the average speed of the detected vehicle and stores this information. The setup is able to detect and measure speed of the vehicle along the single lane irrespective of the vehicle direction.

The experiment was performed on a single lane as shown in Figure 12.d. Four readings each at approximate speed of 20km/hr, 30 km/hr and 40 km/hr were obtained and compared with the corresponding readings from a GPS device kept inside the vehicle. Figure 13 shows the comparison between these two readings for each detected vehicle. Table 1 gives the relative error between the line sensor measurement and GPS reading. A certain percentage of the error could be attributed to the precision of GPS device.

Flow Classification

A typical mobile object like a vehicle would move in different directions along fixed lanes. In order to classify and assign flow attributes to these vehicles the algorithm must extract certain vehicle features. Classical two dimensional image processing proposes a variety of classification techniques based on statistical measures and transforms for feature extraction (Shapiro & Stockman, 2001). Statistical feature measurements

Table 1. Speed Measurement

Vehicle Count	1	2	3	4	5	6	7	8	9	10	11	12
GPS Speed (Km/hr)	21.2	21.7	9.7	23.3	29.2	34.1	32.9	29.8	40.2	38.7	43.0	43.5
Line Sensor Speed (Km/hr)	22.3	21.34	18.73	21.88	33.01	32.94	32.2	30.66	40.4	41.53	42.61	41.65
% Relative Error	5.21	1.66	4.92	6.09	13.05	3.4	2.13	2.89	0.5	7.31	0.91	4.25

Figure 13. Speed Detection with GPS and LineSensor

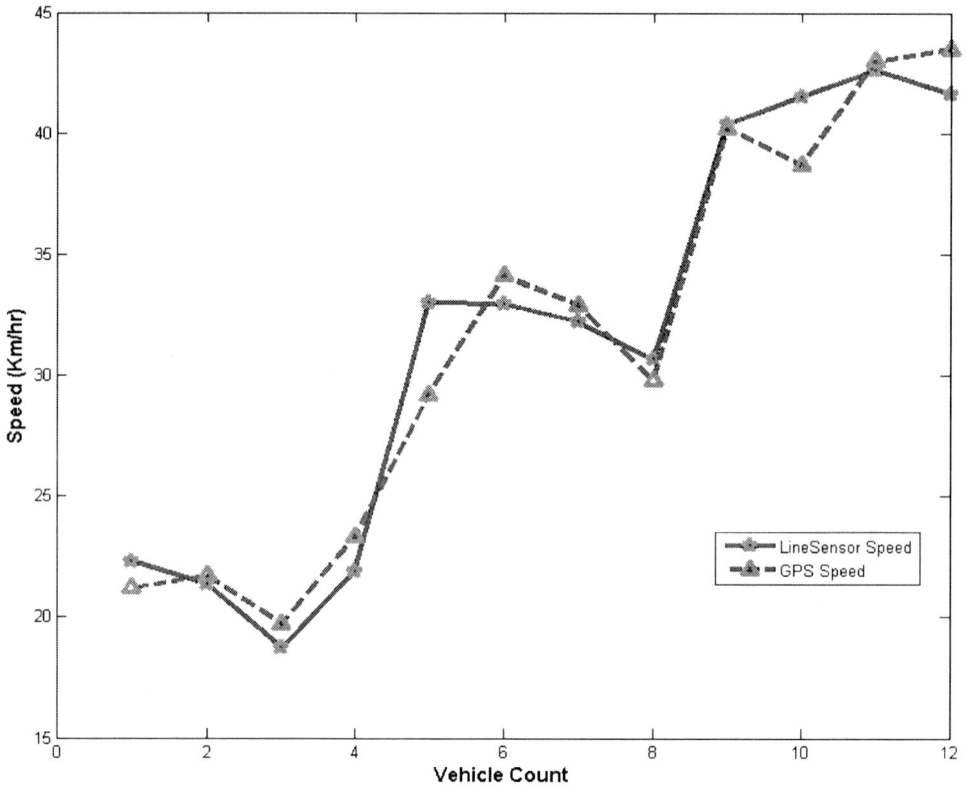

are comparatively less computationally complex than transformation based features.

Due to resource constraints, the line sensor based image algorithms can process only limited number of sampled lines containing the moving vehicle. Hence it is imperative for these algorithms to maintain a history of past measurements rather than storing the sample line itself.

Figure 14 represents a four lane highway where the two line sensor devices are placed at "S" meters apart. The line sensor device records the maximum height of foreground pixels and width (total number of detected lines). The line

Figure 14. Multi Lane Vehicle Flow Classification

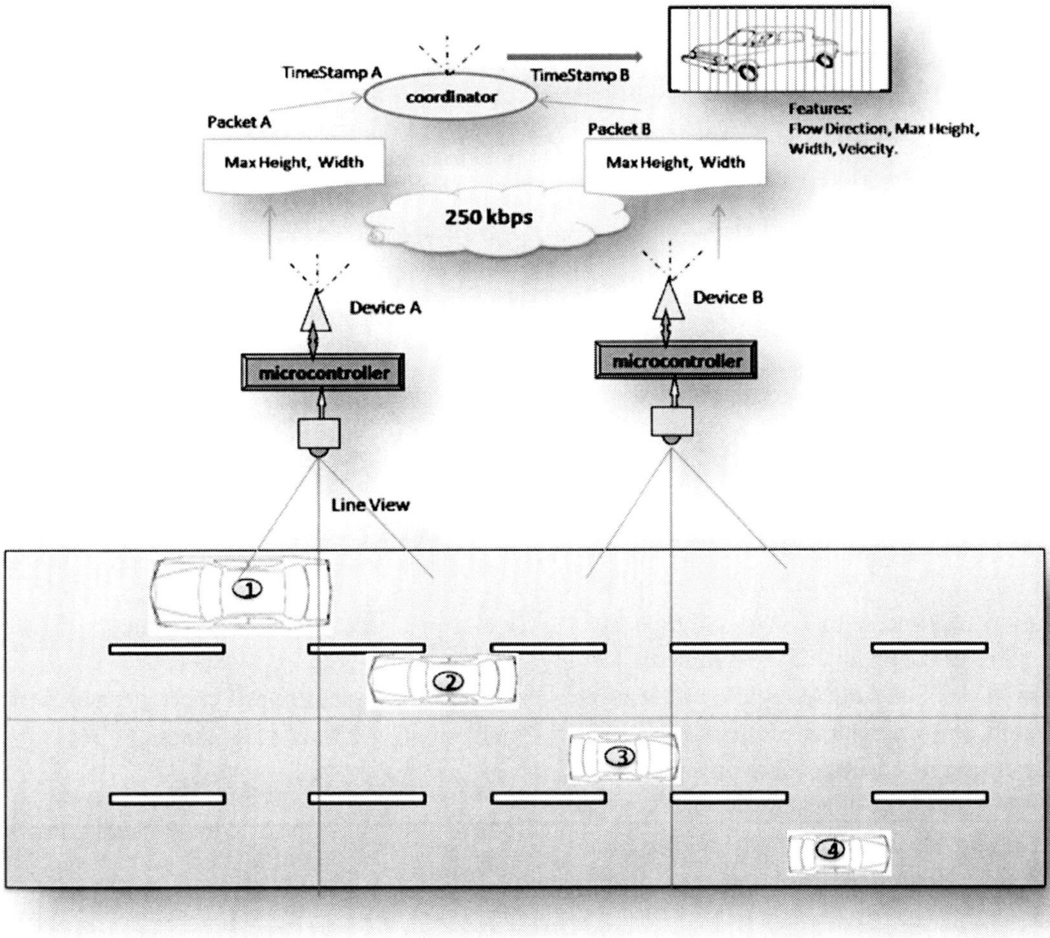

sensor device communicates this information along with the timestamp of the detected object. As shown in the figure, the two devices are in charge of detecting at most four vehicles along the four-way highway. Two of these lanes move in one direction while the other in the opposite direction. This scenario complicates the simple problem of velocity measurement, since the speed detection algorithm would raise false alarms due to detection of a different vehicle from another lane.

In order to avoid these false alarms, the detection algorithm must calculate a minimal feature vector in order to recognize the detected vehicle at the corresponding end device. This feature vector must require minimal amount of storage and computation cost.

The flow classification algorithm is based on a feature vector consisting on the timestamp of the detected object (computed at the coordinator node), height and width of the vehicle in terms of connected pixels and lines. The detected width of the object depends on the vehicle width, its speed and the frame rate of the sampled line. Hence, both devices must operate at the same sampling rate. The height of the detected object depends on the actual vehicle height and the distance of the lane from the end device. Assuming that the vehicles have high correlation in terms of their size, the farther the distance of the lane, the shorter is the height of the detected vehicle, as shown in the figure by vehicle 1,2,3 and 4. This height, width, timestamp feature vector are used to classify every moving vehicle along the four lanes and recognize the direction, lane and velocity of the vehicle along with its approximate size.

CONCLUSION

We presented the problem of deploying visual resource constrained wireless sensor network. The traditional techniques of image processing and distribution are inefficient in WSN. Hence, there is a need for low bandwidth solutions using reduced data size images able to reveal only the relevant information from the target scene. We propose a wireless line sensor technique proving its advantages over the traditional two-dimensional images in MWSN. A suite of simple image processing algorithms operating on line images are ideal for resource constrained wireless sensor nodes. The algorithms described in this chapter can be used for applications of object detection, approximate recognition, counting, speed and flow measurement.

REFERENCES

Akyildiz, I. F., & Melodia, T. Chowdhury, & K. R. (2007, March). *A Survey on Wireless Multimedia Sensor Networks Computer Networks, 51*(4), 921-960. ISSN 1389-1286.

Campbell, J., Gibbons, P. B., Nath, S., Pillai, P., Seshan, S., & Sukthankar, R. (2005, November). IrisNet: An Internet-scale architecture for multimedia sensors. In *Proceedings of the 13th Annual ACM international Conference on Multimedia*, Singapore.

Chen, P. W., Ahammad, P., Boyer, C., Huang, S. I., Lin, L., Lobaton, E. J., & Sastry, S. (2008). CITRIC: A low-bandwidth wireless camera network platform. In *Proceedings of the International Conference on Distributed Smart Cameras.*

Chitnis, M., Liang, Y., Zheng, J. Y., Pagano, P., & Lipari, G. (2009, October 9). Wireless line sensor network for distributed visual surveillance. In *Proceedings of 6th ACM PE-WASUN*, Tenerife, Canary Islands, Spain.

CMOS Image Sensor with Image Processing. (2010 October 20). *HV7131GP*. Retrieved from www.globaltec.com.hk/ databook/ hynix/ Hyca3_V20.pdf.

CMUcam3 datasheet version 1.02 (2007, September). *Carnegie Mellon University*, Pittsburgh, PA.

Coleri, S., Cheung, S. Y., & Varaiya, P. (2004, September). *Sensor networks for monitoring traffic*. 42nd Annual Allerton Conference on Communication, Control, and Computing, U. of Illinois.

E.R.I.K.A. (2010 October 20). Retrieved from http://www.evidence.eu.com/ content/ view/ 27/ 254/.

Feng, W., Kaiser, E., Feng, W., & Baillif, M. L. (2005). Panoptes: Scalable low-power video sensor networking technologies. *ACM Trans. Multi. Comp. Comm. Appl.*, *1*(2), 151–167. doi:10.1145/1062253.1062256

Hengstler, S., & Aghajan, H. (2006, March). *WiSNAP: A wireless image sensor network application platform*. 2nd Int. Conf. on Testbeds and Research Infrastructures for the Development of Networks and Communities (TridentCom).

Hengstler, S., Prashanth, D., Fong, S., & Aghajan, H. (2007). MeshEye: A hybrid-resolution smart camera mote for applications in distributed intelligent surveillance. *6th International Symposium on Information Processing in Sensor Networks (IPSN)* (pp. 360-369). Cambridge, MA.

Hu, N., Gao, T., Camhi, J., Lee, C., Rosario, D., & Aghajan, H. (2008, November). *Smart node: Intelligent traffic interpretation*. ITS World Congress.

Huang, Y. K., Pang, A. C., & Hung, H. N. (2008). *An adaptive GTS allocation scheme for IEEE 802.15.4. IEEE Transactions on Parallel and Distributed Systems*. Piscataway, NJ, USA: IEEE Press.

Itoh, S., Kawahito, S., & Terakawa, S. (2006, May). A 2.6 mW 2 fps QVGA CMOS one-chip wireless camera with digital image transmission function for capsule endoscopes. *Proceedings of the IEEE Int. Symp. Circuits Syst. (ISCAS).*

Kleihorst, R., Abbo, A., Schueler, B., & Danilin, A. (2007, September). Camera mote with a high-performance parallel processor for real-time frame-based video processing (pp. 109–116). ICDSC.

Kleihorst, R., Schueler, B., Danilin, A., & Heijligers, M. (2006). *Smart camera mote with high performance vision system.* ACM SenSys. Workshop on Distributed Smart Cameras (DSC), Boulder, CO.

Koubaa, A., Alves, M., & Tovar, E. (2006, July). I-GAME: *An implicit GTS allocation mechanism in IEEE 802.15.4 for time-sensitive wireless sensor networks*. Euromicro Conference on Real-Time Systems (ECRTS 2006), Dresden (Germany).

Kulkarni, P., Ganesan, D., Shenoy, P., & Lu, Q. (2005). SensEye: A multi-tier camera sensor network. In *Proceedings of the 13th Annual ACM International Conference on Multimedia*, Singapore.

Liu, J., & Fang, Y. (2006). Urban traffic control system based on wireless sensor networks. In *Proceedings of the International Conference on Information Acquisition* (pp. 295-300). Washington, DC., USA: IEEE Computer Society.

Microchip MRF24J40 (2010, October 20). Retrieved from http://www.microchip.com/ wwwproducts /Devices.aspx? dDocName=en027752.

Microchip PIC32MX360F512L (2010, October 20). Retrieved from http://www.microchip.com/ wwwproducts /Devices.aspx? dDocName=en532441.

Pagano, P., Nastasi, C., & Liang, Y. (2008, June). The Multivision problem for wireless sensor networks: A discussion about node and network architecture. International Workshop on Cyber-Physical Systems Challenges and Applications (CPS-CA08). In *Proceedings of the DCOSS 2008 Conference*, Santorini island, Greece.

Rahimi, M., Baer, R., Iroezi, O., Garcia, J., Warrior, J., Estrin, D., & Srivastava, M. (2005, Nov). Cyclops: In situ image sensing and interpretation in wireless sensor networks. In *Proceedings of the ACM Conf. on Embedded Networked Sensor Systems (SenSys)*, San Diego, CA.

Real Time Network Simulator. (2010, October 20). Retrieved from http://rtns.sssup.it/ RTNSWebSite/ RTNS.html.

SAFIRE. (Situational awareness for firefighters) (2010, October 20). *CERT University of California Irvine.* Retrieved from http://cert.ics.uci.edu/ SAFIRE/.

Sensor Andrew. (2010, October 20). Retrieved from http://www.ices.cmu.edu/ censcir/ sensor-andrew/.

Shapiro, L., Stockman, G. (n. d.). *Computer Vision*. Upper Saddle River, NJ: Prentice Hall. ISBN-10: 0130307963.

Situational Awareness from Multimodal Input (SAMI). (2010, October 20). *University of California, Irvine*. Retrieved from http://sami.ics.uci.edu/.

Takex Takenaka System Co. Ltd. (2010, October 20). *Line scan camera.* Retrieved from http://www.takex-system.co.jp/product/line/eng/line.html.

The Flex board. (2010, October 20). Retrieved from http://www.evidence.eu.com/.

Wireless Multimedia Sensor Testbed. (2010, October 20). Retrieved from http://www.ece.gatech.edu/research/labs/bwn/WMSN/testbed.html.

Zheng, J. Y., & Sinha, S. (2007). Line cameras for monitoring and surveillance sensor networks, *ACM Conference Multimedia, 7*, 433-442, Augsburg, Germany.

KEY TERMS AND DEFINITIONS

ED: End-device

ITS: Intelligent Transport System is a distributed sensing system responsible to gather information on vehicles and urban traffic monitoring.

Line Sensor: An imaging technique using one dimensional image vector to detect moving objects. **MWSN:** Multimedia Wireless Sensor Network is a special class of WSN sensing and distributing visual data of the target location.

VM: Visual monitoring refers to general class of camera networks irrespective of computation power and communication medium.

WLSN: Wireless Line Sensor Network

WSN: Wireless Sensor Network is a distributed network of low cost and resource constrained sensor nodes in charge of monitoring and reporting the sensed environment.

ENDNOTE

1 Italian Patent submission number FI-2010A00005F, property of Scuola Superiore Sant'Anna, Pisa, Italy and Consorzio Nazionale Interuniversitario per le Telecomunicazioni (CNIT), Italy.

Chapter 14
FPGA Technology for Implementation in Visual Sensor Networks

Wai Chong Chia
The University of Nottingham (Malaysia Campus), Malaysia

Wing Hong Ngau
The University of Nottingham (Malaysia Campus), Malyasia

Li-Minn Ang
The University of Nottingham (Malaysia Campus), Malaysia

Kah Phooi Seng
Sunway University, Malaysia

Li Wern Chew
The University of Nottingham (Malaysia Campus), Malaysia

Lee Seng Yeong
The University of Nottingham (Malaysia Campus), Malaysia

ABSTRACT

A typical configuration of Visual Sensor Network (VSN) usually consists of a set of vision nodes, network motes, and a base station. The vision node is used to capture image data and transmit them to the nearest network mote. Then, the network motes will relay the data within the network until it reaches the base station. Since the vision node is usually small in size and battery-powered, it restricts the resources that can be incorporated onto it. In this chapter, a Field Programmable Gate Array (FPGA) implementation of a low-complexity and strip-based Microprocessor without Interlocked Pipeline Stage (MIPS) architecture is presented. In this case, the image data captured by the vision node is processed in a strip-by-strip manner to reduce the local memory requirement. This allows an image with higher resolution to be captured and processed with the limited resources. In addition, parallel access to the neighbourhood image data is incorporated to improve the accessing speed. Finally, the performance of visual saliency in using the proposed architecture is evaluated.

DOI: 10.4018/978-1-61350-153-5.ch014

INTRODUCTION

In the Wireless Sensor Network (WSN), a set of sensor nodes capable of wireless communication is distributed in different places to harvest the desired information. Then, the captured information is transmitted to the nearest network mote. The network mote will relay the information within the network until it reaches the base station. Initially, the WSN is used for harvesting simple information such as temperature, pressure, humidity, or location of an object (Akyildiz, Melodia & Chowdhury, 2006). This is due to the reasons that the sensor node is small in size and battery-powered. Because the nodes are battery-operated, power consumption is always a critical issue restricting the (a) type and the number of processing unit, (b) peripheral, and (c) memory that can be incorporated onto it.

Recently, the advancement in miniature technology such as the production of camera with low power consumption (Soro & Heinzelman, 2009) resulted in the extension of visual sensor node (visual node). This leads to the Visual Sensor Network (VSN), which is capable to capture image or video data. As a result, the amount of data that need to be processed by the visual node has significantly increased. It is also required for the data to be compressed prior to transmission due to limited bandwidth. This can help to achieve better quality in video streaming (Akyildiz, Melodia & Chowdhury, 2008) as well as to reduce the power consumption (Bhargava, Kargupta & Powers, 2003). It has been shown that the power consumed by transmitting data is much higher than processing them on-board (Raghunathan, Ganeriwal & Srivastava, 2004). Therefore, the function of the processing unit has become increasingly important. However, the trade-off between performance and power consumption should be taken into consideration in the selection of a processing unit for the visual node.

Throughout years of development, many commercial products have been produced to target different applications. Here, the processing unit incorporated onto the sensor node can range from a small microcontroller to a microprocessor. Ultimately, the type of processing unit to be used depends on the amount of data and the complexity of the processing. For example, a more powerful processing unit is preferred when the visual node needs to process a considerably large amount of image data. Majority of these processing units are developed based on the Reduced Instruction Set Computing (RISC) architecture, such as the ATMEGA128L microcontroller, PXA and ARM7 microprocessor. In this case, only a limited set of simple instructions that each can be executed efficiently with the help of pipelining is adopted (Luker & Prasad, 2001). Likewise, the Microprocessor without Interlocked Pipeline Stage (MIPS) architecture that served as the basis in this chapter is also a type of RISC architecture.

In this chapter, a Field Programmable Gate Array (FPGA) implementation of a low-complexity and strip-based MIPS architecture targeted at the visual node is presented. This architecture is designed in favor of the strip-based processing (Chew, Chia, Ang & Seng, 2009). The basic idea of the strip-based processing is to divide the captured image into number of strips and carry out the processing in a strip-by-strip manner as shown in Figure 1. In this case, the captured image is first stored into the external memory. Then, a single strip will be transferred to the local memory, one at a time for processing. The procedure is repeated until the entire image has been processed. This method helps to reduce the requirement on local memory and allows image with higher resolution to be processed with limited resources.

The proposed architecture also allows the neighbourhood image data to be accessed in parallel. In fact, image processing is usually repetitive and involves the use of neighbourhood image data. Hence, parallel access significantly reduces the total numbers of read operation required to obtain all the neighbourhood data. This is very helpful in many applications, such as the

Figure 1. The basic of strip-based processing

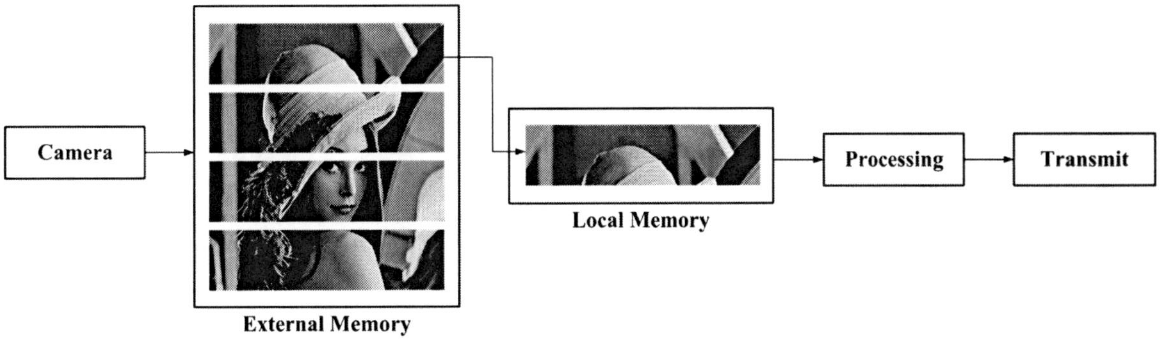

visual saliency, and watershed. The proposed architecture is realized using the RC250 development board (Mentor Graphics, 2010) that incorporated with an Altera Stratix II FPGA (Altera, 2007) for fast prototyping. In this case, the Handel-C programming language that natively designed for better support in parallelism is used to implement the proposed architecture.

Literature Review

In this section, the basics of the FPGA and MIPS architecture will be briefly described. This includes the operation and the recent development in using the FPGA and MIPS architecture for VSN. Furthermore, the 4-connectivity approach proposed in (Yeong, Ngau, Ang & Seng, 2009) that is used to achieve parallel access of neighbourhood image data will be explained as well. This is followed by the background knowledge on the visual saliency, which the performance is evaluated using the proposed architecture.

FPGA Architecture

In most cases, a FPGA is constructed by a two-dimensional array of Logic Array Block (LAB) that surrounds the Input/Output Element (IOE). The LABs are adopted to implement the desired logic function, whereas the IOEs will be used to interface the FPGA with other peripheral. Other than these, Digital Signal Processing (DSP) and Random Access Memory (RAM) are often incorporated into the FPGA as well to improve the performance, such as in the Altera Stratix II FPGA shown in Figure 2. Here, whenever there is a need to perform complex computation, the DSP can be used to replace the LABs. This helps to improve the processing speed and reduce the LABs usage. In addition, the on-board RAM also allows the data to be retrieved in a shorter time compared to data accessing from the external memory. The design that is to be implemented using the FPGA will be mapped onto these resources.

The FPGA has been widely used in the development of VSN. In the work by (Liu, Xie, Chen, Chen & Li, 2009), an asynchronous event-driven processor is implemented using the FPGA. This work takes into consideration that a sensor node will be in the idle mode most of the time. Hence, the power consumption in the idle mode can be significant in long term. Due to this reason, an asynchronous processor is proposed to replace the use of a global clock. The asynchronous processor will reside in the FPGA together with other modules that used to perform any kind processing. Once an event is detected, the modules required to complete the processing are awaken by the asynchronous processor. Therefore, the FPGA is not necessary to maintain the global clock that is used to time the control of the modules.

Figure 2. The simplified architecture of the Altera Stratix II FPGA

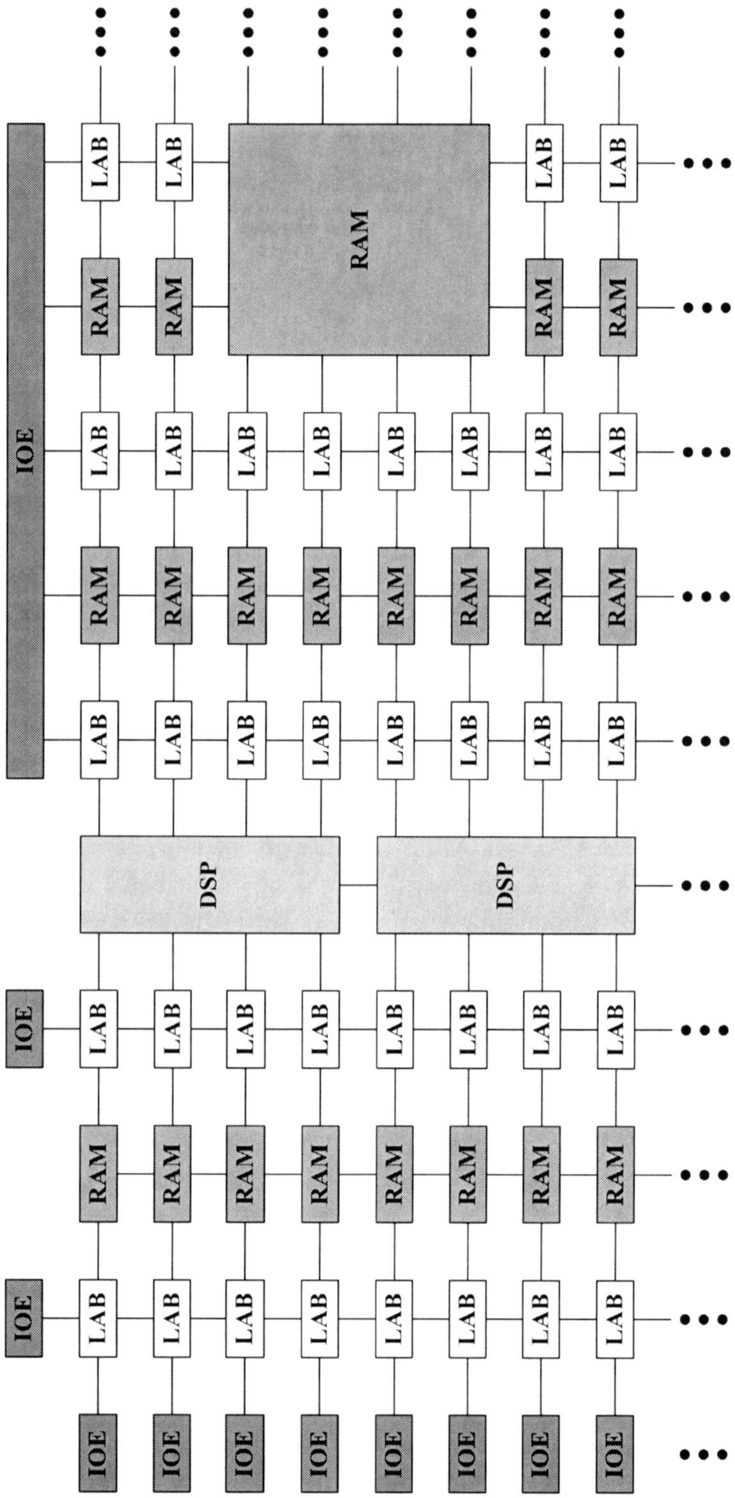

In another work proposed by (Hu, Liu, Pan, Zeng & Meng, 2009), a similar approach is also adopted. Different from the previous approach that focused on the asynchronous processor, this work focuses on using a microcontroller to aid the FPGA in image processing. In this case, the tasks that need to be executed by the sensor node are divided into two parts. The wireless communication is handled by a microcontroller, whereas the image processing will be processed using the FPGA. It should be noted that the microcontroller resides inside the FPGA along with the modules used for image processing. The microcontroller is realized using the Nios II soft-core processor developed by Altera.

Generally, the asynchronous processor described previously can be considered as a soft-core processor. Due to the rapid development, the soft-core processor has become increasingly important in FPGA. Soft-core processors are usually compact and can be easily expanded to improve the performance or interface with other modules (Kateeb & Mandepanda, 2009). This allows them to be implemented onto the FPGA without any trouble. On contrary to the soft-core processor, the work proposed by (Cao, Ji & Hu, 2005) uses a hard-core processor to aid the FPGA in image processing. In this case, the FPGA is fully utilized to perform the image processing. This can be useful in the case where the resources on the FPGA are not sufficient to implement the soft-core processor and the modules needed to carry out the desired processing.

In contrast to the approaches discussed above, (Kwok & Kwok, 2006) have proposed to reconfigure the FPGA on the sensor node prior to performing the desired processing. This is based on the concept that a specific architecture might not be able to execute all sort of processing efficiently (Ratha & Jain, 1999). It is able to perform well in certain operations, but not in all. Hence, it is wise to reconfigure the FPGA to the architecture that is able to carry out the desired processing efficiently. However, reconfiguring the FPGA takes time, and it will affect the operation of the sensor node. In order to overcome this problem, (Garcia, Ross & George, 2009) have proposed to reconfigure the FPGA partially. In this case, the modules that do not need to be reconfigured can maintain their operation without any hindrance.

MIPS Architecture

An ordinary five stages 32-bits MIPS architecture described in (Patterson & Hennessy, 2005), which serves as the basis of the proposed architecture, is simplified and shown in Figure 3. The five stages are usually denoted as Instruction Fetch (IF), Instruction Decode (ID), Execution (EX), Memory Access (MEM), and Write Back (WB). Starting from the IF stage, an instruction code is first loaded out from the instruction memory and stored into the IF/ID pipeline register. Since the instruction memory usually contains a list of instruction code, the order of executing the instruction code will be controlled by the Program Counter (PC). In this case, the value hold by the PC is supplied as the read address of the instruction memory. After fetching an instruction code from the instruction memory, the value of the PC is increased by four if the branch is not taken, or replaced by the branch address otherwise.

As suggested by the name, the width of the instruction code is 32 bits. It should be noted that the registers and memory located in the ID and MEM stage respectively are of width 32 bits as well. Based on the structure of the instruction code, they can be categorized into (a) R-format, (b) I-format, or (c) J-format, as summarized in Table 1. It can be seen that an instruction code is divided into number of fields, which can consist of an opcode, a few operands, a shift/immediate value, and a function code. In this case, the opcode will be used to specify the structure of the instruction code. The use of other fields will be explained in more detail while going through the remaining operation of the MIPS architecture.

Figure 3. The simplified architecture of an ordinary five stages MIPS

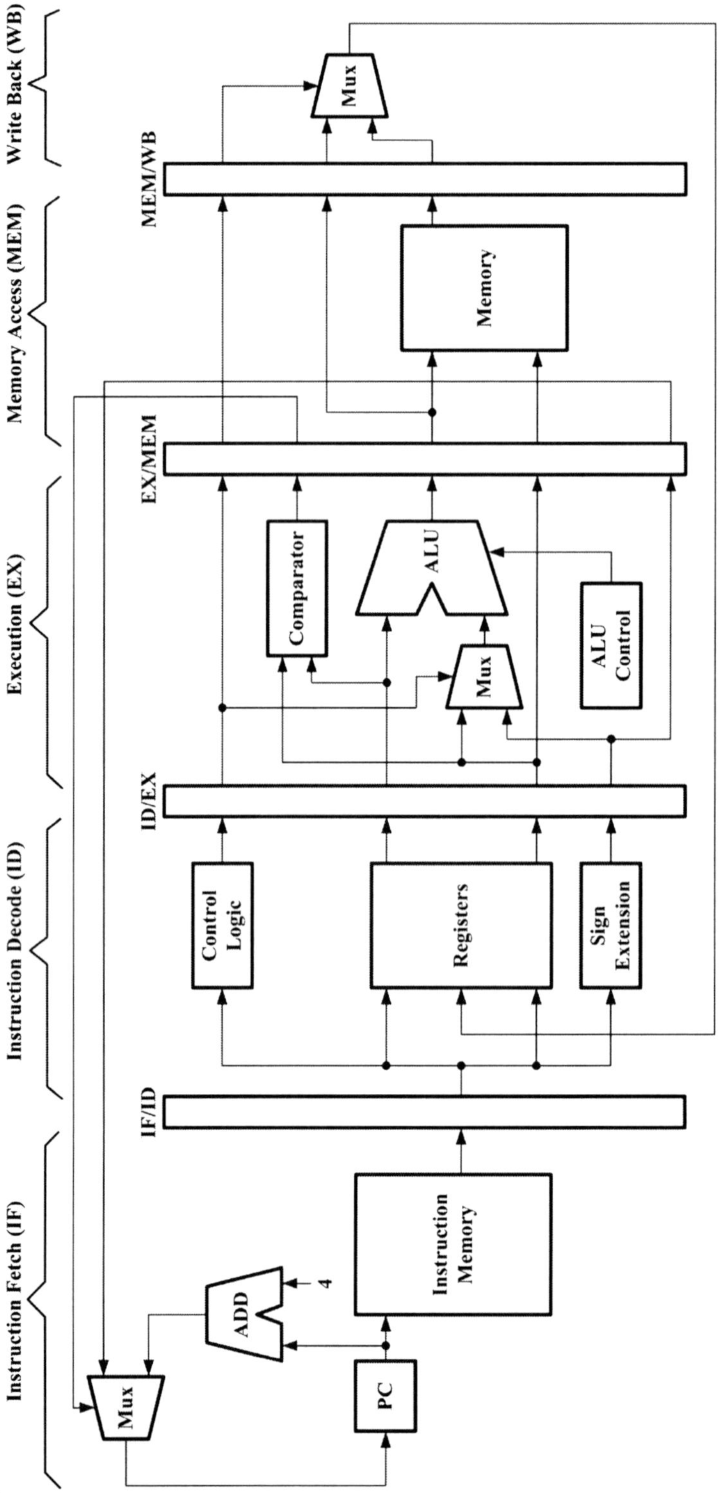

Table 1. The structure of the instruction code in R-format, I-format and J-format

Bits Allocation	6 Bits	5 Bits	5 Bits	5 Bits	5 Bits	6 Bits
R-Format	Opcode	Operand 1	Operand 2	Operand 3	Shift Value	Function Code
I-Format	Opcode	Operand 1	Operand 2	Immediate Value / Branch Address		
J-Format	Opcode	Immediate Branch Address				

When arriving at the ID stage, the instruction code stored in the IF/ID pipeline register will be decoded. The structure of the instruction code is determined by referring to the opcode stated earlier, and the information in other fields will be extracted accordingly. At the same time, the opcode is used to generate the appropriate control signal that is needed to control the multiplexers and the operation of the remaining stages. If the instruction code is in the R-format, two data will be loaded out from the registers, and the two read address are specified by operand 1 and 2 respectively. Then, these data are stored into the ID/EX pipeline register for processing in the next stage. The type of processing to be carried out will be controlled by the function code. Finally, the result obtained after the processing is written back into the registers with the write address supplied from operand 3.

Whereas for the I-format, the role of operand 1, 2, and 3 in the R-format will be taken up by operand 1, immediate value/branch address, and operand 2 in the I-format respectively. In this case, the type of processing to be carried out in the next stage is usually included into the opcode. It is also important to note that the width of the immediate value/branch address is only 16 bits. Therefore, it is required to pad it to 32 bits through the sign extension before storing into the ID/EX pipeline register. The J-format is usually used when there is a need to perform an absolute jump in the instruction code. Hence, it only required an opcode to control the multiplexers, and an immediate branch address to specify the location where the PC should go.

In the EX stage, the data obtained from the registers or the padded immediate value which have been stored into the ID/EX pipeline register will be fed to the Arithmetic Logic Unit (ALU). The input to the ALU is selected by the two multiplexers that controlled by the control signal generated from the opcode in the ID stage. As stated earlier, the type of processing that is to be performed by the ALU is selected by the function code. After the processing is completed, the result will be stored into the EX/MEM pipeline register. In this case, this result can serve as the read/write address of the data memory in the MEM stage, or pass through to the MEM/WB pipeline register for being written back to the registers in the ID stage. At the same time, the comparator will compare the two data obtained from the registers. A branch decision is generated and stored into the EX/MEM pipeline register as well. It will be used to determine whether a branch should be taken or not.

Generally, the development of the MIPS processor can be categorized into four types, which respectively focus on (a) improving the efficiency of code execution or performance, (b) extension for specific application, (c) reduction in complexity, and (d) reduction in power consumption. In the work by (Tyson, Romas, Intan & Adiono, 2009), a new branch mechanism that removes the stalls after a branch instruction is proposed. In the ordinary MIPS architecture, two clock cycles are needed to determine whether a branch should be taken or not before the branch can take place. Hence, it will slow down the processing. Furthermore, the MIPS architecture is reduced from the ordinary five stages to four stages. This can help to improve the efficiency of code execution and reduce the complexity. Another similar but slightly different work is also proposed by (Hasan, Rahman, Hasan, Hasan & Ali, 2010) to improve the branch mechanism. However, the later work remains using the ordinary fives stages MIPS architecture. The basic idea of these two approaches is to pre-calculate the branch address and stored into a specific memory.

On the other hand, (Zulkifli, Yudhanto, Soetharyo & Adiono, 2009) also proposed a MIPS architecture that can reduce the number of stalls. Instead of reading out one instruction code at a time, the proposed architecture can extract three instruction codes simultaneously, and detect whether stalls are required. If the condition is true, the sequence of executing the code will be altered to avoid the need of inserting the stalls. In addition, dedicated multiplier and modular are incorporated into the architecture to improve the processing of cryptography. Alternatively, the work proposed by (Ramdas, Ang & Egan, 2004) has focused on introducing some additional instruction code in favor of human face detection. In general, the two proposed architecture is more application-specific.

Recently, the development on the processor for personal computer has been focusing not only the architecture itself, but also extending the number of cores. In the work proposed by (Ammari & Jemai, 2008), a similar approach is adopted to speed up the decoding process of motion JPEG. In this case, the core is implemented using the MIPS architecture and multiple number of cores are been joint together. It has also been proven that it is possible to improve the performance by using more than one core. In addition, (Faraji, Didehban & Zarandi, 2010) have also proposed a dual-core processor that uses the MIPS architecture. However, the performance has not been evaluated, since the main target is to discuss about the fault tolerance in a dual-core processor.

Different from the work described above, (Gautham, Parthasarathy & Balasubramanian, 2009) have focus on reducing the power consumption of the MIPS architecture. This is achieved by eliminating all unwanted transition. In this case, the data will be passed through to the next stage if the current stage does not perform any processing on the incoming data. Indirectly, this also slightly improves the performance by reducing the architecture from the ordinary five stages to four stages when there is a pass through. Otherwise, the operation of the MIPS architecture will remains the same, as it was five stages.

In addition, the MIPS architecture has also been realized as a soft-core processor for VSN in (Kateeb & Mandepanda, 2009). The soft-core processor can coexist and interface with other resources in the same chip easily. Moreover, it is more flexible when compared to the hard-core processor. However, the processing speed of the soft-core processor is slower and limited by the chip where it situated. Another compact MIPS architecture proposed by (Li & Li, 2004) that uses very few resources has been realize as a soft-core processor as well. This architecture is primarily targeted for implementation under the hardware constrained environment.

The 4-Connectivity Approach for Parallel Access to Neighbourhood Image Data

Generally, an image is constructed by a two-dimensional array of data, which is also known as pixel. The effect of image processing is achieved by processing these pixels. Most of the image processing cannot be completed by using only a single pixel. It usually involves its neighbourhood, such as in the case of edge detection, image coding, etc. However, the conventional approach of accessing neighbourhood data from a memory bank sequentially is slow and redundant. The 4-connectivity approach is proposed by (Yeong, Ngau, Ang & Seng, 2009) to resolve the issue. This approach is named after its capability in providing a 4-way connection that links the Center (C) pixel to its four neighbourhoods, which is located to the North (N), South (S), East (E) and West (W).

In the conventional approach, the image data is usually being stored into a single memory bank, which is large enough to fit all the data. Then, the image data will be accessed sequentially, since only a single access is allowed within a clock cycle. This problem is overcome by the 4-connectivity approach that evenly distributes the image data into multiple numbers of memory bank (multibank memory). In this case, eight memory banks are needed to achieve the 4-way connection. The image data that has been stored into the eight memory banks can be accessed in parallel to retrieve all the neighbourhood data within a clock cycle. Furthermore, it should be noted that this approach does not increase the memory usage. The same amount of memory is used to store the image data. It is also important to clarify that in the remaining of the chapter, multibank memory is used to represent the eight memory banks. In other words, each multibank memory contains eight memory banks. The arrangement of the components and the multibank memory when storing and accessing the image data is illustrated in Figure 4(a) and (b) respectively.

Figure 4. The arrangement when (a) filling up and (b) accessing the image data from the multibank memory. In this case, all the non-active components are color in light gray.

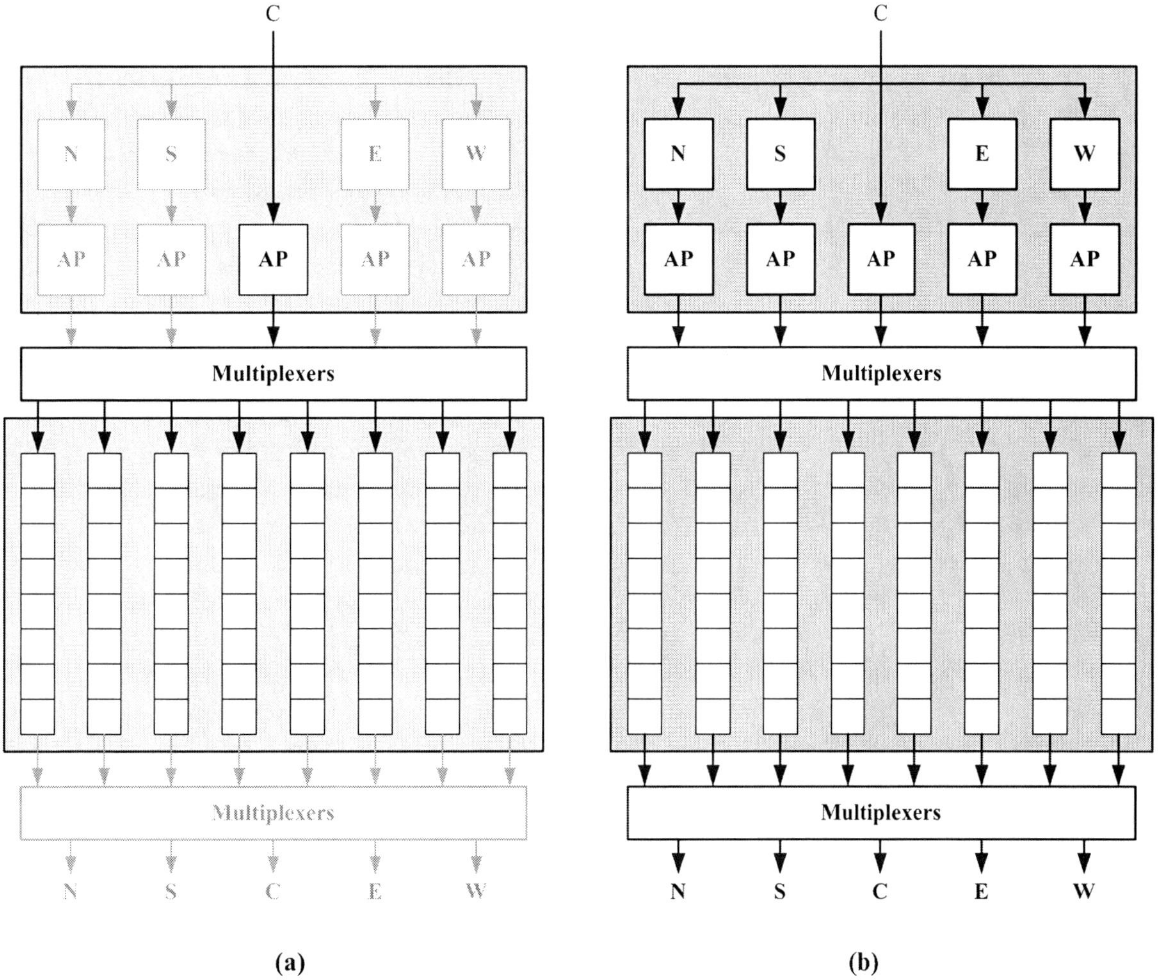

On one hand, it is necessary to first obtain the bank number and address where the incoming image data should be stored when filling up the multibank memory. This will be done by the Address Processor (AP) using the Boolean equations shown in (1) and (2). The bank number and address can be easily computed using the actual *row* and *column* number that indicates the location of the data in the original image. All the calculation is carried out in binary format. For example, *row*[0] represents the least significant bit of the *row* number in binary format. If the row number is five, the equivalent in binary format is "101". Therefore, *row*[0] will take the value of "1", since the lease significant bit (right most) in "101" is "1". Whereas the value of *row*[1] will be "0", because it refers to the first bit in "101". The bit location is started to count from right to left, starting from zero. In the remaining of the chapter, the [*x*] format will be used to represent the bit location at the *x* position of a particular signal or data.

$$B[2] = \{row[0] \cdot column[0]'\} + \{column[0] \cdot row[0]'\}$$

$$B= row[1] \oplus row[0] \oplus column[1]$$

$$B[0] = row[0] \quad (1)$$

$$A = row\ [\log_2(\max\_row) - 1{:}\ 1]\ @\ column[\log_2(\max\_column) - 1{:}\ 2] \quad (2)$$

By referring to the Boolean equations (1) and (2), the symbol •, +, ⊕, ', and @ are used to represent the operation of logical AND, logical OR, logical XOR, bit inversion, and bit concatenation respectively. In this case, *B*[2:0] is the bank number in binary format, which is obtained by performing logical AND, OR, and XOR on bit 0 and 1 of the *row* and *column* number. Here, A is the address where the data is been placed in the bank selected by *B*[2:0]. It is computed by concatenating the bit(s) from ($\log_2$(*max_row*)-1) to 1 of the *row* number with the bit(s) from ($\log_2$(*max_column*)-1) to 2 of the *column* number. In this case, *max_row* and *max_column* are the resolution of the original image in the *row* and *column* respectively.

Similar to the process of filling up the multibank memory with image data, it is also necessary to provide the *row* and *column* number of the C pixel when retrieving the image data. At the same time, the *row* and *column* number of the N, S, E, and W pixel can be computed from the C pixel. Then, the *row* and *column* number for each pixel will pass through their own AP separately to get the respective bank number and address. Once the bank number and address for the C, N, S, E, and W pixel are known, the image data will be loaded out from the multibank memory in parallel.

Visual Saliency

In the Human Visual System (HVS), abundant visual information captured by the retina is transferred by means of optical nerves to the visual cortex in the brain for processing. Although the eye is capable of capturing and delivering massive amount of visual information to the brain in a short period of time, the brain itself has a limited processing capacity to handle the entire amount of received information. Therefore, to avoid bottlenecking and to promote efficient utilization of processing resources; the brain selects only relevant information to be processed at a time. A literature on an early computational HVS model with stimulus selection via selective filters for processing can be found in (Broadbent, 1958). With the brain having no prior knowledge on the content of the visual information captured; in order for the information to achieve a chance of being process, it has to have a quality that makes it stand out from the rest of the information. This quality, rather subjective in nature; is called salience. When referring to unique visual information where no prior information is available, the term "bottom-up visual saliency" is normally used.

Since the introduction of the HVS model, researches in the field of psychophysics and vision have shown dedicated interest in understanding and developing the mechanism of the HSV, which eventually led to the recent biologically-plausible but fully computational visual saliency model (Itti & Koch, 2000). Interested readers are referred to this paper for in-depth discussion and algorithms. Wolfe (Wolfe, 1998) suggested that features (distinct property) play an important role in providing an object, a region, or any visual information for the matter; the quality of being unique. Among the features humans are able to detect efficiently are color, orientation, intensity, texture, scale, motion, onset, offset, cues, to mention a few. Back in 1985, the idea of a saliency map was brought about to provide a measure of salience from all contributing features. According to Koch and Ullman (Koch & Ullman, 1985), the saliency map is a topographical map containing information from the features available to provide a global measure of salience. Similarly, Treisman presented the saliency map as a master map consisting of conspicuous location based on the local salience of each separable feature maps (Treisman, 1985).

Up to now, there are various models of the bottom-up visual saliency, ranging from biologically-plausible to purely computational ones; computational methods from pixel based computations to region based; and computational domains in spatial or transform. In this work, the biologically-plausible model with pixel based computation is emphasized due to its popularity and solid establishment. Taking the Itti and Koch model (Itti & Koch, 2000) for example, the generic operation of a biologically plausible model is as follows. First, the input image is captured and divided into individual channels (red, green, and blue in the case of RGB color space). Then, features are generated from these channels: intensity and opponent colors. Once these are obtained, they undergo a series of linear filtering and sub-sampling giving rise to image pyramids; each with decreasing image resolution and size as the level goes higher towards the top of the pyramid. These image pyramids are important for analyzing unique features at various scales (Burt, 1988). These pyramids are also called Gaussian pyramids since they are convolved using Gaussian kernels. The orientation features are computed from the intensity pyramid. Normally, the number of features used is limited to three or four features to avoid the model being bogged down with excessive computations.

The next step of the operation is the center-surround (CS) process. This important process can be found in the HVS and is used to emphasize fine details at lowers scales of the image pyramid and coarse details at higher scales (Itti, Koch & Neibur, 1998). The CS process is performed by resizing the image pyramids to a standard size and subtracting images from subsequent levels from the image at the current level. The CS results are summed across scale and normalized according to feature, giving the conspicuity map for each feature. The saliency map is found by linearly combining all the conspicuity maps. The step after the saliency map is task dependent and is usually customized according to applications. Current available applications which utilizes the bottom-up visual saliency include object detection (Itti & Koch, 2000), mobile robot (Frintrop, 2007), image compression (Yu & Lisin, 2009), and human tracking on a mobile platform (Montabone & Soto, 2010), to name a few.

Documented works of applications for the wireless VSN using bottom-up visual saliency are currently unavailable, which could be due to the immature area of wireless multimedia technology in the past years; high cost of miniature hardware; and difficulties in implementing a saliency model on a custom hardware with restricted resources. Nevertheless, the area of VSN is currently expanding and visual saliency can be used in possible applications such as surveillance tracking, suspicious activity monitoring, traffic control, smart visual systems, and human detection.

Towards Hardware Implementation

Although there are many bottom-up visual saliency models which successfully to an extent, capable of detecting and highlighting conspicuous objects; most of them consume a large amount of memory. This is one main reason such models are difficult to be implemented on small portable hardware. The large amount of memory consumption comes from the storing of image pyramids and image maps before the final saliency map can be computed. There are other purely computational models which does not require the image pyramid (Hou & Zhang, 2007) (Achanta, Hemami, Estrada, & Susstrunk, 2010) but they each have issues with boundary representation or detection performance. Another factor to consider is the complexity of the algorithm itself. Complex operations such as domain transforms and the use of advance mathematical operators, for example; exponent, logarithm, trigonometry; should be avoided if the model is implemented on a basic processor.

Here, the bottom-up visual saliency is to be implemented on a custom strip-based MIPS. To complement the strip-based MIPS, the strip-based visual saliency model of (Ngau, Ang, & Seng, 2009a) will be used. In their works, it is documented that the total memory consumption for the Itti and Koch model (Itti & Koch, 2000) can be reduced up to more than 90% for large input images. However, only the intensity feature will be used in this implementation as the strip-based MIPS capture input is in grayscale format. With the strip-based method, implementation of a saliency model with large memory consumption on a portable hardware with limited resources is made possible since a small portion of the images is processed at an instance.

For implementation, the Gaussian image pyramid is replaced with a multi-resolution pyramid constructed with recursive discrete wavelet transform (DWT) decomposition. The multi-resolution pyramid is more favorable for hardware implementation since a more efficient computation method is available. In conventional DWT, a convolution based filter bank method is used. The Gaussian pyramid computation shares similar traits; whereby it is constructed by convolving separable Gaussian kernels. Instead of the convolution method, a lifting based filtering method can be used for computational efficiency and memory savings (Acharya & Tsai, 2004). For the DWT filter, the reversible Le Gall 5/3 spline filter (Gall & Tabatabai, 1988) is used. The filter indexing terms are in the 2^n format, which will be an added advantage, since division and multiplication operations can be performed by bit-wise shifting. In addition, orientation features can be directly obtained from the DWT sub-bands if required, without having to use complex mathematical operations for orientation computations.

Discrete Wavelet Transform (DWT)

The DWT is a kind of transformation that transforms an image from spatial domain to frequency domain. In this case, the image will be decomposed into number of sub-bands, each containing different frequency components of the original image. An example of applying the DWT on an image is shown in Figure 5.

The number of sub-bands that are created after the DWT is determined by the level of decomposition (p). Generally, the number of sub-bands that are created with p level of decomposition is ($3p$+1). After one level of decomposition, the image is decomposed into four sub-bands, which are labeled as the LL1, LH1, HL1, and HH sub-bands. The second level of decomposition will be applied on the LL1 sub-band to obtain the LL2, LH2, HL2, and HH2 sub-bands. This process is repeated for additional level of decomposition.

After the first level of decomposition, most of the energy will be concentrated in the LL1 sub-band. The LL1 sub-band contains the low frequency component that preserves the outline of the image. It can be observed that most of the outline in the original image still can be seen in

Figure 5. The process of applying the discrete wavelet transform on an image

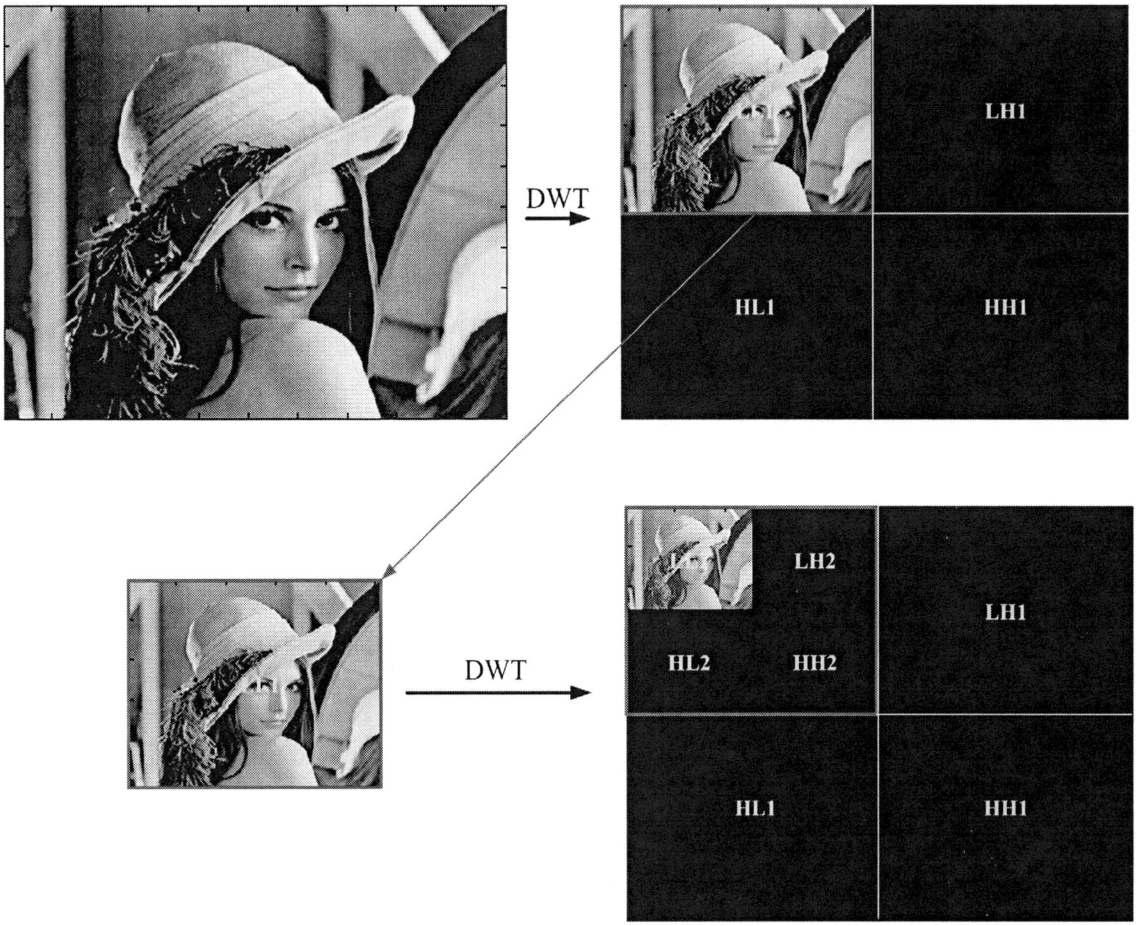

the LL1 sub-band. In contrast, the LH1, HL1, and HH1 sub-band consist of the higher frequency component, which is insensitive to human eye. When the level of decomposition increases, more and more information will be compacted to the LL sub-band. Meanwhile, more details will be isolated from the low frequency component as well.

Although there are different types of DWT, the focus of the discussion in this chapter is on the LeGall 5/3 DWT, which is a part of the visual saliency. The DWT can be performed using the convolution scheme or the lifting scheme. However, the latter scheme is easier to implement when come to hardware. Hence, it has become increasingly popular. The filtering process of different type of DWT is the same, except the filter length and the factor of multiplication or division. They all first started with the filtering process on the horizontal direction follow by the vertical direction.

The lifting scheme of the LeGall 5/3 DWT is built based on the split, prediction, and updating steps. First, the input sequence $X[n]$ is split into odd and even components for the horizontal filtering process. In the prediction phase, a high-pass filtering is applied to the input signal that results in the generation of the detailed coefficient $H[n]$. In the updating phase, a low-pass filtering is applied to the input signal that leads to the generation of the approximation coefficient $L[n]$. Likewise, for the vertical filtering, the split, prediction, and updating processes are repeated for both the H[n] and L[n] coefficients.

The lifting implementation of the 5/3 DWT filter used in JPEG2000 (Archarya & Tsai, 2004) can be illustrated as shown in Figure 6. The detailed coefficients H[n] need to be obtained prior to the computation of the approximation coefficients L[n]. Generally, the illustration can be described by using equations (3) and (4). As can be seen from equations (3) and (4), three computation operations—addition, subtraction, and shift—are needed in the implementation of a lifting-based LeGall 5/3 DWT. All these operations can be implemented easily when compared to multiplication and division. Therefore, the LeGall 5/3 DWT is very suitable for hardware implementation.

$$H[2n+1] = X[2n+1] - [(X[2n] + X[2n+2]) / 2] \quad (3)$$

$$L[2n] = X[2n] + [(H[2n-1] + H[2n+1] + 2) / 4] \quad (4)$$

STRIP-BASED MIPS PROCESSOR

The proposed low-complexity strip-based architecture with the capability to access the neighbourhood image data in parallel is shown in Figure 7. In fact, it is an extension and modification from the ordinary five stages MIPS architecture discussed previously. In order to reduce the number of clock cycles that is required to write back the result into the registers, the proposed architecture is reduced from the ordinary five stages to four stages. In this case, the EX stage and the MEM stage are combined and will be denoted as the EXM stage. Furthermore, the data memory in the MEM stage is replaced by the multibank memory and connected as describe in the 4-connectivity approach.

In addition, three separate multibank memories are incorporated instead of one to pipeline the process of transferring and processing the data. If only one multibank memory is used, the process of transferring and processing the image data have to be done sequentially as shown in Figure 8(a). This can slow down the strip-based processing. As mentioned previously, the entire image is first stored into the external memory. The image data will be transferred to the local memory (multibank memory) and being processed in a strip-by-strip manner. In this case, no processing can be done when the transferring takes place. By having three separate multibank memories that can be accessed simultaneously, it is possible to pipeline the pro-

Figure 6. The relationship of the input sequence in generating the H[n] and L[n] coefficients for LeGall 5/3 DWT

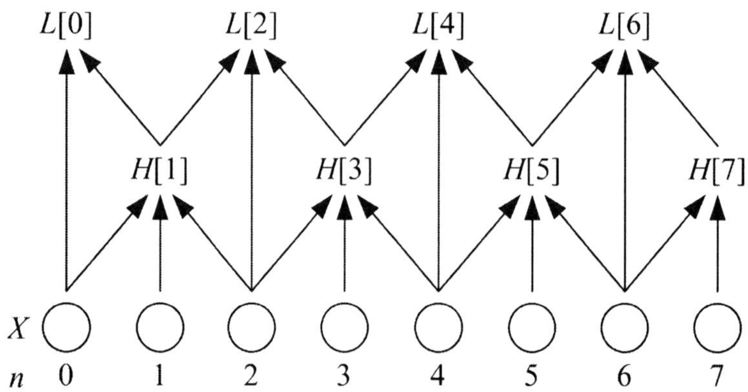

Figure 7. The simplified architecture of the proposed strip-based MIPS architecture

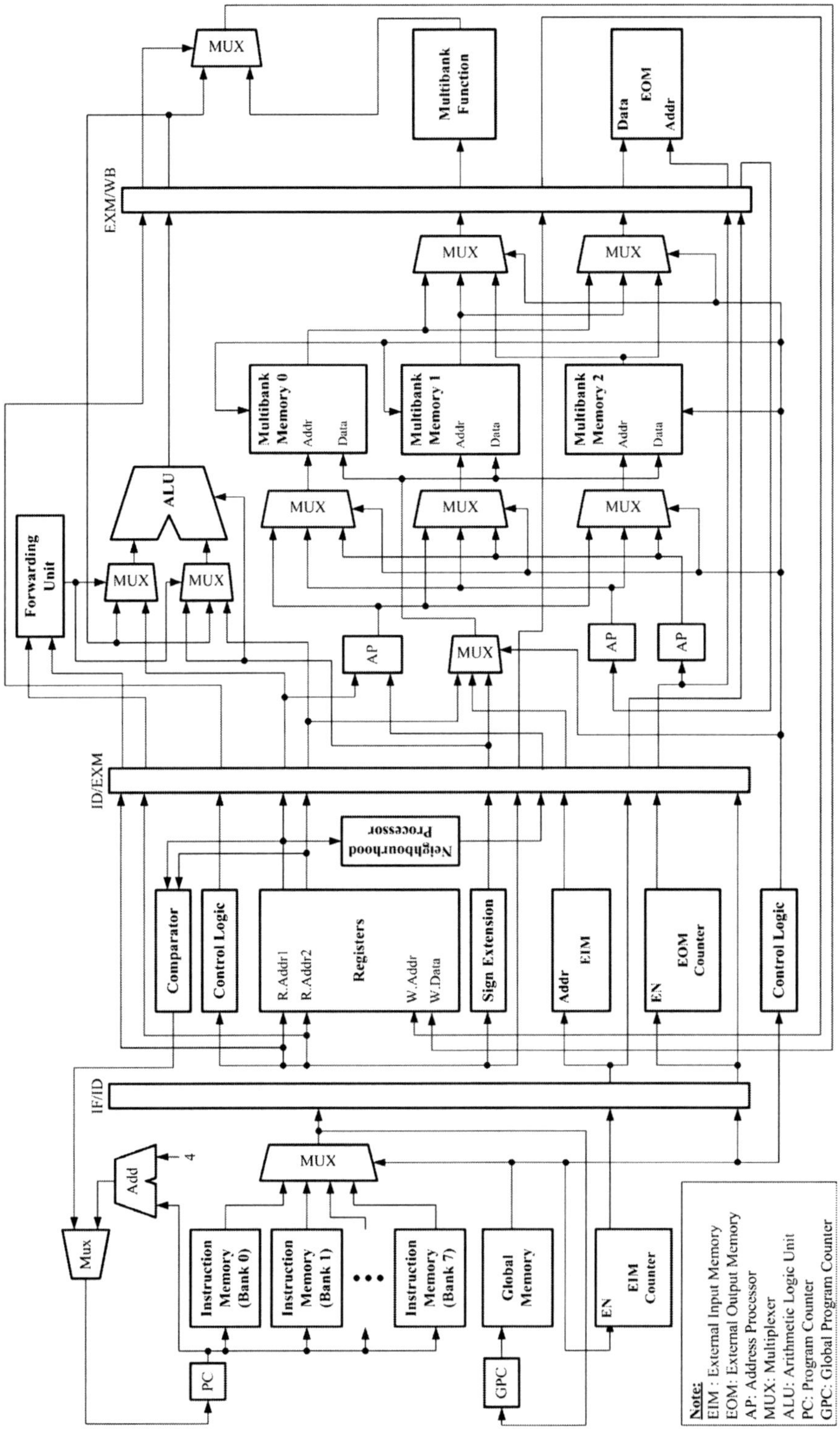

Figure 8. The difference in using the strip-based processing technique (a) without and (b) with pipelining

Time Slot

1	2	3	4
External Input Memory ↓ Local Memory	Process Data in Local Memory	Local Memory ↓ External Output Memory	External Input Memory ↓ Local Memory

(a)

Time Slot

1	2	3	4
External Input Memory ↓ Local Memory (Bank 0)	External Input Memory ↓ Local Memory (Bank 1)	External Input Memory ↓ Local Memory (Bank 2)	External Input Memory ↓ Local Memory (Bank 0)
	Process Data in Local Memory (Bank 0)	Process Data in Local Memory (Bank 1)	Process Data in Local Memory (Bank 2)
		Local Memory (Bank 0) ↓ External Output Memory	Local Memory (Bank 1) ↓ External Output Memory

(b)

cess of transferring and processing the data as shown in Figure 8(b).

Different from the ordinary MIPS architecture that stores the entire instruction codes into a single instruction memory, the proposed architecture allows the user to breakdown the instruction codes and stores them into different instruction memory. This is for the ease of reuse, maintain, and debug the instruction codes. Due to the limited type of MIPS instruction, reusing a section of instruction codes can be tedious. It is often necessary to take few clock cycles to check the condition before determine whether a branch should be taken. Moreover, it can become much more complicated when a section of the instruction codes is reused for many times.

Assuming that the instruction codes are divided into section A, B, C, and they are required to be process in the order of A→B→A→C. In this case, it is required to determine in the end of section A whether it should branch to section B or section C. This is unlike the case in high-level programming language where a macro can be adopted to define and reuse a section of codes easily. This problem can be resolved by breaking down the instruction codes into smaller section and stores them into different instruction memory. The instruction memory can be selected using a global code that sits inside the global memory. After all the instruction codes within an instruction memory have been executed, the second global code can be loaded out to select another instruction memory to continue the operation.

In addition, the global code is also used to control the pipelining process of transferring the data from the external input memory to the multibank memory, and from the multibank memory to the external output memory. The way of using the global code and the operation of the proposed strip-based architecture will be explained in more detail in a stage-by-stage manner as follows.

Instruction Fetch (IF)

Depending on the value hold by the global memory counter, the global code that determines the operation of the entire architecture is first obtained from the global memory. Similar to the instruction code, it is also of width of 32 bits, and the usage of each of the bits is summarized in Table 2. Primarily, the global code is used to control the access of the instruction memories and multibank memories. After all the instruction codes within the selected instruction memory have been executed, a special instruction code is inserted to indicate the end of instruction codes. This will terminate the process and increase the value of the global memory counter. Then, the next global code is obtained from the global memory again to continue the operation.

In addition, the global code is also being used to initiate the process of transferring image data from the external input memory to the multibank memory, and from the multibank memory to the external output memory. This can be done by enabling the external input and output memory counter. It can be seen in Figure 7 that the enable signal used to select the multibank memory is not stored into the pipeline register. All the signals generated from the global code will remain the same as long as the global code is unchanged. The process of transferring will stop when the selected multibank is fully filled up with a strip of image data.

Instruction Decode (ID)

Similar to the ordinary MIPS architecture, the instruction code is decoded and the data will be loaded out from the registers as specified by the operands. Before these data are stored into the ID/EXM pipeline register, the equality of the data is evaluated. This is to determine whether a branch should be taken when executing a branch instruction. Since the branch mechanism is shifted one stage in front, this reduces the number of stalls from two to one. In the ordinary MIPS architecture, the branch will only take place after two clock cycles.

When accessing the multibank memory, the instruction code in the I-format is adopted. In this case, the first operand is point to the read address of the registers that contains the *row* and *column* number of the C pixel. Once the *row* and *column* number for the C pixel is determined, the *row* and *column* number of the N, S, E, and W pixels will be computed and stored into the ID/EXM pipeline register as well.

If observed carefully, it can be noticed that there are boundary issues where certain neighbours do not exist. However, the non-existent pixels can be easily identified when computing their *row* and *column* number. For example, assuming that the C pixel is set to coordinate (0, 0) where the N and W pixels are not existed in this case. When com-

Table 2. The usage of each of the bits in the global code to control the operation of the MIPS processor

Bits	Usage
31:29	Select line for the multiplexer where the inputs are connected to all the instruction memories.
28	Not Used.
27:26	Generate enable signal to select the multibank utilized by the MIPS processor.
25	Not Used.
24:23	Generate enable signal to select the multibank that transferring data to the external output memory.
22	Enable the external output memory counter that drives the external output memory.
21	Not Used.
20:19	Generate enable signal to select the multibank that accepting data from the external input memory.
18	Enable the external input memory counter that drives the external input memory.
17:0	Not used.

puting the *row* and *column* value for the N and W pixels that involve a subtraction of one in the *row* and *column* number respectively, an overflow will occur. This causes the most significant bit of the *row* and *column* number to become a '1'. Hence, the most significant bit can be used to identify the validity of the N and W pixels.

Whereas for the S and E pixels, their validity have to be confirmed using a slightly different way. First, the *row* and *column* number for the S and E pixels are determined from the C pixel. Then, their validity can be known by checking the bit *row*[$\log_2$(*max_row*)] and *column*[$\log_2$(*max_column*)] of the computed *row* and *column* number respectively. Consider that the *row* and *column* number are started to count from zero, these value should be less than *max_row* and *max_column* respectively for a valid pixel. Hence, if the bit *row*[$\log_2$(*max_row*)] and *column*[$\log_2$(*max_colunn*)] is '1', it implies that the particular pixel is located outside the image, which is not valid. Finally, the validity of the N, S, E, and W pixels will be stored into the ID/EXM pipeline register as well. The validity is important when processing the C, N, S, E, and W pixels.

Execution and Memory Access (EXMEM)

The reason of combining the EX stage with the MEM stage is to reduce the number of clock cycles taken to write back the result produced by the ALU. The main drawback of doing this is the multibank memory can no longer be accessed using indirect addressing. In other words, the multibank memory can only be accessed using direct addressing where the address has to be the output from the registers. This problem can be resolved by first compute the address and stored it into the registers prior to accessing the memory.

The bank number and address for accessing the multibank memory will be computed in this stage, by using the *row* and *column* number of the C, N, S, E, and W pixels stored in the ID/EXM pipeline register. This will be done by using a separate AP. Then, the eight outputs from the multibank memory will be stored into the EXM/WB pipeline register. Although all the eight outputs are stored into the EXM/WB pipeline register, three out of the eight outputs are actually invalid. The output for C, N, S, E, and W pixels can be chosen from the eight outputs easily by using a multiplexer. However, this is only carried out in the next stage.

In addition, it should be noted that the priority of accessing the three multibank memories is given to the processing of instruction codes. In other words, if the multibank memory is selected for processing and transferring at the same time, the access will be given to the processing. The steps taken to access the multibank memory under different operation is summarized into Table 3.

Write Back (WB)

The result to be written back to the register specified by the operand is selected by the multiplexer. If the instruction code only uses the ALU to perform the computation on the input data, the result from the ALU will be chosen. Otherwise, the output from the multibank memory is selected. In this case, the output can be referred to the unprocessed or processed data. In the former case where a normal load word instruction is executed, the value hold by the C pixel will be directly selected as the output. This is different from the latter case where the output is the outcome of processing the C, N, S, W, and W pixels.

It should be noted that the width of the data to be written back to the registers is only 32 bits. Although it is possible to squeeze the value hold by the five pixels into 32 bits, but the value of a pixel can only range from 0 to 64, which is not practical for image processing. This problem is resolved by process the five pixels directly and use the outcome as the output. Hence, six new instructions, which are designed for processing the data obtained from the multibank memory, have been incorporated into the proposed architecture. These

Table 3. The steps taken to access the multibank memory. The "Read" column is for retrieving data from the multibank memory. Whereas, the "Write" column is for writing data into the multibank memory

Operation	Read	Write
External Input Memory ↓ Multibank Memory	Not Available	(1) Counter value is converted to the bank number and address for multibank memory. (2) The bank number and address is treated as the C pixel as shown in Figure 5(a). (3) The data is written into the specified bank and address.
Multibank Memory ↓ External Output Memory	(1) Counter value is converted to the bank number and address for multibank memory. (2) The bank number and address is treated as the C pixel and fed to all the eight banks of memory. (3) Eight data are obtained but only the one correspond to the C pixel is chosen.	Not Available
Load Word (Read) / Store Word (Write)	(1) The *row* and *column* number specified by the first operand is converted to the bank number and address for multibank memory. (2) The bank number and address is treated as the C pixel and fed to the eight banks of memory. (3) Eight data are obtained but only the one correspond to the C pixel is chosen.	(1) The *row* and *column* number specified by the first operand is converted to the bank number and address for multibank memory. (2) The bank number and address is treated as the C pixel as shown in Figure 5(a). (3) The data is written into the specified bank and address.
Multibank Function	(1) Output from Register 1 is converted to the bank and address for multibank memory. (2) The coordinate for the N, S, E, W pixels are determined and convert to the bank and address for multibank memory. (3) The addressing is same as the operation shown in Figure 5(b).	Not Available

instructions are designed in favor of the visual saliency, which can be used to find the maximum or minimum value, and performing low-pass or high-pass filtering on the neighbourhood data.

When finding the maximum value among the neighbourhood data, the boundary issue is overcome by forcing the value of the non-existent pixel to zero. Whereas, the most significant bit of the non-existent pixel will be set to '1' when finding the minimum value. The main idea of doing this is to change the value of the non-existent pixel so that it is always smaller or larger than other pixels in the front and latter case respectively.

Data Transferring from External Input or Output Memory

On the RC250 development board, two external memories denoted as the PL1RAM0 and PL1RAM1 will serve as the external input and output memory in the proposed architecture respectively. These two external memories take one clock cycle to load the address and another clock cycle to read or write the data. In other words, two clock cycles are needed to complete the write or read operation. However, they support the use of pipelining that can improve the efficiency in transferring data with proper planning. It is possible to load the new address into the external memory while performing the read or write operation.

In this section, the prefix of *IF/ID*, *ID/EX* and *EXM/WB* is used to indicate in which pipeline register that the data is being stored. For example, *IF/ID Enable* and *EXM/WB Count* imply that the enable signal and the counter value are stored in the *IF/ID* and *EXM/WB* pipeline register respectively. The propagation of all these data can be referred to Figure 7.

The process of transferring data from the external input memory to the multibank memory with respect to the clock cycle is summarized into Table 4. First, the external input memory counter is enabled when *global code[18]* is set to '1'. Meanwhile, the read address of the external input memory is loaded with the value hold in the *IF/ID Count*. In the first clock cycle, the value inside the *IF/ID Count* will be updated. When the data is being reading out, the new value hold by the *IF/ID Count* is set as the new read address of the external input memory. In the second clock cycle, the data will be written into the multibank memory where the write address is fed by the value in the *EXM/WB Count*. At the same time, the value in the *IF/ID Count* and *ID/EX Count* will be updated simultaneously as well. Then, the same process will continue in the third clock cycle and so on. The main concern is to time the read and write operation precisely, so that the data from the external input memory can be written into the multibank memory continuously.

On the other hand, the process of transferring data from the multibank memory to the external output memory with respect to the clock cycle is summarized into Table 5. The external output memory counter is enabled when the *global code[22]* is 1. In the first clock cycle, the write address of the external memory is loaded with the value hold in the *ID/EX Count*. Then in the second clock cycle, the enable signal in *IF/ID Enable* will propagate to the *ID/EX Enable*. The writing process is initiated when the *ID/EX Enable* hold a value of 1. This will write the data in the *EXM/WB Data* to the external output memory where the write address has been set on the first clock cycle. In the third clock cycle, the value in the *IF/ID Count* and *ID/EX Count* is updated again, so does the write address of the external output memory. When come to the fourth clock cycle, new data will be written into the new address of the external memory. From this point onwards, the counter value will increase and propagate

Table 4. The process of transferring data from external input memory to multibank memory

Clock Cycle	**0**	**1**	**2**	**3**
Enable Signal – Use to enable the counter to generate the address.				
Global Code [18]	1	1	1	1
IF/ID Enable	0	1	1	1
Count Value – The count value from the counter that propagates down the pipeline register.				
IF/ID Count	0	1	2	3
ID/EX Count	0	0	1	2
EXM/WB Count	0	0	0	1
RAM Operation – The operation of the data transferring process.				
Set Address [Set (IF/ID Count)]	Set (0)	Set (1)	Set (2)	Set (3)
Read Data [RAM → Data]	X	RAM(0) → Data	RAM(1) → Data	RAM(2) → Data
Transfer [Data → MM (EXM/WB Count)]	X	X	Data(0) → MM(0)	Data(1) → MM(1)
Note:				
X : Don't Care				
MM : Multibank Memory				

Table 5. The process of transferring data from the multibank memory to the external output memory

Clock Cycle	0	1	2	3	4
Enable Signal – Use to enable the counter to generate the address.					
Global Code [22]	1	1	1	1	1
IF/ID Enable	0	1	1	1	1
ID/EX Enable	0	0	1	1	1
Count Value – The count value from the counter that propagates down the pipeline register.					
IF/ID Count	0	0	1	2	3
ID/EX Count	0	0	0	1	2
EXM/WB Count	0	0	0	0	1
Data – The data extracted from the multibank memory its propagation in pipeline register.					
Read Data [MM (ID/EX)]	Data (0)	Data (0)	Data (0)	Data (1)	Data (2)
EXM/WB Data	Data (0)	Data (0)	Data (0)	Data (0)	Data (1)
RAM Operation – The operation of the data transferring process.					
Set Address [Set (ID/EX Count)]	Set (0)	Set (0)	Set (0)	Set (1)	Set (2)
Transfer [EXM/WB Data→RAM]	X	X	Data(0)→RAM(0)	Data(0)→RAM(0)	Data(1)→RAM(1)
Note:					
X	: Don't Care				
MM	: Multibank Memory				

accordingly, which allows the read and write operation to be carried out smoothly.

MIPS Instruction Set

The list of MIPS instructions incorporated into the proposed MIPS architecture is summarized into Table 6. Generally, most of the instructions listed below are similar to those appear in the ordinary MIPS architecture, except those multibank instructions. The multibank instructions are created to improve the processing speed of the visual saliency and DWT. On one hand, the instruction FMAX and FMI are used to find the maximum and minimum value among the neighbourhood pixels, which is a common process in visual saliency. On the other hand, the instruction HP-R, LP-R, HP-C and LP-C are used to perform the row and column filtering in the LeGall 5/3 DWT. It should be noted that all these instructions are capable to identify the non-existent pixels around the boundary using the method discussed previously. This reduces the time taken to identify the boundary manually using the instruction codes.

VISUAL SALIENCY MODEL

A simplified version of the strip-based visual saliency model in the works of Ngau et al. (Ngau, Ang, & Seng, 2009) is implemented as an application for the strip-based MIPS presented in this chapter. The simplified version is much similar to the actual, but with reduced features and the normalization step is skipped. The model uses a strip-based approach whereby the captured input is first partitioned into even sized strips. Each strip will then be processed independently

Table 6. The list of MIPS instructions that is incorporated into the proposed MIPS architecture

Operation	Code	Format	Example	Remark
Arithmetic Function				
Addition	ADD	R	\$R1, \$R2, \$R3	\$R3 = \$R2 + \$R1
Subtraction	SUB	R	\$R1, \$R2, \$R3	\$R3 = \$R2 - \$R1
Add Immediate	ADDI	I	\$R1, \$R2, Value	\$R2 = \$R1 + Value
Add Shift	AS	R	\$R1, \$R2, \$R3	\$R3 = (\$R1 + \$R2) >> 1
Add Shift Shift	ASS	R	\$R1, \$R2, \$R3	\$R3 = (\$R1 + \$R2 + 2) >> 2
Logical Function				
Bit-wise AND	AND	R	\$R1, \$R2, \$R3	\$R3 = \$R2 & \$R1
Bit-wise OR	OR	R	\$R1, \$R2, \$R3	\$R3 = \$R2 | \$R1
Shift Right	SRL	R	\$R1, X, \$R3	\$R3 = \$R1 >> 1
Shift Left	SLL	R	\$R1, \$R2, \$R3	\$R3 = \$R1 << 1
Set on Less Than	SLT	R	\$R1, \$R2, \$R3	$\$R3 = \begin{cases} 1, \text{ If } (\$R2 > \$R1) \\ 0, \text{ Otherwise} \end{cases}$
Multibank Addressing				
Load Word	LW	I	\$R1, \$R2, X	\$R2 = Multibank Memory[\$R1]
Store Word	SW	I	\$R1, \$R2, X	Multibank Memory[\$R1] = \$[R2]
Multibank Function*				
Find Maximum	FMAX	I	\$R1, \$R2, X	\$R2 = Max(N / S / E / W / C)
Find Minimum	FMIN	I	\$R1, \$R2, X	\$R2 = Min(N / S / E / W / C)
High-Pass Filter Row	HP-R	I	\$R1, \$R2, X	\$R2 = C – (W + E) / 2
Low-Pass Filter Row	LP-R	I	\$R1, \$R2, X	\$R2 = C + (W + E + 2) / 4
High-Pass Filter Column	HP-C	I	\$R1, \$R2, X	\$R2 = C – (N + S) / 2
Low-Pass Filter Column	LP-C	I	\$R1, \$R2, X	\$R2 = C + (N + S + 2) / 4
Note:				
X	: The field is not used in that instruction.			
*	: The output of register \$R1 will serve as the address for the center pixels in the multibank function.			

through a recursive saliency algorithm. The strip size is user specific and can be varied according to the available memory resources. The output will be better defined with larger strip size since the loss of information at the strip boundaries can be minimized. Once all the strips are processed, they are recombined to form the final saliency map to be output. Then, the whole process is repeated with the next captured input. The next paragraph onwards will cover the saliency algorithm used in the implementation.

For each grayscale strip, the lifting based 5/3 DWT filter is applied to construct the multi-resolution pyramid. For a level of DWT decomposition, there will be four output sub-bands. Each sub-band will have a size down-sampled by a factor of two of the input strip. The LL sub-band (approximate coefficients, A) is a coarser representation of the input strip. This LL sub-band is used to compute the intensity feature. The other three sub-bands, HL (vertical details, V); LH (horizontal detail, H); and HH (diagonal detail, D) can be used as the orientation feature. However, in this simplified version, they are not used and can be discarded to free additional memory. The LL sub-band is further decomposed twice, giving a three-level pyramid with level $j \in \{1,2,3\}$.

Once the multi-resolution pyramid is constructed, the center-surround (CS) process is applied. In many biological plausible models, this process is performed by designating an image at level j (center) and subtracting images from subsequent levels (surround) from the center image. The original process will require an image pyramid with eight levels, which will significantly increase the amount of memory required. Instead of having additional pyramid levels, another version of the CS process presented by Tsapatsoulis (Tsapatsoulis & Rapantzikos, 2006) can be used. This version of CS simulates the center-surround in the retina ganglion cells in the form of morphological close-minus-open (CMO). The CMO operation is described in (5).

$$CM^{-j}(m,n) = Y_A^{-j}(m,n) \bullet se - Y_A^{-j}(m,n) \circ se \quad (5)$$

In (5), $CM^{-j}(m,n)$ and $Y_A^{-j}(m,n)$ are the conspicuity map and the LL sub-band at level j respectively; se is a cross structuring element with four neighbor connectivity; $\bullet$and $\circ$ denotes morphological opening and closing respectively.

The conspicuity maps at level j are then interpolated to the size of the input strip and summed across scale. The final saliency map is found by averaging the summed conspicuity maps with the number of levels and then normalizing it to the range of [0, 255] (grayscale). The normalization step, as mentioned earlier in the section, is skipped to keep the implementation simple. However, the normalization step can be performed at the end-user level for viewing purposes. The saliency map for each strip is gradually recombined to form the full saliency map.

For the visual saliency implementation on the strip-based MIPS, the grayscale capture is of size 128 x 128. The strip size is chosen to be 16 x 128, giving a total of eight strips per captured input.

Implementation of the Discrete Wavelet Transform

With the used of 4-connectivity approach, the processing speed of the lifting scheme LeGall 5/3 DWT can be greatly improved. In this case, the coefficients at the odd coordinates are filtered before the even ones. Using the 4 x 4 pixels example shown in Figure 9, the sequences of filtering in the horizontal direction will be done for coefficient '6' followed by '5' and then '8' followed by '7'.

Here, with the proposed 4-connectivity methodology, when the pixel '6' is read from the memory, the four neighbours '2', '5', '7' and '10' are also read at the same time. For row filtering, only coefficient '5' and '7' are needed. As shown in the pseudo-code below that comparing the traditional method where each of the coefficient '5' and '7' are needed to be loaded besides the '6', the number of instructions for the proposed work with 4-connectivity methodology is reduced. Instead of taking five clock cycles to load and process the coefficients '5', '6' and '7', only two clock cycles are needed to load all three of the coefficients and process them with the proposed work. This reduction can lead to very huge improvement in the processing speed of the DWT implementation. It should be noted that the process of filtering using the proposed 4-connectivity methodology is applied to both the odd and even components in row filtering as well as for the column filtering process (Exhibit 1 and 2).

Implementation of the Visual Saliency Model

The visual saliency algorithm using the strip-based approach discussed in the previous section is executed sequentially as a set of MIPS instructions.

Figure 9. An example of applying the 4-connectivity approach to speedup the filtering process

	Col 0 (even)	Col 1 (odd)	Col 2 (even)	Col 3 (odd)
Row 0 (even)	1	**2**	3	4
Row 1 (odd)	**5**	**6**	**7**	8
Row 2 (even)	9	**10**	11	12
Row 3 (odd)	13	14	15	16

Exhibit 1. Pseudo Codes for Row Filtering using Traditional Methodology:

```
Load coefficient '5' into Register 0.
Load coefficient '6' into Register 1.
Load coefficient '7' into Register 2.
Perform the computation of ('5' + '7' + 2) / 2 and store into Register 0.
Subtract content of Register 0 from Register 1 and store into Register 2.
*The Register 2 now stored the row filtered value of coefficient '6'.
```

Exhibit 2. Pseudo Codes for Row Filtering using the 4-Connectivity Methodology:

```
Load coefficient '6' and its 4 neighbours, and perform the computation of '6'
- [('5' + '7' + 2) / 2].
Store the result into Register 0.
* The Register 0 now stored the row filtered value of coefficient '6'.
```

In this section, the focus will be on the algorithm after the multi-resolution pyramid construction. The discrete wavelet transform algorithm is handled by a separate set of MIPS instructions. The pseudo-code for the algorithm in terms of MIPS implementation is shown on the following page (Exhibit 3 and 4).

The most prominent advantage of implementing the visual saliency on the strip-based MIPS can be seen in the significant reduction of clock cycles required to perform the center-surround process (CS). As explained in the previous section, the CS process is performed by means of morphological operations whereby the opening result is subtracted from the closing result. For the morphological closing, dilation is first applied to the image and followed by erosion. Similarly, in the morphological opening, the erosion is applied first followed by dilation.

In this MIPS implementation, the dilation operation takes the location (m, n) as the center (C) pixel, and compares the value at this location

Exhibit 3. Pseudo-code for strip-based visual saliency approach

```
Input grayscale image
Divide input image into eight strips images
Set main counter to zero
While main counter is less than nine
        Input a strip image
        Perform three-level DWT on strip image
        Store LL sub-bands from scale j: one to three
        Set scale j to one
        While scale j is less than four
                Perform morphological closing on LL sub-band at scale j
                Perform morphological opening on LL sub-band at scale j
                Subtract opening result from closing result (CMO)
                Resize CMO result for scale j to size of strip image
        Sum CMO result for all j scales
        Average the CMO result
Store the average CMO result to output memory
Get next input strip image
Output saliency map
```

with the values of the surrounding west (W), north (N), east (E), and south (S) pixels to find a maximum. It then stores the maximum value at another memory location with respect to the location (*m*, *n*) for the dilation result. On the other hand, the erosion operates in the same way as the dilation, with the difference of finding the minimum between the C, W, N, E, and S pixels instead of a maximum. The pseudo-code for the morphological closing as a MIPS implementation is also shown in addition to the pseudo-code for the saliency algorithm. The pseudo-code for the morphological opening is similar to the morphological closing with the exception of performing erosion first and subsequently followed by dilation.

In the original model, bilinear interpolation is used to resize the conspicuity maps. However, due to the limited type of instructions; a simplified interpolation similar to the nearest neighbor interpolation is used keep the algorithm simple and the number of clock cycles low. The interpolation is performed in sequence of up-sampling by factor of two. Hence, the conspicuity map at level three would require three passes to reach the actual size of the strip.

SIMULATION RESULTS

Since multiple levels are used in the visual saliency model, the morphological close-minus-open (CMO) operation covers a significant amount of the total processing time of the algorithm. A reduction in clock cycles for the CMO alone would allow a tremendous speed up in processing time. The strip-based MIPS processor has the capability of simultaneously accessing all four neighbors' values when given the center address. This simultaneous data access lies in the multibank memory design. Consider the case of a dilation operation in the visual saliency implementation. The dilation operation requires accessing to the four neighbor values through a load word instruction that uses the address of the center pixel as the

Exhibit 4. Pseudo-code for morphological closing at scale j for MIPS implementation

```
Store starting address for scale j LL sub-band values into register $s1
Store starting address for dilation result destination into register $s2
Store number of rows of scale j LL sub-band into register $t2
Store number of columns of scale j LL sub-band into register $t3
Set register $t4 as row counter with initial value of zero
Set register $t5 as column counter with initial value of zero
While $t4 is less than $t2
        While $t3 is less than $t5
                Invoke compare-word-max instruction for address in $s1
                Store maximum neighbourhood result into register $s3
                Store value in $s3 into Memory[$s2]
                Increment counter value in $t5 by one
                Increment address in $s1 by one
Increment address in $s2 by one
        Increment counter value in $t4 by one
        Increment address in $s1 to one row below
Increment address in $s1 to one row below
Store starting address for dilation result into register $s1
Store starting address for erosion result destination into register $s2
Reset value in register $t4 to zero
Reset value in register $t5 to zero
While $t4 is less than $t2
        While $t3 is less than $t5
                Invoke compare-word-min instruction for address in $s1
                Store minimum neighbourhood result into register $s3
                Store value in $s3 into Memory[$s2]
                Increment counter value in $t5 by one
                Increment address in $s1 by one
Increment address in $s2 by one
        Increment counter value in $t4 by one
        Increment address in $s1 to one row below
Increment address in $s2 to one row below
```

core for maximum value comparison. In the case where instructions are processed sequentially and all five pixels (four neighbors and center) have to be loaded individually into registers for comparison, the total clock cycle consumed will be five.

With the multibank memory access offered by the strip-based MIPS, all five values can be accessed simultaneously for comparison in a single clock cycle. A single strip alone has three CMO operations with 8,064 center addresses to be loaded into registers. In a non-multibank access, accessing all four neighbors along with the center would require 40,320 clock cycles for a single strip; and for a complete image, it would take up 322,560 clock cycles. If the neighbors' and the center values are to be accessed using the multibank method, it would only require 64,512 clock cycles for an entire image. This would result in a savings of 80% clock cycles compared to the one with non-multibank access.

Simulation results for single grayscale inputs are presented in Figure 10. Figure 10(b) shows the results from the original algorithm using only the intensity feature. Figure 10(c) and (d) each shows the implemented saliency algorithm using non strip-based and strip-based approaches respectively. From the results of the implementation, it can be seen that most of the distortion comes from the simplified interpolation, as the method does not provide a smooth transition across pixel values like the bilinear interpolation. As for the loss of information at the strip boundaries due to the strip-based approach, horizontal line artifacts are observed. These artifacts are due to predicted values along the boundaries as the actual neighboring values only available in the subsequent and previous strips. Distortions due to these artifacts are less obvious when large images or large strips are used. However, if the horizontal artifacts are of concern, they can be removed using an overlapping strip-based method proposed in (Ngau, Ang, & Seng, 2009b). Other than the noticeable distortion, the detection result works as well as the original algorithm. The operating results of the visual saliency implementation on the strip-based MIPS are summarized in Table 7.

The implemented visual saliency model on the strip-based MIPS is able to achieve a frame rate of 10.06 fps with the discrete wavelet transform and visual saliency instructions. It is an improvement of the Itti and Koch model (Itti & Koch, 2000) with better output resolution, more defined object boundaries, and more accurate highlighting of salient objects. Compared to the implementation by Han and Zhou (Han & Zhou, 2007) of 12.2 fps for an image with resolution of 352 x 288 on a 2.4 GHz Pentium 4 processor; the obtained result of 10.06 fps running at the speed of 24.576 MHz did not fall far behind from the former implementation on a central processing unit (CPU). In the implementation on a Field-Programmable Gate Array (FPGA) by Kapre et al. (Kapre, Walther, Koch, & DeHon, 2004), 30 fps was achieve with the Itti and Koch model; however, no processor clock rate was documented. It is worth noticing that the implemented visual saliency algorithm operates pixel-based with available neighbor values; therefore, the number of clock cycles is the same for any image of the same size.

CONCLUSION

In this book chapter, a strip-based MIPS processor incorporated with the 4-connectivity approach for parallel access to neighbourhood data is presented. The strip-based MIPS processor is designed in favor of strip-based processing and implementation for VSN. In this case, the transferring and processing of data can be pipelined to improve the efficiency when compared to performing these tasks sequentially. In addition, the performance of implementing the visual saliency model using the proposed MIPS processor is also demonstrated. When the MIPS processor is operating at 24.576 MHz, a processing rate of 10.06fps, which includes the execution of DWT and visual saliency, is achieved.

Table 7. The simulation results on performing the visual saliency using the proposed MIPS processor.

Input Size of 128 x 128		
Number of Instructions (DWT + Saliency)	Number of Clock Cycles (DWT + Saliency)	Frames Per Second (fps)
457 + 1456	696824 + 1746416	10.06

Figure 10. A comparison of the visual saliency outcome of the non strip-based and strip-based method

REFERENCES

Achanta, R., Hemami, S., Estrada, F., & Susstrunk, S. (2009). Frequency-tuned salient region detection: *IEEE International Conference on Computer Vision and Pattern Recognition*.

Akyildiz, I. F., Melodia, T., & Chowdhury, K. R. (2006). A survey on wireless multimedia sensor networks. *Computer Networks: The International Journal of Computer and Telecommunications Networking*, *51*(4), 921–960.

Akyildiz, I. F., Melodia, T., & Chowdhury, K. R. (2008). Wireless multimedia sensor networks: Applications and testbeds, In *Proceedings of the IEEE* (pp. 1588-1605).

Altera Corporation. (2007). *Stratix II device family data sheet.* Retrieved September 11, 2010, from http://www.altera.com/ literature/ hb/ stx2/ stx2_ sii5v1_ 01.pdf

Ammari, A. C., & Jemai, A. (2009). Multiprocessor platform-based design for multimedia. *Journal of IET Computers and Digital Techniques*, *3*(1), 52–61. doi:10.1049/iet-cdt:20070168

Archaya, T., & Tsai, P.-S. (2004). *JPEG2000 Standard for image compression: Concepts, algorithms and VLSI architectures*. New York, USA: Wiley-Interscience.

Bhargava, R., Kargupta, H., & Powers, M. (2003). Energy consumption in data analysis for on-board and distributed applications, In *Proceedings of the ICML Workshop on Machine Learning Technologies for Autonomous Space Applications*, Washington, DC, USA.

Broadbent, D. (1958). *Perception and communication* (p. 338). London, New York: Pergamon Press. doi:10.1037/10037-000

Burt, P. (1988). Attention Mechanism for Vision in a Dynamic World, In *Proceedings of the 9th International Conference on Pattern Recognition*.

Cao, Z.-Y., Ji, Z.-Z., & Hu, M.-Z. (2005). An image sensor node for wireless sensor networks, In. *Proceedings of the International Conference on Information Technology: Coding and Computing*, *2*, 740–745.

Chew, L. W., Chia, W. C., Ang, L. M., & Seng, K. P. (2009). Very low-memory wavelet compression architecture using strip-based processing for implementation in wireless sensor networks, *EURASIP Journal on Embedded Systems*, 2009.

Faraji, I., Didehban, M., & Zarandi, H. R. (2010). Analysis of transient faults on a MIPS-based dualcCore processor, In *Proceedings of the International Conference on Availability, Reliability and Security* (pp. 125-130), Krakow, Poland.

Frintrop, S. (2007). *Visual robot localization and mapping based on attentional landmarks*. (LNCS 4667, pp. 456-459). Berlin/Heidelberg, Germany: Springer-Verlag.

Gall, D. L., & Tabatabai, A. (1988). Sub-band coding of digital images using symmetric short kernel filters and arithmetic coding techniques. *Proceedings of IEEE International Conference on Acoustic, Speech, and Signal Processing*, *2*, 761–764.

Garcia, R. -Ross, A. G., & George, A. D. (2009). Exploiting partially reconfigurable FPGAs for situation-based reconfiguration in wireless sensor networks, In *Proceedings of the IEEE Symposium on Field Programmable Custom Computing Machiens* (pp. 243-246).

Gautham, P., Parthasarathy, R., & Balasubramanian, K. (2009). Low-power pipelined MIPS processor design, In *Proceedings on the 12th International Symposium on Integrated Circuits* (pp. 462, 465), Suntec, Singapore.

Graphics, M. (2010). *RC250: High performance SOPC platform*. Retrieved June 15, 2010, from http://www.mentor.com/ products/ fpga/ handel-c/ rc-series- platforms/ upload/rc250.pdf

Han, B., & Zhou, B. (2007). High speed visual saliency implementation on GPU, In *Proceedings on IEEE International Conference on Image Processing* (pp. 361 – 364), San Antonio, USA.

Hasan, M. R., Rahman, M. S., Hasan, M., Hasan, Md. M., & Ali, M. A. (2010). An improved pipelined processor architecture eliminating branch and jump penalty, In *Proceedings of the 2nd International Conference on Computer Engineering and Applications* (pp. 621 – 625), Bali Island, Indonesia.

Hou, X., & Zhang, L. (2007). Saliency detection: A spectral residual approach, In *Proceedings of the IEEE Conference on Computer Vision and Recognition* (pp. 1-8).

Hu, X., Liu, Z., Pan, Y., Zeng, Z., & Meng, M. Q.-H. (2009). A novel FPGA-based wireless vision sensor node, In *Proceedings of the IEEE International Conference on Automation and Logistics* (pp.841-846), Shenyang, China.

Itti, L., & Koch, C. (2000). A saliency-based search mechanism for overt and covert shift of visual attention. *Journal of Vision Research*, *40*(10-12), 1489–1506. doi:10.1016/S0042-6989(99)00163-7

Kapre, N., Walther, D., Koch, C., & DeHon, A. (2004). Saliency on a chip, *The neuromorphic engineer* (pp. 1 – 2).

Kateeb, A. E., & Mandepanda, A. (2009). Embedded Soft Processor for Sensor Networks, *Proceedings of the International Conference on Network-Based Information Systems* (pp. 268 – 272), Indianapolis, Indiana, USA.

Koch, C., & Ullman, S. (1985). Shifts in selective visual attention: Towards the underlying neural circuitry. *Human Neurobiology*, *4*, 219–227.

Kwok, T. T.-O., & Kwok, Y.-K. (2006). Computation and energy efficient image processing in wireless sensor networks based on reconfigurable computing, In *Proceedings of the International Conference Workshops on Parallel Processing* (pp.43-50), Columbus, OH.

Li, X., & L, T. (2004). ECOMIPS: An economic MIPS CPU design on FPGA, In *Proceedings on the 4th IEEE International Workshop on System-on-Chip for Real-Time Applications* (pp. 291 – 294), Banff, Alberta, Canada.

Liu, Y., Xie, G., Chen, P., Chen, J., & Li, Z. (2009). Designing an asynchronous FPGA processor for low-power sensor networks, In *Proceedings of the International Symposium on Signals, Circuits and Systems* (pp. 1-6), Iasi, Romania.

Luker, J. D., & Prasad, V. B. (2001). RISC system design in an FPGA, In *Proceedings of the 4th IEEE Midwest Symposium on Circuits and Systems* (pp. 532 - 536), Dayton, OH.

Montabone, S., & Soto, A. (2010). Human detection using a mobile platform and novel features derived from a visual saliency mechanism [Amsterdam, The Netherlands: Elsevier.]. *Image and Vision Computing*, *28*(3), 391–402. doi:10.1016/j.imavis.2009.06.006

Ngau, C. W. H., Ang, L.-M., & Seng, K. P. (2009a). Low memory implementation of saliency map using strip-based method, In *Proceedings of the International Visual Informatics Conference* (pp. 712-726), Kuala Lumpur, Malaysia.

Ngau, C. W. H., Ang, L.-M., & Seng, K. P. (2009b). Low memory strip based visual saliency algorithm for hardware constrained environment. *International Journal of Computer and Network Security*, *1*(2).

Patterson, D. A., & Hennessy, J. L. (2005). *Computer organization and design – The hardware / software interface*. San Francisco, CA: Morgan Kaufmann.

Raghunathan, V., Ganeriwal, S., & Srivastava, M. (2004). Energy efficient wireless packet scheduling and fair queuing. *ACM Transactions on Embedded Computing System*, *3*(1), 3–23. doi:10.1145/972627.972629

Ramdas, T., Ang, L. M., & Egan, G. (2004). FPGA implementation of an integer MIPS processor in Handel-C and its application to human face detection, In *Proceedings of the IEEE 10 Conference on TENCON* (pp. 36 - 39), Chiang Mai, Thailand.

Ratha, N. K., & Jain, A. K. (1999). Computer vision algorithms on reconfigurable logic arrays. *IEEE Transactions on Parallel and Distributed Systems*, *10*(1), 29–43. doi:10.1109/71.744833

Soro, S., & Heinzelman, W. (2009). A survey of visual sensor networks. *Advances in Multimedia, 2009*(640386). doi:10.1155/2009/640386

Treisman, A. (1985). Preattentive processing in vision [San Diego, CA: Academic Press Professional, Inc.]. *Computer Vision Graphics and Image Processing*, *31*(2), 156–177. doi:10.1016/S0734-189X(85)80004-9

Tsapatsoulis, N., & Rapantzikos, K. (2006). *Wavelet based estimation of saliency maps in visual attention algorithms* (LNCS 4132, pp. 538-547). Berlin/Heidelberg, Germany: Springer-Verlag.

Tyson, R. A. L., Intan, S., & Adiono, T. (2009). A pipelined double-issue MIPS based processor architecture, In *Proceedings on the International Symposium on Intelligent Signal Processing and Communication Systems* (pp.583 - 586), Kanazawa, Japan.

Wolfe, J. (1998). In Pashler, H. (Ed.), *Visual search, in attention* (pp. 13–74). London, UK: University College London.

Yeong, L. S., Ngau, C. W. H., Ang, L. M., & Seng, K. P. (2009). Efficient processing of a rainfall simulation watershed on an FPGA-based architecture with fast access to neighbourhood pixels, *EURASIP Journal on Embedded Systems*, 2009.

Yu, S. X., & Lisin, D. A. (2009). *Image compression based on visual saliency at individual scales* (LNCS 5875, pp. 157-166). Berlin/Heidelberg, Germany: Springer-Verlag.

Zulkifli, M., Yudhanto, Y. P., Soetharyo, N. A., & Adiono, T. (2009). Reduced stall MIPS architecture using pre-fetching accelerator, In *Proceedings of the International Conference on Electrical Engineering and Informatics* (pp. 611-616), Selangor, Malaysia.

ADDITIONAL READING

Dandamudi, S. P. (1995). *Fundamentals of computer organization and design*. Berlin/Heidelberg, Germany: Springer-Verlag.

Ghanbari, M. (2003). *Standard codecs: Image compression to advance video coding*. UK: MPG Books Limited.

Hennessy, J., & Patterson, D. A. (2007). *Computer architecture: A quantitative approach*. San Francisco, CA: Morgan Kaufmann.

Li, J., Tian, Y., Huang, T., & Gao, W. (2009). A dataset and evaluation methodology for visual saliency in video, In *Proceedings of the IEEE International Conference on Multimedia and Expo* (pp. 442- 445), New York, USA.

MacLean. W. J. (2005). An evaluation of the suitability of FPGAs for embedded vision systems, In *Proceedings of the IEEE Computer Society Conference on Computer Vision and Pattern Recognition* (pp. 131- 137), San Diego, CA, USA.

Navabi, Z. (2007). *Embedded core design with FPGAs*. New York, NY: McGraw-Hill.

Rabaey, J. M. (2000). Silicon platforms for the next generation wireless systems – What role does reconfigurable hardware play? In *Proceedings of the Roadmap to Reconfigurable Computing* (pp. 277- 285), London, UK.

Sohraby, K., Minoli, D., & Znati, T. F. (2007). *Wireless sensor networks: Technology, protocols, and applications*. Hoboken, NJ: John Wiley & Sons. doi:10.1002/047011276X

Chapter 15
Applications for Wireless Visual Sensor Networks:
The Digital Zoo

Johannes Karlsson
Umeå University, Sweden

Tim Wark
CSIRO, Australia

Keni Ren
Umeå University, Sweden

Karin Fahlquist
Umeå University, Sweden

Haibo Li
Umeå University, Sweden

ABSTRACT

In this chapter we will describe our work to set up a large scale wireless visual sensor network in a Swedish zoo. It is located close to the Arctic Circle making the environment very hard for this type of deployment. The goal is to make the zoo digitally enhanced, leading to a more attractive and interactive zoo. To reach this goal the sensed data will be processed and semantic information will be used to support interaction design, which is a key component to provide a new type of experience for the visitors. In this chapter we will describe our research work related to the various aspects of a digital zoo.

INTRODUCTION

The Swedish zoo "Lycksele Djurpark" is the northernmost zoo in the country. It is specialized in housing nordic animals and have animals like the bear, elk, musk-ox and wild boar to mention some. The zoo is home to more than 400 animals, divided into 30 species and covers an area of 47 hectares. It has been one of the biggest tourist attractions in northern Sweden. Due to the weather conditions and the sparsely populated location of the zoo, it is difficult to attract visitors to the zoo during winter. The opening season is for that

DOI: 10.4018/978-1-61350-153-5.ch015

Figure 1. Work packages in the Digital Djurpark project. A wireless visual sensor network is used to capture information about the animals. This information is processed and interaction techniques are used to present the information for visitors.

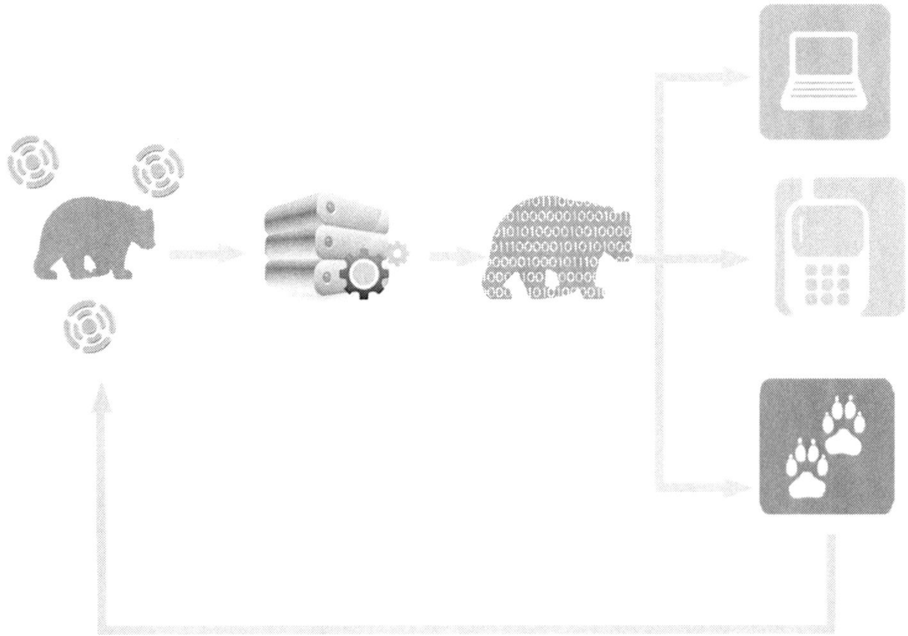

reason from the end of May to the end of September with some additional openings around holidays like Christmas and Easter. Although the zoo is closed during the winter the animals' normal life continues as usual and they are outdoors. Consequently, it is challenging for the zoo to make a profit and to develop further.

The goal of the project "Digital Djurpark" (Digital Zoo) funded by the EU regional development fund is to make the zoo digital and more attractive. The approach is to deploy a wireless sensor network covering the whole zoo. This network will consist of around 100 wireless sensors, which are able to configure themselves into a working network. The sensors will be used to collect data such as video, pictures, sound and temperature. The sensed data will be processed and semantic information will be used to support visitor interaction design, which is a key aspect of the new experience envisaged for the visitors. Figure 1 shows an overview of the working packages included in the project.

In this chapter we will describe our work on creating the digital zoo, based on the proposed large scale testbed using wireless visual sensor networks. This will give us a unique opportunity to test long term wireless senor network deployments in an environment where there are a lot of events to detect. In the following sections we will describe the hardware platforms, sensors, computer vision system used in the network, and interaction design.

Hardware Platforms

To collect information such as audio, video and position from individual animals a heterogeneous set of wireless sensors are used. At some locations we have access to the mains power supply and can use more powerful cameras that are able to perform advanced computer vision functions and

Figure 2. Illustration of camera stack formed by a Fleck™ sensor node, TI DSP or AD Blackfin board and CMOS camera

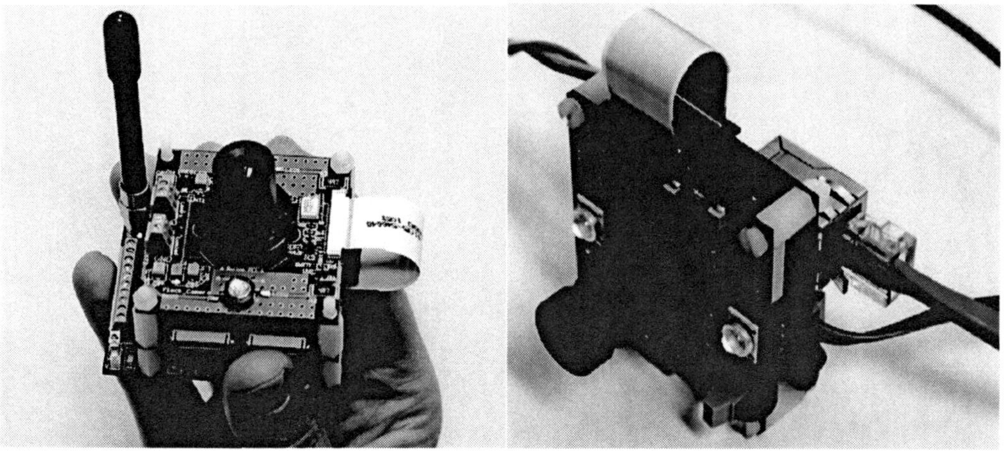

deliver high-quality video streaming. At other locations we need to use low-power cameras that can operate using only harvested solar energy. These sensors will not be at continuous full operation. Instead they will be triggered on specific events and will process the collected media information only when some interesting events are detected.

Low-Power Wireless Camera System

One of the central challenges for image processing in sensor networks is that a single image contains a relatively large amount of raw data while sensor network nodes have limited computing power and memory. The typical Wireless Sensor Network (WSN) "mote-class" nodes have 8 kbytes of RAM which is far less than the 77 kbytes required for even a small QVGA (320x240) resolution image. Various approaches have been investigated in the literature (Diao, Ganesan, Mathur, Shenoy 2007; Yick, Mukherjee, Ghosal 2008; Xia, 2009) to tackle this problem. Our approach to this dilemma is to augment the basic (mote-class) sensor node with an image sensor and high-performance processor, which we treat as one subsystem, as shown in Figure 2. By selective use of the high-performance co-processor, the system can run as a low-power node for most of its operations, but access much more powerful (and expensive) resources only when needed. Our low-power image processing platform consists of a "mote-class" platform known as Fleck™ (Corke, Sikka, 2006) and a high performance co-processor for image processing.

The Fleck™, consists of an Atmega1281 microcontroller with 8 Kbyte of RAM running at 8 MHz, a Nordic NRF905 radio transceiver, a real-time clock, 1Mbyte of flash memory and a temperature sensor. For high-performance co-processor we use the Analog-Devices 600 MHz ADSP-BF537 Blackfin processor. The Blackfin belongs to the family of 32-bit RISC processors and has no memory management. We use the BlueTechnix BF537E board which comprises the CPU with 132 KB SRAM, 32 MB external SDRAM and 4 MB flash memory. The Blackfin has an SPI bus that is connected to the Fleck for inter-processor communications. Finally, the image processing platform has a camera daughter board that contains a sensor chip, lens holder and two switchable ultra-bright LEDs for illumination. The sensor is an OV7640 Color CMOS device from OmniVision Technologies Inc., which outputs VGA (640 x 480) or QVGA (320 x

Figure 3. Our high-performance camera platform based on the Gumstix Overo™ Fire computer-on-module

240) images at a maximum rate of 30 fps or 60 fps respectively.

High Performance Camera

To capture information emanating from the animals we have also developed a camera platform based on the Gumstix Overo™ Fire computer-on-module and a custom image sensor board, as shown in Figure 3. The Gumstix Overo™ Fire consists of a Texas Instrument Omap 3530 application processor with ARM Cortex-A8 CPU and a C64x+ digital signal processor (DSP) core, both running at 600 MHz. The board has 256 MB of RAM and 256 MB of flash memory. For communication an IEEE 802.11g wireless LAN is used as a wireless mesh network. The board measures only 17 mm x 58 mm x 4.2 mm which enabled us to build a really small and powerful camera platform.

The camera board contains an OmniVision OV9710 image sensor supporting 1280x800 at 30 fps. The image sensor also has a good low light sensitivity of 3.3V per lux-second.

For this platform we use a multi-hop wireless ad-hoc network on 802.11 wireless links. Since this platform has high computational capacity, as well as high communication capacity, it provides the ability to send high-quality real-time video. This platform has however a rather high power consumption, typically a few watts, and therefore will in general be connected to the mains power supply.

Sensor Network Deployment

The Lycksele zoo is located in northern Sweden close to the Arctic Circle. During winter the sun does not comes up over the horizon for many days and during this period it is dark the whole days. The temperature is often below minus 30 degrees Celsius and can occasionally drop well below minus 40 degrees Celsius. The ground is covered by more then one meter of snow from November to May. We are deploying a large scale wireless visual sensor network that must operate under these harsh conditions.

The zoo covers an area of 47 hectares and is home to 400 animals divided into 30 species. Therefore there is a rich set of events to detect, given the large numbers of active animals. This is a great setting for us to test a large scale, long term deployment of wireless visual sensor networks. We plan to cover the entire zoo and use around 100 sensors.

The current deployment in the zoo consists of 25 video transmitting nodes. The nodes used in this network are based on our high-performance wireless LAN platform. They are distributed so as to cover the entire zoo, see Figure 4. At a central location in the zoo, close to the bear den, we have a fiber connection to the Internet and this is the location of the sink node in the wireless network. The nodes configure themselves into a working ad-hoc network and send all collected multimedia information back to the sink node. The network is rather large and to reach some nodes four hops are needed from the sink node. This deployment has been in place and running for over one year in the zoo.

Energy Analysis

In order to ensure continuous operation of cameras over extended periods of time, cameras will be deployed with a combination of rechargeable battery energy storage and photo-voltaic solar cells. To assist with design decisions and developing research questions around adaptive energy protocols, we have seeked to build a thorough understanding of the amount of solar energy which is available to harvest throughout the year.

In order to understand the trends in daily and yearly solar radiation (independent of weather changes) we have utilized a standard atmospheric model (Jeong, Jiang, Culler, 2008) which relates solar radiation as a function of the position on the surface of the planet and day of year. An example of this is shown in Figure 5 for the latitude of Umea, Sweden. As expected, peak solar radiation occurs during the months of June/July and during the middle of the day.

We have also analyzed the patterns from historical solar radiation data from the area of the zoo deployment, shown in Figure 6(a) for a period of three years. As expected, yearly solar radiation patterns follow a similar trend to that predicted by the solar atmospheric model. In addition to the yearly trends, key periods of low solar radiation can be observed due to local weather conditions.

Given this data, we also calculate what the effective duty-cycle of the camera would be using the energy model for our platform as described in (Wark, Corke, Liu, Moore, 2008). As shown in Figure 6(b) (assuming that only enough energy storage exists to store a day's energy) we show that in order for a camera to maintain continuous operation under the solar radiation levels occurring at Umea, the camera needs to run at a maximum duty cycle of 18 percent in the longest days of the year down towards zero in the shortest days. This results in an average duty cycle of around 5% over the entire year.

Ongoing research will investigate how the use of additional stored energy, smart triggering strategies (e.g. low-power passive IR sensors) and adaptive protocols can be used to both increase the average duty-cycle as well as bias the active periods towards periods the maximize the overall utility of the cameras.

Computer Vision

The Computer vision added to the system, concerned with the theory behind artificial systems that extract information from videos and images make the camera into an artificial intelligence machine

Figure 4. An overview of the zoo and the deployed sensor nodes

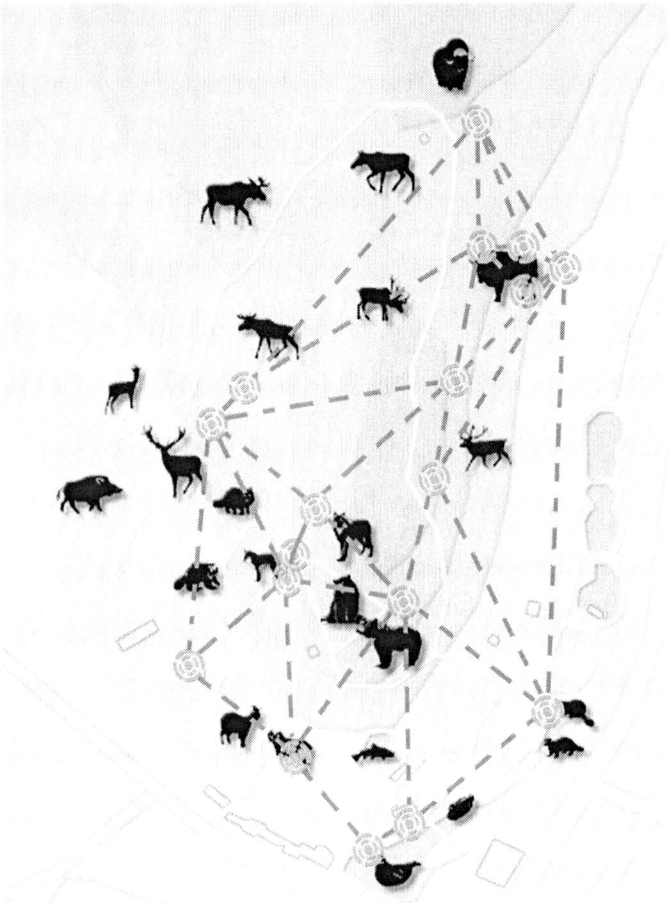

that not only can "see", but also "understand". In the Digital Djurpark project, the computer vision system analyzes the videos captured from the ad-hoc camera system. Examples of applications of computer vision include systems that:

- Detect the motion of the animals.
- Separate the foreground and background. Retrieve the location information of the animals.
- Track moving animals
- Understand the behavior of and control the animals.

Detection of motion focuses on video which contains animal events, avoids long video time with a blank background without animals. Motion detection works by comparing the intensity of pixels between new image and a reference image. If there is no motion and no noise, the difference between the images is zero. If there has been a change between the pictures the result will be different. To avoid the noise from being regarded as motion, a threshold is used which means that there has to be a certain number of pixels with changed intensity for a change to be regarded as motion. With this simple strategy, motion video can be recorded from live camera.

Foreground and background detection from a video stream is an essential task in video processing. In the real environment of the zoo, there is complex background appearing in the video, for

Figure 5. Normalized daily solar radiation at Umea (as derived from solar atmospheric model) as a function of day of year and time of day

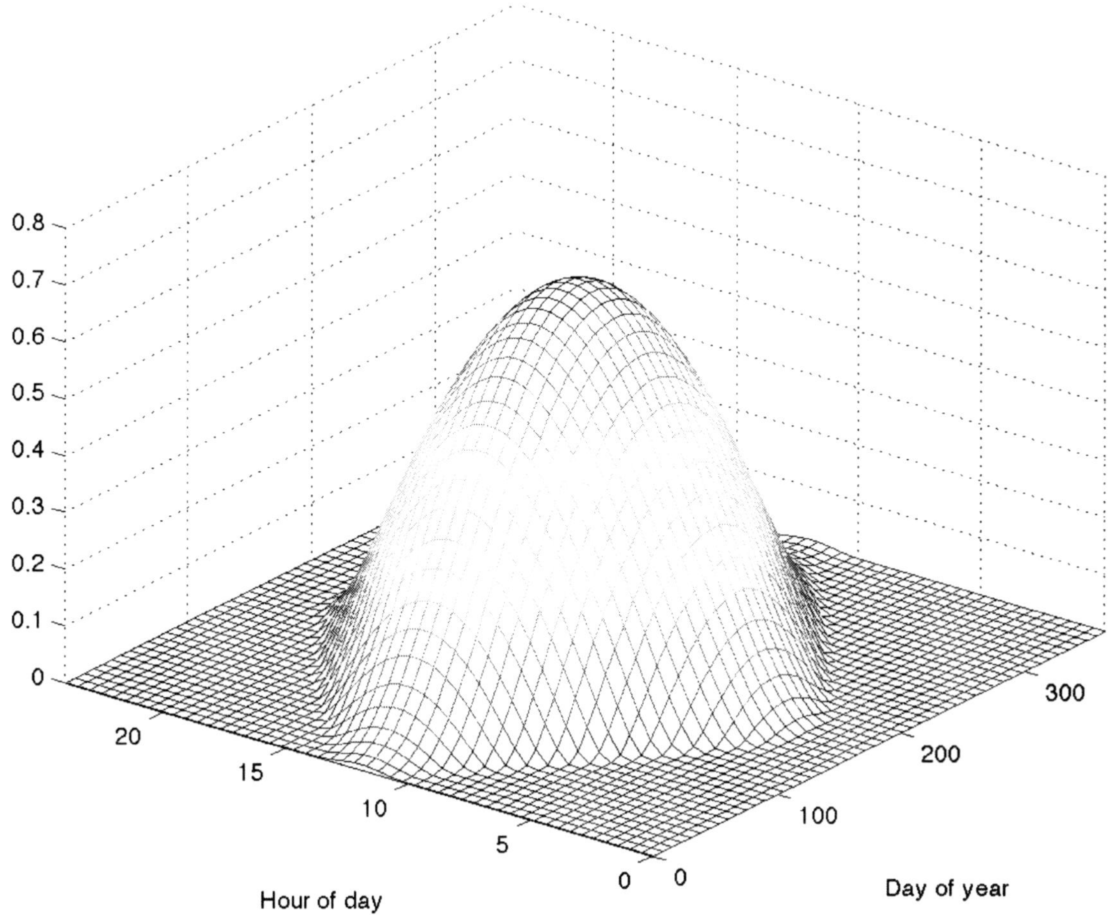

example, background that contains shadows, moving objects, e.g., wavering tree branches, flickering water surfaces, small mugs near the cameras and changing illumination. Foreground and background detection in challenging environments has previously been investigating by for example Li (2003). A Bayes decision rule for background/foreground classification from selected feature vectors is formulated by Arulampalam (2002). With the rule, different types of background object will be segmented from foreground objects by choosing a feature vector. Foreground objects are extracted by fusing the classification results from both stationary and moving pixels. Learning strategies for the gradual and sudden "one-off" background changes are proposed to adapt to various changes in the background throughout the video. The convergence of the learning process is proven and a formula to select a proper learning rate is also derived. See Figure 7. for an example of foreground and background detection from a camera in the zoo.

The location information of the animals is estimated with semi-continuous tracking and labeling of animals by combining occasional RFID readings with tracking of animals from the camera sensor network. Since most animals living in the zoo are required to have passive RFID implant

Figure 6(a). Historical data of daily solar radiation at Umea over three years; (b) Estimate of duly cycle required for continuous operation of camera nodes during different times of year (assuming 900mW camera and 300mA solar panel at 3.4V with sufficient storage for each days solar energy only)

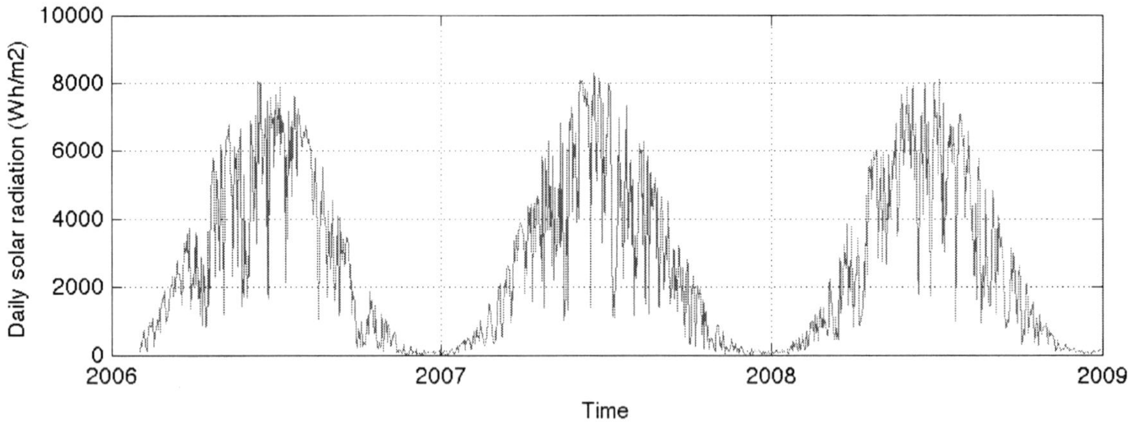

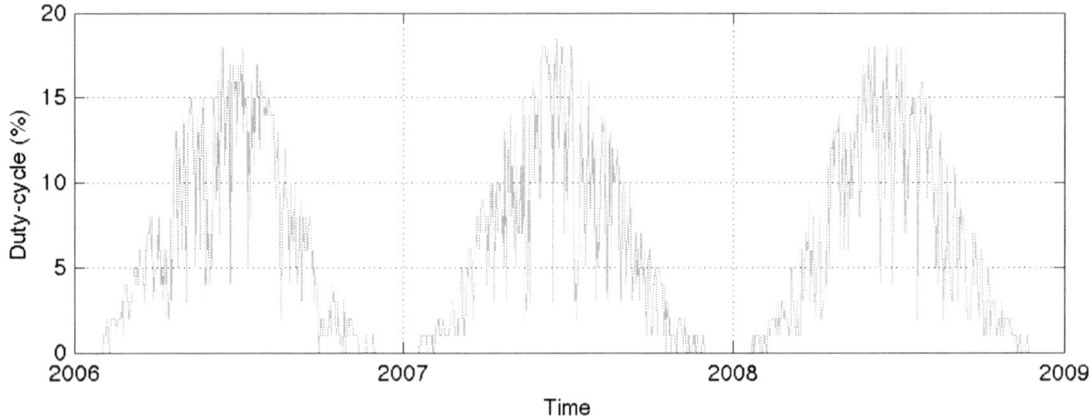

Figure 7. Motion detection system for the wireless visual sensor network in the zoo

Figure 8. Overview of the animal tracking system

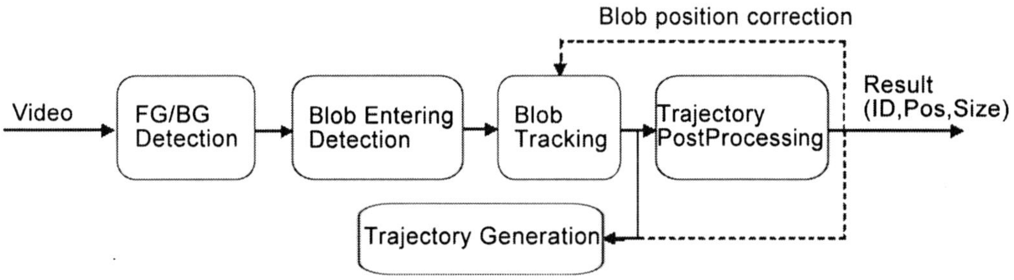

we can make use of that for identification of individual animals. The reading distance for passive RFID chips is, however, rather limited. Depending on size of the implant, the design of the reader and the radio standard used the reading range can vary from a few centimeters up to about one meter. For this reason, the RFID readers will be placed at strategic locations like drinking or feeding locations. Other good locations are gates or tunnels where the animal can actually pass through a loop antenna. The location of RFID is estimated depending on the position relative to the camera's view. Once the animal stands near the RFID readers, the name will be captured and by tracking individual animals, the individual's location can be calculated and sent to the control center. This application also can be used in multiple cameras working by using Scale-invariant feature transform (or SIFT).

For the tracking the moving animals' part blob tracking is employed, see Figure 8. In blob tracking large groups of connected foreground pixels are considered as a blob. The foreground and background detection perform segmentation for each pixel in every frame of a video sequence. Detection of entering blobs uses the foreground pixels to detect new blob object entering to the scene. Once the blob pixels have been determined, the blobs can be tracked. Generally there is a trade-off between speed/accuracy that needs to be chosen between on a case by case basis. With the blob tracking, the system performs a saving function to collect the position of all blobs and save the blob trajectory. A trajectory post-processing is optional to smooth the trajectory. After the blob tracking module, a file is created to report the blob IDs, positions and pixel size. The output video draws an ellipse around the animals in the frame which shows where the tracked animals are. By combining occasional RFID readings with tracking of animals through the camera sensor network, the system writes the name above the individuals to explain who they are.

Understanding the behavior of the animals, and controlling the animals will be achieved by a computer vision system. The visual sensor uses blob based motion tracking to retrieve the ID of each animal, and its position in the frame. This information is also stored and analyzed for animal behavior estimation. For a simple case, the feeding place position is related to eating while a long-time static position is related to the resting. The behavior training can be applied to a computer system in order to understand the animals.

INTERACTION DESIGN

With the wireless sensor network in the zoo it is possible to give the visitors of the zoo a new experience. The idea is to give people an experience that will give them a chance to get to know the animal on a more personal level.

Figure 9. The two applications: interactive map and the social media connection

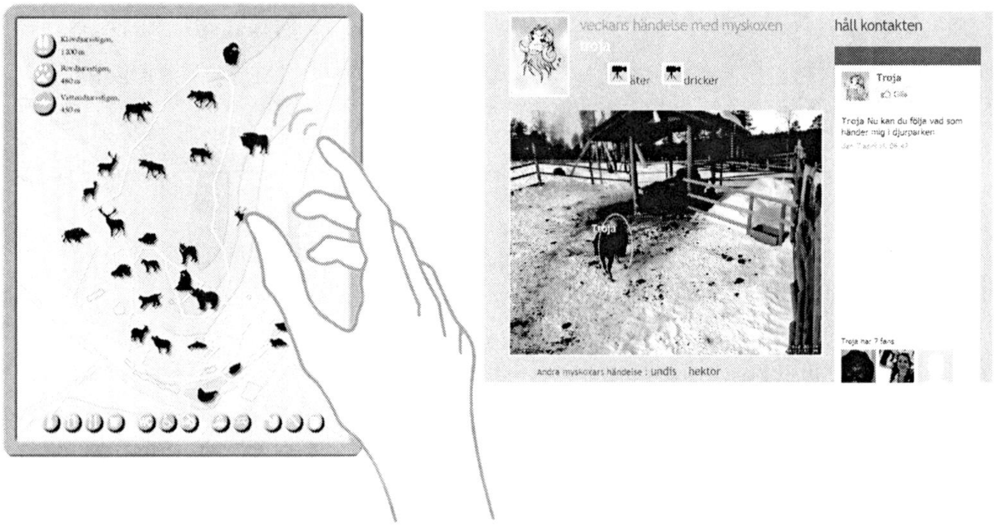

With the applications made, based on the wireless sensor network information, the concept "zoo visitor" becomes a wider meaning. The visitor would not only be someone in the zoo at Lycksele but also a user of the applications, applications that can be located anywhere where there is an internet connection. The goal is to make the visitor an active user of the collected information, the visitors/users becomes "interactors". The information collected with the wireless sensor network is in form of "moving pictures". These moving pictures are both video clips and live stream videos from the animals in the zoo.

So with the following in mind, letting the visitor get to know the animals and to make the visitors interactors, applications is made based on the information from the wireless sensor network.

Application: Interactive Map

To get a strong physical connection with the actual zoo, "Lycksele Zoo", an interactive map over the zoo is made as an application (Figure 9). This interactive map is made in two versions: one for the web and one for being in the actual zoo. These two maps looks the same but they differ in interaction. The one made for the web is made to be used at a computer outside the zoo as part of a web site. The interaction is then made through the keyboard and the computer mouse. The other map, which is made for use inside the zoo, is made for a touch screen concept. This means that the user has her hands for direct interactions with the map. When designing the interaction for the touch screen it is important to think of the fact that the "rollover" function with the mouse is nearly never used. The reason is that a movement from one place to another on the screen is usually made in the air instead of dragging the mouse over the screen. The interaction used in this application is based on the fact that we can see it a model–world metaphor (Hutchins, 1987). The map becomes something that the user has seen before and something that she can interact directly with. The touch screen concept is made so the user only can use the map, the possibilities to use anything else on the computer is prevented. The map and the touch screen becomes a whole concept of its own. This is the first step of making the computer invisible for the interactor and at the same time become ubiquitous as Streitz and Nixon talks about (2005). The visitor can only interact with the map and the map is starting to become an artifact that can give information to the user without forcing the visitor to bring some artifact with her.

Except for the interaction differences the two maps gives the same information to the visitor/user. It is the location of the maps that make the interaction and use of the actual information different. At home at the computer the visitor will get to know the zoo through the interactive map by getting an idea of the location and information like entrance fee and starting getting to know the animals. The visitor in "Lycksele djurpark" would probably be more interested in information like where to find different facilities (restrooms, restaurants and so on) and were to find different animals. In both the versions of the interactive map it is possibility to see the animals live in real time inside the zoo.

Application: Social Media Connection

In the next application the focus is even more on the interaction together with the fact that the visitor should be able to follow and get to know the animal on a more personal level. The fact that we are able to identify and track the animals in the video clips is used here. The goal is also to involve people in a more active way with the animals but also with other visitors that had the same interest for a specific animal.

The decision is made to use today's way of connecting and interacting with others: social media. By using social media it is possible to use an already existing social network together with interactions and information. In this way it will be possible to spread the concept in a natural way by using friend-to-friend connection and at the same time create new connection based on the same interest for a specific subject, in this case the animal.

The result is an application that showed the newest video clips from the animals in the zoo. The visitor/user can choose which animal to watch depending on her interest. From the video clip that the user picked she is then able to see more video clips for a specific individual. This is done by automatically identifying each individual by name, tracking them in the video clip and making them clickable in the video clip.

Each individual animal then has her own video clips that would show the newest everyday events for the individuals live. Together with these video clips it is possible for the user to use her social connections that she already has within social media so a Facebook connection was made. The goal with the Facebook connection is to make the visitor/user and the animal individual have a contact that can be seen as a social contact. By using Facebook connections the visitor will be able to comment directly on what she is seeing and experience. The individual animal is at the same time able to keep in contact with its visitor through Facebook, by telling when something new is happening to it live.

The information that the visitor/user can get from this application is all based on the wireless sensor network. The video clips are all created automatically since we can register activity in front of the camera. Each clip then generate an appropriate comment that is posted on Facebook. The comment on Facebook works as the "personal" connection and information from the individual animal.

New Applications

By using sensors like GPS together with camera-sensors it would be possible to make an augmented reality application (Figure 10). This would be an application used in a handheld artifact in the actual zoo. The information given to the visitors would then depend on where they would be in the zoo. One scenario could be that the visitor would walk around in the zoo and come to a fence where there are no animals in sight. The handheld artifact the visitor could get information by holding up

Figure 10. The augmented reality application

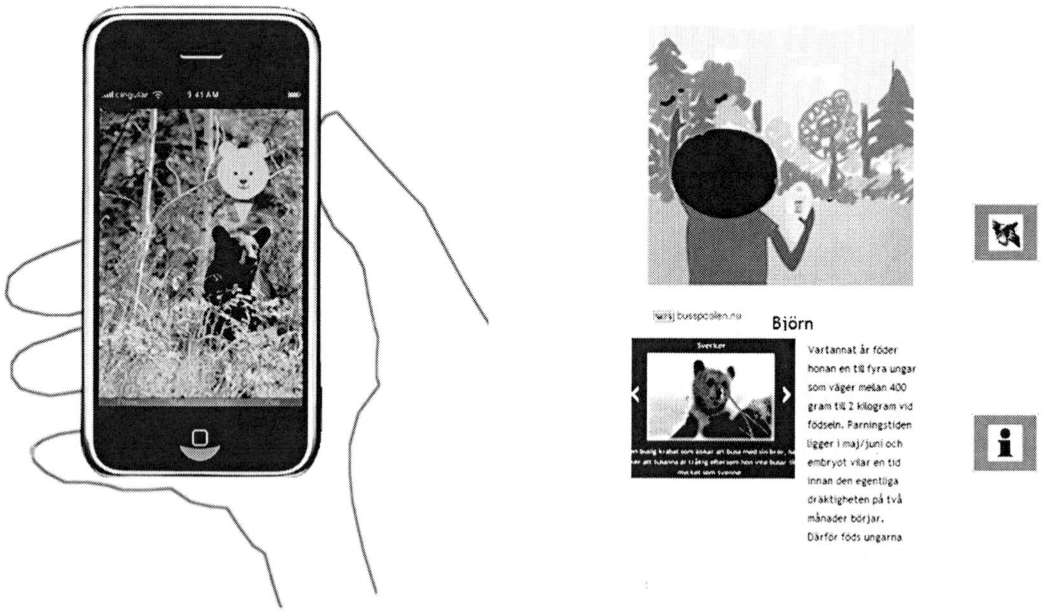

the artifact towards the fens. With the techniques based on the wireless sensor network together with the GPS could then tell the visitor were to look in the fence but also give facts about the animal in the handheld artifact. Based on the discussions that Janlert and Stolterman makes (2008), the artifact used here should be of the type that has an external simplicity and an internal complexity. All this so that the user, who is the visitor in the park, doesn't have to lose the focus on the main activity: visiting the zoo and watching animals.

FUTURE RESEARCH DIRECTIONS

One long term goal is to further analyze the captured data from the wireless visual sensor network installed in the zoo to not only detect an animals motion, but also understand the animals behaviour. This will require multi-camera tracking of the animals to extend the time each animal is tracked.

The sensor network deployment will also be extended in the zoo, from the current number of about 25 nodes to at least 100 nodes. More types of wireless multimedia sensor nodes will also be developed and installed. One interesting approach is to use compressive sampling at the nodes to create extremely simple and low-cost vision sensor nodes.

One interesting challenge for the sensor network is the Arctic climate at the deployment site. For the low-power devices this will cause special requirements. During winter virtually no energy could be harvested, on the other hand a lot can be harvested during summer. This in combination with the very cold weather during winter, leading to difficulties to store energy, will lead to very challenging conditions. To handle this, the nodes must be highly adaptive.

For interaction design more types of sensed data will be explored, for example sound, temperature and heart beat. To enable this more types of sensors will be deployed in the zoo.

CONCLUSION

We have presented our work on creating a digital zoo using a wireless visual sensor network. This is an interesting testbed where algorithms for networking and computer vision can be tested under realistic conditions in a large-scale and long-term deployment. The current deployment shows that a visual sensor network can be used to capture multimedia information from a large area under very hard weather conditions.

To process the sensed data several computer vision algorithms has been developed and implemented in the system. The main focus has been to detect and track animals and to separate the foreground from the background. We have shown that it is possible to use single camera tracking of animals in the zoo, even though it is a relatively challenging environment.

In this work we are dealing with all aspects of creating a digital zoo, from collecting the data to how it should be presented. We have investigated and demonstrated several applications where we use interaction design to connect the sensed data with the user. This includes applications both for the Internet and for physical visitors in the zoo.

REFERENCES

Arulampalam, M., Maskell, S., Gordon, N., & Clapp, T. (2002). A tutorial on particle filters for online nonlinear/non-Gaussian Bayesian tracking. *IEEE Transactions on Signal Processing*, *50*(2), 174–188. doi:10.1109/78.978374

Corke, P., & Sikka, P. (2006). *Results from the farm*, In 3rd Workshop on Embedded Networked Sensors (EmNets 2006).

Diao, Y., Ganesan, D., Mathur, G., & Shenoy, P. (2007). *Re-thinking data management for storage-centric sensor networks*, In 3rd Biennial Conference on Innovative Data Systems Research (CIDR).

Hutchins, E. (1987). *Metaphors for interface design (Tech. rep.)*. Ft Belvoir, VA: Defense Technical Information Center.

Janlert, L.-E., & Stolterman, E. (2010). Complex interaction. *ACM Transactions on Computer-Human Interaction*, *17*(2), 1–32. doi:10.1145/1746259.1746262

Jeong, J., Jiang, X., & Culler, D. (2008). Design and analysis of micro-solar power systems for wireless sensor networks, In *Proceedings of the International Conference on Networked Sensing Systems (INSS)* (pp. 182-188).

Li, L., Huang, W., Gu, I. Y. H., & Tian, Q. (2003). Foreground object detection from videos containing complex background, In *Proceedings of the 11th ACM International Conference on Multimedia (MULTIMEDIA '03)* (pp. 2-10). New York, NY, USA: ACM.

Streitz, N., & Nixon, P. (2005). Introduction. *Communications of the ACM*, *48*(3), 32–35. doi:10.1145/1047671.1047700

Wark, T., Corke, P., Liu, J., & Moore, D. (2008). Design and evaluation of an image analysis platform for low-power, low-bandwidth camera networks, *ImageSense '08 Workshop* in conjunction with *ACM Sensys '08* (pp. 7-11).

Xia, F. (2009). Wireless sensor technologies and applications. *Sensors (Basel, Switzerland)*, *9*, 8824–8830. doi:10.3390/s91108824

Yick, J., Mukherjee, B., & Ghosal, D. (2008). Wireless sensor network survey. *Computer Networks*, *52*(12), 2292–2330. doi:10.1016/j.comnet.2008.04.002

ADDITIONAL READING

Ahmad, J., Khan, H., & Khayam, S. (2009). *Energy efficient video compression for wireless sensor networks* (pp. 629–634).

Akyildiz, I. F., Melodia, T., & Chowdhury, K. R. (2007). A survey on wireless multimedia sensor networks. *Computer Networks*, *51*(4), 921–960. doi:10.1016/j.comnet.2006.10.002

Akyildiz, I. F., Melodia, T., & Chowdhury, K. R. (2008). Wireless multimedia sensor networks: Applications and Testbeds. *Proceedings of the IEEE*, *96*(10), 1588–1605. doi:10.1109/JPROC.2008.928756

Charfi, Y., Wakamiya, N., & Murata, M. (2009). Challenging issues in visual sensor networks. *Wireless Communications, IEEE*, *16*(2), 44–49. doi:10.1109/MWC.2009.4907559

Chow, K.-Y., Lui, K.-S., & Lam, E. (2006). *Balancing image quality and energy consumption in visual sensor networks* (p. 5).

Fahlquist, K., Karlsson, J., Ren, K., Liu, L., Li, H., Rehman, S. U. & Wark, T. (2010). *Human animal machine interaction: Animal behavior awareness and digital experience*, in ACM Multimedia 2010 - Brave New Ideas.

Feng, W.-C., Kaiser, E., Feng, W. C., & Baillif, M. L. (2005). Panoptes: Scalable low-power video sensor networking technologies. *ACM Trans. Multimedia Comput. Commun. Appl.*, *1*(2), 151–167. doi:10.1145/1062253.1062256

Gu, Y., Tian, Y., & Ekici, E. (2007). Real-time multimedia processing in video sensor networks. *Signal Processing Image Communication*, *22*(3), 237–251. doi:10.1016/j.image.2006.12.013

Han, B., Wu, F., & Wu, D. (2010). Image representation by compressive sensing for visual sensor networks. *Journal of Visual Communication and Image Representation*, *21*(4), 325–333. doi:10.1016/j.jvcir.2010.02.007

He, Y., Lee, I., & Guan, L. (2009). Distributed algorithms for network lifetime maximization in wireless visual sensor networks. *IEEE Transactions on Circuits and Systems for Video Technology*, *19*(5), 704–718. doi:10.1109/TCSVT.2009.2017411

He, Z., Liang, Y., Chen, L., Ahmad, I., & Wu, D. (2005). Power-rate-distortion analysis for wireless video communication under energy constraints. *IEEE Transactions on Circuits and Systems for Video Technology*, *15*(5), 645–658. doi:10.1109/TCSVT.2005.846433

Karlsson, J., Wark, T., Valencia, P., Ung, M., & Corke, P. (2007). *Demonstration of image compression in a low-bandwidth wireless camera network* (pp. 557 -558).

Lee, I., Shaw, W., & Fan, X. (2009). Wireless multimedia sensor networks, in Subhas Chandra Misra; I. Woungang, & S. Misra (Eds.), Guide to wireless sensor networks (pp. 561-582). London: Springer.

Lobaton, E., Vasudevan, R., Bajcsy, R., & Sastry, S. (2010). A distributed topological camera network representation for tracking applications. *IEEE Transactions on Image Processing*, *19*(10), 2516–2529. doi:10.1109/TIP.2010.2052273

Mainwaring, A., Culler, D., Polastre, J., Szewczyk, R., & Anderson, J. (2002). Wireless sensor networks for habitat monitoring, In *Proceedings of the 1st ACM International Workshop on Wireless Sensor Networks and Applications (WSNA '02)* (pp. 88-97). New York, NY, USA: ACM.

Misra, S., Reisslein, M., & Xue, G. (2008). A survey of multimedia streaming in wireless sensor networks. *IEEE Communications Surveys Tutorials*, *10*(4), 18–39. doi:10.1109/SURV.2008.080404

Naumowicz, T., Freeman, R., Heil, A., Calsyn, M., & Hellmich, E. Brдndle, A.,...Schiller, J. (2008). Autonomous monitoring of vulnerable habitats using a wireless sensor network, In *Proceedings of the workshop on Real-world wireless sensor networks (REALWSN '08)* (pp. 51-55). New York, NY, USA: ACM.

Obraczka, K., Manduchi, R. & Garcia-Luna-Aveces, J. (2002). Managing the information flow in visual sensor networks (Vol. 3, pp. 1177 – 1181).

Politis, I., Tsagkaropoulos, M., Dagiuklas, T., & Kotsopoulos, S. (2008). Power efficient video multipath transmission over wireless multimedia sensor networks. *Mobile Networks and Applications*, *13*(3-4), 274–284.

Soro, S., & Heinzelman, W. (2009). A survey of visual sensor networks. *Advances in Multimedia*, *2009*, 1–22. doi:10.1155/2009/640386

Stojkoska, B., & Davcev, D. (2009). *Web interface for habitat monitoring using wireless sensor network* (pp. 157–162).

Szewczyk, R., Mainwaring, A., Polastre, J., Anderson, J., & Culler, D. (2004). An analysis of a large scale habitat monitoring application, In *Proceedings of the 2nd International Conference on Embedded Networked Sensor Systems (SenSys '04)* (pp. 214-226). New York, NY, USA: ACM.

Wark, T., Corke, P., Karlsson, J., Sikka, P., & Valencia, P. (2007). Real-time image streaming over a low-bandwidth wireless camera network (pp. 113 -118).

Wark, T., Corke, P., Sikka, P., Klingbeil, L., Guo, Y., & Crossman, C. (2007). Transforming agriculture through pervasive wireless sensor networks. *IEEE Pervasive Computing / IEEE Computer Society [and] IEEE Communications Society*, *6*(2), 50–57. doi:10.1109/MPRV.2007.47

Wark, T., Swain, D., Crossman, C., Valencia, P., Bishop-Hurley, G., & Handcock, R. (2009). Sensor and actuator networks: Protecting environmentally sensitive areas. *IEEE Pervasive Computing / IEEE Computer Society [and] IEEE Communications Society*, *8*(1), 30–36. doi:10.1109/MPRV.2009.15

Yu, C., & Sharma, G. (2010). Camera scheduling and energy allocation for lifetime maximization in user-centric visual sensor networks. *IEEE Transactions on Image Processing*, *19*(8), 2042–2055. doi:10.1109/TIP.2010.2046794

Zahariadis, T., & Voliotis, S. (2007). *Open issues in wireless visual sensor networking* (pp. 335–338).

KEY TERMS AND DEFINITIONS

Artifact: Man-made entity whether it is material objects or immaterial procedures.

Bayes Decision Rule: In decision theory, Bayes decision rule is used to minimize the posterior expected value of a loss function. From a Bayesian perspective, the tracking problem becomes the mechanism for updating knowledge about the target state by using extra information from new data..

Blob: In the area of computer vision, "blob" refers to visual modules that are aimed at points and/or regions in the image that are either brighter or darker than the surrounding.

Feature Vector: In pattern recognition and machine learning, 'feature vector' refers to an n-dimensional vector of numerical features that represent some object.

Interactors: The participants involved in an interaction.

Model-World Metaphor: A metaphor describing the world as a model world that directly can be perceived and that directly can be affected by action.

Social Media: The type of webpages that connect people and letting them share different contents with each other.

Ubiquitous: Being everywhere at the same time.

Chapter 16
Visual Sensor Network Processing and Preventative Steganalysis

Julien Sebastien Jainsky
Texas A&M University, USA

Deepa Kundur
Texas A&M University, USA

ABSTRACT

In this chapter, we discuss the topic of security in wireless visual sensor networks. In particular, attention is brought to steganographic security and how it can be discouraged without challenging the primary objectives of the network. We motivate the development and implementation of more lightweight steganalytic solutions that take into account the resources made available by the network's deployment and its application in order to minimize the steganalysis impact on the WVSN workload. The concept of preventative steganalysis is also introduced in this chapter as a means to protect the network from the moment it is deployed. Preventative steganalysis aims at discouraging any potential steganographic attacks by processing the WVSN collected data such that the possibility of steganography becomes very small and the steganalysis leads to high rate of success.

INTRODUCTION

Wireless visual sensor networks (WVSNs) provide unprecedented opportunities to improve surveillance and assisted-living applications. Such broadband and often mobile networks when combined with visual data acquisition provide keen insights for tactical battlefield scenarios, geriatric patient monitoring amongst other applications. This inherent application space makes issues of security and privacy a significant concern while their often aggressive design (maximizing both efficiency and visual utility) makes engineering of security and privacy challenging.

DOI: 10.4018/978-1-61350-153-5.ch016

From a technological perspective, security and privacy may be addressed through an effective marriage between signal processing, communications and applied cryptography. It is imperative that such solutions be studied during system specification and creation rather than applied as an afterthought, which can lead to superficial protection requiring repeated upgrade. In this chapter, we propose to introduce and study the novel yet important problem in WVSN security of *preventative steganalysis*.

BACKGROUND

Most well known measures to protect WVSNs, to date, have focused on the problem of providing privacy in vision-rich systems. Lo *et al.* (Lo, Wang & Yang, 2005) introduce an automated homecare monitoring system for the elderly named *UbiSense* where image processing is conducted directly at the camera to convert visual data directly into abstractions that reveal no personal information and hence protect the privacy of the monitored individuals. Fidaleo *et al.* (Fidaleo, Nguyen & Trivedi, 2004) introduce the *Networked Sensor Tapestry (NeST)* architecture designed for the secure sharing, capture, and distributed processing and archiving of multimedia data. They introduce the notion of "subjective privacy" in which processing of raw sensor data is conducted to remove personally identifiable information; thus the behavior, but not the identity of an individual under surveillance is conveyed. The resulting data, approved for public viewing, is communicated in a network that employs the secure socket layer protocol and client authorization for network-level protection. Wickramasuriya *et al.* (Wickramasuriya, Datt & Mehrotra, 2006) present a privacy preserving video surveillance system that monitors subjects in an observation region using video cameras along with motion sensors and RFID tags. The motion detectors are used to trigger the video cameras on or off, and the RFIDs of the subjects provide authorization information in order to specify which individuals are entitled to privacy and hence have their visual information masked through image processing. More recently, Kundur *et al.* (Kundur, 2008) (Kundur, Luh, Okorafor & Zourntos, 2008) present the HoLiSTiC (Heterogeneous Lightweight Sensornet for Trusted Visual Computing) framework for WVSN security that exploits secure protocols in a hierarchical directional link communication network to achieve broadband low power communications. A decentralized visual secret sharing approach is used to preserve privacy.

Research has also emerged with the goal of assuring the authenticity of the data collected by sensor networks. When nodes are corrupted and provide false information, the entire network's legitimacy is compromised. The authentication of each node allows for the network to remain trusted. Several proposed solutions utilize common cryptographic concepts to provide such security. Feng *et al.* (Feng & Potkonjak, 2003) introduce a paradigm to cryptologically embed signatures into the collected data via watermarking techniques. Their objective is to efficiently watermark the data while introducing as little distortion as possible. Zheng et al. (Zheng, Li, Lee & Anshel, 2006) propose to offer authenticity assurance using a public key cryptographic scheme. A derivable public key scheme is used which has the effect of simplifying the cryptography and reducing the need for key storage, therefore making it more suitable for large scale sensor networks. Because these methods still increase the workload of the WSN, Martinovic *et al.* (Martinovic, Pichota & Schmitt, 2009) propose a novel paradigm that relies on the properties of wireless communications to provide authentication capabilities. They focus their study on taking advantage of frequency jamming to detect attacks and strengthen the WSN's security. Energy is always a concern when dealing with WSN where the nodes forming the network have very limited resources. As a potential solution, Blaß *et al.* (Blaß, Wilke & Zitterbart, 2008) develop *Extended Secure Aggregation for Wireless Sensor Networks* (*ESAWN*). ESAWN finds a trade-off between decreasing the energy consumption of the network via data aggregation and providing authentication mechanisms although the authors agree that the security created is weaker compared to mechanisms that are not driven by energy preservation.

These existing approaches for WVSN protection all focus on protecting the *overt* data acquisition and communications systems. Fundamental questions however arise regarding the possibility of *covert* approaches for networking leading to possible breaches in both security and privacy. In this book chapter, we propose to study the possibility of, implications to and mitigation approaches for covert networking in the context of WVSNs. The high degree of collaboration, resource sharing and data redundancy in WVSNs promote data hiding and hence it necessary to study methods to overcome such covert channels.

PREVENTATIVE STEGANALYSIS FOR WVSNS

Covert Communications and Steganalysis

Covert channels have traditionally been employed for stealthy communication of sensitive information from high security areas to low security areas. This often undetectable information leakage has the potential to breach both security and privacy of a given system. A determined attacker, on a more general scale, can exploit various covert channels that exist in a WVSN to create a covert network to promote widespread information leakage. For example, an entire covert network can be build upon an overt WVSN camera network by exploiting camera nodes that have been corrupted by one or more attackers as shown in Figure 1.

The mechanics of covert communications depend on the medium used for stealthy information transfer. In a WVSN, the visual information provides a rich environment to piggyback covert data on overt images through a process of data hiding called steganography. A steganographic system usually involves two entities sharing information: the sender and the receiver. The sender can pass on sensitive information to the receiver by hiding data, usually via the use of a secret key shared by both parties. The overt host information is called the *cover-media*. After embedding (i.e., hiding) the covert information within the cover-media, the resulting signal is called the *stego-media* and is typically perceptually identical to the cover-media hence making it difficult for another party to identify the covert communications. For instance if the media represents still or moving images then the stego-image would have to be visually identically to the cover-image. WVSNs are of particular interest to a steganographer because of the potential for high bandwidth covert communications by exploiting the redundancy of the visual data for hiding large volumes of information. Thus, defending against WVSN steganography is a priority.

Steganalysis is the process of detecting (and potentially annihilating) the presence of steganography by employing various signal processing and statistical tools. Several classifications of steganalysis exist: The steganalysis can be reactive or proactive and it can involve an active or passive warden scenario. A reactive steganalysis approach targets one specific type of steganography whereas proactive steganalysis targets a class of steganographic techniques. In active or passive warden scenarios, the role of the "warden" is that of the steganalyst. A passive warden/steganalyst only eavesdrops on overt communications to potentially detect any covert activity. In contrast, an active warden has the freedom to apply trans-

Figure 1. Creation of Attacker's Covert Network

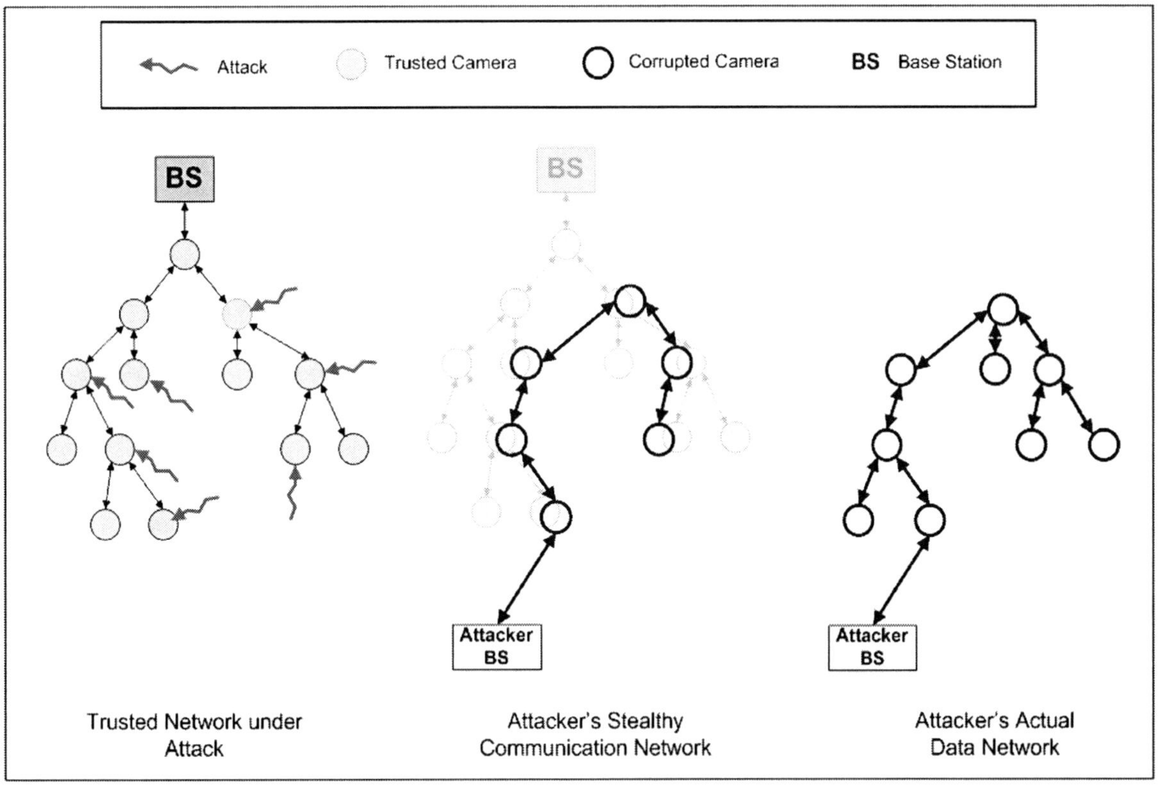

formations on the communicated information before it reaches the receiver potentially removing any hidden data if any thus limiting or preventing covert transmission. These concepts can be used in many applications but are currently unsuitable to defend against WVSN steganography because with a passive warden, regardless of whether the steganalysis is reactive or proactive, only the presence of steganography is identified once the cover data has been sent through the communication channel. Thus, even though covert communications could be flagged as contaminated thanks to the steganalyst, the exchange of information between the malicious sender and the receiver has already occurred. This is where the passive steganalyst falls short: the detection of steganography can only happen after the attacker manages his or her malicious objective. In the case of the active warden, the lossy transformation usually used by the steganalyst compromises the integrity of the cover data which, in many cases, contains critical information. In a WVSN where the collection of data and its subsequent interpretation is of the utmost importance, corrupting the raw information is not an option. For example, if the WVSN works toward tracking intruders in a zone of interest, any lossy transformation due to an active warden might also affect the network's primary function by raising the rates of false-positive or false-negative detections.

One question arises as to whether the inherent processing in a WVSN can be designed and subsequently exploited to serve the joint role of enhancing the capabilities of a steganalysist (thus discouraging any data hiding) while providing much needed data interpretation and network aggregation. It is possible that certain visual processing could serve to facilitate the intended overall

WVSN goal, while ensuring to a high degree that any data formerly embedded or posthumously embedded will have a chance of detection by a steganalyst. We call this concept preventative steganalysis.

The goal of preventative steganalysis is to provide security against covert communications in WVSNs by ensuring that any potential cover-media within the network has statistical characteristics such that any previously hidden data has close to no chance of being undetected by a covert receiver and any possible future data to be hidden has limited opportunity to be communicated imperceptibly within that cover-media. Stated more specifically, the presence of steganography within the cover-media has a detection probability $1\text{-}\varepsilon$ where $\varepsilon \rightarrow 0$ thus discouraging any potential attacker from conducting data hiding.

Towards Preventative Steganalysis via Distributed Data Processing

One general strategy to discourage steganalysis is to reduce the entropy of the cover-media. This has the effect of reducing the uncertainty of the cover information and hence detecting any statistical deviations due to data hiding. By definition, the entropy, or uncertainty, $H(X)$ of a random variable X is a measure of its randomness and is given by (Cover & Thomas, 1991):

$$H(X) = -\sum_{i=1}^{n} p(x_i)\log_2(p(x_i))$$

where $X = \{x_1, x_2, \ldots, x_n\}$ and $p(x_i)$ is the probability of occurrence of x_i.

In the case of images, defining the entropy objectively is more complex because the purely statistical content of the image data does not correspond exactly to visual cues; thus entropy is considered to be an estimate of the visual uncertainty within an image. For this study, we chose a first-order estimate of the entropy which can be defined as:

$$H_{im} = N * \left[-\sum_{i=1}^{255} p(l_i)\log_2(p(l_i)) \right]$$

where N is the total number of pixels in the image and $p(l_i)$ is the probability that gray level i occurs in the image examined. The image entropy is expressed in bits.

From the well-known *data processing inequality*, it is known that any type of processing on a data set cannot increase the entropy of the signal. The inequality states that for random variables X, Y, Z forming a Markov Chain X →Y→Z, the mutual information between X and Z is less than or equal to that between X and Y (Cover & Thomas, 1991):

$$I(X;Y) \geq I(X;Z)$$

We can rewrite this equation for the specific case of the X →X→f(X) Markov Chain where f(X) is a function of X representing the processing on X. Therefore we have:

$$I(X;X) \geq I(X;f(X))$$

$$\Leftrightarrow H(X) - H(X|X) \geq H(f(X)) - H(f(X)|X)$$

$$\Leftrightarrow H(X) \geq H(f(X))$$

Since $H(X|X)$ and $H(f(X)|X)$ equal zero (there is no uncertainty about X or f(X) when X is known). Therefore data processing within a WVSN can never increase the entropy of potential cover-images.

Moreover if the process f is non-invertible (i.e. one-way) the processing strictly reduces the entropy. Most steganographic schemes embed covert data within noisy or irrelevant parts of a cover-media in order to maintain covertness both perceptually and statistically. For instance, "noisy" image characteristics often visually represent high texture areas that can be modified to embed stego-data through slight color variations; the changes can be applied to a greater extent enabling a higher capacity of data hiding than visually flat areas consisting of almost one color. It is therefore well known that by reducing the image entropy of the cover-media, the image irrelevancies and noise are reduced and steganography becomes more difficult because the available covert communication bandwidth is reduced.

It is thus clear that one-way data processing inherent to WVSNs (for example for privacy preservation as highlighted above) will help prevent steganography. Consider the example of a camera network for identifying the presence of an intruder in a safe area. The usual setup would be for the cameras in that safe area to forward the data to a specific base station which would, through various processes such as color conversion, contour extraction and pattern recognition, decide whether content resembling a human figure is present in the zone under surveillance. If the raw data is sent to the base station through the multihop network, there exists much room for an attacker to hide information within this network as the raw data is being communicated throughout. However if the cameras themselves directly extract the contours of the image and then directly communicate the extracted contours to the base station, it becomes more difficult for an attacker residing within the multihop network to hide information. Furthermore, if the cameras conduct all necessary processing only forwarding to the base station the decision of whether or not an intruder is present, there is close to no room for data hiding. Following the logic of this example, we assert that there is an opportunity to discourage and mitigate against steganography within a WVSN by distributing the signal processing effectively throughout the overall network taking into account practical network constraints.

Sensor networks are comprised of nodes often with limited computational and communication capabilities. Low computing power and short battery life are two of the most critical limitations of sensor nodes. In a WVSN, using cameras as sensors, these limitations become even more significant due the volume of data being acquired. Thus, we assert that when security design is concerned a simple yet efficient solution is required for WVSNs. This is especially true since information security is not the primary function of the network and is often considered a "tax" on the overall system. In that mindset, techniques to encourage security that employ existing system components and processing have the potential to provide protection without possible overhead. We look at how network topology and the distribution of integral processing affect steganographic security. In addition, since we do not want the steganalysis to add unnecessary load to or interfere with the WVSN primary purpose, it is wise to first utilize the resources at hand and explore how they can be shaped in order to help the steganalysis without disturbing the WVSN. Thus, we choose to study how the necessary processing resources can be allocated for the purpose of resilience against steganography and covert communications.

Steganographic security is not the primarily objective of a WVSN which means that in the quest for preventative steganalysis, priority must be given to essential network properties such as its application-based functionality and lifespan. An important goal is to achieve a certain level of security without compromising the network's priorities. By distributing the processing related to the existing network's function, we aim to create a certain level of steganographic security without interfering with the WVSN goals. What a distributed approach is intended to do is break the overall WVSN objective into tasks that are then separately assigned amongst select nodes. For example, in the case of an intruder target-tracking surveillance WVSN, a goal of the network is to

Figure 2. Illustration of Image Processing Steps for Surveillance Application (source image 'office_4. jpg' from the MATLAB library)

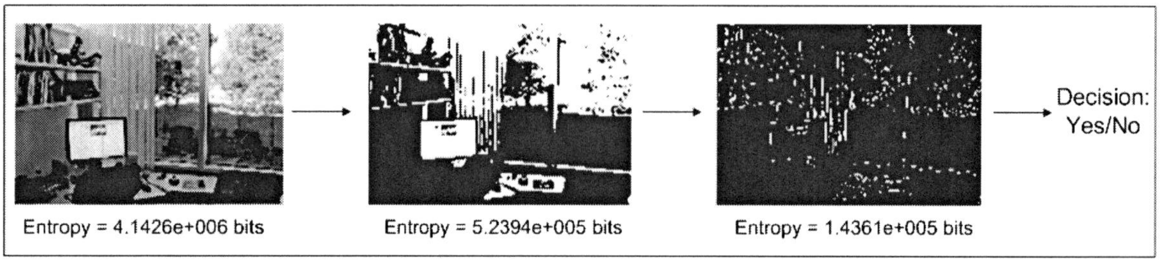

determine the presence of a person or an intruder in the monitored area. This goal could be divided into the following three possible steps as illustrated in Figure 2:

- **Step 1.** Convert the raw data into a black and white image;
- **Step 2.** Compute the contours of the black and white image;
- **Step 3.** Using one or more pattern recognition algorithms, determine the presence of an intruder.

Figure 2 also shows the respective image entropy after each processing step and demonstrates how processing the data greatly reduces entropy and therefore covert embedding capacity.

The reader should note that performing each one of these intermediate steps would eliminate part of the data that is irrelevant to the application such that the volume of cover-media that can mask the steganography is decreased. Therefore the entropy and the associated embedding capacity are also reduced. Once this embedding capacity further decreases, the data will become less attractive to the attacker for steganographic purposes. In Figure 2, the last step of the processing results in a decision on the presence of intruders that can be represented as binary data, i.e. '0' and '1' or 'yes' and 'no'. By achieving this level of data reduction, the bandwidth of effective steganography has reduced and preventative steganalysis has succeeded.

Distributed Data Processing and Network Topology

As discussed, preventative steganalysis can be achieved via breaking up network processing into multiple data processing steps and effectively distributing it. We next explore how this preventative steganalysis can be implemented in a WVSN and its associated costs. To do so, we consider three data processing architectures as applied to a variety of network configurations. The three data processing schemes are:

- **Central data processing:** in which, the base station is responsible for all of the processing; the nodes only transfer the raw data from the recording node to the base station;
- **Uniform distributed data processing:** here, the nodes take some or all of the data processing; Each node, starting with the initial data sensing node, performs one and only one step of data processing if its battery level allows it;
- **Greedy distributed data processing:** similarly, the nodes take some or all of the data processing. However, each node starting with the recording node tries to perform as many of the processing steps

as its battery allows. We assume that the batteries are rechargeable after a period of time (for example, a day if solar recharging is employed) so that nodes that deplete power can be effectively revived within a period of time.

Even if preventative steganalysis can be achieved in theory, when it comes to practical applications to a WVSN, several challenges arise. Our preventative steganalytic solution requires the application of a distributed data processing approach, however, in a WVSN, by decentralizing the processing of the raw data, which takes place at the base station in most cases, the overall computational load on the network is intuitively increased and its lifespan could become shorter. Therefore, we also study the effect of providing steganographic security on the battery usage in a WVSN.

We consider a network as being comprised of a set of nodes connected through wireless communication links; two nodes are considered to be neighbors if they are one-hop (i.e., a direct communication link) apart. From the perspective of communication of sensed data from a source node to the base station we consider the information to be routed through neighboring nodes in a multi-hop fashion; along such a route a non-source node within this route usually has predecessor and successor neighbor(s) that aid in routing information from a source node to the base station. A successor node is defined as being one-hop further from the base station whereas a predecessor is said to be one-hop closer to the base station. Many types of node placements forming a given topology are possible. Whether the nodes are organized in a straight line or are all gathered around the base station, the network follows a communication topology. We define a terminal node as one with no successors at the "edge" of the communication network. The reader should note that the term "node" does not include reference to the base station. Four common communication topologies are considered for the tests in this work (Figure 3):

- **Horizontal topology:** every node is only one-hop away from the base station; equivalently, every node is a termination node.
- **Fixed Tree topology:** every node has only one predecessor and two successors;
- **Random Tree topology:** every node has only one predecessor but can have multiple successors;
- **Vertical or straight line topology:** every non-termination node has only one predecessor and only one successor.

We chose to consider several topologies, which depend on the observation environment of the WVSN and on the network's application. For example, in a controlled environment the nodes can be placed in a specific pattern whereas in a hostile environment, the sensors may be randomly deployed. Also if the application requires the monitoring of a large area then a more widespread topology would be preferred whereas for specific monitoring of an airport gate for example, a horizontal topology might be better.

Intuitively, the topology of the WVSN naturally influences the life of the network. A WVSN is alive as long as it can perform its original function whereas when a WVSN cannot fulfill the goal it was built for, it is said to be dead. Because a network is but a set of nodes, its lifespan depends directly on the life of its cameras: by extension, the network dies when its nodes die. However, all the nodes do not have to be dead in order for the WVSN to die. As expressed earlier, the network dies when it cannot perform its original application. This scenario can occur for example when all the nodes closest to the base station have depleted batteries. In such a case, even if the cameras further away are still alive, no transmission can reach the base station and therefore the network becomes

Figure 3. Examples of Communication Network Topologies. BS represents the network base station. (a) Horizontal, (b) Fixed Tree, (c) Random Tree, (d) Vertical.

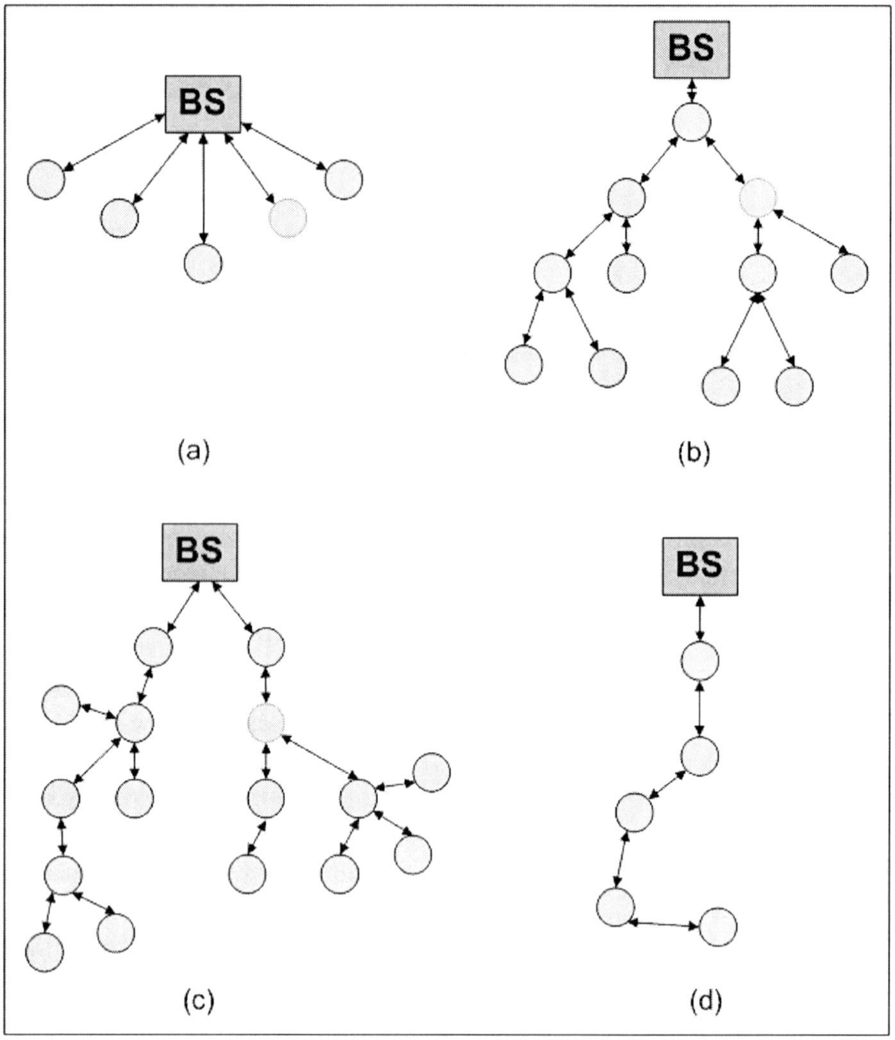

useless; it has died. We define the life of the network, or lifespan, from a performance perspective as the time from the moment all the nodes are fully charged until so many of the network's nodes die that the WVSN cannot record any event, i.e., all paths to the base-station are broken. For example, a tree topology which presents a bottleneck close to the base station requires using the same nodes (the ones one-hop from the base station) for every data collecting process. Since these nodes will be used often, their batteries will deplete sooner and therefore the network's lifespan will be shorter compared to a more horizontal topology where there is no bottleneck at the base station.

In a similar way, it is intuitive that the network's topology can influence its steganographic security. For example, nodes spread in a wider area using a rather vertical distribution create more node links and therefore more potential breaches exploitable by an attacker. On the other hand, a shorter more horizontal network where all nodes would only be one-hop away from the base station, presents little interest and little opportunity for an attacker. The WVSN's topology is therefore an important parameter that is considered in our investigation.

Empirical Studies

We study this matter in more detail and perform tests for a given communication topology using the following assumptions:

- The WVSN's application is the monitoring of a predefined observation area and is used to detect the presence of intruders. The data is processed in three steps as illustrated in Figure 4.
- Each node has a limited battery level of *Bt* units.
- Each data processing step requires the use of *k* units of battery.
- The transmission of raw data requires *Btrans* units of battery and each processing step reduces the power necessary for data transmission by 25% as illustrated in Figure 4.
- Each processing step increases the steganographic security level of the data by 1 unit (cf. Figure 4). The weakest security level is 0, the highest is 3.

This model conveniently facilitates a focused study on steganalytic aspects of WVSNs while providing insight into interaction of security with communication, processing and battery power. Figure 4 illustrates the data processing chain that we consider for an intruder target-tracking surveillance WVSN. Each processing step reduces the amount of information in the images recorded by eliminating any irrelevant data: Step 1 eliminates the variations in grey levels, Step 2 eliminates the texture of potential objects present in the image, Step 3 eliminates any visual information since only the decision on the presence of an intruder remains. As mentioned previously the data processing chain chosen reduces the amount of information to be conveyed to the base station which also reduces its entropy and the volume of cover-media that can mask the steganography at the same time. With a smaller entropy, and therefore a smaller redundancy, the data offers a reduced embedding capacity which leads to an increase in steganalytic security. Figure 4 points out also that the decrease in data volume for each data processing step reduces the amount of energy necessary for transmission to the next node.

Furthermore, in order to determine the effects of our preventative steganalysis approach on the overall WVSN, we also measure the success of the WVSN in detecting intrusion events. The overall surveillance process is illustrated in Figure 5. It consists of six main steps. The WVSN is monitoring an area of interest (Step 1). When an event is detected (Step 2), the camera records an image of the scene (Step 3) and forwards it to the base station (Step 4). Once the data arrives at the base station (Step 5) the network registers a successful record of event (Step 6). A recording is unsuccessful when the data does not arrive to the base station because of broken links. A broken link occurs if a node along the path to the base station is unresponsive because it has died due to lack of battery power or because the transmission of the data is blocked. Along the rate of successful recordings, we look at the average level of security achieved for each transmission. The level of security corresponds to the amount of processing steps the data has gone through: the more processing steps have occurred, the higher the security level of the data transmission. By looking at both the security levels achieved and the percentage of successful recordings of event, we can better understand how the increase in steganographic security can influence the WVSN

Figure 4. Data Processing, Security Level and Transmission Battery Usage Models

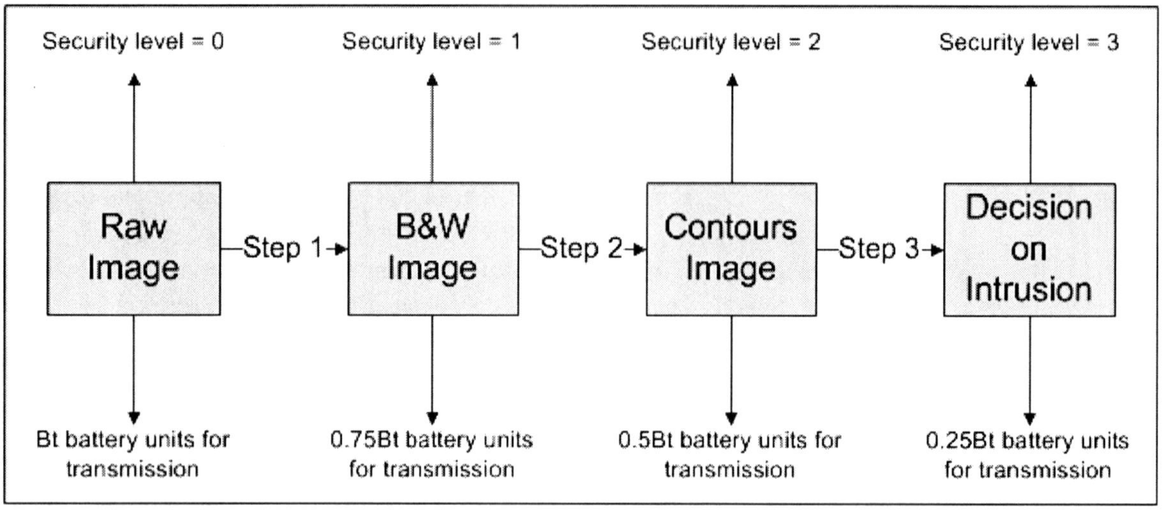

Figure 5. Overall Surveillance Process Diagram

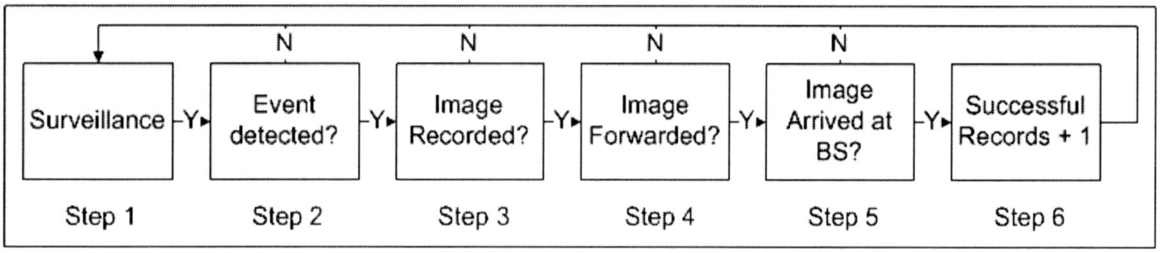

performance. For example, distributing the data processing at the node level can make the nodes die faster which means the number of broken links would increase rapidly.

The tests show the comparison of three data processing schemes in the three different network topologies cited earlier. For the simulations, we used a WVSN composed of 200 cameras. Each camera has a battery of 10 units to transmit the data to the base station and process it if necessary. Once the battery is depleted, the camera is virtually disconnected from the network and every successor becomes at the same time obsolete. Also, we set the amount of battery necessary for each data processing step to be $k=0.5$ units and the amount for transferring the raw data (full image) to be $Btrans=1$ unit.

The results from the simulations are presented in Figures 6, 7, 8 and 9.

Without any form of processing occurring at the camera, no measure of steganographic security exists; the average transmission security level is zero for these cases. However, as suspected, graphs in Figure 6 (b), Figure 7 (b), Figure 8 (b) and Figure 9 (b) prove that the use of a distributed data processing scheme adds a significant amount of security to the WVSN without adding more workload on its primary function by reducing the data entropy: the average transmission security level fluctuates between 1

Figure 6. Horizontal WVSN of 200 cameras with Bt = 10 units of battery/node. (a) Rate of successful records of events. (b) Average transmission security level.

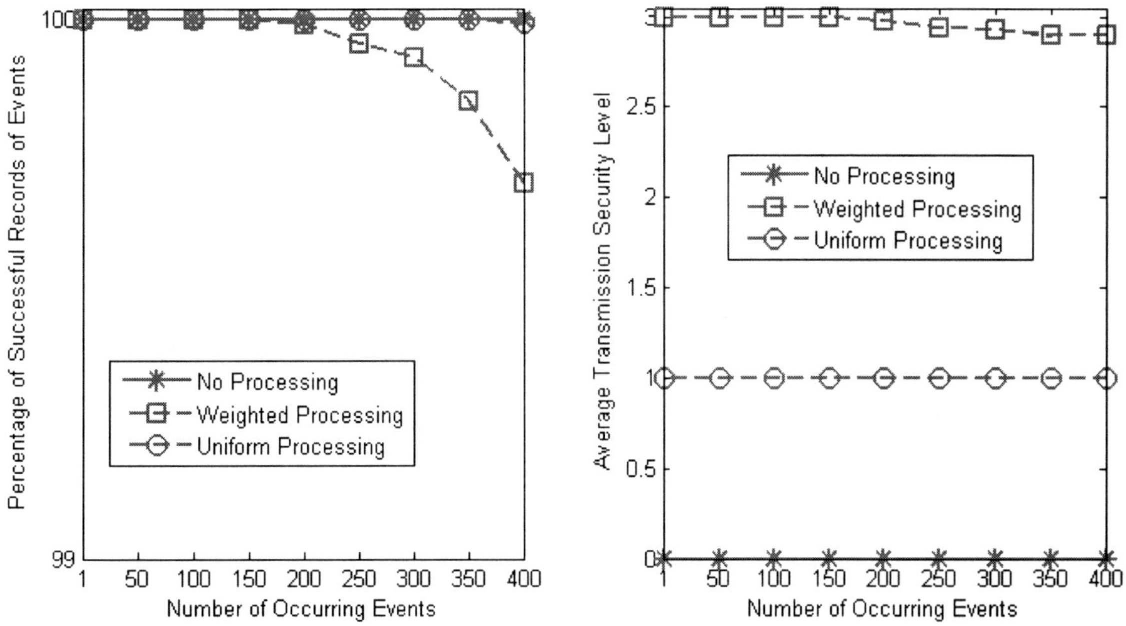

Figure 7. Vertical WVSN of 200 cameras with Bt = 10 units of battery/node. (a) Rate of successful records of events. (b) Average transmission security level.

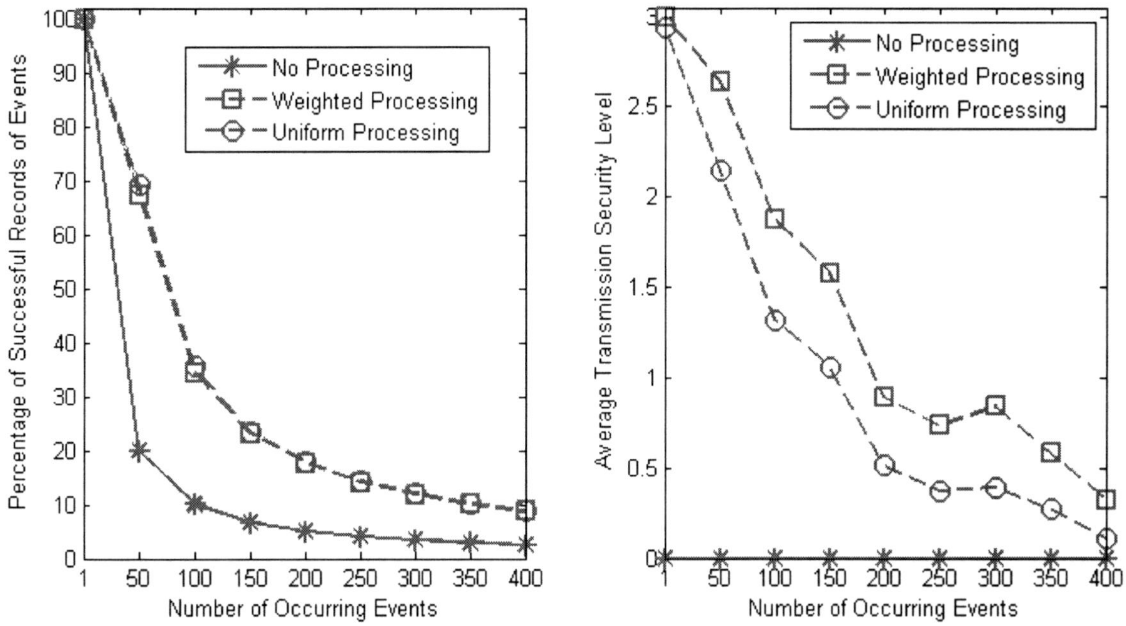

Figure 8. Fixed Tree WVSN of 200 cameras with Bt = 10 units of battery/node. (a) Rate of successful records of events. (b) Average transmission security level.

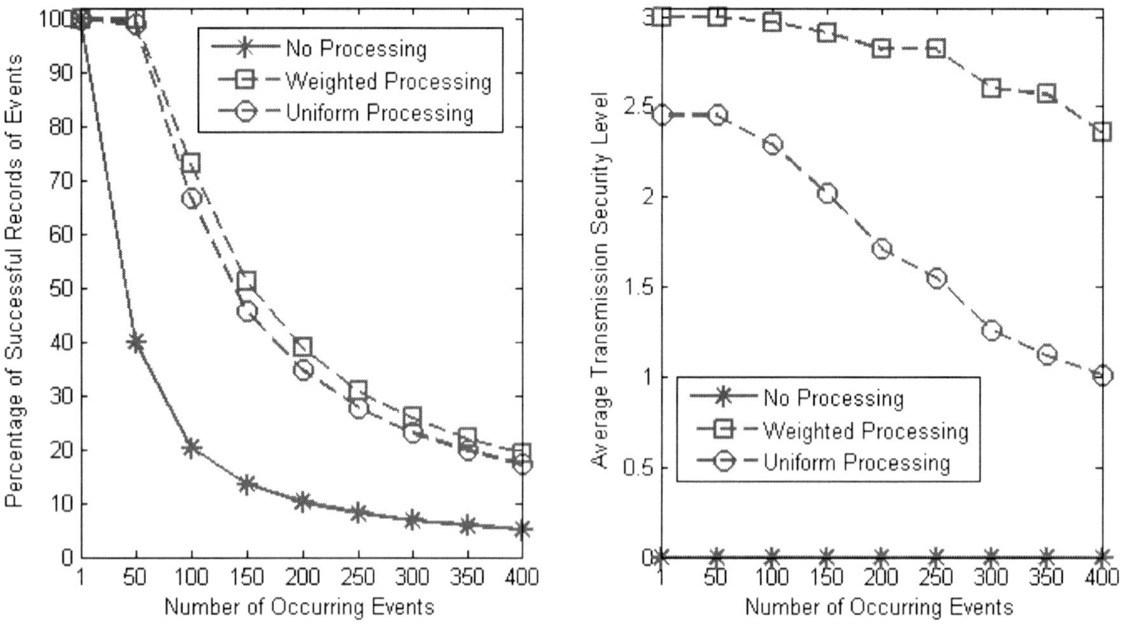

Figure 9. Random Tree WVSN of 200 cameras with Bt = 10 units of battery/node. (a) Rate of successful records of events. (b) Average transmission security level.

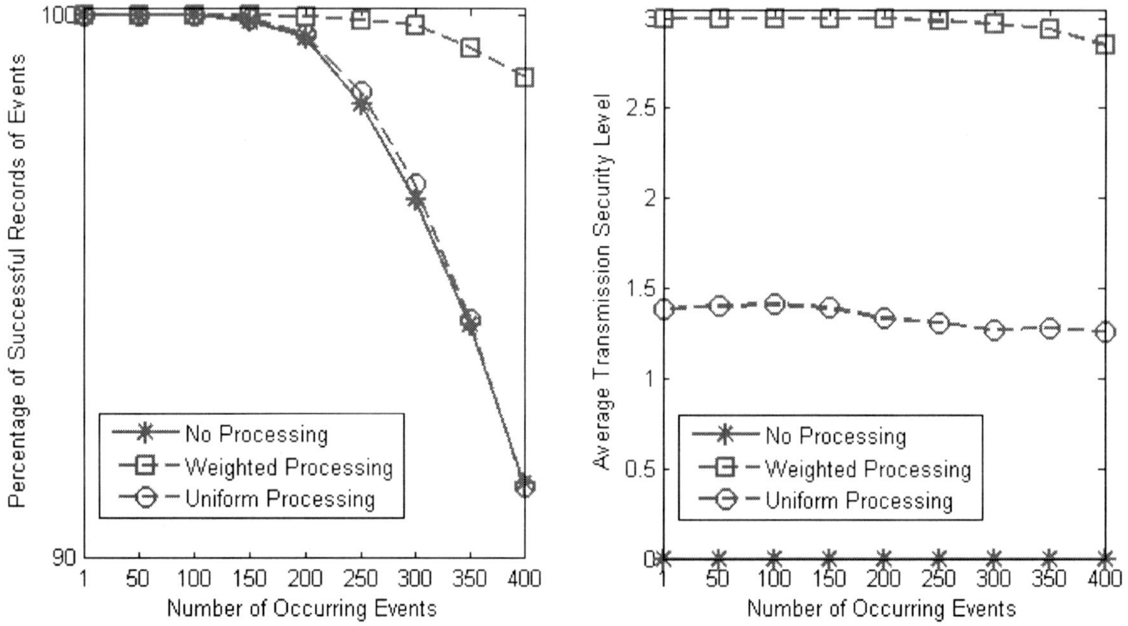

and 3 in most cases except for the Fixed-Tree topology where it fluctuates between 0.25 and 3.

In addition, the distributed scheme helps liberate the network's bottleneck, formed by the nodes format close routing proximity to the base station, of heavy battery drainage by reducing the amount of data to be forwarded and therefore reducing the amount of battery power needed to transmit the information along. Our empirical results show that without distributed processing assignments, the percentage of successful records drops faster than in the other studied cases except in the case of the horizontal network. This is because in more vertical topologies, the nodes closest to the base-station are solicited with every data transmission toward the base station. Furthermore, without any kind of data processing at the nodes' level, the full raw data is handled. This results in the workload of select nodes being high. By distributing the data processing, the amount of data handled by the nodes closest to the base-station is reduced therefore the workload is lightened and Figure 6 (a), Figure 7 (a), Figure 8 (a) and Figure 9 (a) prove that the number of successful events records is increased.

Moreover, the comparison between the results obtained for the Uniform and the Greedy architectures in the figures above confirm that the sooner the data is processed, the more secure the rest of the communication channels along the transmission path will be.

Additionally, these results show that choosing a more horizontal network topology can give some added steganographic security. However a purely horizontal network as shown in Figure 2 does not cover a great deal of geographic space which limits the potential applications where such a network would be needed.

Overall, these simulations and their results provide us with insights and useful information to create a better, more efficient and more secure network against steganography: A horizontally spread network with a distributed scheme will improve the lifespan of the network and discourage attackers from corrupting the network by reducing the entropy of the data as early as possible on the communication path.

FUTURE RESEARCH DIRECTIONS

With the current state of WVSN development, research should focus on simple security solutions that make the best out of the network's parameters. Preventative steganalysis presents the advantage of proactivity over standard steganalysis. With regular steganalysis, there is a constant battle between the steganalyst and the attacker: on one hand the steganalyst offers a solution against a specific attack, and on the other hand, the attacker creates a steganographic method that defeats the solution developed by the steganalyst. This leads to efficiency compromises in the overt network. It is well known that covert communications is often limited at the expense of overt communications performance. The study of preventative steganalysis aims to determine how necessary overt communications mechanisms can be effectively harnessed to aid in mitigating covert communications. This new concept has potential for a rich set of research problems. We assert that future research should focus on providing more innovative and efficient solutions that provide a steganalytic alternative without excessively compromising WVSN efficiency. A more rigorous formulation on how distributed systems can analyze steganographic mechanisms is a good research avenue to pursue.

It is also interesting to study how the topology of the WVSN can help the steganalysis. We have in this chapter approached four common topologies in a passive way. Modifying or rather adapting the network's topology in an active manner could lead to the development of intelligent camera configurations that would minimize the potential for steganography. Future work would look at determining optimally topologies that aid in preventative steganography.

CONCLUSION

Steganography and steganalysis are of great interest in an era where digital communication and networks are used ubiquitously and security and privacy are an integral part of information acquisition and exchange. With the development of WVSNs, attack strategies will increase and therefore protecting the network against non-trusted parties will become a significant WVSN design goal. As for now, it is important to provide holistic means for security and privacy that will not interfere with the network's application nor will not shorten its lifespan. This chapter shows that simple solutions are possible.

However, sensor networks are widely studied and new technology arises on a regular basis. It is of great importance for the steganalyst to follow the WVSN's evolution to adapt the steganalysis to the technology available. And although a more proactive, more global steganalytic solution that would fit a large panel of WVSN and applications is preferred, only by focusing on the network at hand, by studying its environment, topology and purpose, can a custom and more efficient steganographic security be found.

REFERENCES

Blaß, E., Wilke, J., & Zitterbart, M. (2008, September). Relaxed authenticity for data aggregation in wireless sensor networks, In *Proceedings of the 4th International Conference on Security and Privacy in Communication Networks*, Istanbul, Turkey (pp. 1-10).

Cover, T. M., & Thomas, J. A. (1991). *Elements of information theory*. New York, NY: John Wiley & Sons. doi:10.1002/0471200611

Feng, J., & Potkonjak, M. (2003, June). Security in sensor networks: Watermarking techniques, In *Proceedings of the SPIE, Security and Watermarking of Multimedia Contents V,* (Vol. 5020, pp. 391-402).

Fidaleo, D. A., Nguyen, H.-A., & Trivedi, M. (2004, October). The networked sensor tapestry (NeST): A privacy enhanced software architecture for interactive analysis of data in video-sensor networks, In *Proceedings of the ACM 2nd International Workshop on Video Surveillance & Sensor Networks*, New York, USA (pp. 46–53).

Kundur, K., Luh, W., Okorafor, U. N., & Zourntos, T. (2008, January). Security and privacy for distributed multimedia sensor networks, In. *Proceedings of the IEEE Special Issue on Distributed Multimedia*, *96*(1), 112–130.

Lo, B. P. L., Wang, J. L., & Yang, G.-Z. (2005, May). From imaging networks to behavior profiling: Ubiquitous sensing for managed homecare of the elderly, In *Adjunct Proceedings of the 3rd International Conference on Pervasive Computing*, Munich, Germany (pp. 101–104).

Martinovic, I., Pichota, P., & Schmitt, J. B. (2009, March). Jamming for good: A fresh approach to authentic communication in WSNs, In *Proceedings of the 2nd ACM cCnference on Wireless Network Security*, Zurich, Switzerland (pp. 161-168).

Wickramasuriya, J., Datt, M., Mehrotra, S., & Venkatasubramanian, N. (2004, October). Privacy protecting data collection in media spaces, In *Proceedings of the 12th Annual ACM International Conference on Multimedia*, New York, USA (pp. 48–55).

Zheng, J., Li, J., Lee, M. J., & Anshel, M. (2006, December). A lightweight encryption and authentication scheme for wireless sensor networks. *International Journal of Security and Networks*, *1*(3), 138–146. doi:10.1504/IJSN.2006.011773

ADDITIONAL READING

Budhia, U., & Kundur, D. (2004, April). Video steganalysis using collusion sensitivity, In *Proceedings SPIE, Sensors, Command, Control, Communications, and Intelligence (C3I) Technologies for Homeland Security and Homeland Defense*, Orlando, Florida, USA (Vol. 5403, pp. 210-221).

Budhia, U., Kundur, D., & Zourntos, T. (2006, December). Digital video steganalysis exploiting statistical visibility in the temporal domain, In *Proceedings of the IEEE Transactions on Information Forensics and Security* (pp. 502-516).

Chiasserini, C.-F., & Magli, E. (2002, September). Energy consumption and image quality in wireless video-surveillance networks, In *Proceedings of the 13th IEEE International Symposium on Personal, Indoor and Mobile Radio Communications*, Lisbon, Portugal (Vol. 5, pp. 2357-2361).

Chu, M., Reich, J., & Zhao, F. (2004, February). Distributed attention in large scale video sensor networks, In *Proceedings of the Institute of Defence and Strategic Studies*, London, UK (pp. 61-65).

Fridrich, J., Goljan, M., & Du, R. (2001, August). Steganalysis based on JPEG compatibility, In *Proceedings SPIE Multimedia Systems and Applications IV* (pp. 275—280).

Fridrich, J., Goljan, M., & Du, R. (2001, October). Reliable detection of LSB steganography in color and grayscale images, In *Proceedings of the 2001 Workshop on Multimedia and Security: New challenges*, Ottawa, Ontario, Canada (pp. 27-30).

Grieco, L. A., Boggia, G., Sicari, S., & Colombo, P. (2009, October). Secure wireless multimedia sensor networks: A survey, In *Proceedings of the 2009 3rd International Conference on Mobile Ubiquitous Computing, Systems, Services and Technologies* (pp. 194-201).

Guerrero-Zapata, M., Zilan, R., Barceló-Ordinas, J., Bicakci, K., & Tavli, B. (2010, September). The future of security in wireless multimedia sensor networks: A position paper. *Telecommunication Systems*, *45*(1), 77–91. doi:10.1007/s11235-009-9235-0

Jainsky, J. S., Kundur, D., & Halverson, D. R. (2007, September). Towards digital video steganalysis using asymptotic memoryless detection, In *Proceedings of the 9th Workshop on Multimedia & Security,* Dallas, Texas, USA (pp. 161-168). New York, NY: ACM.

Johnson, N. F., Duric, Z., & Jajodia, S. (2001). *Information hiding: steganography and watermarking: attacks and countermeasures. Advances in information security* (*Vol. 1*, p. 137). Dordrecht, The Netherlands: Kluwer Academic Publishers.

Katzenbeisser, S., & Petitcolas, F. A. (2000). *Information hiding techniques for steganography and digital watermarking*. Norwood, MA: Artech House, Inc.

Kipper, G. (2003). *Investigator's guide to steganography*. Boca Raton, FL: CRC Press, Inc. doi:10.1201/9780203504765

Luh, W., Kundur, D. & Zourntos, T. (2007, January). A novel distributed privacy paradigm for visual sensor networks based on sharing dynamical systems, *EURASIP Journal on Advances in Signal Processing Special Issue on Visual Sensor Networks* (pp. 218-218).

Lyu, S., & Farid, H. (2002). Detecting hidden messages using higher-order statistics and support vector machines, In *Proceedings of the 5th International Workshop on Information Hiding* (pp. 340-354).

Memon, N., Avcibas, I., & Sankur, B. (2001, October). Steganalysis based on image quality metrics, *IEEE 4th Workshop on Multimedia Signal Processing*, Cannes, France (pp. 517-522).

Perrig, A., Stankovic, J., & Wagner, D. (2004, June). Security in wireless sensor networks. *Communications of the ACM, 47*(6), 53–57. doi:10.1145/990680.990707

Perrig, A., Szewczyk, R., Wen, V., Culler, D., & Tygar, J. D. (2002, September). Spins: Security protocols for sensor networks. *Wireless Networks, 8*(5), 521–534. doi:10.1023/A:1016598314198

Sullivan, K., Madhow, U., Chandrasekaran, S., & Manjunath, B. S. (2005, January). Steganalysis of spread spectrum data hiding exploiting cover memory, In *Proceedings IST/SPIE 17th Annual Symposium on Electronic Imaging Science Technology*, San Jose, CA, USA (pp. 38-46).

Sullivan, K., Madhow, U., Chandrasekaran, S., & Manjunath, B. S. (2006, June). Steganalysis for Markov cover data with applications to images, In. *Proceedings of the IEEE Transactions on Information Forensics and Security, 1*, 275–287. doi:10.1109/TIFS.2006.873595

Wayner, P. (2002). *Disappearing cryptography: Information hiding: Steganography and watermarking* (2nd ed.). San Francisco, CA: Morgan Kaufmann Publishers Inc.

Westfeld, A., & Pfitzmann, A. (1999, September). Attacks on steganographic systems, In *Proceedings of the 3rd International Workshop on Information Hiding* (pp. 61-76).

Xiao, Y. (2007). *Security in Sensor Networks*. Boca Raton, NY: CRC Press, Inc.

Yu, Y., Prasanna, V. K., & Krishnamachari, B. (2007). *Information processing and routing in wireless sensor networks*. Singapore: World Scientific Publishing Co., Inc.

Zhang, Y., Lee, W., & Huang, Y. (2003, September). Intrusion detection techniques for mobile wireless networks. *Wireless Networks*, 545–556. doi:10.1023/A:1024600519144

Zhu, S., Setia, S., & Jajodia, S. (2003, October). LEAP: Efficient security mechanisms for large-scale distributed sensor networks, In *Proceedings of the 10th ACM Conference on Computer and Communications Security*, Washington, DC., USA (pp. 62-72).

KEY TERMS AND DEFINITIONS

Embedding Capacity: Measure of the amount of information that can be covertly hidden in a cover-media.

Entropy: Measure of uncertainty of a random variable.

Information Theory: Theory of communications developed by Claude E. Shannon.

Mutual Information: Measure of the degree of dependence between two random variables.

Preventative Steganalysis: Use and redistribution of essential networking processes to protect potential cover-media from information embedding prior to occurrence of steganographic attacks.

Steganalysis: Science of detecting steganography.

Steganography: Art of embedding hidden information in covert media.

About the Contributors

Li-minn Ang received his Ph.D and Bachelor degrees from Edith Cowan University, Australia in 2001 and 1996 respectively. He is currently an Associate Professor at the University of Nottingham Malaysia Campus. His research interests are in the fields of visual information processing, intelligent processing techniques, hardware architectures, reconfigurable computing, and engineering education. He has authored or coauthored a number of journal and conference papers in these areas. Currently he leads the research activities at the Visual Information Engineering Research Group at the University of Nottingham Malaysia Campus. He has received research grants from the Malaysian government and industry for his research activities and has served as a reviewer and committee member for a number of journals and conferences. Dr. Ang is a Senior Member of the IEEE and a Fellow of the Higher Education Academy (UK).

Kah Phooi Seng received her Ph.D and Bachelor degrees from the University of Tasmania, Australia in 2001 and 1997 respectively. She is currently a Professor of Computer Science at Sunway University, Malaysia. Her research interests are in the fields of intelligent visual processing, biometrics and multi-biometrics, artificial intelligence, and signal processing. She has authored or coauthored a number of journal and conference papers in these areas. Dr. Seng has received research grants from the Malaysian government and industry for her research activities and has served as a reviewer for journals and conferences.

* * *

Ruth M. Aguilar-Ponce received a Bachelor degree in Computer Science from the Autonomous University of Puebla, Mexico in 1997. She obtained the M.S. in Electronics from the National Institute for Research in Optics, Astrophysics and Electronics, Mexico in 1999. She joined the Center for Advanced Computer Studies at the University of Louisiana at Lafayette, where she obtained the M.S in Computer Engineering in 2004 and the Ph.D. in 2007. Currently, she is an assistant professor at the University of San Luis Potosi, Mexico. She is a junior member of the National Research System in Mexico since 2009. Her research interests include sensor networks, communication protocols, system on a chip, and signal and image processing.

Alfonso Alba received the B.S. degree in Electronic Engineering (1999) and the M.Sc. in Applied Mathematics (2001) from the Autonomous University of San Luis Potosi (UASLP), Mexico. In 2001, he also became a full-time professor at the Faculty of Science, UASLP. In 2007, he obtained the Ph.D.

degree in Computer Science from the Center for Research in Mathematics (CIMAT, Guanajuato) and became a member of the Academic Committee of the graduate program in Electronic Engineering at UASLP. Dr. Alba is currently a fellow member of the National Research System. His current research interests include biomedical signal analysis and visualization, pattern recognition, and computer vision.

Samir Brahim Belhouari received a diploma in telecommunication from Institute of Telecommunications of Oran (ITO), Algeria. He received Master degree from School of Toulouse (ENSEEIHT), France, during which he has developed a new method of multiple accesses in mobile networks by clock change. He has received a Ph.D. degree in Mathematical Sciences with an emphasis on Stochastic Processes from Federal Polytechnic School of Lausanne, Switzerland. During which he has been working at the Laboratory of Stochastic Processes (IMA-PrSt) at (EPFL). He has worked at the Integrated Systems Laboratory (LSI) of EPFL with main responsibility of applying advanced techniques of routing for data transmission over the internet (IP packet-Internet Protocol). Currently he is working at the University of Technology PETRONAS, Electrical and Electronic Engineering Department as S. Lecturer and his research interest is in Image processing and Pattern Recognition.

Li Wern Chew received her Master of Science and Bachelor degrees from the University of Nottingham Malaysia Campus and the University of Adelaide, Australia in 2007 and 2004 respectively. She is currently pursuing her Ph.D at the University of Nottingham Malaysia Campus. Her research interests are in the fields of signal, image, vision processing, and hardware architectures.

Wai Chong Chia received his Master of Science and Bachelor degrees from the University of Nottingham Malaysia Campus and the University of Nottingham Trent in 2008 and 2006 respectively. He is currently pursuing his Ph.D at the University of Nottingham Malaysia Campus. His research interests are in the fields of image and vision processing, hardware architectures, and wireless sensor network.

Mangesh Chitnis received his Bachelors of Engineering in Information Technology in 2003 from Mumbai University and later Masters in Information Technology in 2006 from Scuola Superiore Sant'Anna, Italy. In 2004 he was hired by Infosys Technologies, India to work on Enterprise based Applications using J2EE. Since January 2007, he has been pursuing his PhD on Wireless Sensor Networks at Retis Lab in Scuola Superiore Sant'Anna, Italy. His main research interest is in the area of deterministic and stochastic bandwidth estimation for multimedia wireless sensor networks.

Shung Han Cho received B.E. (Summa Cum Laude) with specialization in Telecommunications from both the department of Electronics Engineering at Ajou University, Korea and the department of Electrical and Computer Engineering at Stony Brook University - SUNY, NY in 2006. He was a recipient of Award for Academic Excellence in Electrical Engineering by College of Engineering and Applied Sciences at Stony Brook University. He received M.S. in Electrical and Computer Engineering from Stony Brook University with Award of Honor in recognition of outstanding achievement and dedication in 2008. He is currently pursuing his Ph.D. degree in the department of Electrical and Computer Engineering at Stony Brook University. He was a recipient for International Academic Exchange Program supported by Korea Research Foundation (KRF) in 2005. He was a member of Sensor Consortium for Security and Medical Sensor Systems sponsored by NSF Partnerships for Innovation from 2005 to

2006. His research interests include collaborative heterogeneous signal processing, distributed digital image processing and communication, networked robot navigation and communication, heterogeneous system modeling and evaluation.

Yinhao Ding received the B.Sc. degree in Electronic Information Science and Technology from Jilin University, Changchun, China, in 2006, and received the M.Eng. degree in telecommunications engineering from the University of Adelaide, Adelaide, SA, Australia, 2009. He is currently working toward the Ph.D. degree in the School of Electrical & Electronic Engineering, University of Adelaide, Adelaide, SA, Australia. His research interests include wireless sensor networks and compressive sensing technique.

Abdelrahman Elamin received his first degree from university of Khartoum Sudan in electrical and electronic engineering. He has received a master degree from IT university of Copenhagen, Denmark in Multimedia technology in 2005, where he studied speech coding and recognition and also developed speech enhancement system. Currently he is doing PhD at University of Technology PETRONAS, Malaysia in distributed video coding. His research interests in multimedia communication and application.

Karin Fahlquist is a Ph.D. student at Digital Media Lab, Umeå University, Sweden, since 2008. She is currently involved in the EU project "Digital Zoo". Her research area is interaction design in media technology. Before her Ph.D studies she worked six years as a teacher in the media technology area in the Department of Applied Physics and Electronics (TFE), Umeå University, Sweden.

Jesús García is currently an Associate Professor at the Universidad Carlos III de Madrid, member of the Computer Science Department since year 2000 and belonging to the Applied Artificial Intelligence research Group (GIAA). He received his B.S. in Telecommunications Engineering from Universidad Politécnica de Madrid in 1996, and his Ph.D. degree from the same university in 2001. He has participated in several European projects related to sensor fusion (funded by EC and Eurocontrol) and participated and coordinated diverse national research projects. His main interests are computational intelligence, sensor and information fusion, surveillance systems, machine vision, traffic management and autonomous vehicles.

Brahim Hadjou is a professor in the department of Electrical and Computer Engineering (Engineering faculty) at Sherbrooke University. He obtained his Ph.D. from Sherbrooke University in 1996, and Bachelor degree from Constantine University (Algeria) in 1986. His research domain is mainly related to the application of mathematics in robotics and power-limited applications.

Sangjin Hong received the B.S and M.S degrees in EECS from the University of California, Berkeley. He received his Ph.D in EECS from the University of Michigan, Ann Arbor. He is currently with the department of Electrical and Computer Engineering at Stony Brook University. Before joining Stony Brook University, he has worked at Ford Aerospace Corp. Computer Systems Division as a systems engineer. He also worked at Samsung Electronics in Korea as a technical consultant. His current research interests are in the areas of multimedia wireless communications and digital signal processing systems, reconfigurable VLSI Systems and optimization. Prof. Hong is a Senior Member of IEEE and a member of EURASIP journal editorial board. Prof. Hong served on numerous Technical Program Committees for IEEE conferences.

Julien S. Jainsky received his engineering degree in electrical and computer engineering in 2005 from CPE Lyon in France. He is currently a Ph.D. candidate in the Wireless Communication Group (WCL) of the Department of Electrical and Computer Engineering at Texas A&M University under the advisorship of Drs. Deepa Kundur and Don Halverson. His research interests include security in multimedia, sensor networks and smart grids.

Varun Jeoti received his Ph.D. degree from Indian Institute of Technology Delhi India in 1992. He worked on several sponsored R&D projects in IIT Delhi and IIT Madras during 1980 to 1989 developing Surface Acoustic Wave Pulse Compression filters, underwater optical receivers etc.. He was a Visiting Faculty in Electronics department in Madras Institute of Technology for about 1 year during 1989 to 1990 and joined Delhi Institute of Technology for next 5 years till 1995. He moved to Electrical & Electronic Engineering (E&E Engg) department of Universiti Sains Malaysia in 1995 and joined E&E Engg of Universiti Teknologi PETRONAS in 2001. His research interests are in signal processing and wireless communication relating to wireless LAN and MAN technologies, cognitive radio, maritime communication, distributed video coding and security.

Johannes Karlsson is a researcher at the Department of Applied Physics and Electronics (TFE), Umeå University, Sweden. He is currently the project leader for the EU project "Digital Zoo". He has worked on several industrial and academic projects in the field of wireless sensor networks ranging from large scale networks of wireless fire detectors to environmental multimedia sensor networks. His current research focus is around multimedia signal processing and transportation in wireless sensor networks and energy management in large scale environmental sensor networks. He received the Ph.Lic degree in signal processing from Umeå University, Sweden, in 2006. The thesis dealt with real-time video communication over wireless ad-hoc networks.

Ahmed Khoumsi is a professor in the department of Electrical and Computer Engineering (Engineering faculty) at Sherbrooke University. He obtained his Ph.D. and Bachelor degrees from Toulouse University in 1984 and 1988, successively. His research interests include information technology and telecommunication, web services, service interactions, event discrete systems, etc.

Deepa Kundur received the B.A.Sc., M.A.Sc., and Ph.D. degrees all in electrical and computer engineering in 1993, 1995, and 1999, respectively, from the University of Toronto, Canada. From September 1999 to December 2002 she was an Assistant Professor and held the title of Bell Canada Junior Chair-holder of Multimedia in the Department of Electrical and Computer Engineering at the University of Toronto. In January 2003, she joined the Department of Electrical and Computer Engineering at Texas A&M University, where she is currently an Associate Professor. Dr. Kundur is the author of over 100 technical publications in the field of information security, multimedia and signal processing and communication systems. Her research interests include cybersecurity of the electric smart grid and security of multimedia sensor networks. She is an appointed member of the NERC Smart Grid Task Force and an elected member of the IEEE Information Forensics and Security Technical Committee.

K. S. Kwak received the B.S. degree from Inha University, Korea in 1977, and the M.S. degree from the University of Southern California in 1981 and the Ph.D. degree from the University of California at San Diego in 1988, respectively. From 1988 to 1989 he was a Member of Technical Staff at Hughes Network Systems, San Diego, California. From 1989 to 1990 he was with the IBM Network Analysis Center at Research Triangle Park, North Carolina. Since then he has been with the School of Information and Communication, Inha University, Korea as a professor. He had been the chairman of the School of Electrical and Computer Engineering from 1999 to 2000 and the dean of the Graduate School of Information Technology and Telecommunications from 2001 to 2002 at Inha University, Inchon, Korea. He is the current director of Advanced IT Research Center of Inha University, and UWB Wireless Communications Research Center, a key government IT research center, Korea. He has been the Korean Institute of Communication Sciences (KICS)'s president of 2006 year term. In 1993, he received Engineering College Young Investigator Achievement Award from Inha University, and a distinguished service medal from the Institute of Electronics Engineers of Korea (IEEK). In 1996 and 1999, he received distinguished service medals from the KICS. He received the Inha University Engineering Paper Award and the LG Paper Award in 1998, and Motorola Paper Award in 2000. His research interests include multiple access communication systems, mobile communication systems, UWB radio systems and ad-hoc networks, high-performance wireless Internet.

Haibo Li is a full Professor in Signal Processing in the Department of Applied Physics and Electronics (TFE), Umeå University, Sweden. He received the technical doctor degree in Information Theory from Linköping University, Sweden, in 1993. His doctoral thesis dealt with advanced image analysis and synthesis techniques for low bitrate video. Dr. Li got the "Nordic Best Ph.D. Thesis Award" in 1994. In 1997 Dr. Li was awarded the title of "Docent in Image Coding". From 1990 to 1993 he was a teaching assistant of digital video at Linköping University. After graduation Dr. Li joined the technical faculty of Linköping University first as an Assistant Professor and then promoted to an Associate Professor in 1998. During his period at Linköping University he developed advanced image and video compression algorithms, including extremely low bitrate video compression, 3D video transmission, and tele-operation and tele-presence. After joining Umeå University, he is now directing the Digital Media Lab, Umeå Center for Interaction Technology (UCIT), Umeå University, and working on advanced Human, Thing and Information interaction techniques. Prof. Li has been chairing sections at relevant international conferences and was actively involving in MPEG activities in low bitrate video compression. He has contributed to several EU projects, like VIDAS, SCALAR, INTERFACE, MUCHI and INTRO. He has published more than 100 technical papers including chapters in books and holds six international patents as the first inventor in multimedia area.

Cheng-Chew Lim received the B.Sc. degree in electrical and electronic engineering and the Ph.D. degree from Loughborough University, Leicestershire, U.K. He is currently with the University of Adelaide, Adelaide, SA, Australia, as an Associate Professor and the Head of the School of Electrical & Electronic Engineering. He is an Associate Editor of the *Journal of Industrial and Management Optimization*. His research interests include wireless sensor networks and system-on-chip verification and very large scale integration architectures for array processors for matrix-based computations.

James Llinas brings over 35 years of experience in multisource information processing and data fusion technology to his research, teaching, and business development activities. He co-authored the first integrated book on Multisensor Data Fusion, and has lectured internationally for over 30 years on this topic. Dr. Llinas is also the co-editor of the "Handbook of Multisensor Data Fusion (2001). Dr. Llinas is a Technical Advisor to the Defense Department's Joint Directors of Laboratories Data Fusion Group, the only US DoD technology oversight group for Data Fusion, a position he has held for 20+ years. Dr. Llinas is a recipient of the Joseph Mignona Award (1999), given to one person per year by the Data Fusion Group as signifying lifelong contributions to the field of Data Fusion. His experience in applying this technology to different problem areas ranges from defense applications to non-defense applications to include intelligent transportation systems, medical diagnostics, and condition-based maintenance, among others.

Giuseppe Lipari is Associate Professor of Computer Engineering (scientific sector ING-INF/05) at Scuola Superiore Sant'Anna. He is part of the RETIS lab of the Center of Excellence for Information Information, Communication and Perception Engineering. He is IEEE member since many years, and associate editor of IEEE Transactions on Computers. His research interests are in real-time systems, real-time operating systems, scheduling algorithms, embedded systems, wireless sensor networks. He is involved in many EU research projects (FRESCOR, RI-MACS, ARTIST 2) and italian projects (ART-DECO, SensorNet).

Mansouri Majdi was born November 11, 1982 in Kasserine, Tunisia. He received the Dipl.-Ing. in 2006 from High School of Communications of Tunis (SUP'COM) and the M.S. degrees in 2008 from High school of Electronic, Informatique and Radiocommunications in Bordeaux (ENSEIRB). He is PhD student at Troyes University of Technology, France. Since October 2008. His current research interests include statistical signal processing and Wireless Sensors Networks. Mansouri is the author of over 20 papers. Mansouri Majdi, Ouachani Ilham, Snoussi Hichem and Cédric Richard received Best Poster Award for "Adaptive Quantized Target Tracking in Wireless Sensor Networks" in SensorNets 2009: The First International School on Cyber-Physical and Sensor Networks.

Abdelhamid Mammeri is a lecturer in the department of Electrical and Computer Engineering (Engineering faculty), and a researcher in the MOIVRE laboratory (Science faculty) at Sherbrooke University. He obtained his Ph.D. from Sherbrooke University (canada) in 2010, Master degree from UCL University (Belgium) in 2002, and Bachelor degree from Batna University (Algeria) in 1996. His research domain is related to the signal and image processing on power-limited applications, such as visual sensor networks.

S. Mehta received the B.E. and M.S degrees both in Electronics Engineering from Mumbai University, Mumbai, India, and Ajou University, Korea in 2002 and 2005, respectively. He is currently pursuing the Ph.D. degree in Telecommunication engineering from the Inha University, Korea. His research interests are in performance analysis of wireless networks and RFID systems.

José M. Molina received a degree in Telecommunication Engineering from Universidad Politécnica de Madrid in 1993 and a Ph.D. degree from the same university in 1997. He joined Universidad Carlos

III de Madrid in 1993 where currently he is a Full Professor at Computer Science Department. He is the Head of Computer Science Department and leads the Applied Artificial Intelligence Group (GIAA, http://www.giaa.inf.uc3m.es) involved in several research projects related with ambient intelligence, surveillance systems and context based computing. His current research focuses in the application of soft computing techniques (Multiagents Systems, Evolutionary Computation, Fuzzy Systems) to surveillance, ambient intelligence, air traffic management and e-commerce. He is the author of more than 50 journal papers and 100 conference papers.

Yunyoung Nam received B.S, M.S. and Ph.D. degree in computer engineering from Ajou University, Korea in 2001, 2003, and 2007 respectively. He was a research engineer in the Center of Excellence in Ubiquitous System from 2007 to 2009. He was a post-doctoral researcher at Stony Brook University in 2009, New York. He is currently a research professor in Ajou University in Korea. He also spent time as a visiting scholar at Center of Excellence for Wireless & Information Technology (CEWIT), Stony Brook University - State University of New York Stony Brook, New York. He was a recipient the Presidential Award for Excellence in Graduate School of Information and Communication in 2004 and 2007. He earned the Best Paper Award at KDC 2006. He is selected for inclusion in the 2011 Edition of "Who's Who in America." His research interests include multimedia database, ubiquitous computing, image processing, pattern recognition, context-awareness, conflict resolution, wearable computing, and intelligent video surveillance.

Christopher Wing Hong Ngau received his Bachelor degree from the University of Nottingham Malaysia Campus in 2008. He is currently pursuing his PhD at the University of Nottingham Malaysia Campus. His research interests are in the fields of image, vision processing, hardware architectures, and wireless sensor network.

Ilham Ouachani was born in Paris, France, in 1976. She received the diploma degree in electrical engineering from The Ecole supérieure de Communication de Tunis (Sup'Com), Tunisie in 2000. Then, she received the DEA degree from ParisTech Paris, and the Ph.D. in signal processing from the University of Paris-sud, Orsay, France in 2001 and 2005, respectively. Between 2005 and 2006, she was a teaching assistant in with the University of Paris-sud and the Conservatoire National des Arts et Métiers de Paris. Since september 2006, she is associate professor at the Institut Supérieur d'Informatique et des Techniques de Communication, in Hammam. Sousse, Tunisia. She has won the AUF grant and she has spent 6 months as visiting scientist at the LM2S laboratory in the University of Technology of Troyes, France, in 2009. Her research interests include MIMO systems, capacity calculation, OFDM, relay channels, digital communication in wireless sensor networks, She is author of several research papers in journals and international conferences.

Paolo Pagano received his M.S. degree in Physics in 1999 from Trieste University (I). In 2003 he received his Ph.D. degree in High Energy Physics from Trieste University having worked for the COMPASS collaboration at CERN (CH). In 2004 he was hired by HISKP at Bonn University (D). In 2006 he received a Master in Computer Science from Scuola Superiore Sant'Anna in Pisa (I). In the same year he joined the REal-TIme System (RETIS) laboratory of the Scuola as associate researcher to work in the domain of Wireless Sensor Networks. In 2009 he joined the CNIT (National Inter-University Consortium for Telecommunications) and was assigned to the Sant'Anna Research Unit as leader of the Real-Time Networks team at the RETIS lab.

Miguel Á. Patricio received his B.S. degree in Computer Science from Universidad Politécnica de Madrid in 1991, his M.S. degree in Computer Science in 1995, and his Ph.D. degree in Artificial Intelligence from the same university in 2002. He has held an administrative position at the Computer Science Department of Universidad Politécnica de Madrid since 1993. He is currently an Associate Professor at Escuela Politécnica Superior of Universidad Carlos III de Madrid and a research fellow of the Applied Artificial Intelligence Group (GIAA). He has carried out a number of research projects and consulting activities in the areas of automatic visual inspection systems, texture recognition, neural networks, and industrial applications

Matteo Petracca received his M.S. degree in Telecommunication Engineering in 2003 and the Ph.D. degree in Information and System Engineering in 2007, both from the Politecnico di Torino, Turin, Italy. From January 2008 to November 2009 he was a post-doc researcher at the Politecnico di Torino working on multimedia processing and transmission over wired and wireless packet networks. He is currently a post-doc researcher in Real-Time Networks at Scuola Superiore Sant'Anna in Pisa.

Keni Ren is a project researcher in Digital Media Lab in the Department of Applied Physics and Electronics (TFE), Umeå University, Sweden. She received the master degree in Interactive System Engineering from Umeå University in 2010. Her master thesis is about face recoganation technique by using Support vector machines. Her currently areas of focus are around object detection, classification, object tracking and Human, Thing and Information interaction techniques.

Cédric Richard was born January 24, 1970 in Sarrebourg, France. I received the Dipl.-Ing. and the M.S. degrees in 1994 and the Ph.D. degree in 1998 from the University of Technology of Compiegne, France, all in Electrical and Computer Engineering. From 1999 to 2003, he was an Associate Professor at the University of Technology of Troyes, France. From 2003 to 2009, he was a Full Professor at the Institut Charles Delaunay (CNRS FRE 2848) at the UTT, and the supervisor of a group consisting of 60 researchers and Ph.D. In winter 2009 and autumn 2010, he was a Visiting Researcher with the Department of Electrical Engineering, Federal University of Santa Catarina (UFSC), Florianopolis, Brazil. Cédric Richard is a junior member of the Institut Universitaire de France since october 2010. Since September 2009, Cédric Richard is a Full Professor at Fizeau Laboratory (CNRS UMR 6525, Observatoire de la Cote d'Azur), University of Nice Sophia-Antipolis, France. His current research interests include statistical signal processing and machine learning. Prof. Cédric Richard is the author of over 100 papers. He was the General Chair of the XXIth francophone conference GRETSI on Signal and Image Processing that was held in Troyes, France, in 2007, and of the IEEE International Workshop on Statistical Signal Processing that will be held in Nice in 2011. Since 2005, he is a member of the board of the federative CNRS research group ISIS on Information, Signal, Images and Vision. He is a member of GRETSI association board and of the EURASIP society, and Senior Member of the IEEE. Cédric Richard serves as an Associate Editor of the IEEE Transactions on Signal Processing since 2006, and of the EURASIP Signal Processing Magazine since 2009. In 2009, he was nominated liaison local officer for EURASIP, and member of the Signal Processing Theory and Methods (SPTM) Technical Committee of the IEEE Signal Processing Society.

Juan Gómez Romero is currently a Visiting Lecturer at the Computer Science department of Universidad Carlos III of Madrid. He received a degree in Computer Science from Universidad of Granada (Spain) in 2004, and a Ph.D. in Intelligent Systems from the same university in 2008. He worked as an early stage researcher in the Approximate Reasoning and Artificial Intelligence group (ARAI) of Universidad of Granada, and as a post-doctoral researcher in the Applied Artificial Intelligence Group (GIAA) of Universidad Carlos III, where he continues his work as a research fellow. His research interests focus on computable Knowledge Representation and Reasoning formalisms (ontologies, fuzzy ontologies) and their applications to High-Level Information Fusion and Cognitive Vision Systems.

Claudio Salvadori was born in Arezzo, Italy in 1978. He received the Laurea degree in Telecomunication Engineering in 2006 from the Università degli Studi di Siena, Siena, Italy. Since November 2009 he is a PhD student in Wireless Sensor Network at Scuola Superiore Sant'Anna, Pisa, Italy.

Luca Santinelli is a Ph.D. candidate at the Scuola Superiore Sant'Anna of Pisa. He graduated in Electronic Engineering (specialization in Computer Science) at the Polytechnic University of Marche in 2003.From 2003 to 2007 he worked for the MTSGroup as software developer in the R&D department. He specialized on microcontroller and control systems. From 2007 he is doing his Ph.D. at the RETIS Laboratory on real-time systems. His main research interests include real-time operating systems, dynamic scheduling algorithms, quality of service control, real-time controls, resource reservation algorithms, dynamic resource allocation, energy aware scheduling, probabilistic real-time and artificial intelligence.

Nalin Sharda gained B.Tech. and Ph.D. degrees from the Indian Institute of Technology, Delhi. Presently he teaches and leads research in innovative applications of computer technologies at the School of Engineering and Science, Victoria University, Australia. Dr. Sharda's publications include the Multimedia Information Networking textbook, and around 120 papers and handbook chapters. Dr. Sharda has invented Movement Oriented Design (MOD) paradigm for the creation of effective multimedia content based experience, and applied it to e-Learning and other applications. Nalin has led e-Tourism projects for the Australian Sustainable Tourism CRC, and is currently guiding research in the innovative applications of ICT systems to sustainable living. Dr. Sharda has been invited to present lectures and seminars in the Distinguished Lecturer series of the European Union's Prolearn program. He has presented over fifty seminars, lectures, and Key Note addresses in Austria, Australia, Finland, Germany, Hong Kong, India, Malaysia, Pakistan, Japan, Singapore, Sweden, Switzerland, UAE, and USA.

Fethi Smach received a Master degree in Computer science from University of Sfax, Tunisia in 2003, and the Ph.D. degree in image processing from the University of Burgundy, Dijon, France in 2007. Since 2008 he is research engineer at CEA-List Paris. Between August 2009 and august 2010 he was postdoctoral researcher at INRIA TEMICS research group. He is currently research engineer at ActivNetworks. His fields of interest are signal and image processing, and in particular 2D and 3D image and video coding for video transmission over wireless networks, algorithms for pattern recognition and real-time implementation of those, classification algorithms. Besides the academic side, he is interested with hunting.

Hichem Snoussi was born in Bizerta, Tunisia, in 1976. He received the diploma degree in electrical engineering from the Ecole Superieure d'Electricite (Supelec), Gif-sur-Yvette, France, in 2000. He also received the DEA degree and the Ph.D. in signal processing from the University of Paris-Sud, Orsay, France, in 2000 and 2003 respectively. He has obtained the HdR from the University of Technology of Compiègne in 2009. Between 2003 and 2004, he was postdoctoral researcher at IRCCyN, Institut de Recherches en Communications et Cybernétiques de Nantes. He has spent short periods as visiting scientist at the Brain Science Institute, RIKEN, Japan and Olin Neuropsychiatry Research Center at the Institute of Living in USA. Between 2005 and 2010, he has been associate professor at the University of Technology of Troyes. Since September 2010, he has been appointed a Full Professor position at the same university. He is in charge of the regional research program S3 (System Security and Safety) of the CPER 2007-2013 and the CapSec plateform (wireless embedded sensors for security). He is the principal investigator of an ANR-Blanc project (mv-EMD), a CRCA project (new partnership and new technologies) and a GDR-ISIS young researcher project. He is partner of many ANR projects, GIS, strategic UTT programs. He obtained the national doctoral and research supervising award PEDR 2008-2012.

J.Luis Tecpanecatl-Xihuitl received the B.E. degree in electrical engineering from the Autonomous University of Puebla, Mexico and the M.S. degree in electronics from National Institute for research on Astrophysics, Optic, and Electronics, Mexico in 2000. He obtained the Ph.D. in computer engineering from The Center for Advanced Computer Studies, at University of Louisiana at Lafayette in 2009. Luis joined the University of San Luis Potosi, Mexico in 2009 as an assistant professor and became a member of the Academic Committee of the graduate program in Electronic Engineering. He is a junior member of the Mexican National Research System since 2009. His research interests focus on multi-standard digital receivers, VLSI digital processing systems and image processing.

Jing Teng received the BEng degree in electronic information engineering from the Central South University, China, in 2003, and the PhD degree in systems optimization and security from the University of Technology of Troyes, France, in November 2009. Since 2010, she has been a lecturer at the North China Electric Power University. Her research interests include statistical signal processing and its applications in wireless sensor networks.

Tim Wark is a Research Group Leader and Principal Research Scientist in the Autonomous Systems Laboratory at CSIRO ICT Centre, Australia. He has worked in a variety of projects around sensor networks ranging from wearable body sensors for indoor monitoring, to long-term, environmental sensing. His current areas of focus are around adaptive energy management of sensor networks and in-network processing of multimedia information. He currently leads a group of scientists and engineers developing the next generation of pervasive computing technologies targeting a range of indoor and environmental applications. Tim currently holds Adjunct A/Prof appointments at The Australian National University, University of Queensland and Queensland University of Technology.

Lee Seng Yeong received his Masters of Engineering from the University of Nottingham in 2004. He is currently pursuing his Ph.D in the same university. He is a member of the Visual Information Engineering Research Group. His research interest is in the field of embedded visual processing.

Djemel Ziou is a professor in the department of computer science at Sherbrooke University. He obtained his Ph.D. and Master degrees from Nancy University (France) in 1985 and 1991, successively. He is leading MOIVRE laboratory at Sherbrooke University. He is the Canada research chair (NSERC-Bell) in personal imaging. His research domain concern image processing.

Jiang-Yu Zheng received the B.S. degree in Computer Science from Fudan University, China, in 1983, and the M.S. and Ph.D. degrees in Control Engineering from Osaka University, Japan in 1987 and 1990, respectively. Currently he is a professor at the Dept. of Computer Science, Indiana University Purdue University Indianapolis. He research interests are in the areas of image and video processing, computer vision, virtual reality, multimedia and internet application. Dr. Zheng received 1991 Best Paper Award from the Information Processing Society of Japan for generating the world first *digital panoramic image*. He also received Excellent Paper Award from Japan Society of Art and Science in 2000 for his development of a graphics tool to extract human motion from video. He is the inventor of *route panorama* and *scene tunnel* that is awarded as one of the best papers in ACM VRST04. He is a senior member of IEEE.

Shibao Zheng graduated from Xidian University in 1986 and received B.S. degree. From 1986 to 1999, he was an expert of important project in HDTV of China. From 2000, he worked in Institute of Image Communication and Information Processing of Shanghai Jiao Tong University as a professor. His general research interests include DTV, multimedia in network and ASIC design.

Yi Zhou received the B.S. and M.S. degrees of Electronic Engineering from Dalian Maritime University, Dalian, P.R China, in 2003 and 2006 respectively. He is currently a joint educated Ph.D. student in Signal Processing with the Institute of Image communication and Information Processing of Shanghai Jiao Tong University, Shanghai, P. R. China, and ICD/LM2S, University of Technology of Troyes, France since 2007. His research interests include computer vision, machine learning and visual tracing.

Index

A

B

C

D

J

K

L

M

N

O

P

Q

R

S

T

U

V

W

CPSIA information can be obtained at www.ICGtesting.com
Printed in the USA
BVOW021730201011

273945BV00003BF/5/P

9 781613 501535